Green Energy and Technology

Climate change, environmental impact and the limited natural resources urge scientific research and novel technical solutions. The monograph series Green Energy and Technology serves as a publishing platform for scientific and technological approaches to "green"—i.e. environmentally friendly and sustainable—technologies. While a focus lies on energy and power supply, it also covers "green" solutions in industrial engineering and engineering design. Green Energy and Technology addresses researchers, advanced students, technical consultants as well as decision makers in industries and politics. Hence, the level of presentation spans from instructional to highly technical.

Indexed in Scopus.

More information about this series at http://www.springer.com/series/8059

Yaşar Demirel

Energy

Production, Conversion, Storage,
Conservation, and Coupling

Third Edition

 Springer

Yaşar Demirel
University of Nebraska Lincoln
Lincoln, NE, USA

Solution Manual for Instructors to this book can be downloaded on https://www.springer.com/in/book/9783030561635

ISSN 1865-3529   ISSN 1865-3537   (electronic)
Green Energy and Technology
ISBN 978-3-030-56166-6   ISBN 978-3-030-56164-2   (eBook)
https://doi.org/10.1007/978-3-030-56164-2

This Springer imprint is published by the registered company Springer Nature Switzerland AG
The registered company address is: Gewerbestrasse 11, 6330 Cham, Switzerland

To Zuhal, Selçuk, and Can

Preface

This is the third edition to fill the need for a comprehensive textbook on energy. Importance of energy and its effect on everyday life are undisputable. Consequently, many institutions today offer either energy minor or energy major programs. Also, 'Energy Engineering' is emerging as one of the recent engineering disciplines. This textbook is an undergraduate textbook for students with diverse backgrounds and interested to know more on energy and pursue a degree on energy.

The third edition discusses five major aspects of energy in an introductory manner. Those major aspects are: energy production, conversion, storage, conservation, and coupling, discussed in separate chapters. Basic definitions are given in Chap. 1, and the primary and secondary energy sources are discussed in Chap. 2. Chapter 3 discusses mechanical and electrical energies which are the types of major energies other than heat and work. Chapter 4 discusses the internal energy and enthalpy, and Chap. 5 discusses balance equations, heat of reaction, and heat transfer. After these chapters for introducing the basics and building the infrastructure of energy, Chap. 6 discusses energy production mainly using closed and open cycles of heat engines. Chapter 7 discusses the energy conversion with an emphasis on the ways to improve the energy conversion efficiency. Chapter 8 discusses the energy storage by various means. Chapter 9 discusses the energy conservation and recovery. Chapter 10 briefly introduces energy coupling with examples from biological systems. Chapter 11 focuses on sustainability and life cycle analysis in energy systems to emphasize the implications of the use of energy on the environment, society, and economy. In the newly added Chaps. 12 and 13, 'renewable energy' and 'energy management and economics' are discussed with the latest projections.

Each chapter contains fully solved example problems to support the easy understanding and applications of the topics discussed. At the end of each chapter, enough number of practice problems are listed to provide the students with opportunity towards deep understanding the concepts and aspects of energy. There are 188 fully solved examples problems and 730 practice problems listed at the end of 13 chapters.

It is obvious that the present textbook will mature further in reoccurring editions based on the technological developments and suggestions from the students and colleagues. I want to thank to those who helped me in preparing, developing, and improving this third edition. I especially thank Brad Hailey, Michael Matzen, Nghi Nguyen, Mahdi AlHajji, Hannah Evans, Xiaomeng Wang, Serpil Madden, and Dr. M. A. Abdel-Wahab for their help preparing this new textbook and checking the problems.

I very much want to encourage those who use this textbook to contact me with suggestions and corrections for future editions.

Lincoln, Nebraska, USA   Yaşar Demirel
2020

Contents

Symbols

B	magnetic field
c	cost
c_i	molar concentration of component i
C	Coulomb
°C	degrees Celsius
C_p	constant pressure heat capacity
D	diameter
e	total specific energy
e_p	specific energy potential
F	Faraday constant
°F	degrees Fahrenheit
F	mass force
g	acceleration of gravity
G	Gibbs free energy
h	specific enthalpy, heat transfer coefficient
h_i	partial specific enthalpy of component I
H	enthalpy
i	interest rate, discount rate
I	charge
j	flow
m	mass
M	molar mass
n	number of moles
P	pressure
Pr	Prandtl number
q	heat, degree of coupling
Q	heat, volume flow, charge
$\dot{Q}$	heat flow rate
r	pressure ratio
R	universal gas constant, electrical resistance
s	specific entropy

s_i partial specific entropy of component i
S entropy
t time
T absolute temperature
T   Torque
u specific internal energy, mobility
U internal energy
v specific volume, centre-of-mass or barycentric velocity
V volume, voltage
w_i mass fraction of component i, weight factor
W work
x mole fraction, ratio of forces
X_i thermodynamic force

Greek Symbols

α thermal diffusivity
η efficiency, ratio of flows
ρ density
ν stoichiometric coefficient
ψ electrostatic potential
ε emittance
Ψ energy dissipation function
σ Stefan-Boltzmann constant
λ wavelength
ω angular frequency
ν frequency

Subscripts

e effective
gen generation
i,j,k components
max maximum
min minimum
q heat
r chemical reaction

Introduction: Basic Definitions

1

Introduction and Learning Objectives: This chapter introduces basic definitions, properties, and units, hence builds a vocabulary before discussing the concept of energy and its various aspects. For example, definitions of systems and properties with various units are essential to understand the change of states during energy conversions in various processes. The chapter also discusses heat, work, steam tables, steady-flow, isothermal, and adiabatic processes.

The learning objectives of this chapter are to:

- Understand the basic concepts in all the energy related fields,
- Understand energy, heat, work, and power,
- Understand the state of systems and various processes.

1.1 System

System is a quantity of matter or a region with its boundary in space chosen for study as seen in Fig. 1.1a. *Boundary* separates the system from its surroundings. In *simple systems* matter or region is homogenous with little or no external force fields, such as magnetic and surface tension effects. The boundary may be fixed or moving as seen in Fig. 1.1b. The universe outside the system's boundary is *surroundings*. *Environment* refers to the region beyond the immediate surroundings whose properties are not affected by the process at any point. *Control volume* is any arbitrary region in space through which mass and energy can pass across the boundary as seen in Fig. 1.1c. A control volume may involve heat, work, and material flow interactions with its surroundings.

There are various types of systems with the following brief definitions:

- *Open systems* allow mass and energy exchange across the boundary as seen in Fig. 1.1c.

© Springer Nature Switzerland AG 2021
Y. Demirel, *Energy*, Green Energy and Technology,
https://doi.org/10.1007/978-3-030-56164-2_1

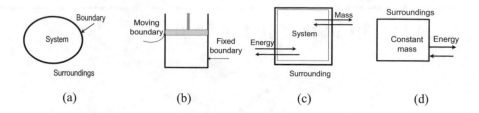

Fig. 1.1 **a** Boundary separates system from its surroundings, **b** a closed system with moving and fixed boundaries, **c** an open system with a control volume; an open system can exchange mass and energy with its surroundings, **d** a closed system does not exchange material with its surroundings, while it can exchange energy

- *Closed systems* consist of a fixed amount of mass, and no mass can cross its boundary but energy can cross the boundary as seen in Fig. 1.1d.
- *Isolated systems* do not exchange mass and energy with their surroundings.
- *Adiabatic systems* do not exchange heat with their surroundings.
- *Isothermal systems* have a uniform temperature everywhere within the system.
- *Isobaric systems* have a uniform pressure everywhere within the system.
- *Isochoric systems* have a constant volume within the system.
- *Steady state systems* have properties, which are independent of time; for example, amount of energy in the systems does not change with time.
- *Unsteady state systems* have properties changing with time.

1.2 Property and Variables

Property is an observable characteristic of a system, such as temperature or pressure. Two types of commonly used properties are *state properties* and *path variables*. The values of *state properties* depend on the state of the system. Temperature, pressure, and volume are examples of state properties. On the other hand, changes in *path variables* depend on the path a system undertakes when it changes from one state to another. For example, work and heat are path dependent variables. Therefore, change of heat in a process that proceeds first at constant pressure and then at constant volume will be different from the one that proceeds first at constant volume and then at constant pressure from state 1 to state 2.

- An *extensive property*, such as volume, depends on the size of the system. If the system consists of several parts, the volume of the systems is the sum of the volumes of its parts, and therefore, extensive properties are additive.
- An *intensive property* does not depend on the size of the system. Some examples of intensive properties are temperature, pressure, and density (specific volume). For a homogeneous system, intensive property can be calculated by dividing an extensive property by the total quantity of the system such as specific volume m^3/kg or specific heat capacity J/g °C.

1.3 Dimensions and Units

Dimensions are basic concepts of measurements such as mass, length, temperature, and time. Units are the means of expressing the dimensions, such as feet or meter for length. Energy has been expressed in several different units. Scientific and engineering quantifications are based on the "International Systems of Units" commonly abbreviated as SI units, which were adopted in 1960 [1, 2]. Table 1.1 shows the seven fundamental measures: length, mass, time, electric current, temperature, amount of substance, and luminous intensity. English system units use length in foot ft, mass in pound lb_m, and temperature in degree Fahrenheit °F. At present, the widely accepted unit of measurement for energy is the SI unit of energy, the joule. Some other units of energy include the kilowatt hour kWh and the British thermal unit Btu. One kWh is equivalent to exactly 3.6×10^6 joules, and one Btu is equivalent to about 1055 joules. The candela is the luminous intensity of a surface of a black body at 1 atm and temperature of freezing platinum. Table 1.2 shows the various units of mass and length.

Table 1.1 Fundamental measures and units

Measure	Symbol	Unit Name & Abbreviation
Length	l	Meter (m), foot (ft)
Mass	m	Kilogram (kg), pound (lb)
Time	t	Second (s)
Current	I	Ampere (A)
Temperature	T	Kelvin (K), Rankine (R)
Luminous intensity	I	Candela (cd)
Plane angle	Φ	Radians (rad)

Table 1.2 Mass and length units and their definitions

Measure	Symbol	Unit Name & Abbreviation
Kilogram	kg	1000 g = 2.204 lb = 32.17 oz
Ounce	oz	28.35 g = 6.25×10^{-2} lb
Pound	lb	0.453 kg = 453 g = 16 oz
Ton, long	ton	2240 lb = 1016.046 kg
Ton, short	sh ton	2000 lb = 907.184 kg
Tonne	t	1000 kg
Ångström	Å	1×10^{-10} m = 0.1 nm
Foot	ft	1/3 yd = 0.3048 m = 12 inches
Inch	in	1/36 yd = 1/12 ft = 0.0254 m
Micron	μ	1×10^{-6} m
Mile	mi	5280 ft = 1760 yd = 1609.344 m
Yard	yd	0.9144 m = 3 ft = 36 in

Table 1.3 Some derived measures and variables

Measure	Symbol	Unit Name & Abbreviation
Area	A	m^2, ft^2
Volume	V	m^3, ft^3
Density	ρ	kg/m^3, lb/ft^3
Pressure	P	Pascal (Pa), psi
Velocity	v	m/s, ft/s
Acceleration	a	m/s^2 ft/s^2
Volume flow	Q	m^3/s, ft^3/s
Mass flow rate	$\dot{m}$	kg/s, lb/s
Mass flux rate	$\dot{m}$	kg/m^2 s, lb/ft^2 s
Force	F	Newton (N), kg m/s^2
Energy	E	Joule (J), Btu (1055.05 J)
Power	P	Watt (W), (W = J/s), Btu/s
Energy flux	E	J/m^2 s, Btu/ft^2 s
Potential difference	ψ	Volt (V)
Resistance	R	Ohm (Ω)
Electric charge	C	Coulomb (C)
Frequency	f	Hertz (Hz)
Electric flux	e	C/m^2

The fundamental measures may be used to derive other quantities and variables called the *derived units*, such as area, volume, density, pressure, force, speed, acceleration, energy, and power. For example, speed is length/time: l/t, acceleration is l/t^2, and force is ml/t^2. Table 1.3 shows some derived variables from fundamental measures, while Table 1.4 shows the prefix system for very small and large magnitudes using metric units.

Table 1.4 Prefixes for SI units

Prefix	Symbol	Factor	Example
Peta	P	10^{15}	Petameter, Pm = 10^{15} m
Tera	T	10^{12}	Terabyte, TB = 10^{12} byte
Giga	G	10^9	Gigajoule, GJ = 10^9 J
Mega	M	10^6	Megawatt, MW = 10^6 W
Kilo	k	10^3	kilojoule, kJ = 10^3 J
Hecto	h	10^2	hectometer, hm = 10^2 m
Deka	da	10^1	decaliter, dal = 10^1 l
Deci	d	10^{-1}	deciliter, dl = 10^{-1} l
Centi	c	10^{-2}	centimeter, cm = 10^{-2} m
Milli	m	10^{-3}	milligram, mg = 10^{-3} g
Micro	μ	10^{-6}	microvolt, μV = 10^{-6} V
Nano	n	10^{-9}	nanometer, nm = 10^{-9} m
Pico	p	10^{-12}	picometer, pm = 10^{-12} m
Femto	f	10^{-15}	femtometer, fm = 10^{-15} m

1.4 Measures of Amounts and Fractions

Common measurements for amount are mass m, number of moles n, or total volume V. Two common fractions are the mass fractions and mole fractions.

- *Mass fraction* is the mass of a substance divided by the total mass of all the substances in the mixture or solution. Mass fraction w_i for a substance i is

$$w_i = \frac{m_i}{m_{\text{total}}} \tag{1.1}$$

- Number of moles n is the ratio of a substance mass m to its molecular weight MW

$$n = \frac{m}{MW} \tag{1.2}$$

- *Mole fraction* x is the ratio of moles of a substance in a mixture or solution divided by the total number of moles in the mixture or solution. Mole fraction x_i for a substance i is

$$x_i = \frac{n_i}{n_{\text{total}}} \tag{1.3}$$

For example, assuming that the mole fraction of nitrogen is 79% and of oxygen is 21% in air, then the mass fractions of nitrogen and oxygen are 76.83% and 23.17%, respectively. Here, the molecular weight of nitrogen is 28.2 g/mol and oxygen 32 g/mol.

1.5 Force

Force F is defined as the product of mass m and acceleration a, according to the Newton's second law

$$F = ma \tag{1.4}$$

The SI unit of force is the newton defined by $(1\ \text{kg})(\text{m/s}^2)$. One kg of mass can be accelerated one m/s^2 by a force of one newton.

The English engineering system of unit of force is the pound force lb$_f$, which can accelerate one pound of mass 32.174 ft/s^2. For the consistency of units, one-pound force is

$$F = \frac{1}{g_c} ma = \frac{1}{g_c} \text{lb}\ 32.174 \frac{\text{ft}}{\text{s}^2} \tag{1.5}$$

where g_c is the proportionality constant

$$g_c = 32.174 \frac{\text{lb ft}}{\text{lb}_f s^2}$$

A pound force and a pound mass are different quantities. If an equation contains both the pound force and pound mass, the proportionality constant g_c is necessary for the equation to have correct dimensions.

Weight is the force of gravity acting on a body and is expressed in newton or in pound force. Since the acceleration of gravity is $g = 9.8$ m/s^2, the weight in the SI unit is

$$F = mg \tag{1.6}$$

In the English engineering system, the weight is

$$F = m \frac{g}{g_c} \tag{1.7}$$

Since the gravitational force is different at different elevations, the weight is a function of elevation. Therefore, the weight for the same mass of a body would vary at different elevations.

1.6   Temperature

The rate of change of entropy when a small amount of energy is added to a system at fixed volume in thermodynamic equilibrium defines temperature

$$\frac{1}{T} = \frac{dS}{dU}\bigg|_V \tag{1.8}$$

Temperature is a physical property that underlies the common notions of hot and cold. Something that feels hotter generally has a higher temperature. Heat flows from a matter with the higher temperature to a matter with the lower temperature. If no net heat flow occurs between two matters, then they have the same temperatures. Statistical physics describe matter as a collection of many particles and define temperature as a measure of the average energy in each degree of freedom of the particles in the matter. For a matter, this energy is found primarily in the vibrations of its atoms. In an ideal monatomic gas, energy is found in the translational motions resulting in changes in position of the particles; with molecular gases, vibrational and rotational motions also provide thermodynamic degrees of freedom [3].

Temperature is measured with thermometers that may be calibrated to a variety of scales as seen in Fig. 1.2a. Table 1.5 summarizes the units, definitions, and conversions of temperature to each other. Common temperature scales are:

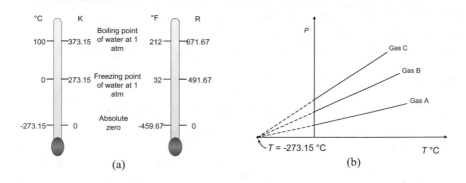

Fig. 1.2 a Comparison of different temperature scales; **b** Using three different gases in a constant-volume gas thermometer, all the straight lines, showing the change of pressure with temperature, will intersect the temperature axis at −273.15°C after extrapolation of measurements at low pressures. This is the lowest temperature that can be obtained regardless the nature of gases. So, the absolute gas temperature scale, called the Kelvin scale, is obtained by assigning a value of 0 K to this lowest temperature of −273.15 °C

Table 1.5 Temperature units and their definitions

Unit	Symbol	Definition	Conversion
Degree Celsius	°C	°C = K − 273.15	°C = (°F − 32)/1.8
Degree Fahrenheit	°F	°F = R − 459.67	°F = °C × 1.8 + 32
Degree Rankine	R	R = °F + 459.67	R = K × 1.8
Degree Kelvin	K	K = °C + 273.15	K = R/1.8

- On the *Celsius scale*, the freezing point of water is 0 °C and the boiling point of water 100 °C (at standard atmospheric pressure). Celsius scale is used in the SI system and named after Anders Celsius.
- On the *Fahrenheit scale*, the freezing point of water is 32°F and the boiling point 212 °F (at standard atmospheric pressure). The. Fahrenheit scale is used in the English engineering units system and named after Daniel Gabriel Fahrenheit.
- *Thermodynamic temperature* scale is independent of the properties of the substances and called the *Kelvin scale* or *Rankine scale*. The temperature on this scale is called the *absolute temperature*.
- *Kelvin* scale is the thermodynamic temperature scale in the SI units. The lowest temperature on the Kelvin scale is −273.15 °C as seen in Fig. 1.2b. The Kelvin temperature scale matches the Celsius scale: 1K = 1 °C. The Kelvin temperature scale is named after Lord Kelvin.
- *Rankine scale* is the thermodynamic temperature scale in the English engineering system units. The lowest temperature on the Rankine scale is −459.67 ° F. Rankine and Fahrenheit temperature scales have the same degree intervals: 1 R = 1°F. The Rankine temperature scale is named after W. Rankine.

Example 1.1 Conversion of temperature units

(a) Convert 27 °C to °F, K, and R.
(b) Express a change of 25 °C in K and a change of 70°F in R.

Solution:

(a) $T(\text{K}) = T(°\text{C}) + 273.15\ °\text{C} = 27\ °\text{C} + 273.15\ °\text{C} = \mathbf{300.15\ K}$

   $T(°\text{F}) = 1.8T(°\text{C}) + 32 = (1.8)(27)°\text{F} + 32°\text{F} = \mathbf{80.6°F}$

   $T(\text{R}) = T(°\text{F}) + 460°\text{F} = 80.6°\text{F} + 460°\text{F} = \mathbf{540.6\ R}$

(b) A change of 25 °C in K is: $\Delta T(\text{K}) = \Delta T(°\text{C}) = 25$ K as Kelvin and Celsius scales are identical.
   A change 70°F in R is: $\Delta T(\text{R}) = \Delta T(°\text{F}) = 70$ R as Rankine and Fahrenheit scales are identical.

1.7 Pressure

Pressure P is the force per unit cross sectional area applied in a direction perpendicular to the surface of an object.

$$P = \frac{F}{A} \tag{1.9}$$

The SI unit for pressure is the Pascal (Pa) defined as 1 Pa = 1 N/m^2 = (1 kg m/s^2)/ m^2. Two common units for pressure are the pounds force per square inch abbreviated as 'psi' and atmosphere abbreviated as 'atm', and they are related by 1 atm = 14.659 psi. The atmospheric pressure is variable, while the standard atmosphere is the pressure in a standard gravitational field and is equal to 1 atm, or 101.32 kPa, or 14.69 psi. Pressure is measured by a manometer. Figure 1.3a shows a basic manometer. Atmospheric pressure is measured by barometer as seen in Fig. 1.3b. The pressure at the bottom of the static column of mercury in Fig. 1.3b is

$$P = \rho g z + P_o \tag{1.10}$$

where P is the pressure at the bottom of the column of the fluid, ρ is the density of the fluid, g is the acceleration of gravity, z is the height of the fluid column, and P_o is the pressure at the top of the column of the fluid.

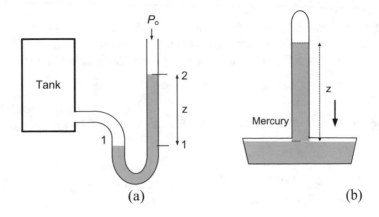

Fig. 1.3 a Basic manometer: $P = \rho g z + P_o$, **b** basic mercury barometer to measure atmospheric pressure. The length or the cross-sectional area of the tube has no effect on the height of the fluid in the barometer

- According to the measuring device, we can measure *relative* or *absolute* pressures. *Gauge pressure*, psig, is the pressure relative to the local atmospheric or ambient pressure. Gauge pressure is measured by an open-end manometer as seen in Fig. 1.3a. If the open end of the manometer is closed and a vacuum is created in that end, then we measure *absolute pressure*, psia. For example, the standard atmosphere is absolute pressure. The relationship between absolute and relative pressures is

$$\text{Absolute pressure} = \text{Gauge pressure} + \text{Atmospheric pressure} \qquad (1.11)$$

- In common usage, pressure is frequently called as *technical atmosphere*, at, (kilogram-force per square centimeter) (kg$_f$/cm^2), and kg$_f$/cm^2 = 9.8×10^4 Pa. Table 1.6 shows some conversion factors among the common pressure units.

Table 1.6 Pressure conversion factors [2]

	kPa	bar	atm	mm Hg	psi
kPa	1	10^{-2}	9.869×10^{-3}	7.50	145.04×10^{-3}
bar	100	1	0.987	750.06	14.503
atm	101.32	1.013	1	760	14.696
mm Hg	0.133	1.333×10^{-3}	1.316×10^{-3}	1	19.337×10^{-3}
psi	6.894×10^3	68.948×10^{-3}	68.046×10^{-3}	51.715	1

1 Pa = 1 N/m^2 = 10^{-5} bar = 10.197×10^{-6} at = 9.8692×10^{-6} atm = 7.5006×10^{-3} torr = 145.04×10^{-6} psi

Example 1.2 Pressure calculations

Consider a column of mercury Hg that has an area of 1 cm^2 and a height of 20 cm. The top of the column is open to atmosphere. The density of mercury at 20 °C is 13.55 g/cm^3. Estimate the pressure exerted by the column of mercury on the sealing plate

Solution:

Data: $A = 1$ cm^2; $z = 20$ cm; $g = 980$ cm/s^2; $\rho = 13.55$ g/cm^3 at 20°C, $P_o = 101.32$ kPa.

$$P = \rho g z + P_o = \left(\frac{13.55\,\text{g}}{\text{cm}^3}\right)\left(\frac{980\,\text{cm}}{\text{s}^2}\right)(20\,\text{cm})\left(\frac{\text{kg}}{1000\,\text{g}}\right)\left(\frac{\text{N s}^2}{\text{kg m}}\right)\left(\frac{100\,\text{cm}}{1\,\text{m}}\right)\left(\frac{\text{m}^2\text{Pa}}{\text{N}}\right)\left(\frac{\text{kPa}}{1000\,\text{Pa}}\right)$$
$$+ 101.32\,\text{kPa} = 26.56\,\text{kPa} + 101.32\,\text{kPa} = 127.88\,\text{kPa}$$

P = 127.88 kPa

In English engineering units, we have the density of mercury

$$\rho = \left(\frac{13.55\,\text{g}}{\text{cm}^3}\right)\left(\frac{62.4\,\text{lb}_m}{\text{ft}^3}\right) = 845\,\text{lb}_m/\text{ft}^3$$

The pressure in English units becomes:

$$P = \rho g z + P_o = \left(\frac{845.5\,\text{lb}_m}{\text{ft}^3}\right)\left(\frac{32.2\,\text{ft}}{\text{s}^2}\right)(20\,\text{cm})\left(\frac{\text{in.}}{2.54\,\text{cm}}\right)\left|\frac{\text{ft.}}{12\,\text{in.}}\right|\left|\frac{\text{s}^2\text{lb}_f}{32.174\,\text{ft lb}_m}\right|\left(\frac{\text{ft}^2}{144\,\text{in}^2}\right)$$
$$+ 14.69\,\text{psi} = 3.85\,\text{psi} + 14.69\,\text{psi} = 18.54\,\text{psi}$$

P = 18.54 psi

In practice, the height of a liquid column is referred to as *head* of liquid. Therefore, the pressure of the column of mercury is expressed as 20 cm Hg, while the pressure on the sealing plate at the bottom becomes 20 cm Hg + 76 cm Hg (1 atm) = 96 cm Hg.

Example 1.3 Pressure conversions

Convert the pressure of 10 GPa to (a) psia, (b) atm, and (c) bar

Solution:

(a) $\dfrac{10\,\text{Gpa}}{1}\left(\dfrac{10^9\text{Pa}}{\text{GPa}}\right)\left(\dfrac{14.69\,\text{psia}}{101.32 \times 10^3\text{Pa}}\right) = 1.45 \times 10^6\text{psia}$

(b) $\dfrac{10\,\text{Gpa}}{1}\left(\dfrac{10^9\text{Pa}}{\text{GPa}}\right)\left(\dfrac{\text{atm}}{101.32\times10^3\text{Pa}}\right)=0.098\times10^6\text{atm}$

(c) $\dfrac{10\,\text{Gpa}}{1}\left(\dfrac{10^9\text{Pa}}{\text{GPa}}\right)\left(\dfrac{\text{bar}}{100.0\times10^3\text{Pa}}\right)=0.1\times10^6\text{bar}$

Example 1.4 Absolute pressure estimations
A gauge connected to a chamber kept under vacuum reads 7.6 psig. The atmospheric pressure at the location is 14.3 psi. Estimate the absolute pressure in the chamber.

Solution:
Absolute pressure shows the actual pressure at a given point. Pressures below atmospheric pressure are called vacuum pressure, and the gauges measuring it indicate the difference between atmospheric pressure and absolute pressure.
$P_{\text{gauge}} = P_{\text{abs}} - P_{\text{atm}}$ (For pressures above atmospheric pressure)
$P_{\text{gauge}} = P_{\text{vacuum}} = P_{\text{atm}} - P_{\text{abs}}$ (For pressures below atmospheric pressure)
$P_{\text{abs}} = P_{\text{atm}} - P_{\text{gauge}} = 14.3\,\text{psi} - 7.6\,\text{psig} = \mathbf{6.7\,psia}$

1.8 Volume

Volume is how much three-dimensional space occupies or contains a substance. The units of volume depend on the units of length. If the lengths are in meters, the volume will be in *cubic* meters, m^3; if they are in feet the volume is in cubic feet, ft^3. Table 1.7 shows common units and their definitions for volume. Table 1.8 shows the conversion factor between the various volume units. Volume related derivative properties are:

- Total volume V_t of a system may be divided by the mass to calculate *specific volume v*. Specific volume is the inverse of density, such as m^3/kg of ft^3/lb

$$\text{specific volume}(v) = \frac{1}{\rho} = \frac{V}{m} \tag{1.12}$$

Table 1.7 Volume units and their definitions [1, 4]

Name of unit	Symbol	Definitions
Barrel (Imperial)	bl (Imp)	36 gal (Imp) = 0.163 m^3
Barrel	bl; bbl	42 gal (US) = 0.158 m^3
Cubic foot	cu ft	0.028 m^3
Cubic inch	cu in	16.387 $\times 10^{-6}$ m^3
Cubic meter	m^3	1 m^3 = 1000 L
Cubic yard	cu yd	27 cu ft = 0.764 m^3
Gallon (U.S.)	gal (US)	3.785 $\times 10^{-3}$ m^3 = 3.785 L
Ounce	US fl oz	1/128 gal (US) = 29.573 $\times 10^{-6}$ m^3
Pint	pt (US dry)	1/8 gal (US dry) = 550.610 $\times 10^{-6}$ m^3
Quart	qt (US)	¼ gal (US dry) = 1.101 $\times 10^{-3}$ m^3
Liter	L	1000 cm^3 = 10^{-3} m^3

Table 1.8 Volume conversion factors [2]

	in^3	ft^3	U.S gal	Liters	m^3
in^3	1	5.787$\times 10^{-4}$	4.329$\times 10^{-3}$	1.639$\times 10^{-2}$	1.639$\times 10^{-5}$
ft^3	1.728$\times 10^3$	1	7.481	28.32	2.832$\times 10^{-2}$
U.S gal	231	0.133	1	3.785	3.785$\times 10^{-3}$
Liters	61.03	3.531$\times 10^{-2}$	0.264	1	1.000$\times 10^{-3}$
m^3	6.102$\times 10^4$	35.31	264.2	1000	1

- Total volume may be divided by number of moles n to calculate *molar volume*

$$V_m = \frac{V_t}{n} \tag{1.13}$$

- *Density* ρ is the ratio of mass per unit volume, such as lb/ft^3, or g/cm^3

$$\rho = \frac{\text{mass}}{\text{volume}} = \frac{m}{V} \tag{1.14}$$

Densities of liquids and solids do not change significantly with pressure at ordinary conditions. However, densities of liquids and solids change with temperature. Densities of gases change with pressure and temperature. For example, the density of ethanol is 0.790 g/cm^3 and the volume of 1 lb of ethanol is (454 g/lb) (cm^3/0.790 g) = 574.7 cm^3/lb.

- *Specific gravity* (sp. gr.) is the ratio of densities of the substance to density of a reference substance

$$\text{specific gravity(sp.gr.)} = \frac{\rho}{\rho_{H_2O\,at\,4°C}} \qquad (1.15)$$

In specific gravity, the reference substance usually is water at 4 °C, with density of 1 g/cm^3, or 62.43 lb/ft^3. The specific gravity of gases frequently is referred to air. Therefore, the data for specific gravity is expressed by both the temperature of the substance considered and the temperature of the reference density measured. For example, sp. gr. = 0.82 at 25°/4° means the substance is at 20°C and the reference substance is at 4 °C. As the density of water at 4 °C is 1 g/cm^3, the numerical values of the specific gravity and density are the same in the SI unit system.

In the petroleum industry the specific gravity of petroleum products is often reported in terms of a hydrometer scale called API *gravity*, and expressed with a standard temperature of 60°F

$$°API\,gravity = \frac{141.5}{sp.gr.\frac{60°F}{60°F}} - 131.5 \qquad (1.16)$$

Concentration usually refers to the quantity of a substance per unit volume. Some common concentrations are:

- *Mass concentration* is the mass per unit volume such as lb solute/ft^3 or g solute/l.
- *Molar concentration* is the moles per unit volume, such as lbmol/ft^3 or mol/l.
- *Parts per million*, ppm, or *parts per billion*, ppb, expresses the concentration of extremely dilute solutions or mixtures

$$ppm = \frac{parts}{10^6 parts} \qquad (1.17)$$

They are dimensionless and equivalent to a mass fraction for solids and liquids, and a mole fraction for gases. For example, about 9 ppm carbon monoxide means that nine parts (mass or mole unit) per million part of air.

- *Parts per million by volume*, ppmv, or *parts per billion by volume*, ppbv.

1.9 Energy

Energy is the capacity to do work. Energy comes in various forms, such as motion, heat, light, electrical, chemical, nuclear, and gravitational. *Total energy* is the sum of all forms of the energy a system possesses. In the absence of magnetic, electrical, and surface tension effects, the total energy of a system consists of the *kinetic, potential, and internal energies*. The *internal energy* of a system is made up of sensible, latent, chemical, and nuclear energies. The sensible internal energy is due to translational, rotational, and vibrational effects of molecules. *Thermal energy* is the sensible and latent forms of internal energy. The classification of energy into different "types" often follows the boundaries of the fields of study in the natural sciences. For example, *chemical energy* is the kind of potential energy stored in chemical bonds, and *nuclear energy* is the energy stored in interactions between the particles in the atomic nucleus. *Microscopic forms of energy* are related to the molecular structure of a system and they are independent of outside reference frames [3].

Hydrogen represents a store of potential energy that can be released by fusion of hydrogen in the Sun. Some of the fusion energy is then transformed into sunlight, which may again be stored as gravitational potential energy after it strikes the earth. For example, water evaporated from the oceans, may be deposited on elevated parts of the earth, and after being released at a hydroelectric dam, it can drive turbines to produce energy in the form of electricity. Atmospheric phenomena like wind, rain, snow, and hurricanes, are all a result of energy transformations brought about by solar energy on the atmosphere of the earth. Sunlight is also captured by plants as *chemical potential energy* in photosynthesis, when carbon dioxide and water are converted into carbohydrates, lipids, and proteins. This chemical potential energy is responsible for the growth and development of a biological cell.

British thermal unit (Btu) is the energy unit in the English system needed to raise the temperature of 1 lb_m of water at 68°F by 1°F. *Calorie* (cal) is the amount of energy in the metric system needed to raise the temperature of 1 g of water at 15 °C by 1 °C. Table 1.9 displays some of the important energy units and their definitions.

1.10 Work

Energy is a conserved and extensive property of every system in any state and hence its value depends only on the state. *Work*, on the other hand, is not a property of a system. Energy may be transferred in the forms of heat or work through the boundary of a system. In a complete cycle of steady state process, internal energy change is zero and hence the work done on the system is converted to heat (|work|=| heat|) by the system. The mechanical work of expansion or compression proceeds with the observable motion of the coordinates of the particles of matter. *Chemical work*, on the other hand, proceeds with changes in internal energy due to changes in the chemical composition (mass action). *Potential energy* is the capacity for

Table 1.9 Some energy units and definitions

Name of unit	Symbol	Definitions
British thermal unit	Btu	$1055 \text{ J} = 5.4039 \text{ psia ft}^3$
Btu/lb$_m$	Btu/lb$_m$	2.326 kJ/kg
Joule	J	$J = mN = 1 \text{ kgm}^2/\text{s}^2$
Calorie	Cal	4.1868 J
kJ	kJ	$\text{kPa m}^3 = 1000 \text{ J}$
kJ/kg	kJ/kg	0.43 Btu/lb$_m$
Erg	Erg	$g \cdot \text{cm}^2/\text{s}^2 = 10^{-7} \text{ J}$
Foot pound force	ft lb$_f$	$g \times \text{lb} \times \text{ft} = 1.355 \text{ J}$
Horsepower hour	hph	$\text{hp} \times \text{h} = 2.684 \times 10^6 \text{ J}$
Kilowatt hour	kWh	$\text{kW} \times \text{h} = 3.6 \times 10^6 \text{ J}$
Quad	Quad	$1 \times 10^{15} \text{ Btu} = 1.055 \times 10^{18} \text{ J}$
Atmosphere liter	Atml	$\text{atm} \times 1 = 101.325 \text{ J}$
kW	kW	3412 Btu/h
Horsepower	hp	2545 Btu/h
Therm	Therm	29.3 kWh
Electronvolt	eV	$\approx 1.602 \; 17 \times 10^{-19} \pm 4.9 \times 10^{-26} \text{ J}$

mechanical work related to the position of a body, while *kinetic energy* is the capacity for mechanical work related to the motion of a body. Potential and kinetic energies are *external energies*, while sensible heat and latent heat are *internal energies*. Table 1.10 lists the conversion coefficients between various units of energy [3].

1.11 Power

The rate of work done is measured by power. The instantaneous power created by a force F on an object moving with velocity $v = dx/dt$ becomes: $P = F (dx/dt) = Fv$. So power is estimated only if the object is in motion. *Power* is the energy exchanged in time. For example, Watt (W) gives rise to one joule of energy in one second of time. Table 1.11 displays some of the power definitions and their conversions. The rate of work that is the work per unit time is called the *power*. In SI units, power is defined as $W = J/s = Nm/s = kg \cdot m^2/s^3$. In English units' power is in Btu/s = 778.1 ft lb$_f$/s = 1.415 hp; mechanical hp = 550 ft lb$_f$/s = 746 W = (electrical hp).

- *Horsepower* is the imperial (British) unit of power. A horsepower is the ability to do work at the rate of 33,000 lb$_f$ ft/min or 550 lb$_f$ ft/s. For example, the total horsepower developed by water falling from a given height is

Table 1.10 Heat, energy, and work conversions

	ft lb	kWh	hph	Btu	Calorie	Joule
ft lb	1	3.766×10^{-7}	5.050×10^{-7}	1.285×10^{-3}	0.324	1.356
kWh	2.655×10^{6}	1	1.341	3.413×10^{3}	8.606×10^{5}	3.6×10^{6}
hph	1.98×10^{6}	0.745	1	2.545×10^{3}	6.416×10^{5}	2.684×10^{6}
Btu	778.16	2.930×10^{-4}	3.930×10^{-4}	1	252	1.055×10^{3}
Calorie	3.086	1.162×10^{-6}	1.558×10^{-6}	3.97×10^{-3}	1	4.184
Joule	0.737	2.773×10^{-7}	3.725×10^{-7}	9.484×10^{-4}	0.2390	1

Table 1.11 Power conversions

	hp	kW	ft lb$_f$/s	Btu/s	J/s
hp	1	0.745	550	0.707	745.7
kW	1.341	1	737.56	0.948	1000
ft lb$_f$/s	1.818×10^{-3}	1.356×10^{-3}	1	1.285×10^{-3}	1.356
Btu/s	1.415	1.055	778.16	1	1055
J/s	1.341×10^{-3}	0.001	0.737	9.478×10^{-3}	1

$$P_{hp} = \frac{\dot{m}gz}{550} \tag{1.18}$$

Or in terms of volumetric flow rate, horsepower is

$$P_{hp} = \frac{\dot{Q}\rho gz}{550} \tag{1.19}$$

where P_{hp} is the horsepower hp, $\dot{m}$ is the mass flow rate (lb/s), z is the head or height (ft), g is the acceleration of gravity (32 ft/s^2), and $\dot{Q}$ is the volumetric flow rate (ft^3/s).

- *The brake horsepower*, The is the amount of real horsepower going to the pump, not the horsepower used by the motor. Due to hydraulic, mechanical, and volumetric losses in a pump or turbine the actual horsepower available for work is less than the total horsepower supplied.

Example 1.5 Power conversions

A 20 kW electric motor is used to pump ground water into a storage tank. Estimate the work done by the pump in (a) Btu/hr, (b) hp, (c) J/s, and (d) ft lb$_f$/h.

Solution:

(a) $(20\,\text{kW}) \left(\dfrac{0.948\,\text{Btu/s}}{\text{kW}} \right) \left(\dfrac{3600\,\text{s}}{\text{h}} \right) = 6.825 \times 10^4\,\text{Btu/h}$

(b) $(20\,\text{kW}) \left(\dfrac{1.341\,\text{hp}}{\text{kW}} \right) = 26.82\,\text{hp}$

(c) $(20\,\text{kW}) \left(\dfrac{1000\,\text{J/s}}{\text{kW}} \right) = 2 \times 10^4\,\text{J/s}$

(d) $(20\,\text{kW}) \left(\dfrac{737.5\,\text{ft lb}_f/\text{s}}{\text{kW}} \right) \left(\dfrac{3600\,\text{s}}{\text{h}} \right) = 53.1 \times 10^6\,\text{ft lb}_f/\text{h}$

1.12 Heat Capacity

The heat capacity C is the amount of heat necessary to raise the temperature of a matter (system) by one degree. When the temperature change occurs at constant pressure or at the constant volume the heat capacity is shown by C_p or C_v, respectively. For a unit amount, such as gram or pound, it is called the specific heat capacity with the units of J/g °C. It is also expressed in molar units as J/mol °C.

1.13 Heat

Heat is part of the total energy flow across a system boundary that is caused by a temperature difference between the system and its surroundings or between two systems at different temperatures. Heat flows from high temperature region to cold temperature region. Therefore, heat is not stored and defined as thermal energy in transit. The unit for heat in the SI system is the joule, J. When heat capacity is constant, such as $C_{p,av}$, the amount of heat changed when a substance changed its temperature from T_1 to another temperature T_2 is

$$q = mC_{p,av}(T_2 - T_1) \qquad (1.20)$$

where q is the heat, and m is the mass. The values of specific heat capacity at constant pressure, C_p, for various substances are shown in Table B3 in Appendix B. The value of heat is positive when transferred to the system and negative when transferred from the system to its surroundings. There is no heat flow through the boundary in an *adiabatic process* as the system and its surroundings have the same temperatures. *Thermal equilibrium* is achieved when two systems in thermal contact with each other cease to exchange energy by heat.

1.14 Internal Energy

Internal energy U of a system or a body with well-defined boundaries is the total of the kinetic energy due to the motion of molecules (translational, rotational, vibrational) and the potential energy associated with the vibrational motion and electric energy of atoms within molecules. Internal energy also includes the energy in all the chemical bonds. From a microscopic point of view, the internal energy may be found in many different forms. For a gas, internal energy comprises of translational kinetic energy, rotational kinetic energy, vibrational energy, electronic energy, chemical energy, and nuclear energy. For any material, solid, liquid, or gaseous, it may also consist of the potential energy of attraction or repulsion between the individual molecules [3].

Internal energy is a state function of a system and is an extensive quantity. One can have a corresponding intensive thermodynamic property called *specific internal energy*, commonly symbolized by the lower-case letter u, which is internal energy per mass of the substance in question. As such, the SI unit of specific internal energy would be the J/g. For a closed system, the internal energy is essentially defined by

$$\Delta U = q + W \tag{1.21}$$

where ΔU is the change in internal energy of a system during a process, q is the heat, and W is the mechanical work. The heat absorbed by a system goes to increase its internal energy plus to some external work.

If an energy exchange occurs because of temperature difference between a system and its surroundings, this energy appears as *heat* otherwise it appears as *work*.

1.15   Enthalpy

Enthalpy H of a system is defined by

$$H = U + PV \tag{1.22}$$

where U is the internal energy, P is the pressure at the boundary of the system and its environment, and V is the volume of the system. The PV term is equivalent to the energy which would be required to make the pressure of the environment remained constant. For example, a gas changing its volume can maintain a constant pressure P. This may be possible, for example, by a chemical reaction that causes the change in volume of a gas in a cylinder and pushes a piston. The SI unit of enthalpy is the joule. The SI unit for specific enthalpy is joules per kilogram. Enthalpy is a state function of a system and is an extensive quantity.

Whenever the enthalpy of a matter is independent of pressure, regardless of the process, or at constant pressure, the change of enthalpy is

$$\Delta H = H_2 - H_1 = C_{p,\mathrm{av}}(T_2 - T_1) \tag{1.23}$$

where the $C_{p,\mathrm{av}}$, is an average value within the temperature interval $(T_2 - T_1)$. ΔH of a system is equal to the sum of non-mechanical work done on it and the heat supplied to it. The increase in enthalpy of a system is exactly equal to the energy added through heat, provided that the system is under constant pressure: $\Delta H = q$. Therefore, enthalpy is sometimes described as the *"heat content"* of a system under a given pressure [3].

1.16   Entropy

Entropy change is determined by the following equation

$$dS = \frac{\delta q_{rev}}{T} \qquad (1.24)$$

where δq_{rev} is the reversible heat flow. In an integrated form, the reversible change of heat flow is estimated by

$$q_{rev} = T\Delta S \qquad (1.25)$$

When a fluid system changes from state A to state B by an irreversible process, then the change of its entropy is $S = S_B - S_A$. Entropy is a state function and an extensive property. The total change of entropy is expressed by: $\Delta S = \Delta S_{boundary} + \Delta S_{internal}$, where $\Delta S_{boundary}$ is the change due to the interaction of a system with its surroundings, and $\Delta S_{internal}$ is the increase due to a natural change within the system, such as a chemical reaction, and is always positive for irreversible changes and zero at equilibrium [1]. For ideal gases ($PV = RT$, and $dV/dT = R/P$), the change in entropy is

$$dS = C_p \frac{dT}{T} - R \frac{dP}{P} \qquad (1.26)$$

For incompressible fluids, entropy change becomes

$$dS = C_{p,av} \frac{dT}{T} \qquad (1.27)$$

Energy exchange accompanied by entropy transfer is the heat transfer, and energy exchange that is not accompanied by entropy transfer is the work transfer.

1.17   State

State is a set of properties that completely describes the condition of that system. At this point, all the properties can be measured or calculated.

- *Standard reference state* for the properties of chemical components is chosen as 25°C (77°F) and 1 atm. Property values at the standard reference state are indicated by a superscript (°).

- *In steady state,* any kind of accumulation, such as mass or energy, within the system is zero. The flows in and out remain constant in time, and the properties of the system do not change in time.
- *Gas* and *vapor states* are often used as synonymous words. Gas state of a substance is above its critical temperature. Vapor, on the other hand, implies a gas state that is not far from its condensation temperature.
- *Thermal equilibrium* means that the temperature is the same throughout the entire system.
- A system is in *mechanical equilibrium* if there is no change in pressure at any point of the system with time.
- In *chemical equilibrium*, the reaction rates in forward and backward directions become equal.
- An ideal gas *equation of state* relates the pressure, temperature and volume of a substance:

$$PV = nRT \tag{1.28}$$

where P is the absolute pressure, T is the absolute temperature, n is the number of moles of a substance and R is called the universal gas constant, which is the same for all substances:

$R = 8.314$ kJ/mol K $= 8.314$ kPa m^3/k mol K
$R = 1.986$ Btu/lbmol R $= 10.73$ psia ft^3/lbmol R

The ideal-gas relation approximately represents the P-V-T behavior of real gases only at low densities (high volume and low pressure). For a constant mass, the properties of an ideal gas at two different states are related to each other by

$$\frac{P_1 V_1}{T_1} = \frac{P_2 V_2}{T_2} \tag{1.29}$$

1.17.1 Saturated Liquid and Saturated Vapor States

A pure substance will start to boil when its vapor pressure is equal to existing atmospheric pressure. This boiling temperature is called the saturation temperature. *Saturated liquid and saturated vapor states* exist at the saturation temperature corresponding to the existing pressure. Consequently, *superheated vapor* exists at a higher temperature and pressure compared with the saturated temperature and pressure. Figure 1.4a shows a typical phase diagram for a pure substance. Between the solid and liquid phases, there is a solid-liquid equilibrium curve. Between liquid and vapor phases there is a liquid-vapor equilibrium curve that exist at the

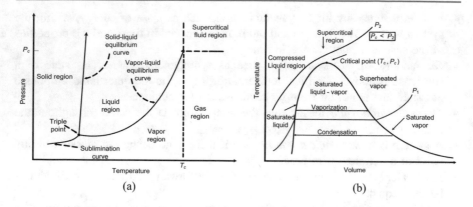

Fig. 1.4 **a** Pressure-temperature diagram of a pure component; **b** temperature-volume diagram of a pure substance. At critical point, saturate liquid line and saturated vapor lines merge. Within the saturation curves, a mixture of vapor and liquid phases exist. Super critical region represents a state of matter for temperature and pressure above P_c and T_c that are the critical pressure and temperature

saturation temperatures and corresponding pressures [4]. At critical point (P_c, T_c) the saturated liquid and vapor states become identical. At triple point, the three phases coexist in equilibrium. The zone where the isobars become flat shows the two-phase region, which is enveloped by saturated liquid and saturated vapor lines. At the supercritical region: $T > T_c$ and $P > P_c$.

The proportions of vapor and liquids phases in the saturated *liquid-vapor* mixture is called the *quality x,* which is the ratio of mass of vapor to the total mass of the mixture

$$\text{quality}(x) = \frac{\text{mass of vapor}}{\text{total mass of mixture}} \qquad (1.30)$$

Average values of specific volume, enthalpy, and internal energy of a saturated liquid-vapor mixture are estimated by

$$V = (1 - x)V_{\text{liq sat}} + xV_{\text{vap sat}} \qquad (1.31)$$

$$H = (1 - x)H_{\text{liq sat}} + xH_{\text{vap sat}} \qquad (1.32)$$

$$U = (1 - x)U_{\text{liq sat}} + xU_{\text{vap sat}} \qquad (1.33)$$

1.17.2 Partial Pressure and Saturation Pressure

Partial pressure is the pressure exerted by a single component in a gaseous mixture if it existed alone in the same volume occupied by the mixture and at the same temperature and pressure of the mixture. Partial pressure depends on the pressure and temperature. On the other hand, *saturation pressure* is the pressure at which liquid and vapor phases are at equilibrium at a given temperature. Saturation pressure depends only on the temperature. When a saturation pressure of a pure component reaches the atmospheric pressure, the liquid starts to boil. For example, water boils at 100 °C when its saturation pressure equals to the atmospheric pressure of 1 atm. The Antoine equation may be used to estimate the saturation pressure

$$\ln P_{sat} = A - \frac{B}{T + C} \quad \left(P \text{ in kPa and } T \text{ in } ^\circ C\right) \tag{1.34}$$

where *A, B*, and *C* are the Antoine constants. Table 1.12 lists the Antoine constants for some substances [4].

Example 1.6 Estimation of saturated vapor pressure
(a) Estimate the saturation vapor pressure of water and acetone 50 °C, (b) Estimate the boiling point temperature of water and acetone at $P = 101.32$ kPa.
Solution
The Antoine constants from Table 1.13:
Water: $A = 16.3872$; $B = 3885.70$; $C = 230.17$
Acetone: $A = 14.3145$; $B = 2756.22$; $C = 228.06$

Table 1.12 Antoine constants [4] for some components: *P* is in kPa and *T* is in °C

Species	A	B	C	Range (°C)
Acetone	14.314	2756.22	228.06	−26–77
Acetic acid	15.071	3580.80	224.65	24–142
Benzene	13.782	2726.81	217.57	6–104
1-Butanol	15.314	3212.43	182.73	37–138
Carbon tetrachloride	14.057	2914.23	232.15	−14–101
Chloroform	13.732	2548.74	218.55	−23–84
Ethanol	16.895	3795.17	230.92	3–96
Ethyl benzene	13.972	3259.93	212.30	33–163
Methanol	16.578	3638.27	239.50	−11–83
Methyl acetate	14.245	2662.78	219.69	−23–78
1-propanol	16.115	3483.67	205.81	20–116
2-propanol	16.679	3640.20	219.61	8–100
Toluene	13.932	3056.96	217.62	13–136
Water	16.387	3885.70	230.17	0–200

Table 1.13 Saturated water in English engineering units

T (°F)	P^{sat} (psia)	V (ft³/lbm)		U (Btu/lb)		H (Btu/lb)		S (Btu/lb R)	
		V_l	V_g	U_l	U_g	H_l	H_g	S_l	S_g
132	2.345	0.01626	149.66	99.95	1053.7	99.95	1118.6	0.1851	1.9068
134	2.472	0.01626	142.41	101.94	1054.3	101.95	1119.5	0.1884	1.9024
136	2.605	0.01627	135.57	103.94	1055.0	103.95	1120.3	0.1918	1.8980
138	2.744	0.01628	129.11	105.94	1055.6	105.95	1121.1	0.1951	1.8937
140	2.889	0.01629	123.00	107.94	1056.2	107.95	1122.0	0.1985	1.8895
142	3.041	0.01630	117.22	109.94	1056.8	109.95	1122.8	0.2018	1.8852
144	3.200	0.01631	111.76	111.94	1057.5	111.95	1123.6	0.2051	1.8810

(a) $T = 50\,°C$ use $\ln P_{sat} = A - \dfrac{B}{T+C}$

Plug the Antoine constants into the equation for water and acetone, respectively,

$P_{w,sat} = 12.33$ **kPa at** $T = 50\ °C$ (This value is very close to the tabulated value in Table F3)

$P_{a,sat} = 82.1$ **kPa at** $T = 50\ °C$

(b) Boiling point, T_b, of water at 101.32 kPa

$$T_b = \frac{B}{A - \ln P} - C$$

$T_{w,b} = 100\,°C$ **and** $T_{a,b} = 56.4\,°C$ **at** 101.33 **kPa.**

1.18   Steam Tables

Steam tables tabulate the properties of saturated and superheated vapor and widely available in English units and SI units (http://www.eisco.co/boilr/steam_tables.pdf). Tables 1.13 and 1.14 show examples of saturated steam tables in English and SI units. Saturated steam is fully defined by either its temperature or pressure. For example, at 140°F liquid water has the value of enthalpy of 107.95 Btu/lb and water vapor of 1122.0 Btu/lb. The difference would be the heat of vaporization at 140°F.

Table 1.14 Saturated water in SI units

T (K)	P^{sat} (kPa)	V (cm³/g)		U (kJ/kg)		H (kJ/kg)		S (kJ/kg K)	
		V_l	V_g	U_l	U_g	H_l	H_g	V_l	V_g
372.15	97.76	1.043	1730.0	414.7	2505.3	414.8	2674.4	1.2956	7.3675
373.15	101.33	1.044	1673.0	419.0	2506.5	419.1	2676.0	1.3069	7.3554
375.15	108.78	1.045	1565.5	427.4	2508.8	427.5	2679.1	1.3294	7.3315
377.15	116.68	1.047	1466.2	435.8	2511.1	435.9	2682.2	1.3518	7.3078
379.15	125.04	1.049	1374.2	444.3	2513.4	444.4	2685.3	1.3742	7.2845
381.15	133.90	1.050	1288.9	452.7	2515.7	452.9	2688.3	1.3964	7.2615
383.15	143.27	1.052	1209.9	461.2	2518.0	461.3	2691.3	1.4185	7.2388

Tables 1.15 and 1.16 show examples of superheated steam tables in English and SI units. The properties of superheated steam are determined by the values of temperature and pressure.

Example 1.7 Energy change during evaporation
A mass of 22 kg of saturated liquid water at 101.3 kPa is evaporated completely at constant pressure and produced saturated vapor. Estimate the temperature of the vapor and amount of energy added to the water.

Table 1.15 Superheated steam in English engineering units

T (°F)	V (ft³/lb)	U (Btu/lb)	H (Btu/lb)	S (Btu/lb R)	V (ft³/lb)	U (Btu/lb)	H (Btu/lb)	S (Btu/lb R)
	P = 250 psia, T^{sat} = 400.97 °F				P = 255 psia, T^{sat} = 402.72 °F			
Sat liq	0.019	375.3	376.1	0.5679	0.019	377.2	378.0	0.5701
Sat vap	1.843	1115.8	1201.1	1.5264	1.808	1116.0	1201.3	1.5247
420	1.907	1125.8	1214.0	1.5413	1.865	1125.1	1213.1	1.5383
440	1.970	1135.9	1227.1	1.5559	1.928	1135.3	1226.3	1.5530
460	2.032	1145.6	1239.6	1.5697	1.989	1145.0	1238.9	1.5669
480	2.092	1154.9	1251.7	1.5827	2.048	1154.5	1251.1	1.5800
500	2.150	1164.0	1263.5	1.5951	2.105	1163.6	1262.9	1.5925

Table 1.16 Superheated steam in SI units

T (°C)	V (cm³/g)	U (kJ/kg)	H (kJ/kg)	S (kJ/kg K)	V (cm³/g)	U (kJ/kg)	H (kJ/kg)	S (kJ/kg K)
	P = 1200 kPa, T^{sat} = 187.96 °C				P = 1250 kPa, T^{sat} = 189.81 °C			
Sat liq	1.139	797.1	798.4	2.2161	1.141	805.3	806.7	2.2338
Sat vap	163.20	2586.9	2782.7	6.5194	156.93	2588.0	2784.1	6.5050
200	169.23	2611.3	2814.4	6.5872	161.88	2608.9	2811.2	6.5630
220	178.80	2650.0	2864.5	6.6909	171.17	2648.0	2861.9	6.6680
240	187.95	2686.7	2912.2	6.7858	180.02	2685.1	2910.1	6.7637
260	196.79	2722.1	2958.2	6.8738	188.56	2720.8	2956.5	6.8523
280	205.40	2756.5	3003.0	6.9562	196.88	2755.4	3001.5	6.9353
300	213.85	2790.3	3046.9	7.0342	205.02	2789.3	3045.6	7.0136

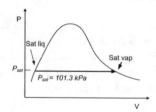

Solution:

From Steam Table: $P_{sat\ liq}$ = 101.3 kPa and $T_{sat\ liq}$ = 100°C.

$H_{sat\ liq}$ = 419.1 kJ/kg and $H_{sat\ vap}$ = 2676.0 kJ/kg

$\Delta H_{vap} = H_{sat\ vap} - H_{sat\ liq}$ = (2676.0-419.1) kJ/kg = 2256.9 kJ/kg

Total amount of heat added = (2256.9 kJ/kg)(22 kg) = **49,651.8 kJ**.

Example 1.8 Energy change during condensation

A mass of 22 kg of superheated steam at 1250 kPa and 220 °C condenses completely to saturated liquid state at constant pressure. Estimate the temperature of the saturated vapor and heat removed from the steam.

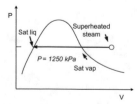

Solution:
From Steam tables: Superheated steam: P = 1250 kPa and T = 220 °C, $H_{\text{superheat vap}}$ = 2861.9 kJ/kg
Saturated steam and liquid at T_{sat} = 189.8 °C
Saturated liquid water at 1250 kPa, and $H_{\text{sat liq}}$ = 806.7 kJ/kg, T_{sat} = **189.8 °C**
Heat of condensation: $\Delta H_{\text{cond}} = H_{\text{sat liq}} - H_{\text{superheat vap}}$ = (806.7−2861.9) kJ/kg
= −2055.2 kJ/kg
Total amount of heat removed = −(2055.2 kJ/kg)(22 kg) = **−45,214.4 kJ**
The removed heat has the negative sign as the heat leaving the system of super-
heated steam after condensation process as sees in the figure above.

Example 1.9 Quality of a saturated liquid and vapor mixture of a steam
A rigid tank contains 15 kg of saturated liquid and vapor water at 85 °C. Only 2 kg
of the water is in liquid state. Estimate the enthalpy of the saturated mixture and the
volume of the tank.

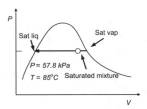

Solution:
From steam tables: Saturated liquid-vapor water mixture at 57.8 kPa and T_{sat} = 85 °C,
Saturated vapor: P = 57.8 kPa, $H_{\text{sat vap}}$ = 2652.0 kJ/kg, V_{vap} = 2828.8 cm³/g
Saturated liquid water at P = 57.8 kPa, $H_{\text{sat liq}}$ = 355.9 kJ/kg, V_{liq} = 1.003 cm³/g
Amount of liquid water = 2 kg
Amount of vapor = 15−2 = 13 kg
Quality x of the saturated mixture: $x = \dfrac{m_{\text{vap}}}{m_{\text{total}}} = \dfrac{13\,\text{kg}}{15\,\text{kg}} = 0.87$
Enthalpy of the mixture, Eq. (1.32):

$$H_{\text{mix}} = (1-x)H_{\text{sat liq}} + xH_{\text{sat vap}} = (1-0.87)355.9\,\text{kJ/kg} + (0.87)2652.0\,\text{kJ/kg}$$
$$= \mathbf{2353.5\,kJ/kg}$$

Volume of the mixture, Eq. (1.31):

$$V_{mix} = (1 - x)V_{sat\,liq} + xV_{sat\,vap}$$
$$= (1 - 0.87)0.001003\,m^3/kg + (0.87)2.828\,m^3/kg = 2.46\,m^3/kg$$

V_{tank} = 15 kg (2.46 m³/kg) = **36.9 m³**

Enthalpy of the saturated mixture depends on the quality of the mixture x and will be

$H_{sat.\ vap} > H_{mix} > H_{sat.\ liq}$.

1.19 Process

Process is any change that a system undergoes from one equilibrium state to another. To describe a process completely, one should specify the initial and final states of the process, as well as the *path* it follows during its progress, and the interactions with the surroundings. There are several types of processes:

- *Adiabatic process* occurs in either open or closed system without exchanging heat with the surroundings. This is an ideal concept as it implies a perfect insulation against the flow of heat, which is impossible in the strict sense.
- In an *isothermal* process, the temperature remains constant. In an *isobaric* process, the pressure remains constant. In an *isochoric* process the volume remains constant.
- *Constant pressure process* is isobaric process (P = constant).
- *Constant volume process* is isochoric process (V = constant).
- *Polytropic process* exists in which the pressure-volume relation is $PV^x = $ constant where x is a constant called the *polytropic index*. In a *batch process*, material is neither added to nor removed from the process during its operation. In a *semi-batch process*, material may be added but product is not removed during the operation.
- In a *steady-flow (uniform-flow) process*, a fluid flows through a control volume steadily and no intensive or extensive properties within the control volume change with time. The fluid properties can change from point to point within the control volume, but at any point, they remain constant during the entire process. Figure 1.5 shows a steady-flow system with mass and energy interactions with the surroundings.
- In a *transient-flow* process, properties change within a control volume with time.
- *Irreversible processes* cannot spontaneously reverse themselves and restore the system to its initial state. *Irreversibility* can be viewed as the wasted work potential or the lost opportunity to do work. Some of the factors causing *irreversibility* are friction, unrestrained expansion, mixing of two gases, heat transfer across a finite temperature difference, electric resistance, inelastic deformation of

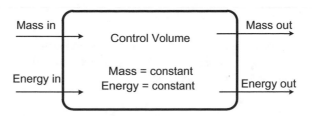

Fig. 1.5 Steady-flow system with mass and energy flow interactions with the surroundings. Mass and energy flows do not change with time in a steady-flow system

solids, and chemical reactions. *Friction* is a familiar form of irreversibility associated with bodies in motion which results from the force that opposes the motion developed at the interface of the two bodies in contact when the two bodies are forced to move relative to each other.

- *Cyclic process*, or series of processes, allows a system to undergo various state changes and return the system to the initial state at the end of the processes. For a cycle, the initial and final states are identical. *Ideal cycle* consists of internally reversible processes.
- *Isentropic process* is an internally reversible and adiabatic process. In such a process the entropy remains constant.
- *Quasi-static, or quasi-equilibrium, process* is a process which proceeds in such a manner that the system always remains infinitesimally close to an equilibrium state. A quasi-equilibrium process can be viewed as a sufficiently slow process that allows the system to adjust itself internally so that properties in one part of the system do not change any faster than those at other parts.

Summary

- *Open systems* allow mass and energy exchange across the boundary.
- *Closed systems* consist of a fixed amount of mass, and no mass can cross its boundary. But energy, in the form of heat or work, can cross the boundary.
- *Isolated systems* do not exchange mass and energy with their surroundings.
- *Steady state systems* have properties, which are independent of time; for example, amount of energy in the systems does not change with time.
- Force F is defined as the product of mass m and acceleration a, according to the Newton's second law, and expressed by

$$F = ma$$

- In the English engineering system, the weight is expressed by: $F = mg/g_c$.
- *Pressure P* is the force per unit cross sectional area applied in a direction perpendicular to the surface of an object. Pressure is:

$$P = F/A$$

- Absolute pressure = Gauge pressure + Atmospheric pressure
- *Parts per million*, ppm, or *parts per billion*, ppb, expresses the concentration of extremely dilute solutions or mixtures:

$$\text{ppm} = \frac{\text{parts}}{10^6 \text{ parts}}$$

- *Standard reference state* for the properties of chemical components is chosen as 25 °C (77°F) and 1 atm. Property values at the standard reference state are indicated by a superscript (°).
- An *equation of state* relates the pressure, temperature, and volume of a substance. At low pressures and high temperature the density of a gas decreases and the volume of a gas becomes proportional to its temperature:

$$PV = nRT$$

where P is the absolute pressure, T is the absolute temperature, n is the number of moles of a substance and R is called the universal gas constant: R = 8.314 kJ/mol K = 8.314 kPa m³/k mol K = 1.986 Btu/lbmol R = 10.73 psia ft³/ lbmol R.
- The Antoine equation may be used to estimate the saturation pressure:

$$\ln P_{\text{sat}} = A - \frac{B}{T+C} \left(P \text{ in kPa and } T \text{ in } °C \right)$$

Problems

1.1. Temperature units are Celsius and Kelvin (K) in the SI units, and Fahrenheit and Rankine (R) in the English engineering system. At what temperature the values of the temperatures in Celsius (°C) and in Fahrenheit (°F) scales become equal?

1.2. In a cooling process, the temperature of a system drops by 36 °F. Express this drop in Celsius (°C), Kelvin, (K), and Rankin (R).

1.3. In a heating process, the temperature of a system increases by 110 °C. Express this increase in Fahrenheit (°F), Kelvin, (K), and Rankin (R).

1.4. The temperature of a steam is 420 °C. Express the temperature of steam in Fahrenheit (°F), Kelvin, (K), and Rankin (R).

1.5. Convert the following units: (a) atm to bar, (b) Btu to J, (c) hp to kW; (d) quad to kJ, (e) lb$_m$ to tonne, (f) quart to liter

1.6. Convert the following units: (a) 4 km to miles, (b) 100 ft³/day to cm³/h, (c) 10 miles to ft, (d) 1000 L to gallons.

1.7. Convert the following units: (a) 40 km to miles, (b) 10 ft³/day to cm³/h, (c) 100 miles to ft, (d) 10 L to gal.

1.8. Convert the following units: (a) 1000 miles to km, (b) 2000 ft³/day to gal/h, (c) 50,000 ft to miles, (d) 1000 gallons to m³.

1.9. Convert the following units: (a) 550 miles to km, (b) 500 ft³/day to gal/min, (c) 10,000 ft to miles, (d) 150 gallons to liters.

1.10. Convert the following units: (a) 1200 miles to meters, (b) 1500 ft³/day to gal/s, (c) 2000 ft to meters, (d) 500 gallons to liters.

1.11. Convert the following units: (a) 200 miles to inches, (b) 2500 ft³/h to gal/s, (c) 600 nm to inches, (d) 300 gallons to in³.

1.12. A heat capacity equation for acetylene is given by

$$C_p = 9.89 + 0.827 \times 10^{-2}T - 0.378 \times 10^{-5}T^2$$

where T is in °F and C_p is in Btu/lbmol °F. Convert the equation for C_p in J/mol °C.

1.13. Parameters in Table B1 require use of temperatures in Kelvin and from that table the molar heat capacity of ethylene in ideal-gas state is

$$\frac{C_p^{ig}}{R} = 1.424 + 14.394 \times 10^{-3}T - 4.392 \times 10^{-6}T^2$$

Develop an equation for C_p^{ig} in J/mol °C.

1.14. Parameters in Table B1 require use of temperatures in Kelvin and from that table the molar heat capacity of propylene in ideal-gas state is

$$\frac{C_p^{ig}}{R} = 1.637 + 22.706 \times 10^{-3}T - 6.915 \times 10^{-6}T^2$$

Develop an equation for C_p^{ig} in cal/mol °C.

1.15. Parameters in Table B1 require use of temperatures in Kelvin and from that table the molar heat capacity of methane in ideal-gas state is

$$\frac{C_p^{ig}}{R} = 1.702 + 9.081 \times 10^{-3}T - 2.164 \times 10^{-6}T^2$$

Develop an equation for C_p^{ig} in cal/mol °C.

1.16. Parameters in Table B2 require use of temperatures in Kelvin and from that table the molar heat capacity of liquid ethanol is

$$\frac{C_p}{R} = 33.866 - 172.60 \times 10^{-3}T + 349.17 \times 10^{-6}T^2$$

Develop an equation for C_p in J/mol °C.

1.17. Parameters in Table B2 require use of temperatures in Kelvin and the molar heat capacity of liquid methanol is

$$\frac{C_p}{R} = 13.431 - 51.28 \times 10^{-3}T + 131.13 \times 10^{-6}T^2$$

Develop an equation for C_p in J/mol °C.

1.18. Parameters in Table B3 require use of temperatures in Kelvin and from that table the molar heat capacity of solid $NaHCO_3$ is

$$\frac{C_p}{R} = 5.128 + 18.148 \times 10^{-3}T$$

Develop an equation for C_p in Btu/lbmol R.

1.19. Experimental values of for the heat capacity, C_p, in J/mol °C of air from temperature, T, interval of 0–900 °C are:

T	0	25	100	200	300	400	500	600	700	800	900
C_p	29.062	29.075	29.142	29.292	29.514	29.782	30.083	30.401	30.711	31.020	31.317

Use the least square method to estimate the values of the coefficients in the following form: $C_p = A + BT + CT^2$

1.20. The heat capacity of ethanol is expressed by $C_p = 61.34 + 0.1572T$ where T is in °C and C_p is in J/mol °C. Modify the expression so that the unit for C_p would be Btu/lbmol R.

1.21. Express 30 GPa in: (a) atmosphere, (b) psia, (c) inches of Hg, (d) mm of Hg.

1.22. A gauge connected to a chamber kept under vacuum reads 4.5 psi. The atmospheric pressure at the location is 14.6 psi. Estimate the absolute pressure in the chamber.

1.23. A gauge connected to a chamber kept under vacuum reads 75 kPa. The atmospheric pressure at the location is 101.2 kPa psi. Estimate the absolute pressure in the chamber.

1.24. A gauge connected to a chamber reads 24.8 psi. The atmospheric pressure at the location is 14.2 psi. Estimate the absolute pressure in the chamber.

1.25. Estimate the atmospheric pressure at a location where the temperature is 25° C and the barometric reading is 755 mmHg.

1.26. A dead-weight gauge uses a piston with diameter, D, of 3 mm. If the piston is pressed by a 50 kg of the weight, determine the pressure reading on the gauge.

1.27. A pressure reading by a dead weight gauge is 30 bar. The piston of the gauge has diameter of 4 mm. Determine the approximate mass in kg of the weights required for the pressure reading.

1.28. The pressure reading on a dead-weight gauge is 3 atm. The piston diameter of the gauge is 0.2 in. Determine the approximate mass in lb_m of the weights required.

1.29. The reading on a mercury open manometer (open to the atmosphere at one end) is 15.5 cm. The local acceleration of gravity is 9.832 m/s². Atmospheric pressure is 101.75 kPa. Determine the absolute pressure in kPa being measured. The density of mercury at 25 °C is 13.534 g/cm³.

1.30. The reading on a mercury open manometer at 70°F (open to the atmosphere at one end) is 10.25 in. The local acceleration of gravity is 32.243 ft/s². Atmospheric pressure is 29.86 in. Hg. Estimate the measured absolute pressure in psia. The density of mercury at 70°F is 13.543 g/cm³.

1.31. At 300 K the reading on a manometer filled with mercury is 65.7 cm. The local acceleration of gravity is 9.785 m/s². What is the pressure in atmosphere that corresponds to this height of mercury?

1.32. The pressure gauge on a tank filled with carbon dioxide reads 45 psi. The barometer reading is 28.5 in Hg. Estimate the absolute pressure in the tank in psia and atmosphere.

1.33. Complete the following table if P_{atm} = 100 kPa and ρ_{Hg} = 13.6 g/cm³

P_{Gauge}(kPa)	P_{abs}(kPa)	P_{abs}(mm Hg)	P_{Gauge}(mm H$_2$O)
5			
	150		
		30	
			30

1.34. Estimate the mass of the air in a room with dimensions of (4)(5)(5)m³ at 20 °C and 1 atm.

1.35. Estimate the temperature of 68.2 kg of carbon dioxide gas in a tank with a volume of 20 m³ and pressure of 250 kPa.

1.36. Estimate the temperature of 12 kg of propane gas in a tank with a volume of 1.6 m³ and pressure of 400 kPa.

1.37. Estimate the temperature of 6.8 kg of hydrogen gas in a tank with a volume of 40 m³ and pressure of 300 kPa.

1.38. Estimate the temperature of 12.6 lb of nitrogen gas in a tank with a volume of 150 ft³ and pressure of 27 psia.

1.39. Estimate the pressure of 2.2 kg of carbon dioxide (CO_2) gas at T = 350 K and V = 0.5 m³.

1.40. Estimate the pressure of 0.6 kg of hydrogen gas at T = 400 K and V = 1.0 m³.

1.41. A tank has been filled with 240 lb of propane (C_3H_8). The volume of the tank is 150 ft³ and the temperature is 80°F. Estimate the pressure of propane using the ideal-gas equation of state.

1.42. Estimate the molar volume and specific volume of nitrogen (N_2) at 20 psia and 180 °F using the ideal-gas equation of state.

1.43. Estimate the molar volume and specific volume of propane C_3H_8 at 15 psia and 100 °F using the ideal-gas equation of state.

1.44. Convert 2.5 μg mol/ml min to lbmol/ft^3 h.

1.45. The current Operational Safety and Healy Administration (OSHA) 8-hour limit for carbon monoxide (CO) in air is 9 ppm. How many grams carbon monoxide/kg air is 9 ppm?

1.46. (a) Estimate the saturation vapor pressure of water and acetone at 40 °C, (b) Estimate the boiling point temperature of water and acetone at 98 kPa.

1.47. (a) Estimate saturation pressures for methanol and ethanol at 40 °C, (b) Estimate boiling points for methanol and ethanol at 101.33 kPa.

1.48. (a) Estimate saturation pressures for 1-propanol at 50 °C, (b) Estimate boiling points for 1-propanol at 101.33 kPa.

1.49. A tank filled with 10 lb of air is heated. The air originally is at 40 psia and 100 °F. After heating, the air pressure becomes 60 psia. Estimate the final temperature of the air.

1.50. A 2.0 m^3 tank filled with hydrogen (H_2) at 200 kPa and 500 K. The hydrogen is cooled until its temperature drops to 300 K. Estimate the amount of hydrogen and the final pressure in the tank after cooling.

1.51. Determine the number of grams of ethanol vapor within a volume of 500 cm^3 at a pressure of 500 kPa and a temperature of 100 °C.

1.52. A cylinder of a fixed volume of 15 m^3 containing CO_2 is initially at a temperature of 100 °C and a pressure of 101 kPa. When the temperature rises to 150°C find the final pressure.

1.53. A piston cylinder system is under isothermal conditions is at a temperature of 15 °C with a volume of 37 cm^3 and a pressure of 120 kPa. The piston moves creating a pressure of 260 kPa within the system. What is the change of volume after the piston has reached its final position?

1.54. Determine the volume occupied by 10 kg of water at a pressure of 10 MPa and the following temperatures: 400 °C and 600 °C.

1.55. Determine the volume occupied by 3.5 kg of water vapor at a pressure of 260 psia and the following temperatures of 480 °F and 1100 °F.

1.56. A mass of 10 kg of saturated liquid water at 101.3 kPa is evaporated completely at constant pressure and produced saturated vapor. Estimate the temperature of the vapor and amount of energy added to the water.

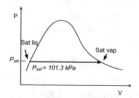

1.57. A mass of 25 lb of saturated liquid water at 3.2 psia is evaporated completely at constant pressure and produced saturated vapor. Estimate the temperature of the vapor and amount of energy added to the water.

1.58. A mass of 100 lb of saturated liquid water at 132 °F is evaporated completely at constant pressure and produced saturated vapor. Estimate the temperature of the vapor and amount of energy added to the water.

1.59. A mass of 10 kg of superheated steam at 1250 kPa and 220 °C condenses completely to saturated liquid state at constant pressure. Estimate the temperature of the saturated vapor and heat removed from the steam.

1.60. A mass of 150 kg of superheated steam at 500 kPa and 200 °C condenses completely to saturated liquid state at constant pressure. Estimate the temperature of the saturated vapor and heat removed from the steam.

1.61. A mass of 25 kg of superheated steam at 470 kPa and 240 °C condenses completely to saturated liquid state at constant pressure. Estimate the temperature of the saturated vapor and heat removed from the steam.

1.62. A mass of 150 lb of superheated steam at 250 psia and 440 °F condenses completely to saturated liquid state at constant pressure. Estimate the temperature of the saturated vapor and heat removed from the steam.

1.63. A mass of 50 lb of superheated steam at 255 psia and 500 °F condenses completely to saturated liquid state at constant pressure. Estimate the temperature of the saturated vapor and heat removed from the steam.

1.64. A rigid tank contains 30 kg of saturated liquid and vapor water at 85 °C. Only 4 kg of the water is in liquid state. Estimate the enthalpy of the saturated mixture and the volume of the tank.

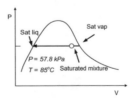

1.65. A rigid tank contains 10 kg of saturated liquid and vapor water at 326.15 K. Only 1.5 kg of the water is in liquid state. Estimate the enthalpy of the mixture and the volume of the tank.

1.66. A rigid tank contains 100 lb of saturated liquid and vapor water at 140°F. Only 8.0 lb of the water is in liquid state. Estimate the enthalpy of the saturated mixture and the volume of the tank.

References

1. Çengel YA, Boles MA (2014) Thermodynamics: an engineering approach, 8th edn. McGraw-Hill, New York
2. Himmelblau DM, Riggs JB (2012) Basic principles and calculations in chemical engineering, 8th edn. Prentice Hall, Upper Saddle River
3. Jaffe RL, Taylor W (2018) The physics of energy. Cambridge Univ. Press, Cambridge
4. Poling BE, Prausnitz JM, O'Connell JP (2001) The properties of gases and liquids, 5th edn. McGraw-Hill, New York

Energy Sources

2

Introduction and Learning Objectives: Energy has far-reaching impact in everyday life, technology, and development. Primary energy sources are extracted or captured directly from the environment, while the secondary energy sources are derived from the primary energy sources, for example, in the form of electricity or fuel. Primary energies are nonrenewable energy (fossil fuels), renewable energy, and waste. Nonrenewable energy resources are coal, petroleum, natural gas, and nuclear, while renewable energy resources are solar, bioenergy, wind, and geothermal. Energy density and the impact of fossil fuels on the global warming are discussed briefly.

The learning objectives of this chapter are to:

- Distinguish various sources of energy,
- Understand energy density, heating value of various fuels, and
- Implication of the energy usage on the environment.

2.1 Energy Sources

Primary and secondary sources of energy are the two main sources of energy, as shown in Fig. 2.1. Primary energy is extracted or captured directly from the environment, while the secondary energy is converted from the primary energy in the form of electricity or fuel. Differences between the primary and secondary energy sources are important in the energy balances to count and record energy supply, transformations, and losses.

© Springer Nature Switzerland AG 2021
Y. Demirel, *Energy*, Green Energy and Technology,
https://doi.org/10.1007/978-3-030-56164-2_2

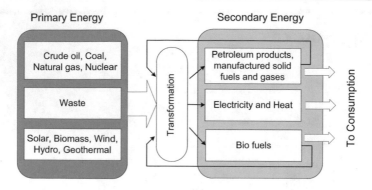

Fig. 2.1 Primary and secondary energy sources. To separate primary and secondary energy is important in energy balances for energy supply, transformation, and losses [13]

2.1.1 Primary Energy Sources

Primary energy is the energy extracted or captured directly from the environment. Three distinctive groups of primary energy are:

- Nonrenewable energy (fossil fuels): coal, crude oil, natural gas, nuclear fuel.
- Renewable energy: hydropower, biomass, solar, wind, geothermal, and ocean energy.
- Waste.

Primary sources of energy consisting of petroleum, coal, and natural gas, and nuclear are the major fossil fuels in primary energy consumption in the world. Projected energy use for power generation in the world shows that coal, and natural gas will still be the dominant energy sources (Fig. 2.2). The principle of supply and demand suggests that as fossil fuels diminish, their prices will rise and renewable energy supplies, particularly biomass, solar, and wind resources, will become sufficiently economical to exploit. Figure 2.2 shows the installed power generation capacity by source with projections [6, 7]. As seen, the shares of coal, nuclear, and oil will decrease, while natural gas and renewable energy will increase. Figure 2.3 shows the world end use energy consumption with projections [7]. Petroleum, natural gas, electricity, and coal will dominate the end-use energy consumption for some time. Figure 2.4 shows the global primary energy consumption by energy source with projections until 2050. Except coal and nuclear energy, the consumptions of other sources will keep increasing where the renewable will have the steepest increase [7].

2.1.2 Secondary Energy Sources

The primary energy is transformed to *secondary energy* in the form of electrical energy or fuel, such as gasoline, fuel oil, methanol, ethanol, and hydrogen. The

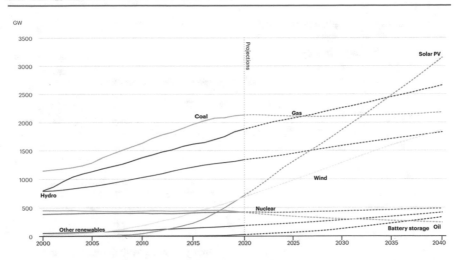

Fig. 2.2 Installed power generation capacity (GW) by source in the Stated Policies Scenario, 2000–2040 [6, 7]

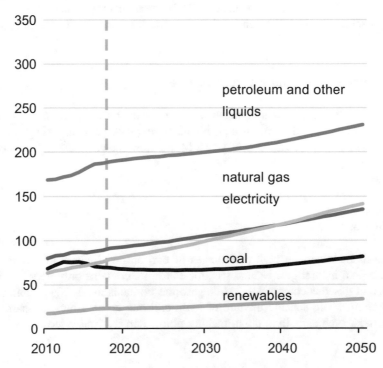

End-use energy consumption by fuel, world
quadrillion British thermal units

Fig. 2.3 World end-use energy consumption by fuel (Quad = 10^{15} Btu); the energy content may be converted to ton of oil equivalent (TOE): 1 TOE = 11630 kWh = 41870 MJ [7]

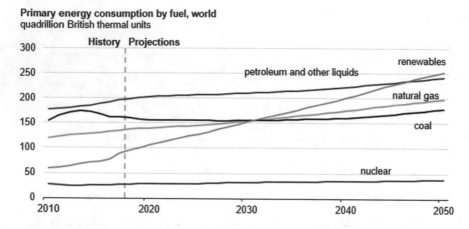

Fig. 2.4 Global primary energy consumption by energy source with projections, quadrillion (10^{15}) Btu [7]

primary energy of *renewable energy* sources, such as sun, wind, biomass, geothermal energy, and flowing water is usually equated with either electrical or thermal energy produced from them. Final energy is often electrical energy and fuel, which is referred to as *useful energy*. The selected four forms of final energy are electrical, thermal, mechanical, and chemical energy. These forms of final energy set a boundary between the energy production and the consumption sectors.

2.2   Nonrenewable Energy Sources

It is generally accepted that nonrenewable energy sources or *fossil fuels* are formed from the remains of dead plants and animals by exposure to heat and pressure in the earth's crust over the millions of years. Major nonrenewable energy sources are:

- Coal
- Petroleum
- Natural gas
- Nuclear

Fossil fuels contain high percentages of carbon and include mainly coal, petroleum, and natural gas. Natural gas, for example, contains only very low boiling point and gaseous components, while gasoline contains much higher boiling point components. The specific mixture of hydrocarbons gives a fuel its characteristic properties, such as boiling point, melting point, density, and viscosity.

2.2.1 Coal

Coals are sedimentary rocks containing combustible and incombustible matters as well as water. Coal comes in various composition and energy content depending on the source and type. Table 2.1 shows some typical properties of various coals. The poorest lignite has less than 50% carbon and an energy density lower than wood. Anthracites have more than 90% carbon, while bituminous coals have mostly between 70 and 75%. Bituminous coal ignites easily and burns with a relatively long flame. If improperly fired, bituminous coal is characterized with excess smoke and soot. Anthracite coal is very hard and shiny and creates a steady and clean flame and is preferred for domestic heating. Furthermore, it burns longer with release of more heat than the other types. For countries with rising oil prices coal may become a cheaper source of energy. It was in the 1880s when coal was first used to generate electricity for homes and factories. Since then, coal played a major role as a source of energy in the industrial revolution.

Coal has impurities like sulfur and nitrogen and when it burns the released impurities can combine with water vapor in the air to form droplets that fall to earth as weak forms of sulfuric and nitric acid as acid rain. Coal also contains minerals, which do not burn and make up the ash left behind in a coal combustor. Carbon dioxide is one of several gases that can help trap the earth's heat and, as many scientists believe, cause the earth's temperature to rise and alter the earth's climate. Because of high carbon content, coals generate more CO_2 per unit of released energy than any other fossil fuel such as crude oil. Sulfur content of coal is also a drawback. Sulfur makes up, typically, about 2% of bitumen coals. However, advanced coal technology can filter out 99% of tiny particles, remove more than 95% of the acid rain pollutants, and reduce the release of carbon dioxide by burning coal more efficiently. Many new plants are required to have flue gas desulfurization, decarbonization, and denitrification units [14] (Tables 2.2 and 2.3).

2.2.2 Petroleum Fractions

Oil is refined and separated into many commodity products from gasoline and kerosene to asphalt and chemical reagents used to make plastics and pharmaceuticals. Figure 2.5 shows a part of a typical refinery processing crude oil to produce various

Table 2.1 Typical properties of various coals [14]

	Anthracite coal	Bituminous coal	Lignite coal
Fixed carbon, weight%	80.5–85.7	44.9–78.2	31.4
Moisture, weight%	2.8–16.3	2.2–15.9	39
Bulk density, lb/ft^3	50–58	42–57	40–54
Ash, weight%	9.7–20.2	3.3–11.7	3.3–11.7
Sulfur, weight%	0.6–0.77	0.7–4.0	0.4

Table 2.2 Typical elemental composition by weight of crude oil [14]

Element	Percent range %
Carbon	83–87
Hydrogen	10–14
Nitrogen	0.1–2
Oxygen	0.1–1.5
Sulfur	0.5–6
Metals	<0.1

Table 2.3 Composition by weight of hydrocarbons in petroleum [14]

Hydrocarbon	Average %	Range %
Paraffins (alkanes)	30	15–60
Naphthenes (cycloalkanes)	49	30–60
Aromatics	15	3–30
Asphaltic	6	remainder

Fig. 2.5 A distillation tower showing the differing weights of various products produced from petroleum

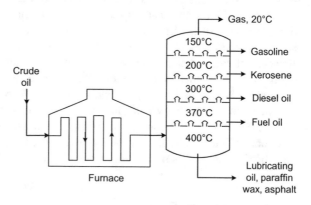

fuels. Approximately 84% by volume of the hydrocarbons present in petroleum is converted into energy-rich fuels, including gasoline, diesel, jet fuel, heating, and other fuel oil and liquefied petroleum gases. The remaining oil is converted to pharmaceuticals, solvents, fertilizers, pesticides, and plastics. Therefore, petroleum is vital to many industries, and thus is a critical concern to many nations (Figs. 2.3 and 2.4).

Some common fractions from petroleum refining are [13, 14]:

- *Liquefied petroleum gas* (LPG) is a flammable mixture of propane (C_3H_8) (about 38% by volume and more in winter) and butane (C_4H_{10}) (about 60% by volume and more in summer) used as a fuel in heating appliances and vehicles. Energy content of liquefied petroleum gas per kilogram is higher for gasoline because of higher hydrogen to carbon ratio. Liquefied petroleum gas emits 81% of the CO_2 per kWh produced by oil and 70% of that of coal. Liquefied petroleum gas has a typical specific heat of 46.1 MJ/kg compared with 43.5 MJ/kg for gasoline.

However, its energy density of 26 MJ/l is lower than either that of gasoline. Pure *n*-butane is liquefied at around 220 kPa (2.2 bar), while pure propane (C_3H_8) at 2200 kPa (22 bar). At liquid state, the vapor pressure of liquefied petroleum gas is about 550 kPa (5.5 bar).

- *Gasoline* is primarily used as a fuel in internal combustion engines. A typical gasoline consists of hydrocarbons with between 4 and 12 carbon atoms per molecule. It consists mostly of aliphatic hydrocarbons obtained by the fractional distillation of petroleum, enhanced with iso-octane or the aromatic hydrocarbons toluene and benzene to increase its octane rating. The specific density of gasoline ranges from 0.71 to 0.77 (6.175 lb/US gal) with higher densities having a greater volume of aromatics. Gasoline contains about 132 MJ/US gal (higher heating value), while its blends differ by up to 4% more or less than the average. The emission of CO_2 from gasoline is around 73.38 g/MJ.

- *Petroleum diesel* contains 8–21 carbon atoms per molecule with a boiling point in the range of 180–360°C (360–680°F). The density of petroleum diesel is about 6.943 lb/gal. About 86.1% of the fuel mass is carbon and it offers a net heating value of around 43.1 MJ/kg. However, due to the higher density, diesel offers a higher volumetric energy density at 128,700 Btu/gal versus 115,500 Btu/gal for gasoline, some 11% higher (see Table 2.7). The CO_2 emissions from diesel are 73.25 g/MJ (like gasoline). Because of quality regulations, additional refining is required to remove sulfur which may contribute to a higher cost.

- *Kerosene* is a thin, clear liquid formed containing between 6 and 16 carbon atoms per molecule, with density of 0.78–0.81 g/cm^3. The flash point of kerosene is between 37 and 65 C (100 and 150 F) and its autoignition temperature is 220 C (428 F). The heat of combustion of kerosene is like that of diesel: its lower heating value is around 18,500 Btu/lb (43.1 MJ/kg), and its higher heating value is 46.2 MJ/kg (19,861 Btu/lb).

- *Jet fuel* is a type of aviation fuel designed for use in aircraft powered by gas-turbine engines. The commonly used fuels are Jet A and Jet A-1 which are produced according to a standardized international specification. Jet B is used for its enhanced cold-weather performance. Jet fuel is a mixture of many different hydrocarbons with density of 0.775–0.840 kg/l at 15°C (59°F). The range is restricted by the requirements for the product, for example, the freezing point or smoke point. Kerosene-type jet fuel (including Jet A and Jet A-1) has a carbon numbers between about 8 and 16; wide-cut or naphtha-type jet fuel (including Jet B), between about 5 and 15.

- *Fuel oil* is made of long hydrocarbon chains, particularly alkanes, cycloalkanes, aromatics, and heavier than gasoline and naphtha. Fuel oil is classified into six classes, numbered 1 through 6, according to its boiling point, composition, and purpose. The boiling point ranges from 175 to 600 °C, and carbon chain length, 9–70 atoms. Viscosity also increases with number, and the heaviest oil must be heated to get it to flow. Price usually decreases as the fuel number increases. Number 1 is similar to kerosene, number 2 is the diesel fuel that trucks and some

Type	Btu/gallon
No. 1 Oil	137,400
No. 2 Oil	139,600
No. 3 Oil	141,800
No. 4 Oil	145,100
No. 5 Oil	148,800
No. 6 Oil	152,400

Table 2.4 Typical heating values of various fuel oils [14]

cars run on, leading to the name 'road diesel'. Number 4 fuel oil is usually a blend of heavy distillate and residual fuel oils. Number 5 and 6 fuel oils are called residual fuel oils or heavy fuel oils. Table 2.4 shows the heating values of various fuel oils per gallon.

Carbon fuels contain sulfur and impurities. Combustion of such fuels eventually leads to producing sulfur monoxides (SO) and sulfur dioxide (SO_2) in the exhaust which promotes acid rain. One final element in exhaust pollution is ozone (O_3). This is not emitted directly but made in the air by the action of sunlight on other pollutants to form *ground level ozone*, which is harmful on the respiratory systems if the levels are too high. However, the *ozone layer* in the high atmosphere is useful in blocking the harmful rays from the sun. Ozone is broken down by nitrogen oxides. For the nitrogen oxides, carbon monoxide, sulfur dioxide, and ozone, there are accepted levels that are set by legislation to which no harmful effects are observed [16].

2.2.3 Natural Gas

Natural gas is a naturally occurring mixture, consisting mainly of methane. Table 2.5 shows the typical components of natural gas. Natural gas is becoming increasingly popular as an alternative power generation fuel. Typical theoretical flame temperature of natural gas is 1960 °C (3562°F), ignition point is 593 °C. The value of gross heating in dry basis is around 37.8 MJ/m^3 and varies between 36.0 and 40.2 MJ/m^3, while the typical caloric value is roughly 1,000 Btu per cubic foot, depending on gas composition.

Natural gas is a major source of electricity production using gas turbines and steam turbines. It burns more cleanly and produces about 30% less carbon dioxide than burning petroleum and about 45% less than burning coal for an equivalent amount of heat produced. Combined cycle power generation using natural gas is thus the cleanest source of power available using fossil fuels, and this technology is widely used wherever gas can be obtained at a reasonable cost.

Liquefied natural gas exists at −161 °C (−258°F). Impurities and heavy hydrocarbons from the gaseous fossil fuel are removed before the cooling process. The density of liquefied natural gas is in the range 410–500 kg/m^3. The volume of the liquid is approximately 1/600 of the gaseous volume at atmospheric conditions [13].

Table 2.5 Typical composition in mole % of a natural gas [13, 14]

Component	Composition %	Range %
Methane	95.2	87.0–96.0
Ethane	2.5	1.5–5.1
Propane	0.2	0.1–1.5
Butane, n-butane	0.03	0.01–0.3
Iso–Pentane, n-pentane, hexane plus	0.01	Trace—0.14
Nitrogen	1.3	0.7–5.6
Carbon dioxide	0.7	0.1–1.0
Oxygen	0.02	0.01– 0.1
Hydrogen	Trace	Trace—0.02
Specific gravity	0.58	0.57–0.62

2.2.4 Nuclear Energy

Nuclear energy plants produce electricity through the fission of nuclear fuel, such as uranium, so they do not pollute the air with harmful gases. *Nuclear fission* is a nuclear reaction in which the nucleus of an atom splits into smaller parts, often producing free neutrons and photons in the form of gamma rays and releasing large amounts of energy. Nuclear fuels undergo fission when struck by free neutrons and generate neutrons leading to a self-sustaining chain reaction that releases energy at a controlled rate in a nuclear reactor [10]. This heat is used to produce steam in a turbine to produce electricity.

Typical fission releases about two hundred million eV (200 MeV) of energy, which is much higher than most chemical oxidation reactions. For example, complete fission energy of uranium-235 isotope is 6.73×10^{10} kJ/kg [10]. The energy of nuclear fission is released as kinetic energy of the fission products and fragments, and as electromagnetic radiation in the form of gamma rays in a nuclear reactor. The energy is converted to heat as the particles and gamma rays collide with the atoms that make up the reactor and its working fluid, usually water or occasionally heavy water. The products of nuclear fission, however, are far more radioactive than the heavy elements that fission as fuel, and remain so for a significant amount of time, giving rise to a nuclear waste problem. More than 400 nuclear power plants operating in 25 countries supply almost 17% of the world's electricity.

Nuclear power is essentially carbon-free. However, the electricity from new nuclear power plants would be relatively expensive, and nuclear energy faces several significant obstacles. The biggest challenges are the disposal of radioactive waste and the threat of nuclear proliferation. New plants would also require long licensing times.

Fission and fusion processes both release large amount of binding energy of nuclei and generate radioactive waste. Nuclear fission is spontaneous process on earth, and slow-neutron-induced fission can be controlled in nuclear power reactors, while fusion power is relatively distant possibility. One fission reaction is

$$^{235}_{92}U + n_{th} \rightarrow {}^{144}_{56}Ba + {}^{90}_{36}Kr + 2n \tag{2.1}$$

The energy appears as kinetic energy beside the fission fragments, neutrons, and γ-rays.

$$\begin{aligned} q &= \Delta({}^{235}U) + \Delta n \rightarrow (\Delta({}^{144}Ba) + \Delta({}^{90}Kr) + 2\Delta n) \\ &= 40.922 + 8.071 - (-73.937 - 74.959 + 2(8.071)) \\ &= 181.747\,\text{MeV} \end{aligned} \tag{2.2}$$

The core of a thermal-neutron reactor is fuel, moderator, and coolant. The coolant transfers the nuclear energy produced by nuclear fission to power generator system, while the moderator slows the neutrons. Most reactors use the water as coolant and moderator. Nuclear reactor designs have active and passive safety systems, such as control rods and coolant pumps as active, while pressure release valves or gravity-driven coolant circulation as passive. Commercial nuclear reactors mainly operate at lower temperatures compared with those operated in fossil fuel power plants. Therefore, turbines in nuclear power stations use mixed water/steam phase and light-water reactors can reach efficiencies of 33–37% that are comparable to coal-fired power plants.

Nuclear force strongly holds protons and neutrons in an atomic nucleus. The ratio of nuclear binding energies to the molecular bond energies is roughly in the magnitude of 1 million. Nuclear fusion in the sun combines four protons into helium nucleus and generates heat that produces solar radiation that drives photosynthesis and creates dynamics of atmosphere and oceans. On the contrary, in the nuclear fission large uranium nuclei become unstable as the electromagnetic repulsion between protons opposes the nuclear binding force and decompose into smaller molecules. Nuclear fission in a nuclear reactor releases large amount of carbon-free energy to produce stream and eventually electric power [10].

2.3 Heating Value of Fuels

The heating value of a fuel is the quantity of heat produced by its combustion at constant pressure and under 'normal' conditions (i.e. to 25° C and under a pressure of 1 atm). The combustion process generates water. The various heating values are:

- The *Higher Heating Value* (HHV) consists of the combustion product of water condensed and that the heat of vaporization contained in the water vapor is recovered. So, all the water produced in the combustion is in liquid state.
- The *Lower Heating Value* (LHV) assumes that the water product of combustion is at vapor state and the heat of vaporization is not recovered.

- *Net heating value* is the same with lower heating value and is obtained by subtracting the latent heat of vaporization of the water vapor formed by the combustion from the gross or higher heating value.
- *The gross heating value* is the total heat obtained by complete combustion at constant pressure including the heat released by condensing the water vapor in the combustion products. Gross heating value accounts liquid water in the fuel prior to combustion, and valuable for fuels containing water, such as wood and coal. If a fuel has no water prior to combustion, then the gross heating value is equal to higher heating value. A common method of relating HHV to LHV per unit mass of a fuel is [4, 15]

$$\text{HHV} = \text{LHV} + \Delta H_{\text{vap}} \left[\left(\text{MW}_{H_2O} n_{H_2O,\text{out}} \right) / \left(\text{MW}_{\text{Fuel}} n_{\text{Fuel,in}} \right) \right] \qquad (2.3)$$

where ΔH_{vap} is the heat of vaporization per mole of water (kJ/kg or Btu/lb), $n_{H_2O,\text{out}}$ is the moles of water vaporized, $n_{\text{fuel,in}}$ is the number of moles of fuel combusted, and MW is the molecular weight. Tables 2.6 and 2.7 show the properties and heating values of some common fuels. The heating value of fossil fuels may vary depending on the source and composition.

Table 2.6 Properties of heating values of some common fuels and hydrocarbons at 1 atm and 20°C; at 25°C for liquid fuels, and 1 atm and normal boiling temperature for gaseous fuels [4, 13–15]

Fuel (phase)	Formula	MW (kg/kmol)	ρ (kg/l)	ΔH_v (kJ/kg)	T_b (°F)	C_p (kJ/kg°C)	HHV** (kJ/kg)	LHV** (kJ/kg)
Carbon (s)	C	12.01	2.000	–		0.71	32,800	32,800
Hydrogen (g)	H_2	2.01	–	–		14.40	141,800	120,000
Methane (g)	CH_4	16.04	–	509	–258.7	2.20	55,530	50,050
Methanol (l)	CH_3OH	32.04	0.790	1168	149.0	2.53	22,660	19,920
Ethane (g)	C_2H_6	30.07	–	172	–127.5	1.75	51,900	47,520
Ethanol (l)	C_2H_5OH	46.07	0.790	919	172.0	2.44	29,670	26,810
Propane(g)	C_3H_8	44.09	0.500	420	–43.8	2.77	50,330	46,340
Butane (l)	C_4H_{10}	58.12	0.579	362	31.1	2.42	49,150	45,370
Isopentane (l)	C_5H_{12}	72.15	0.626	–	82.2	2.32	48,570	44,910
Benzene (l)	C_6H_6	78.11	0.877	433	176.2	1.72	41,800	40,100
Hexane (l)	C_6H_{14}	86.18	0.660	366	155.7	2.27	48,310	44,740
Toluene (l)	C_7H_8	92.14	0.867	412	231.1	1.71	42,400	40,500
Heptane (l)	C_7H_{16}	100.204	0.684	365	209.1	2.24	48,100	44,600
Octane (l)	C_8H_{18}	114.23	0.703	363	258.3	2.23	47,890	44,430
Decane (l)	$C_{10}H_{22}$	142.28	0.730	361		2.21	47,640	44,240
Gasoline (l)	$C_nH_{1.87n}$	100-110	0.72-0.78	350		2.40	47,300	44,000
Light diesel	$C_nH_{1.8n}$	170.00	0.78-0.84	270		2.20	46,100	43,200
Heavy diesel	$C_nH_{1.7n}$	200.00	0.82-0.88	230		1.90	45,500	42,800
Natural gas		~18.00	–	–		2.00	50,000	45,000

*HHV, LHV: higher heating value and lower heating value, respectively; (s): solid; (l): liquid; (g): gas

Table 2.7 Higher heating values (gross calorific value) of some common fuels [4, 13, 14]

Fuel	Higher heating value	
	(kJ/kg)	(Btu/lb)
Anthracite	32,500–34,000	14,000–14,500
Bituminous coal	17,000–23,250	7,300–10,000
Butane	49,510	20,900
Charcoal	29,600	12,800
Coal(anthracite)	30,200	13,000
Coal(bituminous)	27,900	12,000
Coke	28,000–31,000	12,000–13,500
Diesel	44,800	19,300
Ether	43,000	
Gasoline	47,300	20,400
Glycerin	19,000	
Hydrogen	141,790	61,000
Lignite	16,300	7,000
Methane	55,530	
Oils, vegetable	39,000–48,000	
Peat	13,800–20,500	5,500–8,800
Petroleum	43,000	
Propane	50,350	
Semi anthracite	26,700–32,500	11,500–14,000
Wood (dry)	14,400–17,400	6,200–7,500
	(kJ/m^3)	(Btu/ft^3)
Acetylene	56,000	
Butane C$_4$H$_{10}$	133,000	3200
Hydrogen	13,000	
Natural gas	43,000	950–1150
Methane CH$_4$	39,820	
Propane C$_3$H$_8$	101,000	2550
Butane C$_4$H$_{10}$		3200
	(kJ/l)	(Btu/gal)
Gasoline	32,000	115,000
Heavy fuel oil#6	42,600	153,000
Kerosene	37,600	135,000
Diesel	36,300	130,500
Biodiesel	33,500	120,000
Butane C$_4$H$_{10}$	36,200	130,000
Methanol	15,900	57,000
Ethanol	21,100	76,000

1 kJ/kg = 1 J/g = 0.43 Btu/lb$_m$ = 0.239 kcal/kg
1 Btu/lb$_m$ = 2.326 kJ/kg = 0.55 kcal/kg
1 kcal/kg = 4.187 kJ/kg = 1.8 Btu/lb$_m$

Table 2.8 Energy densities of some fuels [13, 14]

Fuel type	Gross (HHV)			Net (LHV)
	(MJ/l)	(MJ/kg)	(Btu/gal)	(Btu/gal)
Conventional gasoline	34.8	44.4	125,000	115,400
High octane gasoline	33.5	46.8	120,200	112,000
LPG (60%Pr. + 40%Bu.)	26.8	46.0		
Ethanol	24.0	30.0	84,600	75,700
Methanol	17.9	19.9	64,600	56,600
Butanol	29.2	36.6		
Gasohol E10 (ethanol 10% vol.)	33.2	43.5	120,900	112,400
Gasohol E85 (ethanol 85% vol.)	25.6	33.1		
Gasoline (petrol)	34.2	46.4		115,500
Diesel	38.6	45.4	138,700	128,700
Biodiesel	33.5	42.2	126,200	117,100
Jet fuel (kerosene based)	35.1	43.8	125,935	
Jet fuel (naphtha)	42.8	33.0	127,500	118,700
Liquefied natural gas (160°C)	22.2	53.6	90,800	
Liquefied petroleum gas	26.8	46.0	91,300	83,500
Hydrogen (liquid at 20 K)	10.1	142.0		130
Hydrogen gas	0.0108	143.0		
Methane (1 atm, 15°C)	0.0378	55.6		
Natural gas	0.0364	53.6		
LPG propane	25.3	49.6		
LPG butane	27.7	49.1		
Crude oil	37.0	46.3		
Coal, anthracite	72.4	32.5		
Coal, lignite		14.0		
Coal, bituminous	20.0	24.0		
Wood		18.0		

2.3.1   Energy Density

Energy density is the amount of energy per unit volume. *Specific energy* is the amount of energy per unit amount. Comparing, for example, the effectiveness of hydrogen fuel to gasoline, hydrogen has a higher specific energy than gasoline but a much lower energy density even in liquid form. Table 2.8 lists energy densities of some fuel and fuel mixtures.

Example 2.1 Energy consumption by a car
An average car consumes 50 gallons gasoline per month. Estimate the energy consumed by the car per year.

Solution:

Assume that gasoline has an average density of 0.72 g/cm^3 and the heating value of 47.3 MJ/kg (Table 2.6).

Data: V = 50 gal/month = 189.25 l/month, 2271.0 l/year (3.785 l = 1 gal)

ρ_{gas} = 0.72 g/cm^3 = 0.72 kg/l

Mass of gasoline: $m_{gas} = \rho V$ = 1635.1 kg/year

Energy consumed per year : E_{gas} = 1635.1 kg/year(47, 300 kJ/kg)

$= \mathbf{77,340,230\,kJ/year}$

$= \mathbf{77,340.2\,MJ/year}$

Example 2.2 Fuel consumption by a low and a high-mileage car

An average daily traveling distance is about 40 miles/day. A car has a city-mileage of 20 miles/gal. If the car is replaced with a new car with a city-mileage of 30 miles/gal and the average cost of gasoline is \$3.50/gal, estimate the amount of energy conserved with the new car per year.

Solution:

Assume: The gasoline is incompressible with ρ_{av} = 0.75 kg/l.

Lower heating value (LHV) = 44,000 kJ/kg; 44,000 kJ of heat is released when 1 kg of gasoline is completely burned, and the produced water is in vapor state.

Fuel needed for the old car: (40 miles/day)/(20 miles/gal) = 2 gal/day

Fuel needed for the new car: (40 miles/day)/(30 miles/gal) = 1.34 gal/day

Old car:

Mass of gasoline: $m_{gas} = \rho_{av}$ (Volume) = (0.75 kg/l)(2.0gal/day)(3.785 l/gal) = **5.7 kg/day**

$$
\begin{aligned}
\text{Energy of gasoline : } E_{gas}(\text{LHV}) &= (5.7\,\text{kg/day})(44,000\,\text{kJ/kg}) \\
&= 250800\,\text{kJ/day}(365\,\text{day/year}) \\
&= 91,542,000\,\text{kJ/year} = \mathbf{91,542\,MJ/year}
\end{aligned}
$$

New car:

Mass of gasoline: $m_{gas} = \rho_{av}$ (Volume) = (0.75 kg/l)(1.34 gal/day)(3.785 l/gal) = **3.8 kg/day**

$$
\begin{aligned}
\text{Energy of gasoline : } E_{gas}(\text{LHV}) &= (3.8\,\text{kg/day})(44,000\,\text{kJ/kg}) \\
&= 167,200\,\text{kJ/day}(365\,\text{days/year}) \\
&= 61,028,000\,\text{kJ/year} = \mathbf{61,028\,MJ/year}
\end{aligned}
$$

The new car reduces the fuel consumption by around 33%
[(91,542–61,028 MJ/year)/91,542 MJ/year], which is significant.

Example 2.3 Daily consumption of natural gas by a city
A city consumes natural gas at a rate of 500×10^6 ft^3/day. The volumetric flow is at standard conditions of 60°F and 1 atm = 14.7 psia. If the natural gas is costing $6/GJ of higher heating value, what is the amount of energy per year for the city.

Solution:
$Q = 500 \times 10^6$ ft^3/day at 60°F and 1 atm = 14.7 psia.
The higher heating value is the heat of combustion of the natural gas when the water product is at liquid state. From Table 2.6, the value of HHV is: 50,000 kJ/kg ($\sim$ 1030 Btu/ft^3 Table 2.7)
Heating value: (1030 Btu/ft^3)(500×10^6 ft^3/day) = 515.0×10^9 Btu/day
(515.0×10^9 Btu/day) (1055 J/Btu) = 543,325 GJ/day
Consumed energy per year: 198,313,625 GJ/year.

Example 2.4 Energy consumed by a car
An average car consumes about 2 gallons (US gallon = 3.785 L) a day, and the capacity of the fuel tank is about 15 gallons. Therefore, a car needs to be refueled once every week. The density of gasoline ranges from 0.72 to 0.78 kg/l (Table 2.6). The lower heating value of gasoline is about 44,000 kJ/kg. Assume that the average density of gasoline is 0.75 kg/l. If the car was able to use 0.2 kg of nuclear fuel of uranium-235, estimate the time in years for refueling.

Solution:
Assume: The gasoline is incompressible with $\rho_{av} = 0.75$ kg/l.
Lower heating value (LHV) = 44,000 kJ/kg; 44,000 kJ of heat is released when 1 kg of gasoline is completely burned, and the produced water is in vapor state.
Complete fission energy of U-235 = 6.73×10^{10} kJ/kg
Mass of gasoline per day: $m_{gas} = \rho_{av} V = $ (0.75 kg/l)(2 gal/day)(3.785 l/gal) = 5.67 kg/day
Energy of gasoline per day: $E_{gas} = m_{gas}$ (LHV) = (5.67 kg/day)(44,000 kJ/kg) = 249,480 kJ/day
Energy released by the complete fission of 0.2 kg U-235:
$E_{U-235} = (6.73 \times 10^{10}$ kJ/kg)(0.2 kg) = 1.346×10^{10} kJ
Time for refueling: (1.346×10^{10} kJ)/(249,480 kJ/day) = **53,952 days = 148 years**
Therefore, the car will not need refueling for about 148 years.

2.4   Renewable Energy Resources

Renewable energy comes from natural resources and are naturally replenished. Major renewable energy sources are:

- Hydroelectric
- Solar energy
- Biomass
- Wind
- Geothermal heat
- Ocean

In its various forms, renewable energy comes directly from the sun, or from heat generated deep within the earth. Figures 2.2, 2.3, 2.4 show the current and projected renewable energy sources. As seen, the generation of electricity from renewables, such as biomass, wind, solar, geothermal, and biofuels is growing steadily. Climate change concerns, high oil prices, and government support are leading to increase in renewable energy usage and commercialization. Renewable energy replaces conventional fuels in four distinct areas: power generation, hot water/space heating, transport fuels, and rural (off-grid) energy services [2, 3, 9]:

- *Renewable power generation* is increasing worldwide. Renewable power generators are spread across many countries, and wind power alone already provides a significant share of electricity in some areas.
- *Solar hot water* contributes a portion of the water heating needs of over 70 million households in many countries.
- *Renewable biofuels* have contributed to a decline in oil consumption in many countries.

New and emerging renewable energy technologies are still under development and include cellulosic ethanol, hot-dry-rock geothermal power, and ocean energy. Renewable energy generally gets cheaper in the long term, while fossil fuels generally get more expensive. Fossil fuel technologies are more mature, while renewable energy technologies are being rapidly improved to increase the efficiency of renewable energy and reduce its cost. In rural and remote areas, transmission and distribution of energy generated from fossil fuels can be difficult and expensive; therefore, producing renewable energy locally can offer a viable alternative [3].

2.4.1 Hydroenergy

Hydroenergy is derived from the force or energy of moving water. Most hydroelectric energy comes from the potential energy of dammed water driving a water turbine and generator. The power extracted from the water depends on the volume and on the difference in height between the source and the water's outflow. This height difference is called the head. The amount of potential energy in water is proportional to the head. To deliver water to a turbine while maintaining pressure arising from the head, a large pipe called a penstock may be used.

One of the major advantages of hydroelectricity is the elimination of fuel. Because there is no fuel combustion, there is little air pollution in comparison with fossil fuel plants and limited thermal pollution compared with nuclear plants. Hydroelectric plants also tend to have longer economic lives than fuel-fired power generation, with some plants now in service which were built 50–100 years ago. Operating labor cost is also usually low, as plants are automated and need few personnel on site during normal operation. The sale of electricity from the station may cover the construction costs after 5–8 years of full operation.

Hydroelectric usually refers to large-scale hydroelectric dams. Micro hydro systems typically produce up to 100 kW of power. Hydro systems without dam derive kinetic energy from rivers and oceans. Ocean energy includes marine current power, ocean thermal energy conversion, and tidal power [6, 9].

2.4.2 Solar Energy

Solar energy is derived from the sun through the form of solar radiation. Solar-powered electrical generation relies on photovoltaics and heat engines. Other solar applications include space heating and cooling through solar architecture, daylighting, solar hot water, solar cooking, and high temperature process heat for industrial purposes. Solar technologies are broadly characterized as either passive solar or active solar depending on the way they capture, convert, and distribute solar energy [5, 6, 9]:

- *Active solar techniques* include the use of solar thermal collectors to harness the energy. The sun is the source of heat and solar collectors collect the solar radiations. A part of the incident energy is lost, and the rest is absorbed by the heat transfer fluid (solar collector fluid). Some active solar techniques include *solar process heat* by commercial and industrial buildings, *space heating/cooling*, and water heating. A typical water heating system includes solar collectors that work along with a pump, heat exchanger, and one or more large storage tanks. The most common collector is called a *flat-plate collector*. Mounted on a roof, it consists of a thin, flat, rectangular box with a transparent cover that faces the sun. Small tubes run through the box and carry the heat transfer fluid, mainly the water or air to be heated. The tubes are attached to an absorber plate, which is painted black to absorb the heat. As heat builds up in the collector, it heats the fluid passing through the tubes. The storage tank then holds the hot liquid. It can be just a modified water heater, but it is usually larger and very well-insulated. Systems that use fluids other than water usually heat the water by passing it through a coil of tubing in the tank, which is full of hot fluid.
- *Passive solar systems* rely on gravity and the tendency of water to naturally circulate as it is heated. *Passive solar techniques* orient buildings to the sun, select materials with favorable thermal mass or light dispersing properties, and design spaces that naturally circulate air.

Nonresidential Solar Collectors

The two main types of solar collectors used for nonresidential buildings are an *evacuated-tube collector* and a *linear concentrator*. They can operate at high temperatures with high efficiency. An evacuated-tube collector is a set of many double-walled, glass tubes and reflectors to heat the fluid inside the tubes. A vacuum between the two walls insulates the inner tube, retaining the heat. Linear concentrators use long, rectangular, U-shaped mirrors tilted to focus sunlight on tubes that run along the length of the mirrors. The concentrated sunlight heats the fluid within the tubes. Solar absorption systems use thermal energy to evaporate a refrigerant fluid to cool the air. In contrast, solar desiccant systems use thermal energy to regenerate desiccants that dry the air, thereby cooling the air [2, 6].

Solar Electric Generating Systems

Solar electric generating system uses parabolic trough collectors to collect the sun's energy to generate steam to drive a conventional steam turbine [2, 6]. The parabolic mirrors automatically track the sun throughout the day. The sunlight is directed to central tube carrying synthetic oil, which heats around 400°C. The heat is used to convert water into steam in order to drive a steam turbine and produce electricity.

Photovoltaic

Solar photovoltaic (PV) converts light into electricity using semiconductor materials. Photovoltaic cell is a *solar cell*, which is a solid-state electrical device that converts the energy of light directly into electricity. Assemblies of cells are known as *solar modules* or *solar panels*. Solar modules are typically deployed as an array of individual modules on rooftops, building facades, or in large scale ground-based arrays. A module consists of many jointly connected solar cells. Most crystalline modules usually consist of 60–72 cells. Photovoltaic cells and modules use various semiconductors; they are of three types: (i) crystalline silicon, (ii) thin-film, and (iii) concentrator. Photovoltaic systems produce direct current, which must be converted to alternating current via an inverter if the output from the system is to be used in the grid.

A major goal is to increase solar photovoltaic efficiency and decrease costs. Current efficiencies for crystalline silicon cells are equal to about 15–20% [2, 5, 9].

2.4.3 Biomass and Bioenergy

Biomass is an organic material made from plants including microorganisms and animals. Plants absorb the sun's energy in photosynthesis and store the energy as biomass. Therefore, biomass is a renewable energy source based on the carbon cycle. Some examples of biomass fuels include wood, crops, and algae. When burned, the chemical energy in biomass is released as heat. Biomass can be converted to other biofuels, such as ethanol and biodiesel. Biomass grown for biofuel

Table 2.9 Lower heating values (LHV) for selected biomass [4]

Product	Moisture (%)	Ash content*(%)	LHV (MJ/kg)
Bagasse sugarcane	18	4	17–18
Coconut husks	5–10	6	16,7
Coffee husks	13	8–10	16,7
Corn stover	5–6	8	17–19
Corncobs	15	1–2	19,3
Cotton husks	5–10	3	16,7
Oil-palm fibers	55	10	7–8
l-palm husks	55	5	7–8
Poplar wood	5–15	1.2	17–19
Rice hulls	9–11	15–20	13–15
Rice straw and husk	15–30	15–20	17–18
Switchgrass	8–15	6	18–20
Wheat straw and husk	7–15	8–9	17–19
Willow wood	12	1–5	17–19

*Approximate

includes corn, soybeans, willow switch grass, rapeseed, sugar beet, palm oil, and sorghum. Cellulosic biomass, such as corn stover, straw, timber, rice husks, can also be used for biofuel production. Anaerobic digestion of biomass produces biogas, while gasification produces syngas, which is the mixture of hydrogen and carbon dioxide to be converted to liquid fuels. Cellulosic ethanol can also be created by a thermo-chemical process, which uses various combinations of temperature, pressure, water, oxygen or air, and catalysts to convert biomass to cellulosic ethanol. Table 2.9 shows lower heating values, moisture, and ash content of some biomass [3].

2.4.4 Carbon Cycle

In the carbon cycle, carbon in various forms is transported between the various components of the earth's biosphere, between the atmosphere, hydrosphere (seas and oceans), lithosphere (rocks, soils, and mineral deposits, including fossil fuels) and biological material including plants and animals. Carbon cycle maintains a state of dynamic equilibrium. Other forms, most notably fossil fuels, can potentially store carbon indefinitely; however, if they are burned the carbon is released, and this makes a net addition to the carbon cycle, raising the total free carbon. If biomass is used without replacement, for example in the case of forest clearance, this too can make a net addition to the carbon cycle. As growing plant absorbs the carbon released by the harvested biomass, sustainable use of biomass makes no direct net contribution [9, 16].

2.4.5 Gross Heating Values of Biomass Fuels

Biomass can be characterized by the *proximate* and *ultimate analyses*:

- The *proximate analysis* determines moisture content, volatile content (when heated to 950 °C), the free carbon remaining at that point, the ash (mineral) in the sample, and the higher heating value based on the complete combustion of the sample to carbon dioxide and liquid water.
- The *ultimate analysis* is the elemental analysis and provides the composition of the biomass in wt% of carbon, hydrogen, oxygen, sulfur, and nitrogen.

Table 2.10 shows measured and estimated gross heating values as well as the proximate and ultimate analyses of some selected fuels, including biomass components, natural biomass (woods, agricultural products), processed biomass, and other solid and liquid fuels [15].

Table 2.10 Proximate and ultimate analyses of biomass fuels in weight percentage [4, 15]

Biomass Name	Fixed Carbon	Volatiles %	Ash %	C %	H %	O %	N %	S %	HHV$_m$ (kJ/g)	HHV$_{est}$ (kJ/g)
Douglas Fir	17.70	81.50	0.80	52.30	6.30	40.50	0.10	0.00	21.05	21.48
Hickory	–	–	0.73	47.67	6.49	43.11	0.00	0.00	20.17	19.82
Maple	–	–	1.35	50.64	6.02	41.74	0.25	0.00	19.96	20.42
Ponderosa Pine	17.17	82.54	0.29	49.25	5.99	44.36	0.06	0.03	20.02	19.66
Poplar	–	–	0.65	51.64	6.26	41.45	0.00	0.00	20.75	21.10
Redwood	16.10	83.50	0.40	53.50	5.90	40.30	0.10	0.00	21.03	21.45
Western Hemlock	15.20	84.80	2.20	50.40	5.80	41.10	0.10	0.10	20.05	20.14
Yellow Pine	–	–	1.31	52.60	7.00	40.10	0.00	0.00	22.30	22.44
White Fir	16.58	83.17	0.25	49.00	5.98	44.75	0.05	0.01	19.95	19.52
White Oak	17.20	81.28	1.52	49.48	5.38	43.13	0.35	0.01	19.42	19.12
Douglas Fir bark	25.80	73.00	1.20	56.20	5.90	36.70	0.00	0.00	22.10	22.75
Loblolly Pine bark	33.90	54.70	0.40	56.30	5.60	37.70	0.00	0.00	21.78	22.35
Peach Pits	19.85	79.12	1.03	53.00	5.90	39.14	0.32	0.05	20.82	21.39
Walnut Shells	21.16	78.28	0.56	49.98	5.71	43.35	0.21	0.01	20.18	19.68
Almond Pruning	21.54	76.83	1.63	51.30	5.29	40.90	0.66	0.01	20.01	19.87
Black Walnut Pruning	18.56	80.69	0.78	49.80	5.82	43.25	0.22	0.01	19.83	19.75
Corncobs	18.54	80.10	1.36	46.58	5.87	45.46	0.47	0.01	18.77	18.44
Wheat Straw	19.80	71.30	8.90	43.20	5.00	39.40	0.61	0.11	17.51	16.71

(continued)

Table 2.10 (continued)

Biomass Name	Fixed Carbon %	Volatiles %	Ash %	C %	H %	O %	N %	S %	HHV$_m$ (kJ/g)	HHV$_{est}$ (kJ/g)
Cotton Stalk	22.43	70.89	6.68	43.64	5.81	43.87	0.00	0.00	18.26	17.40
Corn Stover	19.25	75.17	5.58	43.65	5.56	43.31	0.61	0.01	17.65	17.19
Sugarcane Bagasse	14.95	73.78	11.27	44.80	5.35	39.55	0.38	0.01	17.33	17.61
Rice Hulls	15.80	63.60	20.60	38.30	4.36	35.45	0.83	0.06	14.89	14.40
Pine Needles	26.12	72.38	1.50	48.21	6.57	43.72			20.12	20.02
Cotton Gin Trash	15.10	67.30	17.60	39.59	5.26	36.38	2.09	0.00	16.42	15.85
Cellulose; C6H10O5	–	–	162	44.44	6.17	49.38	–	–	–	17.68
Lignin (Softwood)	–	–	–	63.8	6.30	29.90	–	–	–	26.60
Lignin (Hardwood)	–	–	–	59.8	6.40	33.70	–	–	–	24.93

A relationship between the high heating value (HHV) and the elemental composition is given by

$$HHV(in\ kJ/g) = 0.3491\,C + 1.1783\,H - 0.1034\,O - 0.0211\,A + 0.1005\,S - 0.0151\,N \qquad (2.4)$$

where C is the weight fraction of carbon; H of hydrogen; O of oxygen; A of ash; S of sulfur; and N of nitrogen appearing in the ultimate analysis. This equation represents the experimental data with an average error of 1.5% and can be used in estimating heating values and modeling of biomass processes [15].

Based on chemical functional groups of the fuels, the heating values may vary. When the oxygen percentage is higher in a fuel, the percentages of carbon and hydrogen available for combustion are reduced. This leads to the lower heating values. By using the values of fixed carbon (FC, wt%), the higher heating value of the biomass samples can be estimated by

$$HHV(MJ/kg) = 0.196(FC) + 14.119 \qquad (2.5)$$

The heating values calculated from Eq. (2.5) shows a mean difference of 2.2% between estimated and measured values [10]. Another correlation between the HHV and dry ash content from proximate analysis of biomass (in weight percent) is

$$HHV(MJ/kg) = 19.914 - 0.2324\,Ash \qquad (2.6)$$

Based on the composition of main elements (in wt%) C, H, and O, the heating value is estimated by

$$HHV(MJ/kg) = 0.3137\,C + 0.7009\,H + 0.0318\,O - 1.3675 \qquad (2.7)$$

with more than 90% predictions in the range of $\pm\,5\%$ error.

Example 2.5 Gross heating value estimations
Using data in Table 2.10, estimate the gross heating values in kJ/kg for the biomass redwood from: (a) ultimate analysis, (b) fixed carbon, (c) dry ash content, and (d) carbon (C), hydrogen (H), and oxygen (O) compositions.

Name	Fixed Carbon %	Volatiles %	Ash %	C %	H %	O %	N %	S %	HHV$_m$ (kJ/g)	HHV$_{est}$ (kJ/g)
Redwood	16.10	83.50	0.40	53.50	5.90	40.30	0.10	0.00	21.03	21.45

Solution:

(a) From ultimate analysis

$$HHV(in\,MJ/kg) = 0.3491\,C + 1.1783\,H - 0.1034\,O - 0.0211\,A + 0.1005\,S$$
$$- 0.0151\,N$$

$$HHV(in\,MJ/kg) = 0.3491(53.50) + 1.1783(5.90) - 0.1034(40.3)$$
$$- 0.0211(0.0040) + 0.1005(0.0)$$
$$- 0.0151(0.0010) = 21.44\,MJ/kg = \mathbf{21,440\,kJ/kg}$$

(b) From fixed carbon percentage:

$$HHV(MJ/kg) = 0.196(FC) + 14.119$$

$$HHV\,(MJ/kg) = 0.196(16.10) + 14.119 = 17.3\,MJ/kg = \mathbf{17,300\ kJ/kg}$$

(c) From dry ash content:

$$HHV(MJ/kg) = 19.914 - 0.2324\,Ash$$

$$HHV\,(MJ/kg) = 19.914 - 0.2324\,(0.0040) = 19.914\,MJ/kg = \mathbf{19,914\ kJ/kg}$$

(d) From the main elements (in wt%) C, H, and O

$$HHV(MJ/kg) = 0.3137\,C + 0.7009H + 0.0318\,O - 1.3675$$

$$HHV(MJ/kg) = 0.3137(53.50) + 0.7009(5.9) + 0.0318(40.3)$$
$$- 1.3675 = \textbf{20.83\,MJ/kg} = \textbf{20,830\,kJ/kg}$$

Estimation from Eq. (2.7), used in part (d), is the closest to the measured value of 21.03 MJ/kg (21,030 kJ/kg).

2.4.6 Biofuels

Biological fuels produced from photosynthesis can be categorized into three groups [11]:

- *Carbohydrates*, representing a mixture of mono-di- and polysaccharides (17 kJ/g).
- *Fats*, unsaturated and saturated fatty acids (triglyceride) (39 kJ/g).
- *Proteins*, used partly as fuel source (17 kJ/g).

Carbohydrates are straight-chain aldehydes or ketones with many hydroxyl groups that can exist as straight chains or rings. Carbohydrates such as starch are the most abundant biological molecules, and play numerous roles, such as the storage and transport of energy, and structural components such as cellulose in plants. Triglycerides and fatty free acids both contain long, linear aliphatic hydrocarbon chains, which are partially unsaturated and have a carbon number range. The fuel value is equal to the heat of combustion (oxidation) of fuel. Carbohydrates and fats can be completely oxidized, while proteins can only be partially oxidized and hence has lower fuel values [3].

Some synthetic biofuels are [3]:

- *Bioethanol*: Corn-based ethanol is currently the largest source of biofuel as a gasoline substitute or additive. The gasoline is mixed with 10% ethanol, a mix known as E10 (or gasohol). Only specific types of vehicles named as flexible fuel vehicles can use mixtures with greater than 10% ethanol. E85 is an alternative fuel that contains up to 85% ethanol.
- *Biodiesel*: Biodiesel is most often blended with petroleum diesel in ratios of 2% (B2), 5% (B5), or 20% (B20). It can also be used as pure biodiesel (B100). Biodiesel can be produced from various feedstock and used in regular diesel vehicles without making any changes to the engines [3, 12].
- *Green diesel*: Green diesel is produced by removing the oxygen by catalytic reaction with hydrogen from renewable feedstock containing triglycerides and fatty acids, producing a paraffin-rich product, water, and carbon oxides. Therefore, green diesel has a heating value equal to conventional diesel and is

fully compatible for blending with the standard mix of petroleum-derived diesel fuels. Biodiesel has around 11% oxygen, whereas petroleum-based diesel and green diesel have no oxygen.

Using biomass as a feedstock for liquid fuels production may cut back on waste and greenhouse gas emissions and can offset the use of fossil fuels in heat and power generation [3].

2.4.7 Wind Energy

Earth is unevenly heated by the sun and the differential heating drives a global atmospheric convection system reaching from the earth's surface to the strato-sphere. Most of the energy stored in these wind movements can be found at high altitudes where continuous wind speeds of over 160 km/h (99 mph) occur. To assess the frequency of wind speeds at a location, a probability distribution function is often fitted to the observed data. Wind power is a totally renewable energy source with no greenhouse gas emissions, but due to its unpredictability has problems integrating with national grids. The potential for wind to supply a significant quantity of energy is considerable. Availability of transmission capacity helps large-scale deployment by reducing the cost of delivered wind energy [2, 5, 17].

2.4.8 Geothermal Energy

Geothermal energy is the heat originating from the original formation of the planet, from radioactive decay of minerals, from volcanic activity, and from solar energy absorbed at the surface. The geothermal gradient, which is the difference in tem-perature between the core of the planet and its surface, drives a continuous con-duction of thermal energy in the form of heat from the core to the surface. Geothermal power is cost-effective, reliable, sustainable, and environmentally friendly for district heating, space heating, spas, industrial processes, desalination, and agricultural applications.

Hot water or steam reservoirs deep in the earth are accessed by drilling. Geothermal reservoirs located near the earth's surface maintain a relatively constant temperature of 50°–60°F. The hot water and steam from reservoirs can be used to drive generators and produce electricity. In other applications, the heat produced from geothermal is used directly in heating buildings and industrial plants. As in the case of biomass electricity, a geothermal plant runs 24 h per day, seven days per week and can provide base load power, thus competing against coal plants [2, 5].

2.4.9   Ocean Energy

Systems to harvest electrical power from ocean waves have recently been gaining momentum as a viable technology. The potential for this technology is considered promising. Although the generator is powerful enough to power a thousand homes, the turbine has minimal environmental impact, as it is almost entirely submerged, and the rotors pose no danger to wildlife as they turn quite slowly. Ocean thermal energy conversion uses the temperature difference that exists between deep and shallow waters to run a heat engine [2, 5].

2.5   Thermal Energy

Thermal energy is the sum of energy contained in the relative motion of many of microscopic particles of a macroscopic whole system. The particles randomly move and collide with each other and the wall of container. The kinetic energy of the motion (including rotation and vibration except for monatomic substances) of the particles relative to the fixed wall is thermal energy. Temperature is a relative measure of the amount of thermal energy. Heat is the transfer of thermal energy from a system with higher temperature to a system with lower temperature. At absolute zero temperature, no further energy can be removed from a system without reorganization of chemical or nuclear binding. Therefore, thermal energy only includes the kinetic and potential energies associated with the relative motion, their rotation, and vibrations, as well as the vaporization or condensation energies of molecules [10]. Around 90% of current primary energy consumption at some stage involves thermal energy. For heat added to a system at constant volume without work done, we have internal energy $dq = dU = C_v dT$. For heat added to a system at constant pressure without work done, we have $dq = dH = C_p dT$, where $H = U +PV$. Phase changing by melting or vaporization takes place at constant temperature.

Energy used in a shower: A shower runs around 10 L/min. If a shower takes about 10 min and the water in a shower is heated from 20 to 50°C, the thermal energy used in the shower becomes:

Assume: Density of water is 1 kg/L

$$q = \dot{m} c_p \Delta T = (10\,\text{L/min})(10\,\text{min}/60\text{s})(1\,\text{kg/L})(4.18\,\text{kJ/kgK})(30\,\text{K}) = 209\,\text{kJ}$$

2.6   Hydrogen

Hydrogen is the simplest element. Each atom of hydrogen has only one proton. The sun is basically a giant ball of hydrogen and helium gases. In the sun's core, hydrogen atoms combine to form helium atoms (called fusion process) and gives off *radiant energy*. This radiant energy sustains life on earth as it drives the photosynthesis in plants and other living systems and is stored as chemical energy in fossil fuels.

Hydrogen does not exist on earth as a gas and is found only in compound form with other elements, such as water H_2O and methane CH_4. Hydrogen is produced from other resources, including natural gas, coal, biomass, and even water. The two most common production methods are steam reforming and electrolysis in which the water is split into oxygen and hydrogen. Steam reforming of natural gas is currently the least expensive and most common method of producing hydrogen. Global hydrogen production uses various feedstock, including natural gas, crude oil, coal, and water [5, 9].

Hydrogen has the highest energy content of any common fuel by weight (about three times more than gasoline) but the lowest energy content by volume (see Table 2.9). Hydrogen transports energy in a useable form from one place to another. Like electricity, hydrogen is an energy carrier. Hydrogen burns cleanly producing water, H_2O. When burned in an engine or used in a fuel cell, it is converted to water only. To make hydrogen a renewable fuel, it should use renewable energy, such as wind power or solar power, for production.

There are two primary uses for hydrogen. About half of hydrogen is used to produce ammonia (NH_3) via the Haber process. Ammonia, in turn, is used directly or indirectly as fertilizer. The other half of hydrogen production is used in hydrocracking process to convert heavy petroleum sources into lighter fractions suitable for use as fuels, as well as in other hydrogenation processes. Hydrogen fuel cells generate electricity. They are very efficient. Small fuel cells can power electric cars, while large fuel cells can provide electricity in remote places with no power lines [3, 5].

2.7 Electric Energy

The protons and electrons of an atom carry *electrical charge*. Protons have a positive charge (+) and electrons have a negative charge (−). Opposite charges attract each other. The electrons in an atom's outermost shells do not attract strongly to the protons and can move from one atom to another and create electricity. The amount of electricity a power plant generates, or a customer uses over a period of time is measured in kilowatt hours (kWh), which is equal to the energy of 1,000 watts working for one hour. For example, if you use a 100-W light bulb for 7 h, you have used 0.7 kWh of electrical energy.

Most of the electricity used in the residential sector is for air conditioning, refrigerators, space and water heating, lighting, and powering appliances and equipment. Electricity is the fastest growing form of end-use energy worldwide, as it has been over the past several decades. Electricity is the most well-known energy carrier to transfer the energy in coal, natural gas, uranium, wind power, and other energy sources to homes, businesses, and industry. We also use electricity to transfer the energy in flowing water from hydropower dams to consumers. For many energy needs, it is much easier to use electricity than the energy sources themselves [6, 9].

If the current passes through an electric appliance, some of the electric energy will be converted into other forms of energy (although some will always be lost as heat). The amount of electric energy, E_e, due to an electric current can be expressed in several different ways:

$$E_e = VIt = I^2Rt \tag{2.8}$$

where V is the electric potential difference (in volts), I is the current (in amperes), t is the time for which the current flows (in seconds), and R is the electric resistance (in ohms).

In *alternating current* (AC) the direction of the flow of electrons switches back and forth at regular intervals or cycles. Current flowing in power lines and normal household electricity that comes from a wall outlet is alternating current. The standard current used in the United States is 60 cycles per second (i.e. a frequency of 60 Hz); in Europe and most other parts of the world it is 50 cycles per second (i.e. a frequency of 50 Hz). In *direct current* (DC), on the other hand, electrical current flows consistently in one direction. The current that flows in a flashlight is direct current. One advantage of alternating current is that it is relatively cheap to change the voltage of the current. Furthermore, the inevitable loss of energy that occurs when current is carried over long distances is far smaller with alternating current than with direct current [10].

Example 2.6 Electricity consumption of a laptop computer
A laptop consuming 90 W is used on average 10 h per day. The laptop costs $500 and will be used for four years. Electricity cost is $0.15/kWh. Estimate the total electricity cost in four years for the laptop.

Solution:

$$\text{Cost}_{\text{laptop}} = \frac{\$500}{4\,\text{year}} = \$125/\text{year}$$

$$\text{Cost}_{\text{electricity}} = \left(\frac{\$0.15}{\text{kW h}}\right)\left(\frac{10\,\text{h}}{\text{day}}\right)\left(\frac{365\,\text{days}}{\text{year}}\right)(90\,\text{W})\left(\frac{\text{kW}}{1000\,\text{W}}\right) = \$49.3/\text{year}$$

$$\text{Cost}_{\text{total}} = \text{Cost}_{\text{laptop}} + \text{Cost}_{\text{electricity}} = (\$125/\text{year} + \$49.3/\text{year})(4\,\text{years}) = \mathbf{\$697.2}$$

2.8 Magnetic Energy

There is no fundamental difference between magnetic energy and electric energy: the two phenomena are related by Maxwell's equations. The potential energy of a magnet of magnetic moment m in a magnetic field B is defined as the work of magnetic force (magnetic torque) estimated by

$$E_m = -mB \tag{2.9}$$

Calculating work needed to create an electric or magnetic field in unit volume results in the electric and magnetic fields energy densities. Electromagnetic radiation, such as microwaves, visible light, or gamma rays, represents a flow of electromagnetic energy. The energy of electromagnetic radiation has discrete energy levels. The spacing between these levels is equal to $E = h\nu$, where h is the Planck constant, 6.626×10^{-34} Js, and ν is the frequency of the radiation. This quantity of electromagnetic energy is usually called a photon. The photons which make up visible light have energies of 160–310 kJ/mol [10].

2.9 Chemical Energy

Energy originating from the electromagnetic interactions betweens atoms in molecular levels is stored in chemical bonds and called the chemical energy of a material. The energy stored in a typical chemical bond is in a magnitude of electron volt ($eV = 1.602\ 10^{-19}$ J) that is the energy to move an electron across a one-Volt electric potential difference [10]. Chemical energy results from the associations of atoms in molecules and various other kinds of aggregates of matter. It may be defined as a work done by electric forces that is electrostatic potential energy of electric charges. If the chemical energy of a system decreases during a chemical reaction, the difference is transferred to the surroundings in the form of heat or light. On the other hand, if the chemical energy of a system increases because of a chemical reaction, the difference then is supplied by the surroundings in the form of heat or light. Typical values for the change in molar chemical energy during a chemical reaction range from tens to hundreds of kilojoules per mole. For example, 2,2,4-trimethylpentane (isooctane), widely used in petrol, has a chemical formula of C_8H_{18} and it reacts with oxygen exothermically and produces 10.86 MJ per mole of isooctane

$$C_8H_{18}(l) + 25/2\,O_2(g) \rightarrow 8CO_2(g) + 9H_2O(g) + 10.86\,MJ/mol \qquad (2.10)$$

When two hydrogen atoms react to form a hydrogen molecule, the chemical energy decreases by the bond energy of the H–H. When the electron is completely removed from a hydrogen atom, forming a hydrogen ion, the chemical energy called the ionization energy increases.

Ionization energy measures the energy required to remove electrons from an atom. This energy would be in the range of 5–100 keV and is the least for the outermost electron and highest for the last electrons. To remove a proton or neutron from a nucleus is approximately 8 MeV [10].

2.10 Mass Energy

Mass is a form of energy. Each particle is an excitation of quantum field similar to a single photon of light as a quantum excitation of electromagnetic field. Energy equivalent of a mass by Einstein's formula is $E = mc^2$, where c is the speed of light in a vacuum, $v = 2.997 \times 10^8$ m/s. Converting mass into energy is only possible by bringing a particle in contact with its antiparticle of the same type. Since the solar system does not contain antimatter naturally, mass energy is important only in nuclear reactions and nuclear power [10].

Summary

- *Energy* is the capacity to do work. Energy comes in various forms, such as motion, heat, light, electrical, chemical, nuclear energy, and gravitational. *Total energy* is the sum of all forms of the energy a system possesses.
- The *internal energy* of a system is made up of sensible, latent, chemical, and nuclear energies. The sensible internal energy is due to translational, rotational, and vibrational effects of atoms and molecules.
- *Thermal energy* is the sensible and latent forms of internal energy. The classification of energy into different 'forms' often follows the boundaries of the fields of study in the natural sciences.
- *Primary energy* is the energy extracted or captured directly from the environment. Three distinctive groups of primary energy are: nonrenewable energy, renewable energy, and waste. The primary energy is transformed to *secondary energy* in the form of electrical energy or fuel, such as gasoline, fuel oil, methanol, ethanol, and hydrogen.
- *Nonrenewable energy sources* are formed from the remains of dead plants and animals by exposure to heat and pressure in the earth's crust over the millions of

years. Major nonrenewable energy sources are: coal, petroleum, natural gas, and nuclear.

- *Coals* are sedimentary rocks containing combustible and incombustible matters as well as water. Coal has impurities like sulfur and nitrogen and when it burns the released impurities can combine with water vapor in the air to form droplets that fall to earth as weak forms of sulfuric and nitric acid as acid rain.
- *Petroleum oil* is a naturally occurring flammable liquid consisting of a complex mixture of hydrocarbons of various molecular weights, which define its physical and chemical properties, like heating value, color, and viscosity.
- *Natural gas* is a naturally occurring mixture consisting mainly of methane.
- *Nuclear energy* plants produce electricity through the fission of nuclear fuel, such as uranium. *Nuclear fission* is a nuclear reaction in which the nucleus of an atom splits into smaller parts, often producing free neutrons and photons in the form of gamma rays and releasing large amounts of energy.
- The *higher heating value* (HHV) consists of the combustion product of water condensed and that the heat of vaporization contained in the water vapor is recovered. The *lower heating value* (LHV) assumes that the water product of combustion is at vapor state and the heat of vaporization is not recovered.
- *Net heating value* is the same with lower heating value and is obtained by subtracting the latent heat of vaporization of the water vapor formed by the combustion from the gross or higher heating value. A common method of relating HHV to LHV per unit mass of a fuel is

$$\text{HHV} = \text{LHV} + \Delta H_{vap}\left[\left(\text{MW}_{H_2O}n_{H_2O,\text{out}}\right)/\left(\text{MW}_{\text{Fuel}}n_{\text{Fuel,in}}\right)\right]$$

where ΔH_{vap} is the heat of vaporization per mole of water (kJ/kg or Btu/lb), $n_{H_2O,\text{out}}$ is the moles of water vaporized, $n_{\text{fuel,in}}$ is the number of moles of fuel combusted, and *MW* is the molecular weight.

- *Energy density* is the amount of energy per unit volume.
- *Specific energy* is the amount of energy per unit amount.
- *Renewable energy* comes from natural resources and are naturally replenished. Major renewable energy sources are: hydroelectric, solar energy, biomass, wind, geothermal heat, ocean wave. In its various forms, renewable energy comes directly from the sun, or from heat generated deep within the earth.
- *Hydroenergy* comes from the potential energy of dammed water driving a water turbine and generator. The power extracted from the water depends on the volume and on the difference in height between the source and the water's outflow.
- *Solar energy* is derived from the sun through the form of solar radiation. *Active solar techniques* include the use of solar thermal collectors to harness the energy.

Passive solar systems rely on gravity and the tendency of water to naturally circulate as it is heated.

- *Solar electric generating system* use parabolic trough collectors to collect the sun's energy to generate steam to drive a conventional steam turbine. *Solar photovoltaic* (PV) convert light into electricity using semiconductor materials. Photovoltaic cell is a *solar cell*, which is a solid-state electrical device that converts the energy of light directly into electricity. Assemblies of cells are known as *solar modules* or *solar panels.*

- *Biomass* is an organic material made from plants including microorganisms and animals. Plants absorb the sun's energy in photosynthesis and store the energy as biomass. Biomass fuels are usually characterized by the *proximate* and *ultimate analyses.*

- *Biological fuels* produced from photosynthesis can be categorized into three groups:

- *Bioethanol* is usually corn-based ethanol and currently the largest source of biofuel as a gasoline substitute or additive. The gasoline is mixed with 10% ethanol, a mix known as E10 (or gasohol).

- *Biodiesel* is usually based on soybean oil and is most often blended with petroleum diesel in ratios of 2% (B2), 5% (B5), or 20% (B20).

- *Green diesel* is produced by removing the oxygen by catalytic reaction with hydrogen from renewable feedstock containing triglycerides and fatty acids, producing a paraffin-rich product, water, and carbon oxides. Therefore, green diesel has a heating value equal to conventional diesel and is fully compatible for blending with the standard mix of petroleum-derived diesel fuels.

- Most of the wind energy stored in these wind movements can be found at high altitudes where continuous wind speeds of over 160 km/h (99 mph). The potential for wind to supply a significant quantity of energy is considerable. Availability of transmission capacity helps large-scale deployment by reducing the cost of delivered wind energy.

- *Geothermal energy* is the heat originating from the original formation of the planet, from radioactive decay of minerals, from volcanic activity, and from solar energy absorbed at the surface.

- *Ocean energy* systems harvest electrical power from ocean waves.

- *Hydrogen* doesn't exist on earth as a gas and is found only in compound form with other elements, such as water H_2O and methane CH_4. Hydrogen is produced from other resources including natural gas, coal, biomass, and even water. The two most common production methods are steam reforming and electrolysis in which the water is split into oxygen and hydrogen.

- The protons and electrons of an atom carry an *electrical charge*. Protons have a positive charge (+) and electrons have a negative charge (−). The amount of electric energy, E_e, due to an electric current can be expressed in a number of different ways: $E_e = VIt = I^2Rt$, where V is the electric potential difference (in volts), I is the current (in amperes), t is the time for which the current flows (in seconds), and R is the electric resistance (in ohms).

- There is no fundamental difference between *magnetic energy* and electric energy: the two phenomena are related by Maxwell's equations. The potential energy of a magnet of magnetic moment m in a magnetic field B is defined as the work of magnetic force (magnetic torque) estimated by: $E_m = -mB$.
- *Chemical energy* results from the associations of atoms in molecules and various other kinds of aggregates of matter. It may be defined as a work done by electric forces that is electrostatic potential energy of electric charges.

$$C_8H_{18}(l) + 25/2\,O_2(g) \rightarrow 8CO_2(g) + 9H_2O(g) + 10.86\,MJ/mol$$

Problems

2.1. Why is electrical energy so useful?

2.2. How can the energy in the wind be used?

2.3. How can wind power help conserve our oil supplies?

2.4. How might using wind energy help reduce the air pollution?

2.5. What is the best energy source to convert to electricity?

2.6. Do the white-colored roof tiles keep houses cool?

2.7. How can energy from the sun be used to heat water?

2.8. With the clear advantages of nuclear power, why is it not more commonly used?

2.9. How can using solar energy help reduce pollution in the atmosphere and help conserve our oil supplies?

2.10. Why is the process of photosynthesis so valuable?

2.11. Name some foods that are known to be high energy foods.

2.12. Why are battery-powered vehicles considered to be the transport of the future?

2.13. Why is chemical energy useful to us?

2.14. What other forms of energy can be produced from chemical energy?

2.15. Name three examples of other fuels that contain chemical energy.

2.16. An over used car may consume around 250 gallons of gasoline per month. Estimate the energy consumed by the car per year.

2.17. An over used car may consume around 150 gallons of gasoline per month. Estimate the energy consumed by the car per year.

2.18. A city consumes natural gas at a rate of 500×10^6 ft³/day. The volumetric flow is at standard conditions of 60°F and 1 atm = 14.7 psia. If the natural is costing \$12/GJ of higher heating value what is the daily cost of the gas for the city?

2.19. A city consumes natural gas at a rate of 800×10^6 ft^3/day. The volumetric flow is at standard conditions of 60°F and 1 atm = 14.7 psia. If the natural is costing \$10/GJ of higher heating value what is the daily cost of the gas for the city.

2.20. A car consumes about 6 gallons a day, and the capacity of a full tank is about 15 gallons. The density of gasoline ranges from 0.72 to 0.78 kg/l (Table 2.2). The lower heating value of gasoline is about 44,000 kJ/kg. Assume that the average density of gasoline is 0.75 kg/l. If the car was able to use 0.2 kg of nuclear fuel of uranium-235, estimate the time in years for refueling.

2.21. A car consumes about 3 gallons a day, and the capacity of the full tank is about 11 gallons. The density of gasoline ranges from 0.72 to 0.78 kg/l (Table 2.2). The lower heating value of gasoline is about 44,000 kJ/kg. Assume that the average density of gasoline is 0.75 kg/l. If the car was able to use 0.1 kg of nuclear fuel of uranium-235, estimate the time in years for refueling.

2.22. Using data in Table 2.10 and ultimate analysis, fixed carbon, dry ash content, C, H, and O compositions estimate the gross heating values in kJ/kg for the biomass white oak.

2.23. Using data in Table 2.11 and ultimate analysis, fixed carbon, dry ash content, and C, H, and O compositions only estimate the gross heating values in kJ/kg for the biomass corn stover and wheat straw.

2.24. When a hydrocarbon fuel is burned, almost all the carbon in the fuel burns completely to form CO_2 (carbon dioxide), which is the principal gas causing the greenhouse effect and thus global climate change. On average, 0.59 kg of CO_2 is produced for each kWh of electricity generated from a power plant that burns natural gas. A typical new household uses about 7000 kWh of electricity per year. Determine the amount of CO_2 production in a city with 100,000 households.

2.25. When a hydrocarbon fuel is burned, almost all the carbon in the fuel burns completely to form CO_2 (carbon dioxide), which is the principal gas causing the greenhouse effect and thus global climate change. On average, 0.59 kg of CO_2 is produced for each kWh of electricity generated from a power plant that burns natural gas. A typical new household uses about 10,000 kWh of electricity per year. Determine the amount of CO_2 production in a city with 250,000 households.

2.26. A large public computer lab operates Monday through Saturday. There the computers are either being used constantly or remain on until the next user comes. Each computer needs around 240 W. If the computer lab contains 53 computers and each is on for 12 h a day, during the year how much CO_2 will the local coal power plant have to release to the atmosphere to keep these computers running?

2.27. An average university will have a large public computer lab open Monday through Saturday. There the computers are either being used constantly or remain on until the next user comes. Each computer needs around 240 W. If the computer lab contains 53 computers and each is on for 12 h a day, during the year how much coal will the local coal power plant have to consume to keep these computers running?

2.28. A large public computer lab runs six days per week from Monday through Saturday. Each computer uses a power of around 120 W. If the computer lab contains 45 computers and each is on for 12 h a day, during the year how much CO_2 will the local coal power plant have to release to the atmosphere to keep these computers running?

2.29. If a car consumes 60 gallons gasoline per month. Estimate the energy consumed by the car per year.

2.30. A car having an average 22 miles/gal is used 32 miles every day. If the cost of a gallon fuel is $3.8 estimate the consumed energy per year.

2.31. A car having an average 22 miles/gal is used 32 miles every day. Estimate the yearly energy usage.

2.32. A 150-Watt electric light bulb is used on average 10 h per day. A new bulb costs $2.0 and lasts about 5,000 h. If electricity cost is $0.15/kWh, estimate the consumed energy and its cost per year.

2.33. A laptop consuming 90 W is used on average 5 h per day. If a laptop will be used for five years estimate the total energy.

2.34. A laptop consuming 90 W is used on average 7 h per day. If a laptop will be used for four years estimate the total electricity needed.

2.35. A 20-hP electric motor is used to pump ground water into a storage tank 4 h every day. Estimate the work done by the pump in kW every year.

2.36. A city consumes natural gas at a rate of 250×10^6 ft^3/day. The volumetric flow is at standard conditions of 60°F and 1 atm = 14.7 psia. If the natural gas is costing $6/GJ of higher heating value what is the yearly energy consumed and its cost of the gas for the city.

2.37. A home consumes natural gas at a rate of 4.3ft^3/day to heat the home. The volumetric flow is at standard conditions of 60°F and 1 atm = 14.7 psia. If the natural gas is costing $0.67/MJ of higher heating value what is the yearly use of energy and its cost of the gas for the home?

2.38. A water heater consumes propane, which is providing 80% of the standard heat of combustion. If the price of propane is $2.2/gal measured at 25 °C, what is the amount of energy for heating and its cost in $ per million Btu and in $ per MJ/year?

2.39. An average video games system consumes 170 W of power during game-play. If a person were to play an hour a day for 80% of the year, how many liters of gasoline would the person have burned? (Evaluated at HHV)

2.40. A competitive road cyclist can hold an average of 300 W of power during a four-hour race. During long races they must do this each day for three weeks long race. How many protein bars will the cyclist have to eat at 184 calories per bar to just make up the calories lost during the race?

2.41. Describe the process of how natural gas goes from its natural state to the market?

2.42. Some people like to have background noise when they are falling asleep. Many choose to listen to their television. The television will usually run on about 340 W and will run during the 8 h that you are asleep. With electricity costing $0.20/kWh, calculate how much this will cost you if you do this for five days a week for an entire year.

2.43. What are the advantages and disadvantages of electrical energy in an alternating current?

2.44. What are the advantages and disadvantages of electrical energy flowing in direct current?

2.45. In the search for new sources of energy that are renewable and emit less greenhouse gases, carbon-based biofuels are of major interest. These fuels are still carbon based and must undergo combustion to release the chemical energy. Why is this process being looked at as a reasonable energy source?

2.46. While fixing wiring in a house, an electrician aims to deliver the same amount of electric energy to a devise at the same rate it is currently coming in. The wiring he is replacing has 2 ohms of resistance and runs at 20 amp current. To deliver the same amount of electric energy how many amps will be needed if the resistance is changed to 4 ohms?

2.47. Calculate the yearly dollar savings if you cut down from a daily nine-minute shower to a six-minute shower. The shower volumetric flow rate is 3.2 gpm and the amount of energy used per gallon is 440 Btu and energy costs $0.13/kWh.

2.48. Rank the following carbon-based fuels in the order of lowest to highest gross energy density: diesel, ethanol, conventional gasoline, and kerosene-based jet fuel.

References

1. Alhajji M, Demirel Y (2015) Energy and environmental sustainability assessment of crude oil refinery by thermodynamic analysis. Int. J Energy Research 39:1925–1941
2. Blackburn JO (2014) Renewable Energy. In: Anwar S (ed) Encyclopedia of energy engineering and technology, 2nd edn. CRC Press, Boca Raton
3. Demirel Y (2018) Biofuels. In: Comprehensive energy systems, I. Dincer (ed) Elsevier, Amsterdam, vol. 1, Part B, pp 875–908. (https://www.sciencedirect.com/science/article/pii/B9780128095973001255)
4. Demirbas A (1997) Calculation of higher heating values of biomass fuels. Fuel 76:431–434
5. Dincer I, Midilli A (2014) Green Energy. In: Anwar S (ed) Encyclopedia of energy engineering and technology, 2nd edn. CRC Press, Boca Raton
6. IEA (2019) World energy outlook 2019, IEA, Paris https://www.iea.org/reports/world-energy-outlook-2019

7. IEO (2019) International energy outlook 2019 with projections to 2050. September 2019, U. S. Energy Information Administration, U.S. Department of Energy, Washington, DC 20585. https://www.eia.gov/ieo

8. IRENA (2020a) Renewable power generation costs in 2019. International Renewable Energy Agency, Abu Dhabi

9. IRENA (2020b) Global renewables outlook: energy transformation 2050 (Edition: 2020). International Renewable Energy Agency, Abu Dhabi

10. Jaffe RL, Taylor W (2018) The physics of energy. Cambridge Univ. Press, Cambridge

11. Marks DB (1999) Biochemistry, 3rd edn. Kluwer, Ney York

12. Nguyen N, Demirel Y (2013) Economic analysis of Biodiesel and glycerol carbonate production plant by glycerolysis. J Sustain Bioenergy Sys 3:209–216

13. Øvergaard S (2008) Issue paper: definition of primary and secondary energy. https://unstats.un.org/unsd/envaccounting/londongroup/meeting13/LG13_12a.pdf. Accessed April 2020

14. Parker G (2014) Coal-to-Liquid Fuels. In: Anwar S (ed) Encyclopedia of energy engineering and technology, 2nd edn. CRC Press, Boca Raton

15. Sheng C, Azevedo JLT (2005) Estimating the higher heating value of biomass fuels from basic analysis data. Biomass Bioenergy 28:499–507

16. Vimeux F, Cuffey KM, Jouzel J (2002) New insights into southern hemisphere temperature changes from Vostok ice cores using deuterium excess correction. Earth Planetary Sci Let 203:829–843

17. Wiser R, Bolinger M (2010) DOE EERE, Wind technologies market report. https://www.energy.gov/eere/wind/2019-wind-energy-data-technology-trends. Accessed May 2020

Mechanical Energy and Electrical Energy

3

Introduction and Learning Objectives: This chapter introduces mechanical and electrical energy. Mechanical energy can be broadly classified into potential energy and kinetic energy. Potential energy refers to the energy any object has because of its position in a force field. Kinetic energy is the work required to accelerate an object at a given speed. Mechanical energy due to the pressure of the fluid is known as pressure energy. This chapter discusses various forms of mechanical energy as well as electric energy.

The learning objectives of this chapter are to understand:

- Various forms of mechanical energy, such as potential and kinetic energy,
- Various forms mechanical work, such as boundary work and isentropic work,
- Electric energy and work.

3.1 Mechanical Energy

There are two forces that we can experience: gravitational forces and electromagnetic forces. Both the forces act through space. *Mechanical energy* describes the sum of potential energy and kinetic energy present in the components of a *mechanical system*. Mechanical energy is the energy associated with the motion or position of an object under gravitational force. Mass causes gravitational attraction. Charge and mass obey an inverse square law, which states that a specified physical quantity or strength is inversely proportional to the square of the distance from the source of that physical quantity. The gravitational force law depends inversely upon the square of the distance between two masses, so mass plays a role somewhat like the role charge plays in the force law. Charged particles exhibit non-gravitational forces between them. The force of attraction or repulsion between two electrically charged particles is directly proportional to the product of the electric charges and is inversely proportional to the square of the distance between them. This is known as

© Springer Nature Switzerland AG 2021
Y. Demirel, *Energy*, Green Energy and Technology,
https://doi.org/10.1007/978-3-030-56164-2_3

Coulomb's law. The law of conservation of mechanical energy states that if a body or a system is subjected only to conservative forces the total mechanical energy of that body or system remains constant [1, 2].

The mechanical energy is the form of energy that can be converted to mechanical work completely and directly by a mechanical device such as a turbine. Thermal energy cannot be converted to work directly and completely. The familiar forms of mechanical energy are the kinetic and potential energies. Pressure energy is also another form of mechanical energy due to pressure of a fluid. Therefore, the mechanical energy may be defined by

$$\dot{m}\left(\frac{\Delta P}{\rho} + \frac{\Delta v^2}{2}\left(\frac{kJ/kg}{1000 \ m^2/s^2}\right) + g\Delta z\right) = \dot{W}_{shaft} + \dot{W}_{loss} \qquad (3.1)$$

Pressure energy + Kinetic energy + Potential energy = Mechanical energy
+ Work loss

with the conversion factor for the kinetic energy: $\left(\frac{kJ/kg}{1000 \ m^2/s^2}\right)$, where $\dot{m}$ is the mass flow rate, P is the pressure, ρ is the density, v is the flow velocity, g is the acceleration of gravity, z is the elevation height, W_{shaft} is the net shaft work in per unit mass for a pump, fan, or similar equipment, and W_{loss} represents the work loss due to friction and other irreversibilities. The mechanical energy equation for a pump, a fan, or a turbine can be written in terms of energy per unit mass (ft^2/s^2 or $m^2/s^2 = N \ m/kg$). Equation (3.1) is often used for incompressible flow problems and is called the *mechanical energy equation* [1].

3.2 Kinetic Energy

Kinetic energy KE is the energy that a system or a material possesses because of its velocity relative to the surroundings. The kinetic energy of a flowing fluid relative to stationary surroundings is estimated by

$$KE = \frac{1}{2}mv^2 \qquad (3.2)$$

where m is the mass and v is the average velocity. The value of a change in the kinetic energy occurs in a specified time interval and depends only on the mass and the initial and final values of the average velocities of the material.

$$\Delta KE = \frac{1}{2}m\left(v_2^2 - v_1^2\right) \qquad (3.3)$$

Mass flow rate $\dot{m}$ is related to the density ρ, cross sectional area A, and average velocity v by

$$\dot{m} = \rho A v \tag{3.4}$$

In English engineering units, for unit consistency, g_c ($g_c = 32.174\,\mathrm{lb_m ft/lb_f s^2}$) is included and the unit of kinetic energy becomes

$$KE = \frac{mv^2}{2\,g_c} = \frac{\mathrm{lb_m\,ft^2/s^2}}{\mathrm{lb_m\,ft/lb_f s^2}} = \mathrm{ft\,lb_f} \tag{3.5}$$

Momentum is related to kinetic energy in mechanics and estimated for a particle of mass moving with velocity (dx/dt) by $P = m\,(dx/dt)$. Kinetic energy in terms of momentum is $E = \frac{p^2}{2m}$. Like energy, the total momentum is conserved for an isolated system.

Newton's three laws of motion are: (i) an object in motion experiences no change in its velocity unless an external force acts, (ii) an acceleration a of an object with a mass m under the influence of a force is: $F = ma$, and (iii) for every action there is an equal and opposite reaction.

Air resistance-fluid (air/liquid) resistance causes energy loss of a moving object with a speed v

$$\frac{dE_{loss}}{dt} = c_d \frac{A\rho v^3}{2} \tag{3.6}$$

where ρ is the mass density of the fluid, A is the cross-sectional area of the object, and c_d is the drag coefficient, a dimensionless quantity of order one. The product $c_d A$ is often used and called effective area [3].

Example 3.1 Calculation of the kinetic energy for a flowing fluid
Water with a flow rate of 2.0 kg/s is pumped from a storage tank through a tube of 2.5 cm inner diameter. Calculate the kinetic energy of the water in the tube.

Solution:
Assume the density of water as $\rho = 1000\,\mathrm{kg/m^3}$
Basis is 2.0 kg/s.
Radius of the tube $(r) = 2.5/2 = 1.25\,\mathrm{cm}$
Cross sectional area of the tube:
$A = \pi r^2 = 3.14(1.25\,\mathrm{cm})^2(1\,\mathrm{m}/100\,\mathrm{cm})^2 = 4.906 \times 10^{-4}\,\mathrm{m^2}$
The average velocity of the water: $v = \dfrac{\dot{m}}{\rho A} = \dfrac{2.0\,\mathrm{kg/s}}{(1000\,\mathrm{kg/m^3})\,(4.906 \times 10^{-4}\,\mathrm{m^2})}$
$= 4.076\,\mathrm{m/s}$

The kinetic energy of the water in the tube:

$$KE = \frac{1}{2}mv^2 = \frac{1}{2}(2.0\,\text{kg/s})(4.076\,\text{m/s})^2\left(\frac{\text{kJ/kg}}{1000\,\text{m}^2/\text{s}^2}\right) = 0.0166\,\text{kJ/s} = 16.6\,\text{W}$$

J = N m, where N is the Newton.

Example 3.2 Kinetic energy of a car

A car having a mass of 2750 lb is travelling at 55 miles/h. Estimate: (a) the kinetic energy of the car in kJ, (b) the work necessary to stop the car.

Solution:

Equation: $EK = \frac{1}{2}mv^2$

Data: $m = 2750\,\text{lb}(\text{kg}/2.2\,\text{lb}) = 1250\,\text{kg}$;

$$v = 55\,\text{miles/h} \rightarrow v = \left(\frac{55\,\text{miles}}{\text{hr}}\right)\left(\frac{1609.34\,\text{m}}{\text{mile}}\right)\left(\frac{\text{h}}{3600\,\text{s}}\right) = 24.58\,\text{m/s}$$

(a) $EK = \frac{1}{2}(1250\,\text{kg})(24.58\,\text{m/s})^2\left(\frac{\text{kJ/kg}}{1000\,\text{m}^2/\text{s}^2}\right) = 377.6\,\text{kJ}$

(b) $W = EK = 377.6\,\text{kJ}$.

3.3   Potential Energy

Potential energy exists whenever an object has a position within a force field. The gravitational force near Earth's surface varies with the height h and is equal to the mass m multiplied by the gravitational acceleration $g = 9.81$ m/s^2. When the force field is the earth's gravitational field, then the potential energy of an object is:

$$PE = mgz \tag{3.7}$$

where z is the height above earth's surface or a reference surface. Potential energy is stored within a system and is activated when a restoring force tends to pull an object back toward some lower energy position. For example, when a mass is lifted, the force of gravity will act to bring it back down. The energy that went into lifting the mass is stored in its position in the gravitational field. Therefore, the potential energy is the energy difference between the energy of an object in each position and its energy at a reference position Δz

$$\Delta PE = mg\Delta z \tag{3.8}$$

In English units, the potential energy is expressed as

$$\Delta PE = \frac{mg\Delta z}{g_c} = \frac{\mathrm{lb_m\,ft\,ft/s^2}}{\mathrm{lb_m\,ft/lb_f s^2}} = 1\,\mathrm{lb_f\,ft} \tag{3.9}$$

For unit consistency, g_c ($g_c = 32.174\ \mathrm{lb_m\ ft/lb_f\ s^2}$) is included and the unit of potential energy becomes $\mathrm{lb_f}$ ft. Various types of potential energy are associated with a force [4]:

- *Gravitational potential energy* is the work of the gravitational force. If an object falls from one point to another inside a gravitational field, the force of gravity will do work on the object, such as the production of hydroelectricity by falling water.
- *Elastic potential energy* is the work of an elastic force. Elastic potential energy arises because of a force that tries to restore the object to its original shape, which is most often the electromagnetic force between the atoms and molecules that constitute the object.
- *Chemical potential energy* is related to the structural arrangement of atoms or molecules. For example, chemical potential energy is the energy stored in fossil fuels and when a fuel is burned the chemical energy is converted to heat. Green plants convert solar energy to chemical energy through the process called photosynthesis.
- *Nuclear potential energy* is the potential energy of the particles inside an atomic nucleus. The nuclear particles are bound together by the strong nuclear force.

Example 3.3 Potential energy change of water
Water is pumped from one reservoir to another. The water level in the second reservoir is 30 ft above the water level of the first reservoir. What is the increase in specific potential energy of one pound of water?

Solution:
The problem requires the change of potential energy of 1 $\mathrm{lb_m}$ of water. The difference between the water levels of the two reservoirs is $\Delta z = 30$ ft. Equation (3.8) yields

$$\Delta PE = \left(\frac{32.2\,\mathrm{ft}}{\mathrm{s^2}}\right)\left(\frac{1\,\mathrm{b_f\,s^2}}{32.2\,\mathrm{lb_m\,ft}}\right)(30\,\mathrm{ft})\left(\frac{\mathrm{Btu}}{778.2\,\mathrm{ft\,lb_f}}\right) = \mathbf{0.0385\,Btu/lb}$$

Example 3.4 Energy of an elevator

An elevator with a mass of 1500 kg rests at a level of 3 m above the base of an elevator shaft. The elevator has traveled to 30 m above the base. The elevator falls from this height freely to the base and strikes a strong spring. Assume that the entire process is frictionless. Estimate:

(a) The potential energy of the elevator in its original position relative to the base of the shaft.
(b) The work done traveling the elevator.
(c) The potential energy of the elevator at its highest position relative to the base of the shaft.
(d) The velocity and kinetic energy of the elevator just before it strikes the spring.

Solution:

Assume $g = 9.8 \, \text{m/s}^2$

Data: $m = 1500 \, \text{kg}, z_1 = 3 \, \text{m}, z_2 = 30 \, \text{m}$

(a) $PE_1 = mgz_1 = (1500)(9.8)(3) = \textbf{44100 J} = \textbf{44.1 kJ}$

(b) $$W = \int_{z_1}^{z_2} F dz = \int_{z_1}^{z_2} mg dz = mg(z_1 - z_2) = (1500)(9.8)(30 - 3)$$

$$= \textbf{396.9 kJ}$$

(c) $PE_2 = mgz_2 = (1500)(9.8)(30) = \textbf{441.0 kJ}$

So $W = PE_2 - PE_1 = (441.0 - 44.1) \, \text{kJ} = 396.9 \, \text{kJ}$

(d) $KE = PE_2 = \textbf{441.0 kJ}$

$$KE = \frac{mv^2}{2} \rightarrow v = \sqrt{\frac{2KE}{m}} = \sqrt{\frac{2(441,000)}{1500}} = \textbf{24.25 m/s.}$$

Example 3.5 Mechanical energy of a plane

Consider a Boeing 777 airplane with approximately 350 tons at a cruising speed of 850 km/h. Estimate the kinetic and potential energy.

Solution:

This plane has a kinetic energy:

$$E = \frac{1}{2}mv^2 = \frac{1}{2}(350,000\,\text{kg})[(850\,\text{km/h})(1000\,\text{m/km})(\text{h}/3600\text{s})]^2 = 9.7\,\text{GJ}$$

The same plane flying at an attitude of 10,000 m has the potential energy:

$$E = mgz = (350,000\text{kg})(9.8\,\text{m/s}^2)(10,000\,\text{m}) = 34.3\,\text{GJ}.$$

3.4 Pressure Energy

The pressure unit Pa is defined as: $\text{Pa} = \text{N/m}^2 = \text{N m/m}^3 = \text{J/m}^3$, which becomes energy per unit volume. The product of pressure with volume $P(1/\rho)$ leads to the energy per unit mass. Therefore, the mechanical energy due to the pressure of the fluid is known as pressure energy PV and expressed by $PV = P/\rho$, where ρ is the density. Pressure energy can be converted into mechanical energy completely, for example, by a reversible turbine. Transferring of mechanical energy is usually accomplished through a rotating shaft, and mechanical work is usually referred to as shaft work. A turbine, for example, converts the mechanical energy of a fluid to shaft work. Only the frictional effects cause the loss of mechanical energy in the form of heat. Consider a tank filled with water, as shown in Fig. 3.1.

$$\dot{W}_{\text{max}} = \dot{m}\frac{P}{\rho} = \dot{m}\frac{\rho gz}{\rho} = \dot{m}gz \qquad (3.10)$$

Fig. 3.1 Pressure energy P/ρ can be converted into mechanical energy completely by a reversible turbine

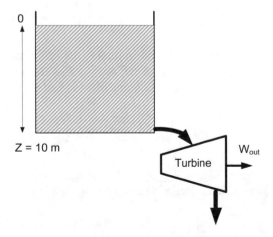

For example, for a water mass flow rate of 10 kg/s, maximum work would be

$$\dot{W}_{\text{max}} = \dot{m}gz = (10\,\text{kg/s})(9.81\,\text{m/s}^2)(10\,\text{m}) = 981\,\text{W}$$

For a flowing fluid, pressure energy is the same with the flow energy using the gauge pressure instead of absolute pressure [1].

Example 3.6 Pressure energy of a hydraulic turbine
Electricity is produced by a hydraulic turbine installed near a large lake. Average depth of the water in the lake is 45 m. The mass flow rate of water is 600 kg/s. Determine the work output of the turbine.

Solution:
Assume: The mechanical energy of water at the turbine exit is small and negligible. The density of water is 1000 kg/m^3.
$W_{\text{out}} = PE = \dot{m}gz = (600\,\text{kg/s})(9.81\,\text{m/s}^2)(45\,\text{m})$
$\qquad = \mathbf{264.9kW}\,(\text{kJ/kg} = 1000\text{m}^2/\text{s}^2)$
The lake supplies 264.9 kW of mechanical energy to the turbine.

3.4.1 Pressure Head

For an incompressible fluid, such as liquids, the pressure difference between two elevations can be expressed as *static pressure difference*: $P_2 - P_1 = -\rho g\,(z_2 - z_1)$. The *pressure head h* is estimated by the static pressure difference [5].

$$h = \frac{P_2 - P_1}{\rho g}\,(\text{in m or ft}) \tag{3.11}$$

The pressure head is related to the height of a column of fluid required to give a pressure difference of $\Delta P = P_2 - P_1$. The mechanical energy equation (Eq. 3.1) can also be written in terms of head and per unit mass.

$$\left(\frac{\Delta P}{\rho g} + \frac{\Delta v^2}{2g}\left(\frac{\text{kJ/kg}}{1000\,\text{m}^2/\text{s}^2}\right) + \Delta h\right) - h_{\text{shaft}} = h_{\text{loss}} \tag{3.12}$$

where $h_{\text{shaft}} = W_{\text{shaft}}/g$ is the net shaft energy head in per unit mass and the head loss due to friction is estimated by

$$h_{\text{loss}} = W_{\text{loss}}/g = \text{loss head due to friction} \tag{3.13}$$

Example 3.7 Pumping water

Water is pumped from an open tank at the level of 0 ft to an open tank at the level of 12 ft. The pump adds 6 hp to the water when pumping a volumetric flow rate of 3 ft^3/s. Estimate the loss energy in head.

Solution:
Data: $Q = 3$ ft^3/s, $W_{shaft} = 6$ hp
Since $v_{in} = v_{out}$, $P_{in} = P_{out}$ and $g = 32.2$ ft/s^2.
Density of water = 62.4 lb/ft^3.
The special conversion factor with symbol $g_c = 32.174$ ft lb$_m$/(s^2 lb$_f$).
From Eq. (3.12): $h_{shaft} = h_{loss} + h_{out}$

$$h_{shaft} = \frac{W_{shaft}}{\rho g Q} = \frac{(6\,hp)\left(\dfrac{550\,ft\,lb}{hp}\right)}{(62.4\,lb/ft^3)(3\,ft^3/s)} = 17.6\,ft$$

where, specific weight of water is 62.4 lb/ft^3 and 1 hp (English horsepower) = 550 ft lb/s

$$h_{loss} = (17.6 - 12)\,ft = \mathbf{5.6\,ft}.$$

Example 3.8 Calculation of the power needed to pump water

A pump draws water from a 10 ft deep well and discharges into a tank, which is 2 ft above the ground level. The discharge rate of water is 0.25 ft^3/s. Calculate the power needed by the pump.

Solution:
Assume that negligible heat dissipation due to friction occurs within the pipe and the pump.
The power needed can be calculated from: $\dot{W} = \dot{m}g(z_{out} - z_{in})$
where g is the standard gravity: $g = 9.806$ m/s$^2 = 32.2$ ft/s^2.
The special conversion factor with symbol $g_c = 32.174$ ft lb$_m$/(s^{2}lb$_f$).
Density of water = 62.4 lb/ft^3
With a basis of lb$_m$/s, the mass flow rate of water is
$\dot{m} = (0.25\,ft^3/s)(62.4\,lb/ft^3) = 15.6\,lb_m/s$
Using the conversion of 1.0 kW = 737.56 lb$_f$ ft/s

$$\text{Power: } W = \frac{(15.6\,lb_m/s\,)(32.2\,ft^2/s\,)(12\,ft)}{32.2\,ft\,lb_m/(s^2\,lb_f)}\left(\frac{1.0\,kW}{737.56\,lb_f\,ft/s}\right) = \mathbf{0.253\,kW} = \mathbf{0.340\,hp}$$

3.5   Surface Energy

Surface energy is a measure of intermolecular forces that occur when a surface is created. The surface energy is the excess energy at the surface of a material compared to the bulk. For a liquid, the surface tension (surface energy density) is defined as the force per unit length. Water has a surface energy density of 0.072 J/m^2 and a surface tension of 0.072 N/m. Raindrops are spherically shaped because of surface tension or surface energy. Surface energy is a characteristic property of a liquid and directly related to the liquid's surface area. High surface area geometries contain high surface energy, while low surface area geometries contain low surface energy levels. Liquids form themselves into spherical shapes to lower their overall energy state since the sphere has the smallest surface area and has the lowest energy state. The driving force behind spherically shaped liquids involves the cohesive forces, which are the attractive forces that a liquid's molecule has to one another [1, 5].

3.6   Sound Energy

Sound energy is the energy produced by an object's vibrations. Sound energy E results from the integral of the acoustic pressure P times the particle velocity v over a surface A, and is given by the integral

$$E = \int (Pv) \cdot dA \qquad (3.14)$$

The sound energy flux is the average rate of flow of sound energy through any specified area A and is usually referred to as acoustic intensity. Sound energy is, therefore, a form of mechanical energy and is related to the pressure of sound vibrations produce. Sound energy is typically not used for electrical power or for other human energy needs because the amount of energy that can be gained from sound is quite small.

Ultrasound is cyclic sound pressure with a frequency greater than the upper limit of human hearing, which is approximately 20 kHz. For a sound wave, the hertz (Hz) is equal to cycles of the sound wave per second. Ultrasound is used typically to penetrate a medium and measure the reflection signals or supply focused energy. The reflection signals can reveal details about the inner structure of the medium. In ultrasonic welding of plastics, high frequency (15–40 kHz) low amplitude vibration is used to create heat by way of friction between the materials to be joined. Also, ultrasonic testing is used to find flaws in materials and to measure the thickness of objects [4].

3.7   Electric Energy

Electricity starts with *charge* that produces electrical forces. When, for example, a light bulb is turned on, charge flows from the wall plug through the bulb and heats up the filament in the bulb generating light. There are two large forces: gravitational forces and electromagnetic forces. Both the forces act through space. Mass causes the gravitational forces and two masses always attract each other. Some particles possess a charge and may be subject to electromagnetic forces. Charge may be positive or negative and two positive charges or two negative charges will repel each other, whereas the positive and negative charges attract each other. Charge is measured in coulombs (C) named after Charles Augustin Coulomb who was the first scientist to formulate the force law for charges. A farad (F) is the charge in coulombs. The Coulomb's law is given by

$$F = k_e \frac{Q_1 Q_2}{r^2} \tag{3.15}$$

where k_e is the coulomb's constant and given by $k_e = 1/(4\pi\varepsilon_0)$ $= 8.987 \times 10^9 \, \mathrm{Nm^2/C^2}$, ε_0 is the electric constant, and in SI system $\varepsilon_0 \sim 8.885419 \times 10^{-12}$ F/m. For the two charges, Q_1 and Q_2, the force between them is proportional to the product of the two charges and inversely proportional to the square of the distance r (in meters) between them. The *electric charge* coulomb (C) is the amount of electricity carried in one second of time by one ampere of current (C = As). One Faraday is 96,485 C. A positive force implies it is repulsive, while a negative force implies it is attractive. Charge comes in discrete sizes. Electrons and protons have the charge of magnitude 1.6×10^{-19} coulombs that are negative for electrons and positive for protons. The charge on an electron is a fundamental constant of nature.

When charge flows in a wire it is called current (I), which is measured in coulombs/second and called amperes (A). For example, for a current of 4 A and time interval of 15 s, the total charge is: (current)(time) = (4 A)(15 s) = 60 °C. Current can also flow through an ionic solution and through the ground. Current is measured by a device called *ammeter* [4].

3.7.1   Electric Potential Energy

Electric potential energy is the work of the coulomb force. Work of intermolecular forces is called *intermolecular potential energy*. There are two main types of electric potential energy: electrostatic potential energy and electrodynamics potential energy (or magnetic potential energy). The *electrostatic potential energy* is the energy of an electrically charged particle in an electric field. It is defined as the work that must be done to move it from an infinite distance away to its present location, in the absence of any non-electrical forces. Consider two point-like objects

A_1 and A_2 with electrical charges Q_1 and Q_2. The work W required to move A_1 from an infinite distance to a distance r away from A_2 is given by:

$$W = \frac{1}{4\pi\varepsilon_0} \frac{Q_1 Q_2}{r} \tag{3.16}$$

Electric potential known as *voltage* is equal to the electric potential energy per unit charge. Voltage is the driving force behind current. When the charged particles are in motion, they generate *magnetic potential energy*. This kind of potential energy is a result of the magnetism having the potential to move other similar objects. Magnetic objects are said to have some *magnetic moment* [4].

Figure 3.2 shows a simple circuit with a battery and two other components. A battery will supply charge at some specified voltage to an applied electrical load. When the charge flows through the load it gives up the potential energy to the load. If the load is a motor then that energy might be transformed into mechanical energy. If the load is a light bulb, the energy is transformed into light and heat.

Here, V_B, represents the voltage across the battery. Considering the circuit in Fig. 3.2, the energy added to a charge Q would be:

- QV_B: when charge moves from the bottom of the battery to the top of the battery the charge gains an amount of energy QV_B
- QV_1: as a charge Q moves from the top of element 1 to the bottom of element 1, the charge loses an amount of energy QV_1
- QV_2: as a charge Q moves from the top of element 2 to the bottom of element 2, the charge loses an amount of energy QV_2.

Like mechanical potential energy, electrical potential energy and voltage are measured from a reference usually chosen as ground. That means the reference is ground itself. A voltage level of 120 V means that the voltage difference between that point and ground is 120 V. To measure voltage differences in circuits a *voltmeter* is used.

Fig. 3.2 Electrical circuit; V_B represents the voltage across the battery

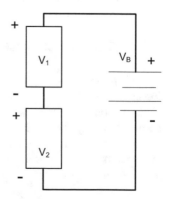

Fig. 3.3 A circuit diagram;
battery voltage, V_b, current I_r
through a resistance R

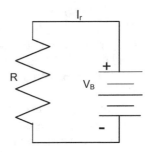

3.7.2 Estimation of Electrical Energy

Figure 3.3 shows a circuit with a resistance R, which determines how much current, will flow through a component. Resistors control voltage and current levels. A high resistance allows a small amount of current to flow. A low resistance allows a large amount of current to flow. Resistance is measured in ohms Ω. One ohm is the resistance through which one volt will maintain a current of one ampere. In a resistor, the voltage and the current are related to each other by Ohm's law

$$I = V/R \text{ or } V_r = RI_r \tag{3.17}$$

Ohm's law defines the relationships between voltage V, current I, and resistance R. If both V and I remain constant during the time t then electrical energy becomes

$$W_e = VIt \tag{3.18}$$

3.7.3 Electric Power

Electric power P is the rate at which electrical energy is transferred by an electric circuit (see Fig. 3.3)

$$P = VI \tag{3.19}$$

where P is the electric power, V is the potential difference, and I is the electric current. Electric power is measured in watts in SI units. The power flowing into the battery is negative, and the power flowing out of the battery is positive. The power delivered to the resistor is the product of the voltage across the resistor and the current through the resistor: VI_r.

In the case of resistive loads (see Fig. 3.3), using the Ohm's law, the following alternative expression for the dissipated power is given by

$$P = I^2R = V^2/R \tag{3.20}$$

where R is the electrical resistance.

3.7.4 Capacitance

A capacitor's ability to store charge is called the capacitance, which is measured by *farad* (*F*). One Farad is the capacitance between two parallel plates that results in one volt of potential difference when charged by one coulomb of electricity (*F* = *C/V*).

In *alternating current* circuits, energy storage elements such as inductance and capacitance may result in periodic reversals of the direction of energy flow. The portion of power flow that, averaged over a complete cycle of the alternating current waveform, results in net transfer of energy in one direction is known as real power or active power. That portion of power flow due to stored energy, which returns to the source in each cycle is known as reactive power, as observed in Fig. 3.4

The relationship between real power, reactive power and apparent power can be expressed by representing the quantities as vectors. Real power is represented as a horizontal vector and reactive power is represented as a vertical vector. The apparent power vector is the hypotenuse of a right triangle formed by connecting the real and reactive power vectors. This representation is often called the *power triangle*. Using the Pythagorean theorem, the relationship among real, reactive, and apparent powers is:

$$(\text{Apparent power})^2 = (\text{Real power})^2 + (\text{Reactive power})^2 \qquad (3.21)$$

Real and reactive powers can also be calculated directly from the apparent power, when the current and voltage are both sinusoids with a known phase angle between them:

$$(\text{Real power}) = (\text{Apparent power})\cos(\theta) \qquad (3.22)$$

$$(\text{Reactive power}) = (\text{Apparent power})\sin(\theta) \qquad (3.23)$$

The ratio of real power to apparent power is called power factor and is a number always between 0 and 1, where the currents and voltages have nonsinusoidal forms, and power factor is generalized to include the effects of distortion [2, 5].

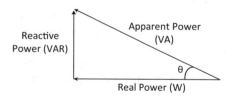

Fig. 3.4 Power triangle and the components of AC power

3.8   Electromagnetic Energy

Electromagnetic force is one of the four fundamental natural forces that are gravity and the strong and weak nuclear forces [4]. Electromagnetic interactions between charge particles and gravitation interactions between bodies are similar to each other. Coulomb's law shows that two charges create large forces equal in magnitude but opposite in direction and same charges repel while opposite-sign charges attract.

$$\mathbf{F}_{12} = -\mathbf{F}_{21} = \frac{1}{4\pi\varepsilon_0}\frac{Q_1 Q_2}{r_{12}^2}\mathbf{r}_{12} \tag{3.24}$$

where Q_1 and Q_2 are the charges (C, coulombs in SI units), r_{12} is the vector separating the two charges, F_{12} is the force exerted by Q_1 on Q_2, and ε_0 is the permittivity of vacuum ($1/4\pi\varepsilon_0 = 8.988 \times 10^9$ Nm^2/C^2).

Electric and magnetic fields that result from charged particles contain electromagnetic energy, which propagates through space as electromagnetic radiation such as visible light (wavelength λ in the range of 400–750 nm). Configuration of electrons and protons can store electromagnetic energy, which can be transmitted through electric circuits. Electromagnetic waves transmit solar energy to earth and become the source of all terrestrial energy sources including fossil fuels, biomass, wind, and wave. Quantum mechanics estimates the electromagnetic energy radiated when a system makes a transition from one quantum energy level to another by Planck's equation $E = h\nu$ where h is the Planck constant ($h = 4.1357 \times 10^{-15}$ eV s or 6.626176×10^{-34} J s) and ν is the frequency. This relation is fundamental to quantum physics. Thermodynamics relates energy and temperature through Boltzmann constant k_B (8.6173×10^{-5} eV/K) [4]

$$E = k_B T = \frac{hc}{\lambda} \tag{3.25}$$

where c is the speed of light. This equation relates the electromagnetic radiation to quantum spectra and thermal processes and define a thermal wavelength

$$\lambda_{\text{th}} = \frac{hc}{k_B T} \tag{3.26}$$

Thermal wavelength sets the scale for the wavelength of thermal radiation from a source at temperature T in which a system has a characteristic magnitude energy of $k_B T$.

Mechanical and electrical energy can be efficiently transformed to one another. However, it is highly inefficient to transform thermal or chemical energy into electromagnetic energy. Storage of electromagnetic energy is challenging yet very important [4]. There is no thermodynamic limit for conversions between mechanical and electromagnetic energy. Solar energy can be converted directly into electromagnetic energy by photovoltaic cells. A 100 W light emitting diode (LED) bulb draws only 16 W from the electric grid and radiates visible light at around 2.6 W [4].

Excitation of some medium can produce wave that can propagate energy over distances without transporting matter. Oscillation periodic in both space and time is the simplest form of wave, such as sine wave $f(x,t) = A \sin(kx - \omega t - \phi)$, where A is the amplitude, $k = 2\pi/\lambda$ is the wave number, λ is the wavelength, $\omega = 2\pi v$ is the angular frequency, v is the frequency, and ϕ is the phase shift. The speed of propagation v is related to λ and v by

$$v = \lambda v = \frac{\omega}{k} \tag{3.27}$$

Excitations at a point may be coupled locally to other close points and waves occur and propagates through the medium in time and space. The energy density in a wave depends on its amplitude quadratically. The one-dimensional wave equation becomes

$$\frac{\partial^2 f(x,t)}{\partial t^2} = v^2 \frac{\partial^2 f(x,t)}{\partial x^2} \tag{3.28}$$

as the wave is a function of $(x - vt)$. The wave equation governs propagation of electromagnetic waves and other typical waves.

For a current I flowing through a voltage difference $V(t)$ consumes power $P(t) = IV(t)$. Electromagnetic energy is converted into thermal energy in a resistor R: $P(t) = V^2(t)/R$ (Joule's law and the dissipation of power is called Joule heating) where $V = IR$ (Ohm's law) [5].

3.8.1 Magnetic Force

Moving charges create magnetic fields, which in turn exerts a force on moving charged particles. A magnetic field is described by a vector $\mathbf{B}(\mathbf{x}, t)$. The magnetic force acting on a moving charged (q) particle with velocity v in magnetic field $\mathbf{B}$ becomes

$$\mathbf{F} = q v \times \mathbf{B} \tag{3.29}$$

Magnetic force on moving charges may describe the principles of motors and generators. In a motor, an external magnetic field exerts force on a loop carrying current producing a torque on the loop driving a rotational motion. In generator moving the loop in the external magnetic field produces forces that drive an external current. Therefore, generators transform mechanical to electric power [4].

3.9 Various Forms of Work

There may be following forms of work [2, 4, 5]:

- Mechanical work
- Boundary work
- Fluid flow work
- Isentropic process work
- Polytropic process work
- Shaft work
- Spring work
- Stirrer work
- Acceleration work
- Gravitational work
- Electrical work.

3.9.1 Mechanical Work

When a force acts on a system through a distance, then energy is transferred as work. Mechanical work is performed when a mechanical force moves the boundary of a system. For example, in a compression work, a piston representing surroundings performs work on the gas (see Fig. 3.5) so that the initial boundary changes and final volume of the gas is less than the initial volume of the gas. The net amount of mechanical work done over a period is

$$W = - \int_{state1}^{state2} \mathbf{F} \cdot \mathbf{dl} \tag{3.30}$$

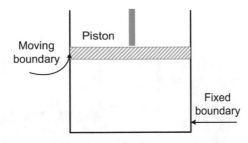

Fig. 3.5 Mechanical work for a mechanical force moving the boundary of a system

where $\mathbf{F}$ is an external force vector acting on the system boundary, and $\mathbf{l}$ is the displacement or pathway vector. Some common types of work are summarized within the next sections:

3.9.2 Boundary Work

The most common form of mechanical work is the boundary work. Boundary work is a mechanism for energy interaction between a system and its surroundings. *Boundary work at constant pressure* is the total boundary work done during the entire constant pressure process and is obtained by adding all the differential works from initial state1 to final state2.

$$W = - \int_{state1}^{state2} P \, dV \tag{3.31}$$

For example, during an expansion process, the boundary work represents the energy transferred from the system to its surroundings. Therefore, for the sign convention adapted here, the expansion work is negative. For a compression process, the boundary work represents the energy transferred from surroundings to the system as seen in Fig. 3.6. Therefore, the compression work is positive according to the sign convention. If the boundary is fixed between states 1 and 2, there would be no work performed on the system. For example, if a gas in a fixed volume container is heated and its initial temperature changed, there would be no work performed associated with the gas because the boundary of the system has not changed. Work calculated by Eq. (3.31) may be any value depending on the pathway taken between states 1 and 2. This means that work is a path dependent function [1, 3].

At constant pressure, Eq. (3.31) becomes

$$W = -P(V_2 - V_1) \tag{3.32}$$

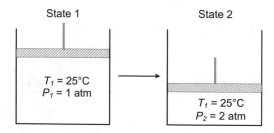

Fig. 3.6 Example of a boundary work: compression at constant temperature

For an ideal gas, *boundary work* at constant temperature for a unit mass is formulated by using $PV = RT = \text{constant} = C$ and $P = C/V$, which may be substituted to Eq. (3.32), and the work is

$$W = -\int_{\text{state1}}^{\text{state2}} P\,dV = -\int_{\text{state1}}^{\text{state2}} \frac{C}{V}\,dV = -C\ln\frac{V_2}{V_1} = -P_1 V_1 \ln\frac{V_2}{V_1} \tag{3.33}$$

The above equation may be modified by using: $P_1 V_1 = P_2 V_2$ as well as $V_2/V_1 = P_1/P_2$. For example, if a piston-cylinder device contains $0.5\ \text{m}^3$ of air at 120 kPa and 100 °C, and the air is compressed to $0.2\ \text{m}^3$ at constant temperature, then the work done would be

$$W = -P_1 V_1 \ln\frac{V_2}{V_1} = -(120\ \text{kPa})(0.5\ \text{m}^3)\ln\left(\frac{0.2}{0.5}\right)\left(\frac{\text{kJ}}{\text{kPa\,m}^3}\right) = 54.97\ \text{kJ}$$

As the work is done by the surrounding on the system the work is positive according to the sign convention adopted in this book.

By using $PV = nRT$ in Eq. 3.31, the isothermal boundary work is

$$W = -\int_1^2 P\,dV = -\int_1^2 \left(\frac{nRT}{V}\right)dV = -nRT\,\ln\left(\frac{V_2}{V_1}\right) = -nRT\,\ln\left(\frac{P_1}{P_2}\right) \tag{3.34}$$

Example 3.9 Expansion and compression work of an ideal gas
Estimate the work during: (a) an expansion of a gas from 1 to $7\ \text{m}^3$ at constant pressure of 120 kPa, (b) a compression of a gas when pressure changes from 100 to 80 kPa and volume changes from 7 to $5\ \text{m}^3$.

Solution:
Assume that the work done by, or on, an ideal gas is computed by taking the product of the pressure and the change in volume.
Data:

(a) $P = 120\ \text{kPa}$, $V_1 = 1\ \text{m}^3$, $V_2 = 7\ \text{m}^3$,

$$W = -P(V_2 - V_1) = -720\ \text{kJ}$$

Notice that the work is negative and done by the gas.

(b) When pressure and volume changes at the same time, we calculate the work underneath the P versus V diagram with two parts: a rectangle and a triangle:

$P_1 = 100\ \text{kPa}$, $P_2 = 80\ \text{kPa}$, $V_1 = 7\ \text{m}^3$, $V_2 = 5\ \text{m}^3$
The "area" of the rectangular region: $W_a = -P_2(V_2 - V_1) = 160\ \text{kJ}$
The "area" of the triangular region: $W_b = (1/2)(P_1 - P_2)(V_2 - V_1) = 20\ \text{kJ}$
Total work done on the gas $= (160 + 20)\ \text{kJ} = \mathbf{180}\ \text{kJ}$.

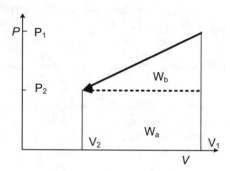

Example 3.10 Isothermal compression work
Estimate the boundary work when $V_1 = 0.5$ m^3 of carbon dioxide at $P_1 = 110$ kPa and $T_1 = 27$ °C is compressed isothermally to $P_2 = 750$ kPa.

Solution:
Assume that the carbon dioxide is ideal gas.
Data: $V_1 = 0.5$ m^3, $P_1 = 110$ kPa, $T_1 = 27$ °C, $P_2 = 750$ kPa
Isothermal system: $\Delta H = \Delta U = 0$
Ideal gas: $PV = nRT$ and $P = \dfrac{nRT}{V}$

$$W = -\int_1^2 P\,dV = -\int_1^2 \left(\frac{nRT}{V}\right) dV = -P_1 V_1 \ln\left(\tfrac{V_2}{V_1}\right)$$

$$W = -P_1 V_1 \ln\frac{V_2}{V_1} = -P_1 V_1 \ln\frac{P_1}{P_2} = -(110\,\text{kPa})(0.5\,\text{m}^3)\ln\left(\frac{110}{750}\right)\left(\frac{\text{kJ}}{\text{kPa}\,\text{m}^3}\right)$$

$$= 105.6\,\text{kJ}$$

The positive sign indicates that the work is done on the system.

3.9.3   Fluid Flow Work

Open systems involve mass flows across their boundaries. The work required to maintain a continuous flow into or out of the control volume is called the fluid-flow work. It is a displacement work done at the moving system boundary. If the fluid pressure is P and the cross-sectional area is A as seen in Fig. 3.7, the force applied on the fluid element becomes

$$F = PA \tag{3.35}$$

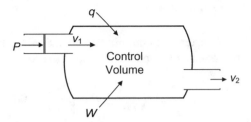

Fig. 3.7 Fluid flow with heat and work interactions between the control volume and surroundings: $\dot{m}\left(\Delta H + \Delta v^2/2\left(\frac{\text{kJ/kg}}{1000\ \text{m}^2/\text{s}^2}\right) + g\Delta z\right) = \dot{q} + \dot{W}_s$ at constant pressure

When the force acts through a distance L, the fluid-flow W_f work becomes

$$W_f = FL = PAL = PV \tag{3.36}$$

In most practical processes, flow can be approximated as steady state and one-dimensional, then the total energy of a fluid, E_{fluid}, entering or leaving a system at steady state is

$$\dot{E}_{\text{fluid}} = \dot{m}\left(\Delta U + PV + \frac{\Delta v^2}{2}\left(\frac{\text{kJ/kg}}{1000\ \text{m}^2/\text{s}^2}\right) + g\Delta z\right) \tag{3.37}$$

The dot over a symbol indicates a quantity per unit time. After the combination of $U + PV$ as enthalpy H ($\Delta H = \Delta U + PV$), Eq. (3.37) becomes

$$\dot{E}_{\text{fluid}} = \dot{m}\left(\Delta H + \left(\frac{\text{kJ/kg}}{1000\ \text{m}^2/\text{s}^2}\right)\frac{\Delta v^2}{2} + g\Delta z\right) \tag{3.38}$$

For fluid flow systems, enthalpy rather than the internal energy is important in engineering since the fluid-flow work is considered within the enthalpy [2, 3]. The general relation between the heat and work for a fluid flow is

$$\dot{m}\left(\Delta H + \left(\frac{\text{kJ/kg}}{1000\ \text{m}^2/\text{s}^2}\right)\frac{\Delta v^2}{2} + g\Delta z\right) = \dot{q} + \dot{W}_s \tag{3.39}$$

3.9.4   Isentropic Process Work

Isentropic process work occurs at isentropic entropy: $S_1 = S_2$. For isentropic process of ideal gas with constants C_v and C_p, we have the following relations

$$S_2 - S_1 = C_p \ln\left(\frac{T_2}{T_1}\right) - R \ln\left(\frac{P_2}{P_1}\right) = 0 \qquad (3.40)$$

Using the following relationships from the ideal gas equation

$$\frac{T_2}{T_1} = \frac{P_2 V_2}{P_1 V_1} \text{ and } C_p - C_v = R \text{ or } R/C_p = (\gamma - 1)/\gamma \qquad (3.41)$$

in Eq. (3.41) yields

$$P_1 V_1^\gamma = P_2 V_2^\gamma \text{ and } \frac{T_2}{T_1} = \left(\frac{P_2}{P_1}\right)^{\frac{\gamma-1}{\gamma}} \qquad (3.42)$$

where γ is the ratio of heat capacities ($\gamma = C_p/C_v$). For an ideal gas, the heat capacity depends only on temperature. The values of R at various units are:

$R = 10.73 \text{ psia ft}^3/\text{lbmol } R = 0.7302 \text{ atm ft}^3/\text{lbmol}$

$R = 83.14 \text{ bar cm}^3/\text{mol K}$

$R = 1.987 \text{ cal/mol K} = 1.986 \text{ Btu/lbmol } R$

$R = 8.314 \text{ J/mol K} = 8.314 \text{ Pa m}^3/\text{mol K} = 8314 \text{ kPa cm}^3/\text{mol K}$

For the isentropic process (reversible, adiabatic), the heat transfer is zero and using the ideal gas relations, the work is estimated by

$$W_s = \dot{m} C_v (T_2 - T_1) = \dot{m}(U_2 - U_1) \qquad (3.43)$$

Example 3.11 Isentropic compression of air
A tank filled with air at 1 bar and 20 °C is compressed adiabatically (isentropic) to 4 bar. Estimate the compression work in kJ/mol. Constant heat capacities for the air are $C_{v,av} = 20.8$ J/mol K and $C_{p,av} = 29.1$ J/mol K.

Solution:
Assume that the air is ideal gas.
Basis: 1 mol of ideal gas.
Data: $T_1 = 20$ °C (293 K); $P_1 = 1$ bar; $P_2 = 4$ bar; $R = 8.314$ J/mol K
$C_{v,av} = 20.8$ J/mol K and $C_{p,av} = 29.1$ J/mol K, $\gamma = C_{p,av}/C_{v,av} = 1.4$

For adiabatic expansion: $T_2 = T_1 \left(\frac{P_2}{P_1}\right)^{\frac{\gamma-1}{\gamma}} = 435\,\text{K}$

$W = C_{v,\text{av}}(T_2 - T_1) = \mathbf{2953.9\,J/mol}$

The positive sign of the work shows that the work is done on the air.

3.9.5   Polytropic Process Work

In *polytropic process work*, pressure and volume are related by

$$P_1 V_1^{\gamma} = P_2 V_2^{\gamma} = C \tag{3.44}$$

and the pressure is

$$P = CV^{-\gamma} \tag{3.45}$$

where γ and C are constants. By using this newly derived pressure in Eq. (3.33), we obtain the polytropic work by [1]

$$
\begin{aligned}
W &= -\int_{\text{state1}}^{\text{state2}} P dV \\
&= -\int_{\text{state1}}^{\text{state2}} CV^{-\gamma} dV = -C \frac{V_2^{-\gamma+1} - V_1^{-\gamma+1}}{1-\gamma} = -\frac{P_2 V_2 - P_1 V_1}{1-\gamma}
\end{aligned} \tag{3.46}
$$

Example 3.12 Calculation of work done by a piston on an ideal gas

A piston does a compression work on an ideal gas. The ideal gas is initially at 400 K and 300 kPa. The volume of the gas is compressed to 0.1 m^3 from 0.3 m^3. Calculate the work done by the piston on the gas by two different paths:

Path A: Isobaric compression at constant pressure $P = 300$ kPa.

Path B: Isothermal compression at constant temperature $T = 400$ K.

Solution:

Assume that the piston is frictionless; has an ideal compression (occurs very slowly), and the ideal gas is the system.

Basis is the amount of ideal gas.

From ideal gas law: $PV = nRT$, where $R = 8.314$ kPa m^3/kmol K

Moles of the gas: $n = PV/(RT)$

$= (300\text{ kPa}) (0.3\text{ m}^3)/[(8.314\text{ kPa m}^3/\text{kmol K})(400\text{ K})] = 0.027$ kmol.

Path A: Isobaric compression at $P = 300$ kPa:
$(V_2 - V_1) = (0.1 - 0.3) \text{ m}^3 = -0.2 \text{ m}^3$;

$$W = -\int_{V_1}^{V_2} P \, dV = -P(V_2 - V_1) = -300 \times 10^3 \text{ Pa} \frac{\text{N}}{\text{m}^2 \text{ Pa}} (-0.2 \text{ m}^3) \frac{\text{J}}{\text{Nm}}$$

$$= 60 \text{ kJ}$$

Path B: Isothermal compression at $T = 400$ K:

$$W = -\int_{V_1}^{V_2} P \, dV = -\int_{V_1}^{V_2} \frac{nRT}{V} dV = -nRT \ln\left(\frac{V_2}{V_1}\right)$$

$$= -(0.027 \text{kmol})\left(\frac{8.314 \text{kJ}}{\text{kmol K}}\right)(400 \text{K}) \ln\left(\frac{0.1 \text{m}^3}{0.3 \text{m}^3}\right) = 98.65 \text{kJ}$$

As work is a path-dependent variable, the work done by the following different paths are different from each other.

Example 3.13 Polytropic expansion of air

Air in a cylinder is at 110 kPa and expands polytropically from 0.01 to 0.05 m^3. The exponent $\gamma = 1.3$ in the pressure volume relation $(PV^\gamma = \text{constant})$ in polytropic process. Estimate the work done during the process.

Solution:
Assume that the air is ideal gas. $PV = RT$ (for one mole of gas).
Data: $V_1 = 0.01 \text{ m}^3$, $V_2 = 0.05 \text{ m}^3$, $P_1 = 110$ kPa, and $\gamma = 1.3$
$P_1 V_1^\gamma = P_2 V_2^\gamma = C = \text{constant}$

$$P_2 = P_1 \frac{V_1^\gamma}{V_2^\gamma} = (110 \text{ kPa})\left(\frac{0.01}{0.05}\right)^{1.3} = 13.6 \text{ kPa}$$

$$W_b = \int_1^2 CV^{-\gamma} dV = \frac{R\Delta T}{\gamma - 1} = \frac{P_2 V_2 - P_1 V_1}{\gamma - 1} \left(\frac{\text{kJ}}{\text{kPa m}^3}\right) = -1.4 \text{ kJ}$$

The negative sign indicates that the work is done on the surroundings.

3.9.6 Shaft Work

Shaft work is the energy transmitted by a rotating shaft and is related to the torque **T** applied to the shaft and the number of revolutions of the shaft per unit time. Shaft work occurs by force acting on a shaft to turn the shaft against a mechanical resistance. A shaft work performed by a pump transferring a liquid body over a distance is positive. On the other hand, the shaft work is negative if a fluid in the system turns a shaft that performs work on the surroundings [5].

A force F acting through a moment arm r produces a torque **T**.

$$\mathbf{T} = Fr \text{ and } F = \frac{\mathbf{T}}{r} \tag{3.47}$$

This force acts through distance of $2\pi rN$, where N is the number of revolutions. The work done during N revolutions becomes

$$W_s = F(2\pi rN) = \frac{\mathbf{T}}{r} 2\pi rN = 2\pi N\mathbf{T} \tag{3.48}$$

A rotating shaft can transmit power. The power transmitted through the shaft is

$$\dot{W}_s = 2\pi \dot{N}\mathbf{T} \tag{3.49}$$

Example 3.14 Estimation of shaft power
Estimate the power transmitted through the shaft of a car when the torque applied is 100 N m. The shaft rotates at a rate of 2000 revolutions per minute.
Solution:
$\dot{N} = 2000/\text{min}, \mathbf{T} = 100\,\text{Nm}$
$$\dot{W}_s = 2\pi \dot{N}\mathbf{T} = 2\pi \left(2000 \frac{1}{\text{min}} \right) \left(\frac{\text{min}}{60\text{s}} \right) (100\text{Nm}) \left(\frac{\text{kJ}}{1000\text{Nm}} \right)$$
$$= 20.9\text{kW} = 28.1\text{hp}$$

3.9.7 Spring Work

The length of a spring changes when a force is applied. For a change of length of dx after applying a force F, the work done becomes

$$\delta W_{sp} = Fdx \tag{3.50}$$

For linear elastic springs, the change in length is proportional to the force applied: $F = kx$, where k is the spring constant with unit kN/m. Therefore, for an initial x_1 and final x_2 displacements the work done becomes

$$W_{sp} = \frac{1}{2}k(x_2^2 - x_1^2)(\text{kJ}) \tag{3.51}$$

Example 3.15 Estimation of spring work
A piston cylinder system is attached to a spring with a spring constant of 110 kN/m. At state 1 there is no displacement on the spring. At state 2, the piston cylinder system performs a compression work and to decrease in volume underneath the piston, spring changes its length by 0.12 m. Estimate the spring work.

Solution:
$k = 110$ kN/m, $x_1 = 0$, $x_2 = 0.12$ m
$W_{sp} = \frac{1}{2}k(x_2^2 - x_1^2) = 792\,\text{J} = 0.792\,\text{kJ}$

3.9.8   Stirrer Work

When a shaft induces stirring of the fluid there occurs a work transfer to the fluid, which is system in this case, as the load. W_{load}. The work transfer W_{sw} is

$$W_{sw} = \int_1^2 mg \, dz = \int_1^2 W_{load} dz = \int_1^2 \mathbf{T} \, d\theta \qquad (3.52)$$

where dz is the vertical drop of the load W_{load} when the induced torque $\mathbf{T}$ has rotated the shaft through an angular displacement by $d\theta$.

3.9.9   Acceleration Work

Acceleration work occurs due to change of velocity of a system under the action of an accelerating force, F_α, which is defined by Newton's Second Law of Motion as $F_\alpha = ma$, where a is the acceleration $a = dv/dt$ and $dl = v \, dt$; hence the acceleration work is=

$$W_a = \int_1^2 F_a dl = \int_1^2 m(dv/dt)(vdt) = m \int_1^2 vdv = (1/2)m\left(v_2^2 - v_1^2\right) \qquad (3.53)$$

where v_1 and v_2 represent the initial and final velocities. The acceleration work represents the change in kinetic energy.

3.9.10   Gravitational Work

Gravitational work is done against gravity. This is simply the change in potential energy. If F_g is the gravitational force acting on the body of mass m, then $F_g = mg$, where g is acceleration due to gravity = 9.81 m/s^2. Then the gravitational work done to raise a mass m from elevation z_1 to z_2 is

$$W_g = \int_1^2 F_g dz = mg \int_1^2 dz = mg(z_2 - z_1) \qquad (3.54)$$

3.9.11 Electrical Work

When an electrical current flows through a resistor, it implies there is work transfer to the system (resistor). For example, the current that flows can drive a motor and hence a mechanical work is performed. *Electrical work* occurs when an electrical current passes through an electrical resistance in the circuit. Electrical work in the rate form is the same as power

$$\dot{W}_e = VI = I^2 R \tag{3.55}$$

where V is the potential difference, I is the electric current, and R is the resistance.

Example 3.16 Estimation of electrical work
A resistance heater within a well-insulated tank is used to heat 12 kg of water. The heater passes a current of 0.6 ampere (A) for 4 h from a 120-V source. Estimate the electric energy used after 4 h of heating.

Solution:
$V = 120$ V, $I = 0.6$ A, $t = 4$ h
Electrical energy received by the water:

$$W_e = VIt = 120\,\mathrm{V}(0.6\,\mathrm{A})(4\,\mathrm{h})(3600\,\mathrm{s/h})\left(\frac{\mathrm{kJ/s}}{1000\,\mathrm{VA}}\right) = 1036.8\,\mathrm{kJ}$$

3.10 Other Forms of Work

Nonmechanical forms of work are formulated in terms of generalized force, F, and generalized displacement, x

$$W_{\mathrm{other}} = Fx \tag{3.56}$$

- *Magnetic work*: (magnetic field strength) (total magnetic dipole moment)
- *Electric polarization work*: (electric field strength) (polarization of the medium).

Polarization of the medium is the sum of the electric dipole rotation moments of the molecules.

Summary

- *Mechanical energy* describes the sum of potential energy and kinetic energy present in the components of a *mechanical system*. Mechanical energy is the energy associated with the motion or position of an object under gravitational force and can be defined by

$$\dot{m}\left(\frac{\Delta P}{\rho} + \frac{\Delta v^2}{2}\left(\frac{kJ/kg}{1000\ m^2/s^2}\right) + g\Delta z\right) = \dot{W}_{shaft} + \dot{W}_{loss}$$

where $\dot{m}$ is the mass flow rate, P is the pressure, ρ is the density, v is the flow velocity, g is the acceleration of gravity, z is the elevation height, W_{shaft} represents net shaft work in per unit mass for a pump, fan, or similar equipment, and W_{loss} represents the work loss due to friction and other nonidealities.
- *Kinetic energy KE* is the energy a system or a material possesses because of its velocity relative to the surroundings. The kinetic energy of a flowing fluid relative to stationary surroundings is estimated by

$$KE = (1/2)mv^2$$

where m is the mass and v is the average velocity. Mass flow rate $\dot{m}$ is related to the density ρ, cross sectional area A, and average velocity v by

$$\dot{m} = \rho A v$$

In English engineering units, for unit consistency g_c ($g_c = 32.174\ lb_m\ ft/lb_f\ s^2$) is included and the unit of kinetic energy becomes

$$KE = \frac{mv^2}{2g_c} = \frac{lb_m\ ft^2/s^2}{lb_m ft/(lb_f s^2)} = ft\ lb_f$$

- *Potential energy* exists whenever an object has a position within a force field. The gravitational force near the earth's surface varies very little with the height h and is equal to the mass m multiplied by the gravitational acceleration $g = 9.81\ m/s^2$. When the force field is the earth's gravitational field, then the potential energy of an object is:

$$PE = mgz$$

where z is the height. In English units, the potential energy is expressed as

$$\Delta PE = \frac{mg\Delta z}{g_c} = \frac{\text{lb}_m\,\text{ft}\,\text{ft/s}^2}{\text{lb}_m\,\text{ft/lb}_f\text{s}^2}$$

With g_c ($g_c = 32.174$ lb$_m$ ft/lb$_f$ s^2) is included, the unit of potential energy becomes lb$_f$ ft.

- *Gravitational potential energy* is the work of the gravitational force.
- *Elastic potential energy* is the work of an elastic force.
- *Chemical potential energy* is related to the structural arrangement of atoms or molecules.
- For an incompressible fluid, such as liquids, the pressure difference between two elevations can be expressed as *static pressure difference*: $P_2 - P_1 = -\rho\,g\,(z_2 - z_1)$.
- The mechanical energy equation (Eq. 3.1) can also be written in terms of head and per unit mass

$$\left(\frac{\Delta P}{\rho g} + \frac{\left(\frac{\text{kJ/kg}}{1000\,\text{m}^2/\text{s}^2}\right)\Delta v^2}{2g} + \Delta h\right) - h_{\text{shaft}} = h_{\text{loss}}$$

where $h_{\text{shaft}} = W_{\text{shaft}}/g$ is the net shaft energy head in per unit mass and the head loss due to friction is estimated by $h_{\text{loss}} = W_{\text{loss}}/g$ = loss head due to friction.

- *Surface energy* is a measure of intermolecular forces that occur when a surface is created.
- *Sound energy* is the energy produced by an object's vibrations.
- *Ultrasound* is cyclic sound pressure with a frequency of approximately 20 kHz.
- *Power* is the energy exchanged in time.
- Electricity starts with *charge* that produces electrical forces.
- *Electric potential energy* is the work of the coulomb force.
- *Electric power P* is the rate at which electrical energy is transferred by an electric circuit: $P = VI$ where P is the electric power, V is the potential difference, and I is the electric current.
- There may be the following forms of work transfer: mechanical work, boundary work, fluid flow work, isentropic process work, polytropic process work, shaft work, spring work, stirrer work, acceleration work, gravitational work, electrical work.

Some common forms of work formulations:

Boundary work	Equation
General	$W = -\int_{\text{state1}}^{\text{state2}} P dV$
Isobaric work P = constant	$W = -P(V_2 - V_1)$
Isothermal T = constant	$W = -P_1 V_1 \ln(V_2/V_1)$
Isentropic work S = constant For ideal gases	$W_s = \dot{m} C_v (T_2 - T_1) = \dot{m}(U_2 - U_1)$
Polytropic work PV^γ = constant	$W = -\dfrac{P_2 V_2 - P_1 V_1}{1 - \gamma}$
Shaft work	$W_s = 2\pi N \mathbf{T}$
Spring work	$W_{sp} = \frac{1}{2} k(x_2^2 - x_1^2)$
Electrical work	$W_e = VIt$

Problems

3.1 Determine the SI units and English units of kinetic and potential energy.

3.2 A one lb$_m$ glass cup is at rest initially, and dropped onto a solid surface: (a) What is the kinetic energy change, (b) What is the sign of potential energy change? (c) What is the work?

3.3 A 50-hp electric motor is used to pump ground water into a storage tank. Estimate the work done by the pump in Btu/h, kW, and J/s.

3.4 A water pumping process consumes 200 hp. Estimate the pump work in Btu/h and in kW.

3.5 A 25-hP electric motor is used to pump ground water into a storage tank. Estimate the work done by the pump in Btu/h, MW, and J/s.

3.6 An 85-hP electric motor is used to pump ground water into a storage tank. Estimate the work done by the pump in Btu/h, kW, and kcal/s.

3.7 A 40-kW electric motor is used to pump ground water into a storage tank. Estimate the work done by the pump in Btu/h, hp, J/s, and ft lb$_f$/h.

3.8 A car having a mass of 1500 kg is initially travelling at a speed of 80 km/h. It slows down at a constant rate, coming to a stop at 90 m.

   (a) What is the change in the car's kinetic energy over the 90 m distance it travels while coming to a stop?
   (b) What is the net force on the car while it's coming to a stop?

3.9 A truck weighing 5000 lb is traveling at 55 miles/h. The brakes are suddenly applied to stop the car completely. Estimate the energy transfer through the brakes.

3.10 A car having a mass of 1000 kg is initially at rest. A constant 1000 N net force acts on the car over 50 m, causing the car to speed up. (a) Estimate the total work done on the car over the 50 m distance it travels. (b) Estimate the speed of the car after 50 m.

3.11   Calculate the kinetic energy (KE) of a 1500-kg automobile at a speed of 30 m/s. If it accelerates to this speed in 20 s, what average power has been developed?

3.12.   A load with mass of 200 kg is pushed to a horizontal distance of 20.0 m by a force of 60.0 N. For a frictionless surface, estimate the kinetic energy of the load when the work is finished, (b) estimate the speed of the load.

3.13.   Estimate the power needed to accelerate a 1400-kg car from rest to a speed of 60 km/h in 10 s on a level road.

3.14.   Estimate the power needed to accelerate a 2000-kg car from rest to a speed of 70 km/h in 5 s on a level road.

3.15.   A car having a mass of 1300 kg is traveling at 55 miles/h. Estimate: (a) the kinetic energy of the car in kJ, (b) the work necessary to stop the car.

3.16.   A car having a mass of 1450 kg is traveling at 65 miles/h. Estimate: (a) the kinetic energy of the car in kWh, (b) the work necessary to stop the car.

3.17.   A car having a mass of 1800 kg is traveling at 40 miles/h. Estimate: (a) the kinetic energy of the car in Btu, (b) the work necessary to stop the car.

3.18.   A car having a mass of 2250 kg is traveling at 65 miles/h. Estimate (a) the kinetic energy of the car in kJ, (b) the work necessary to stop the car.

3.19.   A car weighing 2700 lb is traveling at 60 miles/h. Estimate (a) the kinetic energy of the car in Btu, (b) the work necessary to stop the car.

3.20.   A car having a mass of 3000 lb is traveling at 50 miles/h. Estimate: (a) the kinetic energy of the car in kJ, (b) the work necessary to stop the car.

3.21.   A car having a mass of 4200 lb is traveling at 70 miles/h. Estimate: (a) the kinetic energy of the car in ft lb$_f$, (b) the work necessary to stop the car.

3.22   A box has a mass of 6 kg. The box is lifted from the garage floor and placed on a shelf. If the box gains 200 J of potential energy estimate the height of the shelf.

3.23   A man climbs on to a wall that is 3.4 m high and gains 2400 J of potential energy. Estimate the mass of the man.

3.24   A 1 kg ball is dropped from the top of a cliff and falls with a constant acceleration due to gravity (9.8 m/s^2). Assume that effects of air resistance can be ignored. (a) Estimate the ball's gravitational potential energy change, (b) How much work has been done on the ball?

3.25   Estimate the potential energy of 1000 kg of water at the surface of a lake that is 50 m above a water turbine, which is used to generate electricity.

3.26   Consider a 1400-kg car traveling at 55 km/h. If the car has to climb a hill with a slope of 20° from the horizontal road, estimate the additional power needed for the speed of the car to remain the same.

3.27   Consider a 1200-kg car traveling at 25 km/h. If the car has to climb a hill with a slope of 30° from the horizontal road, estimate the additional power needed for the speed of the car to remain the same.

3.28   Estimate the potential energy of 10 kg of water at the surface of a lake that is 50 m above a water turbine, which is used to generate electricity.

3.29   Estimate the potential energy of 350 lb of water at the surface of a lake that
       is 200 ft above a water turbine, which is used to generate electricity.

3.30   Water is flowing over a waterfall 80 m in height. Assume 1 kg of the water
       as the system. Also assume that the system does not exchange energy with
       its surroundings. (a) What is the potential energy of the water at the top of
       the falls with respect to the base of the falls? (b) What is the kinetic energy
       of the water just before it strikes bottom?

3.31   Estimate the potential energy of 1000 lb of water at the surface of a lake that
       is 100 ft above a water turbine, which is used to generate electricity.

3.32   An elevator with a mass of 3000 kg rests at a level of 3 m above the base of
       an elevator shaft. The elevator has traveled to 30 m above the base. The
       elevator falls from this height freely to the base and strikes a strong spring.
       Assume that the entire process is frictionless. Estimate:

   (a) The potential energy of the elevator in its original position relative to
       the base of the shaft.
   (b) The work done traveling the elevator.
   (c) The potential energy of the elevator in its highest position relative to the
       base of the shaft.
   (d) The velocity and kinetic energy of the elevator just before it strikes the
       spring.

3.33   An elevator with a mass of 2000 kg rests at a level of 2 m above the base of
       an elevator shaft. The elevator has traveled to 20 m above the base. The
       elevator falls from this height freely to the base and strikes a strong spring.
       Assume that the entire process is frictionless. Estimate:

   (a) The potential energy of the elevator in its original position relative to
       the base of the shaft.
   (b) The work done traveling the elevator.
   (c) The potential energy of the elevator in its highest position relative to the
       base of the shaft.
   (d) The velocity and kinetic energy of the elevator just before it strikes the
       spring.

3.34   An elevator with a mass of 4000 lb rests at a level of 5 ft above the base of
       an elevator shaft. The elevator has traveled to 100 ft above the base. The
       elevator falls from this height freely to the base and strikes a strong spring.
       Assume that the entire process is frictionless. Estimate:

   (a) The potential energy of the elevator in its original position relative to
       the base of the shaft.
   (b) The work done traveling the elevator.
   (c) The potential energy of the elevator in its highest position relative to the
       base of the shaft.

    (d) The velocity and kinetic energy of the elevator just before it strikes the spring.

3.35 An elevator with a mass of 5000 lb rests at a level of 10 ft above the base of an elevator shaft. The elevator has traveled to 150 ft above the base. The elevator falls from this height freely to the base and strikes a strong spring. Assume that the entire process is frictionless. Estimate:

    (a) The potential energy of the elevator in its original position relative to the base of the shaft.
    (b) The work done traveling the elevator.
    (c) The potential energy of the elevator in its highest position relative to the base of the shaft.
    (d) The velocity and kinetic energy of the elevator just before it strikes the spring.

3.36 Water is pumped from an open tank at level zero to an open tank at level 14 ft. The pump adds 7.5 hp to the water when pumping a volumetric flow rate of 3.6 ft^3/s. Estimate the loss energy in head.

3.37 Water is pumped from an open tank at level zero to an open tank at a level of 24 ft. The pump adds 16 hp to the water when pumping a volumetric flow rate of 5 ft^3/s. Estimate the loss energy in head.

3.38 Estimate the head equivalent of a pressure difference of 12 psi (lbf/in^2) in ft of water and in ft of mercury.

3.39 Estimate the head equivalent of a pressure difference of 20 psi (lbf/in^2) in ft of water and in ft of mercury.

3.40 Estimate the compression work done by a gas during a process with the following recorded data obtained that relates pressure and volume:

P, kPa	100	200	300	400	500
V, L	1.1	0.8	0.55	0.4	0.3

3.41 A saturated steam is expanded inside a cylinder. The steam is initially at 151.8 °C and 500 kPa. The final pressure is 200 kPa. Determine the work produced by the steam.

P, kPa	500	450	400	350	300	250	200
V, cm^3	374	413	462	524	605	718	885

3.42 Find the work during (a) an expansion of a gas from 1 to 8 m^3 at constant pressure of 120 kPa, (b) a compression of a gas when pressure changes from 140 to 80 kPa and volume changes from 7 to 2 m^3.

3.43 Calculate the boundary work when $V_1 = 1.5$ m^3 of carbon dioxide at $P_1 = 110$ kPa and $T_1 = 27$ °C is compressed isothermally to $P_2 = 750$ kPa.

3.44 A tank filled with an ideal gas at 200 °C and 100 kPa is compressed isothermally to 500 kPa. Estimate the compression work in kJ/mol.

3.45 Air undergoes a constant temperature compression from 100 kPa and 100 °C to 200 kPa. Find the net work done for 4 kg of air.

3.46 A tank filled with air at 6 bar and 20 °C is expanded isothermally to 2 bar. Estimate the expansion work in kJ/mol.

3.47 A tank filled with 10 mol of air at 2 bar and 20 °C is compressed adiabatically (isentropic) to 8 bar. Estimate the compression work in kJ/mol. Constant heat capacities for the air are $C_{v,av} = 20.8$ J/mol K and $C_{p, av} = 29.1$ J/mol K.

3.48 A piston cylinder device contains 2 kg steam at 2 MPa and 450 °C. (a) If the steam is cooled to 225 °C at constant pressure, estimate the compression work the system performs, (b) if the volume of the cylinder at the final state has the saturated vapor, estimate the work and the temperature at the final state.

3.49 Saturated ammonia vapor enters a compressor at −10 °C and leaves at 76.2 °C. Determine the work requirement per 0.4 kg of ammonia for the compressor if the process is adiabatic and reversible. Constant heat capacity is $C_{p,av} = 2.64$ kJ/kg K.

3.50 Air in a cylinder is at 110 kPa and undergoes polytropic expansion from 0.02 to 0.1 m^3. The exponent $\gamma = 1.3$ in the pressure volume relation ($PV^\gamma =$ constant) in polytropic process. Estimate the work done during the process.

3.51 1 kg of air in a cylinder is at 100 kPa and 30 °C. After a polytropic compression the temperature of air reaches to 62 °C. The constant $\gamma = 1.25$ in the pressure volume relation ($PV^\gamma =$ constant) in polytropic process. Estimate the work done during the process.

3.52 Calculate the boundary work when 1.0 m^3 of air at 25 °C is compressed to a final volume of 0.8 m^3 of air at a constant temperature process.

3.53 Calculate the boundary work when 60 ft^3 of nitrogen gas at 80 psia and 150 °F is expanded to 40 psia at a constant temperature process.

3.54 Estimate the boundary work when 500 ft^3 of hydrogen gas at 30 psia and 150 °F is compressed to 150 psia at a constant temperature process.

3.55 Superheated vapor at 45 psia and 500 °F in a piston cylinder system is cooled at constant pressure until 80% of it condenses. Calculate the boundary work per pound of steam.

3.56 Calculate the boundary work when 150 ft^3 of carbon dioxide at 20 psia and 100 °F is compressed to 70 psia at a constant temperature process.

3.57 A piston cylinder device has a 10 kg of saturated vapor at 350 kPa. The vapor is heated to 150 °C at constant pressure. Estimate the work done by the steam.

3.58 Superheated vapor at 800 kPa and 200 °C in a piston cylinder system is cooled at constant pressure until 70% of it condenses. Calculate the boundary work per kg of steam.

3.59 A piston cylinder device contains 0.5 kg of air at 1000 kPa and 400 K. The air undergoes an isothermal expansion to 300 kPa. Estimate the boundary work for this expansion process.

3.60 A piston cylinder device contains 10 lb of air at 20 psia and 500 R. The air undergoes a polytropic compression to 100 psia. Estimate the boundary work for the compression process.

3.61 A piston cylinder device contains 10 lb of air at 100 psia and 500 R. The air undergoes an isothermal expansion to 20 psia. Estimate the boundary work for the expansion process.

3.62 A piston cylinder device contains 0.5 kg steam at 1 MPa and 400 °C. (a) If the steam is cooled to 250 °C at constant pressure, estimate the compression work the system performs, (b) if the volume of the cylinder at the final state is 50% of the initial volume and the pressure in the final state is 0.5 MPa, estimate the work and the temperature at the final state.

3.63 A piston cylinder device has a volume of 0.5 m^3 and contains carbon dioxide (CO_2) at 140 kPa and 100 °C. At the final state pressure is 90 kPa. Estimate the isentropic expansion work and value of temperature at the final state.

3.64 A piston cylinder device has a volume of 0.5 m^3 and contains carbon dioxide (CO_2) at 140 kPa and 400 K. At the final state pressure is 90 kPa. Estimate the isentropic expansion work and value of temperature at the final state.

3.65 A piston cylinder device has a volume of 1.5 m^3 and contains carbon dioxide (CO_2) at 100 kPa and 300 K. At the final state pressure is 200 kPa. Estimate the isentropic compression work and value of temperature at the final state.

3.66 A car creates a torque of 400 Nm. This torque is transmitted through a shaft which rotates at a rate of 3000 revolutions per minute. Estimate the power transmitted through the shaft.

3.67 A car creates a torque of 500 Nm. This torque is transmitted through a shaft which rotates at a rate of 3200 revolutions per minute. Estimate the power transmitted through the shaft.

3.68 A 180-hp car creates a rotation rate of 2600 revolutions/min. Estimate the torque the car creates at this revolution rate.

3.69 A 210-hp car creates a rotation rate of 3000 revolutions/min. Estimate the torque the car creates at this revolution rate.

3.70 Determine the SI system units of (a) electric power; (b) electric charge; (c) electric potential difference, and (d) electric resistance

3.71 A 2-kW resistance heater is used within a water tank of a shower. If you use the shower during half an hour every day, estimate the cost of electricity per month. The unit cost of electricity is $0.10/kWh.

3.72   A 5-kW resistance heater is used within a room with floor heating systems. If you use the heating system during an average of 8 h every day, estimate the cost of electricity per year. The unit cost of electricity is $0.10/kWh.

3.73   A 120-W laptop is used on an average of 12 h every day. Estimate the cost of electricity per year. The unit cost of electricity is $0.15/kWh.

3.74   A resistance heater is used within a well-insulated water tank. The heater passes a current of 1.2 A from a 120-V source. Estimate the electric energy after 4 h of heating.

3.75   A resistance heater is used within a room. The heater passes a current of 0.2 A from a 120-V source. Estimate the electric energy used after 30 min of heating.

3.76   A resistance heater passes a current of 0.4 A (A) from a 120-V source. Estimate the electric energy if the heater is used for 10 h.

References

1. Çengel YA, Boles MA (2014) Thermodynamics: an engineering approach, 8th edn. McGraw-Hill, New York
2. Chattopadhyay P (2015) Engineering thermodynamics, 2nd edn. Oxford University Press, Oxford
3. Himmelblau DM, Riggs JB (2012) Basic principles and calculations in chemical engineering, 8th edn. Prentice Hall, Upper Saddle River
4. Jaffe RL, Taylor W (2018) The physics of energy. Cambridge University Press, Cambridge
5. Moran MJ, Shapiro HN, Boettner DD, Bailey MB (2014) Fundamentals of engineering thermodynamics, 8th edn. Wiley, New York

Internal Energy and Enthalpy

4

Introduction and Learning Objectives: Internal energy is associated with the microscopic and collective random motion of particles constituting the media. Energy of the collective random motion of particles is heat. Heat flows across the boundary of a system. The net energy transfer by heat, work, and mass is equal to the sum of the change in internal, kinetic, and potential energies. Enthalpy is used in many fluid-flow systems. Only changes in internal energy and enthalpy can be calculated. Measurement of heat change by using calorimeter is briefly discussed. Estimations of energy and temperature changes of air using the psychometric chart are briefly discussed. Heat transfer is briefly covered. The concept of energy in fluid flow systems is discussed briefly.

The learning objectives of this chapter are to understand and use:

- Internal energy and enthalpy in various closed and open systems,
- Sensible and latent heats,
- Heat of reaction using the heat of formation,
- Heat of combustion,
- Psychometric chart,
- Heat transfer.

4.1 Internal Energy

Internal energy U of a system or a body with well-defined boundaries is the total of the kinetic energy due to the motion of molecules (translational, rotational, and vibrational) and the potential energy associated with the vibrational motion and electric energy of atoms within molecules. Internal energy also includes the energy in all the chemical bonds [1]. From a microscopic point of view, the internal energy may be found in many different forms. For a gas, internal energy comprises translational kinetic energy, rotational kinetic energy, vibrational energy, electronic

© Springer Nature Switzerland AG 2021
Y. Demirel, *Energy*, Green Energy and Technology,
https://doi.org/10.1007/978-3-030-56164-2_4

energy, chemical energy, and nuclear energy. For any material, solid, liquid, or gaseous, it may also consist of the potential energy of attraction or repulsion between the individual molecules. Internal energy is a state function and an extensive property. One can have a corresponding intensive thermodynamic property called *specific internal energy, u,* which is internal energy per mass of the substance. Table 4.1 shows the different components of internal energy of a system, while Table 4.2 shows the changes in internal energy. Nonmechanical work, such as external electric field, or adding energy through stirring, can change the internal energy.

For a closed system, the internal energy is

$$\Delta U = q + W \qquad (4.1)$$

where ΔU is the change in internal energy of a system during a process, q is the heat, and W is the mechanical work. The heat absorbed by a system goes to increase its internal energy plus to perform some external work. If an energy exchange occurs because of temperature difference between a system and its surroundings, this energy appears as *heat*; otherwise it appears as *work*. When a force acts on a

Table 4.1 Components of internal energy [1.2]

Thermal energy	Sensible heat	Energy change of a system associated with: (a) Molecular translation, rotation, vibration (b) Electron translation and spin (c) Nuclear spin of molecules
	Latent heat	Energy required or released for phase change; change from liquid to vapor phase requires heat of vaporization
Chemical energy	Energy associated with the chemical bonds in a molecule	
Nuclear energy	The large amount of energy associated with the bonds within the nucleus of the atom	

Table 4.2 Physical and chemical processes that can change the internal energy of a system

Transferring energy across the system boundary by	Heat transfer	Energy transfer from a high temperature to low temperature state
	Work transfer	Energy transfer driven by changes in macroscopic physical properties of a system, such as compression or expansion work
	Mass transfer	Energy transfer by mass flowing across a system boundary
Change through internal processes	Mixing	Heat releases upon components mixing that may lead to lower internal energy
	Chemical reaction	Heat required or released during a chemical reaction that changes chemical energy
	Nuclear reaction	Heat released during a nuclear reaction that changes nuclear energy

system through a distance, then energy is transferred as work. Equation (4.1) shows that energy is conserved. According to the sign convention adopted here, the value of heat is positive when transferred to the system and negative when it is transferred from the system to its surroundings. Similarly, the value of work is positive when transferred to the system and negative when transferred from the system to its surroundings. Equation (4.1) in infinitesimal terms is

$$dU = \delta q + \delta W \tag{4.2}$$

where the operator d before the internal energy function indicates that it is an exact differential as it is a state function. On the other hand, the δs show the increments of energy which are not state functions; both the heat and work depend on the path taken between the initial and final states of a system.

The change in internal energy in terms of temperature and volume is

$$dU = C_v dT + [T(\beta/\kappa) - P]dV \tag{4.3}$$

where C_v is the heat capacity at constant volume, β is the thermal expansion coefficient, and κ is the isothermal compressibility. The value of C_v is the amount of heat required to raise the temperature of one gram of a substance by one degree Celsius at a constant volume process. C_v is a temperature-dependent quantity. A gas becomes ideal in the limit as pressure approaches zero, where the interactions between the gas particles will be small enough to neglect. The second term on the right-hand side of Eq. (4.3) may be zero for a constant volume process, for ideal gases, and for incompressible fluids. This term would be approximately zero for low-pressure gases. Whenever the internal energy is independent of volume or at constant volume, the change of internal energy is

$$\Delta U = U_2 - U_1 = \int_{T_1}^{T_2} C_v dT \tag{4.4}$$

If the value of the C_v is constant, such as an average value, $C_{v,av}$, within the temperature interval $(T_2 - T_1)$ considered, then integration in Eq. (4.4) yields

$$\Delta U = U_2 - U_1 = C_{v,av}(T_2 - T_1) \tag{4.5}$$

However, if the C_v is expressed by $C_v = A + BT + CT^2$, we have the change of internal energy in terms of temperature estimated by

$$\Delta U = \int_{T_1}^{T_2} C_v dT = A(T_2 - T_1) + \frac{1}{2}B(T_2^2 - T_1^2) + \frac{1}{3}C(T_2^3 - T_1^3) \tag{4.6}$$

Instead of the absolute value of internal energy, the change in internal energy, $\Delta U = U_{final} - U_{initial}$, is more useful in energy estimations. The U_{final} is the final value of the internal energy, while $U_{initial}$ is the initial value of the internal energy of the system.

4.2   Enthalpy

Enthalpy H of a system is

$$H = U + PV \tag{4.7}$$

where U is the internal energy, P is the pressure, and V is the volume of the system. The SI unit of enthalpy is joule with specific enthalpy in J/g. Enthalpy is a state function and an extensive quantity.

Instead of the absolute value of enthalpy the *enthalpy change* of a system is commonly used in energy estimations. Enthalpy change is defined by: $\Delta H = H_{final} - H_{initial}$, where H_{final} is the final value and $H_{initial}$ is the initial value of the enthalpy of the system. When enthalpy is expressed in terms of T and P, the change of enthalpy is given by

$$dH = C_p dT + (1 - \beta T)V dP \tag{4.8}$$

where C_p is the heat capacity at constant pressure and β is the thermal expansion coefficient. The value of C_p is the amount of heat required to raise the temperature of one gram of a substance by one degree Celsius at a constant pressure process. The value of C_p is a temperature-dependent quantity. The second term on the right-hand side of Eq. (4.8) may be zero for a constant pressure process and approximately zero for low-pressure gases. For ideal gases, the enthalpy of a matter is independent of pressure and a function of temperature only. For low-pressure gases and incompressible fluids, the enthalpy of a matter is approximately independent of pressure. Whenever the enthalpy of a matter is independent of pressure, regardless of the process, or at constant pressure, the change of enthalpy is

$$\Delta H = H_2 - H_1 = \int_{T_1}^{T_2} C_p dT \tag{4.9}$$

If the value of C_p is constant, such as an average value, $C_{p,av}$, within the temperature interval $(T_2 - T_1)$ considered, then integration in Eq. (4.9) yields

$$\Delta H = H_2 - H_1 = C_{p,av}(T_2 - T_1) \tag{4.10}$$

ΔH of a system is equal to the sum of nonmechanical work done on it and the heat supplied to it. The increase in enthalpy of a system is exactly equal to the energy added through heat, provided that the system is under constant pressure: $\Delta H = q$. Therefore, enthalpy is sometimes described as the '*heat content*' of a system under a given pressure.

Estimation of enthalpy by the integral in Eq. (4.9) requires an expression for the temperature dependence of heat capacity

$$C_p = A + BT + CT^2 \tag{4.11}$$

Ideal gas state heat capacities are functions of temperature only designated by C_p^{ig} and C_v^{ig}. The temperature dependence of an ideal gas may be expressed by

$$C_p^{ig}/R = A + BT + CT^2 \tag{4.12}$$

The ratio C_p/R is dimensionless, and the units of C_p are governed by the units of R chosen. In Appendix B, Table B1 lists the parameters used in Eq. (4.12) for various chemical species. The two ideal gas heat capacities, C_p^{ig} and C_v^{ig}, are related by:

$$C_p^{ig} - C_v^{ig} = R \tag{4.13}$$

The departure of real gases from ideal behavior is seldom significant at pressures of several bars, and C_p^{ig} and C_v^{ig} are usually good approximations to their true heat capacities [1].

Figure 4.1 shows the change of enthalpy with pressure and temperature for carbon dioxide. At high pressures, the interactions between the molecules will be considerable and the behavior of carbon dioxide will be nonideal. Figure 4.2 shows the values of enthalpy for water at various temperatures and pressures. The values of enthalpy as high as 1000 bar can be obtained from these pressure–enthalpy diagrams as they represent the nonideal behavior of carbon dioxide and water. Therefore, they are useful in industrial design of compression and expansion processes [2].

Example 4.1 Unit conversions of heat capacity
The parameters in Table B1 require the use of temperatures in Kelvin. Dimensionless heat capacity of propane in ideal gas state in Table B1 is given by $C_p^{ig}/R = 1.213 + 28.785 \times 10^{-3}T - 8.824 \times 10^{-6}T^2$. Develop an equation for C_p^{ig} in J/mol °C.
Solution:
The relation between the two temperature scales: $T\,K = T°C + 273.15$
The gas constant: $R = 8.314$ J/mol K $= 8.314$ J/mol °C

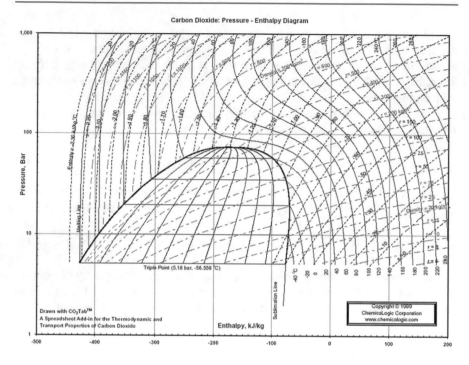

Fig. 4.1 Pressure–enthalpy diagram for carbon dioxide (with permission from Chemicalogic Corporation)

$$C_p^{ig} = R\left(1.213 + 28.785 \times 10^{-3}(T + 273.15) - 8.824 \times 10^{-6}(T + 273.15)^2\right)$$
$$C_p^{ig} = 69.981 + 199.24 \times 10^{-3}T - 73.362 \times 10^{-6}T^2 \text{ in J/mol °C}$$

Example 4.2 Calculation of internal energy change

Calculate the change in internal energy when 0.5 m³ of air at 200 kPa is heated from 25 to 120 °C in a constant volume process.

Solution:

Assume that the system will remain as nearly ideal gas during the process.

Data: $V = 0.5$ m³, $P = 200$ kPa, $T_1 = 25$ °C (298.15 K); $T_2 = 120$ °C (393.15 K), The relation between the two temperature scales: T K $= T$°C $+ 273.15$, $R = 8.314$ Pa m³/mol K

Moles of air: $n_{air} = \dfrac{PV}{RT} = \dfrac{(200 \times 10^3 \text{ Pa})(0.5 \text{ m}^3)}{8.314(\text{Pa m}^3/\text{molK})(298.15 \text{ K})} = 40.3 \text{ mol}$

The heat capacity parameters in Table B1 require the use of temperatures in Kelvin. The heat capacity at constant pressure of air in ideal gas state: (Table B1)

$\dfrac{C_p^{ig}}{R} = 3.355 + 0.575 \times 10^{-3}T \quad (T \text{ in } K)$

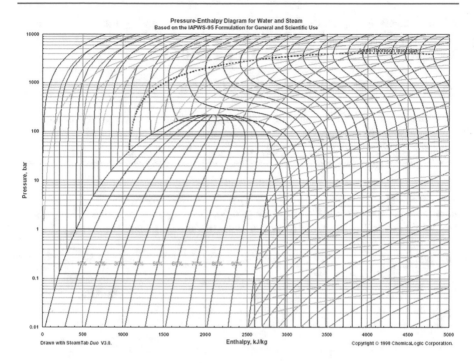

Fig. 4.2 Pressure–enthalpy diagram for water (with permission from Chemicalogic Corporation)

From Eq. (4.13), we have $\frac{C_v^{ig}}{R} = \frac{C_p^{ig}}{R} - 1$ or $C_v^{ig} = C_p^{ig} - R$

$$\frac{C_v^{ig}}{R} = (3.355 + 0.575 \times 10^{-3}T) - 1$$

The gas constant: $R = 8.314$ J/mol K = 8.314 J/mol °C

$$\Delta U = n_{air}R \int_{298K}^{393K} C_v^{ig}dT = (40.3\,\text{mol})(8.314\,\text{J/mol K})$$

$\left(2.355(T_2 - T_1) + \frac{1}{2}0.575 \times 10^{-3}\ (T_2^2 - T_1^2)\right) = \textbf{78.10 kJ}$ The change in internal energy is positive as the heat is transferred to the system (air).

Example 4.3 Calculation of internal energy change during evaporation
One metric ton of liquid water at 373 K is converted to steam at 373 K at 101 kPa. Estimate the change in the internal energy during this phase change.
Solution:
Data: $P_{atm} = 101$ kPa; at 373 K, $\rho_{water} = 1000$ kg/m³, $\rho_{vapor} = 0.6$ kg/m³, and heat of vaporization of water, $\Delta H_{vap} = 2250$ kJ/kg.
When water is converted into steam, its volume expands, and mechanical displacement of the system boundary implies that there is net positive work on the environment.

From the first law of thermodynamics and for unit mass:
$W = P_{atm}(V_{steam} - V_{water}) = P_{atm}(1/\rho_{steam} - 1/\rho_{water}) = 168.323\,kJ/kg$
The heat added to the system is $q = \Delta H_{vap} = 2250\,kJ/kg$

$\Delta U = q - W = 2250 - 168.323 = 2081.677\,kJ/kg$

One metric ton = 1000 kg, and the total change in the internal energy becomes:

$2081.677\,kJ/kg \times (1000\,kg) = 2,081,677\,kJ = \mathbf{2\,081.677\,MJ}$.

Example 4.4 Determination of state properties
Complete the following table for water with the missing state properties.

T (°F)	P (psia)	ΔU (Btu/lb)	State
250	27	550	
213.03		181.16	Saturated liquid
213.03	15	1077.9	
350	30	1125.5	
298	100		Compressed liquid

Solution:
Using the properties from steam tables

T (°F)	P (psia)	ΔU (Btu/lb)	State
250	27	550.0	Saturated mixture
213.03	15	181.16	Saturated liquid
213.03	15	1077.90	Saturated vapor
350	30	1125.50	Superheated vapor
298	100	267.42	Compressed liquid

Example 4.5 Heat value of a saturated liquid and vapor mixture of a steam
A rigid tank contains 15 kg of saturated vapor water at 85 °C. If the vapor partially condenses and 2 kg of the liquid water is formed, estimate the enthalpy of the saturated liquid–vapor mixture and the amount of heat lost through the condensation.

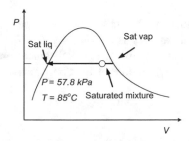

Solution:
Saturated liquid–vapor water mixture at T_{sat} = 85 °C (57.8 kPa)
Saturated vapor: $H_{sat\ vap}$ = 2652.0 kJ/kg, Saturated liquid water: $H_{sat\ liq}$ = 355.9 kJ/kg
Amount of liquid water = 2 kg; Amount of vapor = 15 – 2 = 13 kg

Quality x of the saturated mixture: $x = \dfrac{m_{vap}}{m_{total}} = \dfrac{13\,kg}{15\,kg} = 0.87$

Enthalpy of the liquid–vapor mixture:

$$H_{mix} = (1 - x)H_{sat\ liq} + xH_{sat\ vap} = (1 - 0.87)355.9\,kJ/kg + (0.87)2652.0\,kJ/kg$$
$$= \mathbf{2353.5\,kJ/kg}$$

Heat of condensation: $\Delta H_{cond} = H_{sat\ vap} - H_{sat\ liq}$
= (2652.0 – 355.9) kJ/kg = 2296.1 kJ/kg
Heat lost after forming 2 kg of liquid water: (2 kg)(ΔH_{cond}) = **4592.2** kJ
Enthalpy of the saturated mixture depends on the quality x and will be:
$H_{sat\ vap} > H_{mix} > H_{sat\ liq}$

4.3 Heat

Heat is part of the total energy flow across a system boundary that is caused by a temperature difference between the system and its surroundings or between two systems. Heat flows from high temperature region to cold temperature region. Therefore, heat is not stored and defined as thermal energy in transit. The unit for heat in SI system is joule, J, and the Btu and the calorie are other units. When heat capacity is constant, such as $C_{p,av}$, the amount of heat changed when a substance changed its temperature from T_1 to another temperature T_2 is

$$q = mC_{p,av}(T_2 - T_1) \tag{4.14}$$

where q is the heat, and m is the mass. For processes taking place at constant volume, the specific heat capacity is denoted by C_v. The values of specific heat capacity at constant pressure, C_p, for various substances are shown in Table B3 in Appendix B. For a fluid flow, the rate of heat becomes

$$\dot{q} = \dot{m} C_{p,av} (T_2 - T_1) \tag{4.15}$$

where $\dot{q}$ is the rate of heat transfer (J/s or Btu/s), and $\dot{m}$ is the rate of mass flow (kg/s or lb/s). The value of heat is positive when transferred to the system and negative when transferred from the system to its surroundings. There is no heat flow through the boundary in an *adiabatic process* as the system and its surroundings have the same temperatures [1, 2].

If an amount of mass, m, is heated from T_1 to T_2, it will gain the heat. For example, the energy needed to heat 1.0 kg of water from 0 to 100 °C when the specific heat of water is 4.18 kJ/kg K becomes q = (1.0 kg) (4.18 kJ/kg K) (100–0) °C = 418 kJ. However, if an amount of mass, m, is cooled from T_1 to T_2, it will lose the heat. The value of heat is a path-dependent function. For example, the value of heat at constant pressure process will be different from the value required at a constant volume process.

When there is no work interaction between a closed system and its surroundings, the amount of heat is estimated by

$$q = \int_{T_1}^{T_2} C_p dT \tag{4.16}$$

Using the expression for the temperature dependence of the heat capacity, given in Eq. (4.11), the change of heat between T_1 and T_2 is estimated by

$$q = \int_{T_1}^{T_2} C_p dT = A(T_2 - T_1) + \frac{1}{2}B(T_2^2 - T_1^2) + \frac{1}{3}C(T_2^3 - T_1^3) \tag{4.17}$$

Thermal equilibrium is achieved when two systems in thermal contact with each other cease to exchange energy by heat.

- If two systems are in thermal equilibrium, then their temperatures are the same.
- For an *isolated system*, no energy is transferred to or from the system.
- A *perfectly insulated* system experiences no heat exchange through the boundary.

4.3.1 Sensible Heat

Sensible heat effects refer to the quantity of heat transfer causing the temperature change of a system in which there is no phase transition, no chemical reaction, and no change in composition. For mechanically reversible, constant pressure, and closed systems, the amount of sensible heat transferred would be

$$q = \Delta H = \int_{T_1}^{T_2} C_p \mathrm{d}T \tag{4.18}$$

When $C_{p,\text{av}}$ = constant, we have

$$q = \Delta H = C_{p,\text{av}}(T_2 - T_1) \tag{4.19}$$

According to the sign convention used here, the sign of heat is negative when heat is transferred from the system and positive when heat is transferred to the system.

In a wide variety of situations, it is possible to raise the temperature of a system by the energy released from another system. The calorie is the amount of energy required to raise the temperature of one gram of water by 1 °C (approximately 4.1855 J), and the British thermal unit Btu is the energy required to heat one pound of water by 1°F [1, 2].

4.3.2 Latent Heat

Consider the heat needed for liquefaction of a pure substance from the solid state or vaporization from the liquid state at constant pressure. These heat effects are called the *latent heat of fusion* and *latent heat of vaporization*, respectively. In either case, no change in temperature occurs but the transfer of a finite amount of heat into the system is required. Similarly, a substance state may change from liquid to solid state if a finite amount of heat is transferred from the liquid state to the surroundings. This heat is called the *latent heat of freezing*. The state of a substance changes from vapor to liquid state when a finite amount of heat is transferred from the vapor state to the surroundings. This heat is called the *latent heat of condensation*. There are also heats of transitions accompanying the change of a substance from one solid state to another; for example, when rhombic crystalline sulfur changes to monoclinic structure at 95 °C and 1 bar, the heat of transition is 360 J per gram-atom [1, 2].

The latent heat of vaporization ΔH_{vap} is a function of temperature, and is related to other system properties of an ideal gas at temperature T by the *Clausius-Clapeyron* equation

$$\frac{d \ln P^{\text{sat}}}{dT} = \frac{\Delta H_{\text{vap}}}{RT^2} \tag{4.20}$$

where P^{sat} is the saturation pressure of the pure substance. The value of saturation pressure P^{sat} can be calculated from the Antoine equation, given in Eq. (1.37). The Antoine parameters A, B, and C are listed in Table 1.12. When ΔH_{vap} can be assumed to be constant within the temperature interval of $(T_2 - T_1)$, Eq. (4.20) can be integrated as follows

$$\ln \left(\frac{P_2^{\text{sat}}}{P_1^{\text{sat}}} \right) = -\frac{\Delta H_{\text{vap}}}{R} \left(\frac{1}{T_2} - \frac{1}{T_1} \right) \tag{4.21}$$

Equation (4.21) relates the fluid saturation pressures at two different temperatures to the heat of vaporization. Here the heat of vaporization is assumed to be temperature independent when the temperature difference of $(T_2 - T_1)$ is small [1].

Latent heats may also be measured using a calorimeter, and the experimental values are available at selected temperature for many substances (see Table A1). *Trouton's rule* predicts the heat of vaporization at normal boiling point, ΔH_n, that is at 1 standard atm (101.32 kPa).

$$\frac{\Delta H_n}{RT_n} \simeq 10 \tag{4.22}$$

where T_n is the absolute temperature of the normal boiling point and R is the gas constant. The units of ΔH_n, T_n, and R must be chosen to yield a dimensionless value for the ratio in Eq. (4.22). Another approximate model is [1]

$$\frac{\Delta H_n}{RT_n} \simeq \frac{1.092(\ln P_c - 1.013)}{0.930 - T_n/T_c} \tag{4.23}$$

where P_c and T_c are the critical pressure and critical temperature, respectively. Predictions by this equation are quite satisfactory. For example, application of Eq. (4.23) for water is

$$\frac{\Delta H_n}{RT_n} \simeq \frac{1.092(\ln 220.55 - 1.013)}{0.930 - 0.577} = 13.56$$
$$\Delta H_n = 13.56(8.314)(373.15) = 42.065 \,\text{kJ/mol}$$

where $R = 8.134$ J/mol K, $T_n = 373.15$ K, $P_c = 220.55$ bar, and $T_c = 646.71$ K (Table A3). The experimental value from Table A1 is 40.65 kJ, which is lower by only 3.4%.

The method proposed by Watson (1943) estimates the latent heat of vaporization of a pure substance at any temperature from a known value of the latent heat at a single temperature

$$\frac{\Delta H_2}{\Delta H_1} \simeq \left(\frac{1 - T_{r_2}}{1 - T_{r_1}}\right)^{0.38} \tag{4.24}$$

where T_{ri} is the reduced temperature obtained from the ratio of T_i/T_c. This equation is fairly accurate [1].

4.3.3 Heating with Phase Change

The heating processes for the water between 25 and 200 °C are illustrated on a temperature–volume diagram in Fig. 4.3.

The heating with phase change has the following subprocesses of sensible and latent heating at 1 atm:

Process a: Sensible heating of liquid water between 25 °C (298 K) and 100 °C (373 K)

Process b: Latent heating of vaporization at 100 °C (373 K)

Process c: Sensible heating of water vapor between 100 °C (373 K) and 200 °C (473 K)

The total heat per unit amount of water is calculated by

$$\Delta H_{\text{total}} = \Delta H_a + \Delta H_b + \Delta H_c = \left(\int_{298K}^{373K} C_{p,\text{liq}} dT + \Delta H_{\text{vap}} + \int_{373K}^{473K} C_{p,\text{vap}}^{ig} dT\right) \tag{4.25}$$

Fig. 4.3 Temperature–volume diagram for heating water with phase changes between 25 and 200 °C at 1 atm

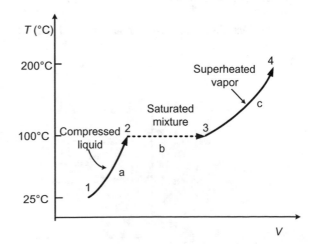

The equations of temperature-dependent heat capacities are needed within the integrals to estimate the sensible heating at liquid and vapor states.

Example 4.6 Calculation of heat of vaporization using Antoine equation and Clausius-Clapeyron equation

Determine the heat of vaporization of benzene at its normal boiling point T_b by using the Antoine equation and compare with the experimental value.

Solution:

$T = T_b = 80.11$ °C $= 353.26$ K (from Table A1), $R = 8.314$ J/mol K

Vapor pressure of benzene from the Antoine equation with the constants from Table 1.12:

A $= 13.7819$, B $= 2726.81$, C $= 217.57$

$$\ln P^{sat} = 13.7819 - \frac{2726.81}{T + 217.57} \quad (P^{sat} \text{ is in kPa and } T \text{ in °C}) \text{ (Eq 1.27)}.$$

The Clausius-Clapeyron equation: $\dfrac{d \ln P^{sat}}{dT} \simeq \dfrac{\Delta H_{vap}}{RT^2}$

Differentiation of the Antoine equation with respect to temperature T yields

$$\frac{d \ln P^{sat}}{dT} = -\frac{2726.81}{(80 + 217.57)^2} \simeq \frac{\Delta H_{vap}}{8.314(353.15)^2}$$

$\Delta H_{vap} = $ **31.95 kJ/mol** or **2.49 kJ/kg** (with the molecular weight of benzene $= 78$ g/mol)

The experimental data from Table A1 is 30.76 kJ/mol and the deviation:

$$\frac{31.95 - 30.76}{30.76} = 0.038 = 3.8\% \text{ (small enough for many engineering calculations)}.$$

Example 4.7 Estimation of change of enthalpy with sensible and latent heat

Estimate the change of enthalpy of a water flowing at a rate of 4.5 kg/s and heated from 25 to 200 °C at atmospheric pressure.

Solution:

Assume that the water will remain in ideal state from the initial and the final temperatures.

The heating processes are shown below on a temperature versus volume diagram:

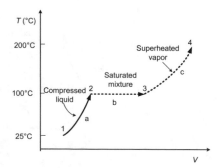

$T_1 = 298$ K, $T_{\text{sat}} = 373$ K, $T_2 = 473$ K

The total heat: $\Delta H_{\text{total}} = \Delta H_a + \Delta H_b + \Delta H_c = \dot{m}\left(\int\limits_{298K}^{373K} C_{p,\text{liq}}dT + \Delta H_{\text{vap}} + \int\limits_{373K}^{473K} C_{p,\text{vap}}dT\right)$

where $\dot{m}$ is mass flow rate of water: 4.5 kg/s, $R = 8.314$ J/mol K,
$MW = 18.02$ g/mol, $R = 0.4615$ kJ/kg K
At $T_{\text{sat}} = 373$ K: $\Delta H_{\text{vap}} = \Delta H_{\text{sat vap}} - \Delta H_{\text{sat liq}} = 2257$ kJ/kg (Steam table)
From Table B1 the dimensionless heat capacities at constant pressure are:

$\dfrac{C_{p,\text{liq}}}{R} = A + BT = 8.712 + 1.25 \times 10^{-3}T$ (T in K)

$\dfrac{C_p^{ig}}{R} = A + BT = 3.47 + 1.45 \times 10^{-3}T$ (T in K)

Process a: $\dot{m}\left(\int\limits_{298K}^{373K} C_{p,\text{liq}}dT\right) = \dot{m}R\left[8.712(373 - 298) + \dfrac{1}{2}1.25 \times 10^{-3}(373^2 - 298^2)\right]$

$= 1422.3$ kJ/s $= 1422.3$ kW

Process b: $\dot{m}(\Delta H_{\text{vap}}) = 10256.3$ kJ/s $= 10256.3$ kW

Process c: $\dot{m}\left(\int\limits_{373K}^{573K} C_{p,\text{vap}}^{ig}dT\right) = \dot{m}R\left[3.47(473 - 373) + \dfrac{1}{2}1.45 \times 10^{-3}(473^2 - 373^2)\right]$

$= 848.0$ kJ/s $= 848$ kW

Total amount of energy (power) needed $= 12426.8$ kJ/s $= \mathbf{12426.8\ kW}$
Using the steam tables
Process a: $\dot{m}(H_2 - H_1) = (4.5\text{ kg/s})(419.06 - 104.8)\text{ kJ/kg} = 1417.5$ kW
Process b: $\dot{m}\Delta H_{\text{vap}} = (4.5\text{ kg/s})(2676.0 - 419.06)\text{ kJ/kg} = 10156.5$ kW
Process c: $\dot{m}(H_4 - H_3)(4.5\text{ kg/s})(2875.3 - 2676.0)\text{ kJ/kg} = 898.8$ kW
Total $= 1417.5 + 10{,}156.5 + 898.8 = \mathbf{12{,}470.8\ kW}$
The results from the steam tables are slightly different from the estimated results
because of the nonideal behavior of water at high temperatures and pressures.

Example 4.8 Estimation of heat of vaporization at another temperature
The latent heat of vaporization of water at 100 °C is 2253.0 kJ/kg. Estimate the
latent heat of water at 150 °C.

Solution:
For water: $T_c = 646.71$ K; $\Delta H_1 = 2253.0$ kJ/kg (Table A1), $T_1 = 373.15$ K,
$T_2 = 423.15$ K

Reduced temperatures to be used in Eq. 4.24: $T_{r1} = 373.15/647.1 = 0.577$ and $T_{r2} = 423.15/647.1 = 0.654$

$$\Delta H_2 \simeq (2253.0\,\text{kJ/kg})\left(\frac{1 - 0.654}{1 - 0.577}\right)^{0.38} = \textbf{2087.4 kJ/kg}$$

The value given in the steam table is 2113.2 kJ/kg (from steam tables)
Deviation: $\frac{2113.2-2087.4}{2113.2} = 0.012$ or 1.2%
Deviation of the estimated value from the tabulated value is low.

Example 4.9 Energy exchange in expansion of air
A closed 5 L cylinder contains air at 100 °C and 2 atm. The cylinder is in the boiling water at normal atmospheric pressure. A piston is released very slowly so the air expands, while absorbing heat from the boiling water until its pressure reaches 1.5 atm. Estimate the final volume of the air and heat absorbed.

(a) Isothermal expansion
(b) Adiabatic expansion

Solution:
As the temperature remains constant and assuming the air as ideal gas
($PV = RT$, for unit mole), we use
$P_1V_1 = P_2V_2 = RT$ = constant; $q = W$
$V_2 = (P_1/P_2)V_1 = (2\ \text{atm}/1.5\ \text{atm})\ 5\ \text{L} = 6.66\ \text{L}$
For an ideal gas expansion $q = P_1V_1\ \ln(V_2/V_1) = (2\ \text{atm})\ (5\ \text{L})$
$\ln(6.66\ \text{L}/5\ \text{L}) = 280.6\ \text{J} = W$
(b) Adiabatic expansion with $\gamma = 1.4$
During the expansion, no heat is added into the air and the work done is
$W = (P_1V_1 - P_2V_2)/(1 - \gamma)$
$(V_2)^\gamma = (P_1/P_2)(V_1)^\gamma =$
$V_2 = V_1\ (P_1/P_2)^{1/\gamma} = (5\ \text{L})\ (2\ \text{atm}/1.5\ \text{atm})^{1/\gamma} = 6.1\ \text{L}$
$W = (P_1V_1 - P_2V_2.)/(1 - \gamma) = (2\ 5 - 1.5\ 6.1)/(1 - 1.4) = 212.7\ \text{J}$

4.3.4 Heat of Reaction

Systems with chemical reactions exchange heat with the surroundings and the temperature may change. These effects result because of the differences in molecular structure and therefore in energy of the products and reactants. The heat associated with a specific chemical reaction depends on the temperatures of reactants and products. A general chemical reaction may be written as follows

$$v_A A + v_B B \rightarrow v_C C + v_D D \tag{4.26}$$

where v_i is the stoichiometric coefficient, which is positive for a product and negative for a reactant.

A standard basis for estimation of reaction heat effects results is when the reactants and products are at the same temperature. For the reaction given in Eq. (4.26), the standard heat of reaction is the enthalpy change when v_A moles of A and v_B moles of B in their standard states at temperature T react to form v_C moles of C and v_D moles of D in their standard states at the same temperature T. Standard state for gases is the pure substance in an ideal gas state at 1 atm. For liquids and solids, the standard state is the real pure liquid or solid at 1 atm. Standard state properties are therefore functions of temperature. Based on this sign convention, a standard heat of reaction is expressed by [2]

$$\Delta H_r^o = \sum_i v_i \Delta H_{fi}^o \tag{4.27}$$

where H_{fi}^o is the *standard heat of formation*. Some values of standard heats of formation are tabulated at 298.15 K or 25 °C in Table C1 as ΔH_{f298}^o.

A formation reaction is defined as a reaction which forms 1 mol of a single compound from its constituent elements with each substance in its standard state at 298.15 K (25 °C). Standard states are: (i) gases (g); pure ideal gas at 1 atm and 25 °C. (b) liquids (l) and solids (s); (ii) pure substance at 1 atm and 25 °C; (iii) solutes in aqueous 1-molal ideal solution at 1 atm and 25 °C. Standard heats of formation for stable elements in their standard states are zero since they are naturally existing. For example, the formation reaction for ammonia is $N_2 + 3H_2 \rightarrow NH_3$ since all the reactants are elements. However, the reaction $H_2O + SO_3 \rightarrow H_2SO_4$ is not a formation reaction since it forms sulfuric acid not from elements but from other compounds.

The summation in Eq. (4.27) is over all products and reactants. For example, for the reaction

$$4HCl(g) + O_2(g) \rightarrow 2H_2O(g) + 2Cl_2(g) \tag{4.28}$$

the stoichiometric coefficients are $v_{HCl} = -4$, $v_{O2} = -1$, $v_{H2O} = 2$, $v_{Cl2} = 2$ and the standard heat of reaction using Eq. (4.27) becomes

$$\Delta H_r^o = \sum_i v_i \Delta H_{fi}^o = v_{H_2O} \Delta H_{fH_2O}^o + v_{Cl_2} \Delta H_{fCl_2}^o + v_{HCl} \Delta H_{fHCl}^o + v_{O_2} \Delta H_{fO_2}^o \tag{4.29}$$

where $\Delta H_{fCl_2}^o = 0$ and $\Delta H_{fO_2}^o = 0$ as they are naturally occurring stable elements.

Heat effects change according to the nature of chemical reactions. Reactions that release heat are called *exothermic reactions* and reactions that require heat are called *endothermic reactions*. For an exothermic reaction at constant pressure, the system's change in enthalpy equals the energy released in the reaction, including

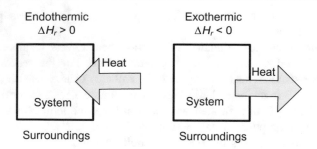

Fig. 4.4 Heat interactions in chemical reactions: endothermic reaction needs energy from outside and exothermic reactions release energy to the surroundings

the energy retained in the system and lost through expansion against its surroundings. In a similar manner, for an endothermic reaction, the system's change in enthalpy is equal to the energy *absorbed* in the reaction. If ΔH_r is positive, the reaction is endothermic, that is heat is absorbed by the system due to the products of the reaction having a greater enthalpy than the reactants. On the other hand, if ΔH_r is negative, the reaction is exothermic, that is the overall decrease in enthalpy is achieved by the generation of heat [2] (Fig. 4.4).

When a heat of reaction is given for a reaction, this applies for the current stoichiometric coefficients. If the stoichiometric coefficients are doubled, the heat of reaction is doubled. For example, consider the following reactions

$$\frac{1}{2}N_2 + \frac{3}{2}H_2 \rightarrow NH_3 \quad \Delta H_r^{\circ} = -46.11 \text{ kJ}$$
$$N_2 + 3H_2 \rightarrow 2NH_3 \quad \Delta H_r^{\circ} = -92.22 \text{ kJ}$$

When chemical reactions are combined by addition, the standard heat of reaction may also be added to estimate the standard heats of the resulting reaction [1, 2]. This is possible since the enthalpy is a state function and, for a given initial and final states, its value is independent of path. Chemical reaction equations considered for this purpose often include an indication of the physical state of each reactant and product; the symbol g, l, or s indicates gas, liquid, or solid states, respectively.

Example 4.10 Estimation of standard heat of reaction
Estimate the heat of reaction at 298.15 K for the reaction
$HCl(g) + 1/4O_2(g) \rightarrow 2/4H_2O(g) + 2/4Cl_2(g)$

Solution:
$HCl(g) + 1/4O_2(g) \rightarrow 2/4H_2O(g) + 2/4Cl_2(g)$

Stoichiometric coefficients ν_i are positive for a product and negative for a reactant. The stoichiometric coefficients for the reaction are $\nu_{HCl} = -1$, $\nu_{O2} = -1/4$, $\nu_{H2O} = 2/4$, $\nu_{Cl2} = 2/4$.

Based on this sign convention, a standard heat of reaction is expressed by $\Delta H_r^o = \sum_i v_i \Delta H_{fi}^o$ where H_{fi}^o is the standard heat of formation. The summation is over all the products and the reactants using the standard heat of formations from Table B3.

$$\Delta H_r^o = (2/4)\Delta H_{fH_2O}^o - \Delta H_{fHCL}^o = (2/4)(-241.82) + (-1)(-92.31) = \mathbf{-28.6\ kJ/mol}$$

The heats of formations for the stable and naturally occurring elements of O_2 and Cl_2 are zero.

Example 4.11 Estimation of standard heats of reaction from standard heats of formation

Estimate standard heats of reaction for the water–gas shift reaction:

$$CO_2(g) + H_2(g) \rightarrow CO(g) + H_2O(g)$$

Solution:

Although this reaction takes place at temperatures well above 25 °C, standard heats of reaction is calculated at 25 °C.

Stoichiometric coefficients v_i are positive for a product and negative for a reactant. The pertinent formation reactions and their heats of formation from Table C1 are:

CO_2: $C(s) + O_2(g) \rightarrow CO_2(g)$ $\Delta H_{f298}^o = -393.51$ kJ/mol, $v = -1$

H_2: $H_2(g)$ $\Delta H_{f298}^o = 0$ because hydrogen is a stable element

CO: $C(s) + \frac{1}{2}O_2(g) \rightarrow CO(g)$ $\Delta H_{f298}^o = -110.52$ kJ/mol, $v = 1$

H_2O: $H_2(g) + \frac{1}{2}O_2(g) \rightarrow H_2O(g)$ $\Delta H_{f298}^o = -241.82$ kJ/mol, $v = 1$

Here all the products and reactants are at their standard states: ideal gas state at 1 atm and 25 °C.

Based on this sign convention, the standard heat of reaction is expressed by Eq. (4.27)

$$\Delta H_r^o = \sum_i v_i \Delta H_{fi}^o = -110.52 - 241.82 - (-393.51) = \mathbf{41.16\ kJ/mol}$$

This result shows that the enthalpy of 1 mol of CO plus 1 mol of H_2O is greater than the enthalpy of 1 mol of CO_2 plus 1 mol of H_2 by 41.16 kJ when all the components are at their standard states. Therefore, the water–gas-shift reaction is an endothermic reaction and requires heat from the surroundings to occur.

4.3.5 Standard Heat of Combustion

Consider the reaction for the combustion of coal: $C(s) + O_2(g) \rightarrow CO_2(g) + \Delta H_r$; here the pure carbon representing the coal is in solid state, while oxygen and carbon dioxide are both in gaseous state. The reactants in a combustion reaction possess greater energy than do the products, and this energy is transferred to the surroundings as heat, so the reaction is exothermic. The *heats of combustion* of fuel may be measured by a flow-calorimeter. The fuel is mixed with air at ambient temperature and mixture flows into a combustion chamber of the calorimeter where reaction occurs. The combustion products enter a water-cooled section and are cooled to the temperature of the reactants. As there is no shaft work produced in the process and potential and kinetic energies are negligible, the change in enthalpy caused by the combustion reaction becomes the heat flowing from the calorimeter

$$q = \Delta H = \text{heat of reaction} \qquad (4.30)$$

For example, 2,2,4-trimethylpentane (isooctane), widely used in petrol, has a chemical formula of C_8H_{18} and it reacts with oxygen exothermically and produces 10.86 MJ per mole of isooctane

$$C_8H_{18}(l) + 25/2O_2(g) \rightarrow 8CO_2(g) + 9H_2O(g) + 10.86\,\text{MJ/mol} \qquad (4.31)$$

Incomplete combustion of petroleum or petrol products results in carbon monoxide. At a constant volume, the heat of combustion of a petroleum product can be approximated by:

$$q_v = 12,400 - 2100d^2 \qquad (4.32)$$

where q_v is measured in cal/g and d is the specific gravity at 60°F (16 °C). Ethane C_2H_6 is an aliphatic hydrocarbon. The complete combustion of ethane releases 1561 kJ/mol, or 51.9 kJ/g, of heat, and produces carbon dioxide and water according to the chemical equation

$$C_2H_6 + 7/2O_2 \rightarrow 2CO_2 + 3H_2O + 1561\ \text{kJ/mol} \qquad (4.33)$$

Gasoline contains about 35 MJ/l (132 MJ/US gal) (higher heating value). This is an average value. Gasoline blends differ, therefore actual energy content varies from season to season and from batch to batch, by up to 4% more or less than the average value [1, 2].

Stable combustion reactions require the right amounts of fuels and oxygen. In theory, there is a theoretical amount of oxygen needed (known as stoichiometric amount) to completely burn a given amount of fuel. In practice, burning conditions are never ideal. Therefore, more air than that of stoichiometric amount is supplied to burn all fuel completely. The amount of air that is more than the theoretical requirement is referred to as *excess air*. Power plant boilers, for example, normally run with about 10–20% excess air. Natural gas-fired boilers may run as low as 5% excess air. Pulverized coal-fired boilers may run with 20% excess air. Gas turbines run very lean with up to 300% excess air. If insufficient amount of air is supplied to the burner, unburned fuel, soot, smoke, and carbon monoxide are exhausted from the boiler. The result is heat transfer surface fouling, pollution, lower combustion efficiency, flame instability and a potential for explosion. To avoid inefficient and unsafe conditions, boilers normally operate at an excess air level.

Example 4.12 Estimation of standard heats of formation

Determine the standard heat of formation for n-pentane: $5C(s) + 6H_2 \rightarrow C_5H_{12}(g)$.

Solution:

Many standard heats of formation come from *standard heats of combustion* measurements in a calorimeter. A combustion reaction occurs between an element or compound and oxygen to form specified combustion products. When organic compounds made up of carbon, hydrogen, and oxygen only, the combustion products are carbon dioxide and water. The state of the water produced may be either vapor or liquid.

The reaction for formation of n-pentane may result from combination of the following combustion reactions in practice. Stoichiometric coefficients v_i are positive for a product and negative for a reactant.

$$5C(s) + 5O_2(g) \rightarrow 5CO_2(g) \qquad \Delta H^\circ_{f298} = (5)(-393.51) \text{ kJ/mol} = \mathbf{-1967.55 \text{ kJ}}$$

$$6H_2(g) + 3O_2(g) \rightarrow 6H_2O(l) \qquad \Delta H^\circ_{f298} = (6)(-285.83) \text{ kJ/mol} = \mathbf{-1714.98 \text{ kJ}}$$

$$5CO_2(g) + 6H_2O(l) \rightarrow C_5H_{12}(g) + 8O_2(g) \qquad \Delta H^\circ_{f298} = 3535.77 \text{ kJ/mol n-pentane}$$

$$5C(s) + 6H_2 \rightarrow C_5H_{12} \qquad \Delta H^\circ_{f298} = \mathbf{-146.76 \text{ kJ/mol}}$$
$$= (-1967.55 - 1714.98 + 3535.77)$$

The net reaction is obtained from the summation of the first three equations by considering that the value of stoichiometric coefficients, v_i, are positive for the products and negative for the reactants. This result is the standard heat of formation of n-pentane listed in Table C1.

Example 4.13 Estimation of standard heats of combustion from standard heats of formation

Estimate standard heats of reaction for the combustion of methane:

$$CH_4(g) + 2O_2(g) \rightarrow CO_2(g) + 2H_2O(l)$$

Solution:

The pertinent formation reactions and their heats of formation from Table C1 are:

CH_4: $C(s) + 2H_2(g) \rightarrow CH_4(g)$ $\Delta H^\circ_{f298} = -74.52$ kJ/mol, $v = -1$

O_2: $O_2(g)$ $\Delta H^\circ_{f298} = 0$ because oxygen is a stable element

CO_2: $C(s) + O_2(g) \rightarrow CO_2(g)$ $\Delta H^\circ_{f298} = -393.51$ kJ/mol, $v = 1$

H_2O: $H_2(g) + \frac{1}{2}O_2(g) \rightarrow H_2O(l)$ $\Delta H^\circ_{f298} = -285.83$ kJ/mol, $v = 2$

Stoichiometric coefficients v_i are positive for a product and negative for a reactant.

Standard heat of reaction:

$$\Delta H_r^o = \sum_i v_i \Delta H_{fi}^o = -393.51 + 2(-285.83) - (-74.52) = \mathbf{-890.65\,kJ/mol}$$

This result shows that the enthalpy of 1 mol of CH_4 plus 2 mol of O_2 is less than the enthalpy of 1 mol of CO_2 plus 2 mol of H_2O by 890.65 kJ when all the components are at their standard states. Therefore, the reaction is an exothermic reaction and emits heat to the surroundings.

4.4 Effect of Temperature on the Heat of Reaction

Standard state of formation enthalpies is function of temperature only

$$dH_{fi}^o = C_{pi}^o dT \tag{4.34}$$

Multiplying by v_i and summing over all products and reactants gives

$$d\Delta H^o = \sum_i dv_i H_{fi}^o = \sum_i v_i C_{pi}^o dT \tag{4.35}$$

The standard heat capacity change of a reaction is

$$\Delta C_p^o = \sum_i v_i C_{pi}^o \tag{4.36}$$

or

$$\Delta C_p^o = \Delta A + \Delta B \times 10^{-3}\,T + \Delta C \times 10^{-6}\,T^2$$

where

$$\Delta A = \sum_i v_i A_i; \quad \Delta B = \sum_i v_i B_i; \quad \Delta C = \sum_i v_i C_i$$

As a result, the effect of temperature on heats of reaction is obtained from

$$\Delta H_{r2}^o = \Delta H_{r1}^o + \int_{T_1}^{T_2} v_i C_{pi}^o dT = \Delta H_1^o + \int_{T_1}^{T_2} \left(\Delta A + \Delta B \times 10^{-3}T + \Delta C \times 10^{-6}T^2\right)dT$$

$$\tag{4.37}$$

where ΔH_{r1}^o is the known value, such as the standard heat of reaction $T_1 = 25\,°C$ and ΔH_{r2}^o is the standard heat of reaction at another temperature T_2.

Industrial reactions are usually not carried out at standard state conditions. In real applications, several reactions may occur simultaneously, actual reactions may also not go to completion, and the final temperature may differ from the initial temperature. Also, some inert components may be present in the reactions. Still, calculations of the heat effects will be based on the principles outlined above [2].

Example 4.14 Estimation of standard heat of reaction at a temperature other than 298 K

Estimate the standard heat of reaction for the methanol synthesis at 800 °C.

$$CO(g) + 2H_2(g) \rightarrow CH_3OH(g)$$

Solution:

By using standard heat of formation data in Eq. (4.27), we have

Stoichiometric coefficients v_i are positive for a product and negative for a reactant.

$\Delta H^{\circ}_{r,298} = -200.66 - (-110.52) = -90.13\,kJ/mol$

Heat of formation for hydrogen is zero as it is naturally occurring stable element. The ideal gas heat capacity parameters for all the components from Table C1 are:

Components	v_i	A	B	C
CO	−1	3.376	0.557×10^{-3}	0.0
H_2	−2	3.249	0.422×10^{-3}	0.0
CH_3OH	1	2.211	12.216×10^{-3}	-3.45×10^{-6}

$\Delta A = (1)(2.211) + (-1)(3.376) + (-2)(3.249) = -7.663$

$\Delta B = [(1)(12.216) + (-1)(0.557) + (-2)(0.422)] \times 10^{-3} = 10.815 \times 10^{-3}$

$\Delta C = [(1)(-3.45) + (-1)(0.0) + (-2)(0.0)] \times 10^{-6} = -3.45 \times 10^{-6}$

$T_1 = 298.15\,K, T_2 = (800 + 273.15)^{\circ}C = 1073.15\,K, R = 8.314\,J/mol\,K$

$$\Delta H^{o}_{r2} = \Delta H^{o}_{r1} + R \int_{T_1}^{T_2} \left(\Delta A + \Delta B \times 10^{-3}T + \Delta C \times 10^{-6}T^2 \right) dT$$

$$= -90.13 + R \int_{298.15}^{1073.15} \left(-7.663 + 10.815 \times 10^{-3}T - 3.45 \times 10^{-6}T^2 \right) dT$$

$$= -13,249.2\,kJ$$

The reaction is exothermic and releases energy. The heat of reaction increased with increased temperature.

4.5 Standard Enthalpy Changes

Standard enthalpy changes are:

- *Enthalpy of formation* is observed when one mole of a compound is formed from its elementary antecedents under standard conditions. The enthalpy change of any reaction under any conditions can be computed using the standard enthalpy of formation of all of the reactants and products.
- *Enthalpy of reaction* is defined as the enthalpy observed when one mole of substance reacts completely under standard conditions.
- *Enthalpy of combustion* is observed when one mole of a substance combusts completely with oxygen under standard conditions.
- *Enthalpy of neutralization* is observed when one mole of water is produced when an acid and a base react under standard conditions.
- *Enthalpy of fusion* is required to completely change the state of one mole of substance between solid and liquid states under standard conditions.
- *Enthalpy of vaporization* is required to completely change the state of one mole of substance between liquid and gaseous states under standard conditions.
- *Enthalpy of sublimation* is required to completely change the state of one mole of substance between solid and gaseous states under standard conditions.
- *Enthalpy of hydration* is observed when one mole of gaseous ions is completely dissolved in water forming one mole of aqueous ions.

4.6 Adiabatic Flame Temperature

There are two types of *adiabatic flame temperatures*: at constant volume and at constant pressure, describing the temperature of a combustion reaction products theoretically reaching if no energy is lost to the outside environment. Constant pressure adiabatic flame temperature results from a complete combustion process that occurs without any heat transfer or changes in kinetic or potential energy. The adiabatic flame temperature of the constant pressure process is lower than that of the constant volume process. This is because some of the energy released during combustion goes into changing the volume of the control system. The constant pressure adiabatic flame temperature for many fuels, such as wood, propane, and gasoline in air is around 1950 °C [2]. Assuming initial atmospheric conditions (1 atm and 25 °C), Table 4.3 lists the adiabatic flame temperatures for various gases for a stoichiometric fuel–air mixture.

Fuel	T_{ad} (°C)	T_{ad} (°F)
Acetylene (C_2H_2)	2500	4532
Butane (C_4H_{10})	1970	3578
Ethane (C_2H_6)	1955	3551
Hydrogen (H_2)	2210	4010
Methane (CH_4)	1950	3542
Natural gas	1960	3562
Propane (C_3H_8)	1980	3596
Wood	1980	3596
Kerosene	2093	3801
Light fuel oil	2104	3820
Medium fuel oil	2101	3815
Heavy fuel oil	2102	3817
Bituminous Coal	2172	3943
Anthracite	2180	3957

Table 4.3 Adiabatic constant pressure flame temperature T_{ad} of common gases/materials [2]

Example 4.15 Maximum flame temperature

Estimate the maximum temperature that can be reached by the combustion of methane with air at constant pressure. Both the methane and air enter the combustion chamber at 25 °C and 1 atm. The combustion reaction is

$$CH_4(g) + 2O_2(g) \rightarrow CO_2(g) + 2H_2O(g)$$

Solution:
Stoichiometric coefficients v_i are positive for a product and negative for a reactant. The heats of formation from Appendix C, Table C1 are:
$\Delta H^\circ_{f298}(CH_4) = -74.52$ kJ/mol, $v = -1$
$\Delta H^\circ_{f298}(O_2) = 0$ because oxygen is an element, $v = -2$
$\Delta H^\circ_{f298}(CO_2) = -393.51$ kJ/mol, $v = 1$
$\Delta H^\circ_{f298}(H_2O) = -241.82$ kJ/mol (vapor state), $v = 2$, $R = 8.314$ J/mol K
Standard heat of reaction:
$\Delta H^o_r = \sum_i v_i \Delta H^\circ_{fi} = -393.51 + (-241.82) - (-74.52) = \mathbf{-802.62 \ kJ/mol}$

Burning one mole of methane releases 802.62 kJ. We have 3.76 mol of N_2 for one mole of O_2 in air. The heat released by the combustion reaction will only be used to heat the products of 1 mol of carbon dioxide, 2 mol of water vapor, and 7.52 mol of nitrogen:

$$\Delta H_{r298}^{o} = \int\limits_{298.15}^{T} C_{p,\text{mix}} dT; \quad C_{p,\text{mix}} = \sum n_i C_{pi}^{o}, \quad \text{by} \quad \frac{C_{pi}^{o}}{R} = A_i + B_i \times 10^{-3} T \quad \text{and}$$

$$\frac{C_{p,\text{mix}}^{o}}{R} = A\prime + B\prime \times 10^{-3} T$$

where n_i is the moles of component i, T the maximum flame temperature in K. Heat capacity parameters are from Table B1:

	CO_2	$H_2O(g)$	N_2
A	5.457	3.47	3.280
B	1.045×10^{-3}	1.450×10^{-3}	0.593×10^{-3}
n_i	1	2	7.52

$$A' = \sum_i n_i A_i = 5.457 + 2(3.47) + 7.52(3.280) = 37.06$$

$$B' = \sum_i n_i B_i [1.045 + 2(1.45) + 7.52(0.593)] 10^{-3} = 8.40 \times 10^{-3}$$

$$802,620\,\text{J/mol} = R \int\limits_{298.15}^{T} C_{p,\text{mix}}^{o} dT = 8.314 [37.06(T - 298.15) + (1/2)8.40 \times 10^{-3}(T^2 - 298.15^2)]$$

$$T = \mathbf{2309\,K}$$

4.7   Air Pollution from Combustion Processes

Emissions from combustion of carbon fuels processes lead to air pollution. The main derivatives of the process are carbon dioxide CO_2, water, particulate matter, nitrogen oxides, sulfur oxides, and some partially combusted hydrocarbons, depending on the operating conditions and the fuel–air ratio. Not all of the fuel will be completely consumed by the combustion process; a small amount of fuel will be present after combustion, some of which can react to form oxygenates or hydrocarbons not initially present in the fuel mixture. Increasing the amount of air in the combustion process reduces the amount of the first two pollutants but tends to increase nitrogen oxides (NO_x) that has been demonstrated to be hazardous to both plant and animal health. Further chemicals released are benzene and 1,3-butadiene that are also particularly harmful. When incomplete burning occurs carbon monoxide (CO) may also be produced [2].

4.8 Heat of Mixing

Upon mixing of pure components, the thermodynamic properties change as shown by the symbol Δ. Figure 4.5 shows the heat of mixing and vapor–liquid equilibrium in ethanol–water mixture. Enthalpy of the mixture is composition dependent. As Fig. 4.5 shows that, at 197.2°F (isotherm), $H_{\text{sat liq}} \cong 155$ Btu/lb, while $H_{\text{sat vap}} \cong 775$ Btu/lb. When the ethanol mass fraction is $w_e = 0.4$ at 197.2°F, the enthalpy of mixture is around 600 Btu/lb.

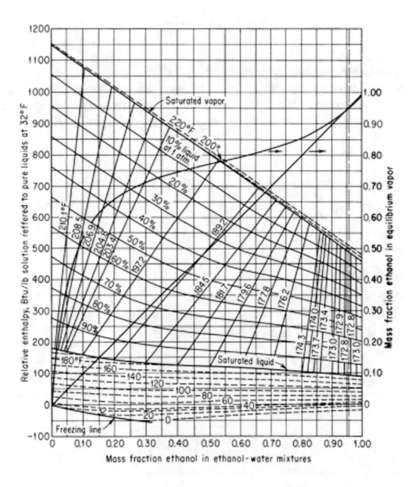

Fig. 4.5 Heat of mixing [1, 2]

The property change of mixing is

$$\Delta H = H - \sum_i x_i H_i \tag{4.38}$$

The partial properties are obtained from the following relationships

$$\Delta H_1 = \Delta H + (1 - x_1) \frac{d\Delta H}{dx_1} \tag{4.39}$$

$$\Delta H_2 = \Delta H - x_1 \frac{d\Delta H}{dx_1} \tag{4.40}$$

Example 4.16 Estimation of partial enthalpies

The heat of mixing (excess enthalpy) for a binary mixture is $H = x_1 x_2 (2ax_1 + ax_2)$, where a is the parameter in J/mol, x_1 and x_2 are the mole fractions of components 1 and 2, respectively. Derive equations for the partial enthalpies of H_1 and H_2 in terms of x_1.

Solution:

The partial properties from Eqs. 4.39 and 4.40:

$$H_1 = H + (1 - x_1) \frac{dH}{dx_1} \qquad H_2 = H - x_1 \frac{dH}{dx_1}$$

With $\sum x_i = 1$: $x_2 = 1 - x_1$, and the equation for the heat of mixing H becomes:

$H = ax_1 - ax_1^3$

The differentiation: $\dfrac{dH}{dx_1} = a - 3ax_1^2$

Therefore, the partial molar excess enthalpies:

$H_1 = ax_1 - ax_1^3 + (1 - x_1)(a - 3ax_1^2) = a - 3ax_1^2 + 2ax_1^3$

$H_2 = ax_1 - ax_1^3 - x_1(a - 3ax_1^2) = 2ax_1^3$

and

$H = x_1 H_1 + x_2 H_2$

4.9 Heat Measurements by Calorimeter

Only the energy change during transition of a system from one state into another can be defined and thus measured. Conventionally, this heat is measured by calorimeter. Figure 4.6 shows a typical flow calorimeter in which the change of temperature is related to the electrical energy supplied through the heater. The heat lost by the system would be equal to the heat gained by the surroundings in an exothermic system. Similarly, heat gained by the system from the surroundings would be equal to heat lost by the surroundings in an endothermic process. In a typical calorimeter, a hot substance (the system) and its surroundings can reach equilibrium at some final temperature. The heat flows from the system (hot) to the surroundings (cold). The heat lost by the system must be equal to the heat gained by the surroundings if there is no heat loss from the surroundings. No calorimeter is perfectly insulating to heat loss. Once the calorimeter has been calibrated to determine the calorimeter constant, we can use it to determine the specific heat effect [2]

Example 4.17 Measurement of heat capacity of a metal in a calorimeter
28.2 g of an unknown metal is heated to 99.8 °C and placed into a calorimeter containing 150.0 g of water at a temperature of 23.5 °C. The temperature can equilibrate, and a final temperature is measured as 25 °C. The calorimeter constant has been determined to be 19.2 J/°C. Estimate the specific heat capacity of the metal.

Solution:
Heat transfer is represented by:
Heat lost by the hot metal = heat gained by the water + heat gained by the calorimeter

$$(28.2 \text{ g})(C_{p,m})(99.8 - 25.0) \text{ °C} = (150.0 \text{ g})(4.184 \text{ J/g °C})(25.0 - 23.5) \text{ °C} + 19.2 \text{ J/°C } (25.0 - 23.5) \text{ °C}$$

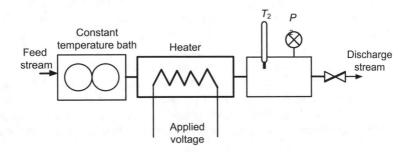

Fig. 4.6 Typical flow calorimeter

Solving for specific heat capacity:

$C_{p,m} = 969\,\text{J}/(2.11 \times 10^3\,\text{g}^\circ\text{C}) = \mathbf{0.459\,J/g\,^\circ C}$

Metals tend to have low specific heat capacities, while water has a relatively high heat capacity.

A *bomb calorimeter* is a well-insulated tank containing a small combustion chamber in a water bath to measure the heat of combustion of a reaction. Bomb calorimeters must withstand the large pressure within the calorimeter as the heat of reaction is being measured. Electrical energy is used to ignite the fuel. The energy released by the combustion raises the temperature of the steel bomb, its contents, and the surrounding water jacket. The temperature change in the water is then accurately measured. This temperature rise, along with a bomb calorimeter constant (which is dependent on the heat capacity of the metal bomb parts), is used to calculate the energy given out by the sample.

In a *differential scanning calorimeter*, conduction heat flows into a sample contained in a small aluminum capsule or a 'pan' and is measured by comparing it to the flow into an empty reference pan. The flow of heat into the sample is larger because of its heat capacity C_p of that sample. The difference in the heat flows induces a small temperature difference ΔT across the slab, which is measured using a thermocouple. Similarly, heat of melting of a sample substance can also be measured in such a calorimeter [2].

4.10 Psychrometric Diagram

Figure 4.7 shows a psychrometric diagram representing physical and thermal properties of moist air in a graphical form. Psychrometric diagrams help in determining environmental control concepts such as humidity of air and change of enthalpy resulting in condensation of the moisture.

- The *dry bulb temperature* T_{db} is the ambient air temperature, which can be measured using a normal thermometer freely exposed to the air but shielded from radiation and moisture. The dry bulb temperature appears as vertical lines in the psychrometric chart.
- The *wet bulb temperature* T_{wb} is indicated by a moistened thermometer bulb exposed to the air flow. Wet bulb temperature can be measured by using a thermometer with the bulb of thermometer wrapped in wet cloth. The adiabatic evaporation of water from the thermometer and the cooling effect is indicated by a 'wet bulb temperature' which is lower than the 'dry bulb temperature' in the air.

The rate of evaporation from a wet bandage on the bulb depends on the humidity of the air. The evaporation is reduced when the air contains more water vapor. The wet bulb temperature and the dry bulb temperature are identical at 100% relative

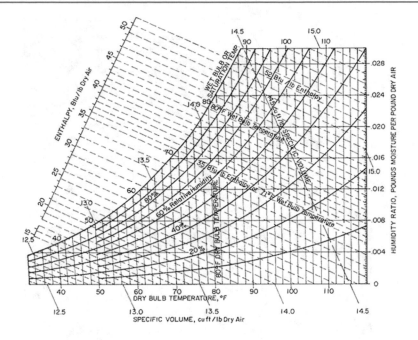

Fig. 4.7 Psychrometric diagram with permission (http://www.engineeringtoolbox.com)

humidity of the air (the air humidity is at the saturation line). Combining the dry bulb and wet bulb temperature in a psychrometric diagram determines the state of the humid air. Lines of constant wet bulb temperatures run diagonally from the upper left to the lower right on the psychrometric diagram.

The *dew point* is the temperature at which water vapor starts to condense out of the air (the temperature at which air becomes completely saturated). Above this temperature the moisture will stay in the air. For example, if moisture condensates on a cold bottle taken from the refrigerator, the dew point temperature of the air is above the temperature in the refrigerator. The dew point is given by the saturation line in the psychrometric diagram [2].

The Mollier diagram is used as a basic design tool for engineers for heating, ventilation, and air conditioning (HVAC) calculations as well as the values of enthalpy of air at various conditions. Figure 4.8 shows the Mollier diagram for water and steam.

Example 4.18 Determination of air properties on a psychrometric chart
For a dry bulb temperature of 74°F and a wet bulb temperature of 67°F:

(a) Estimate the air properties and its enthalpy.
(b) Estimate the heat to be added if the air is heated to 84°F dry bulb temperature at constant humidity ratio.

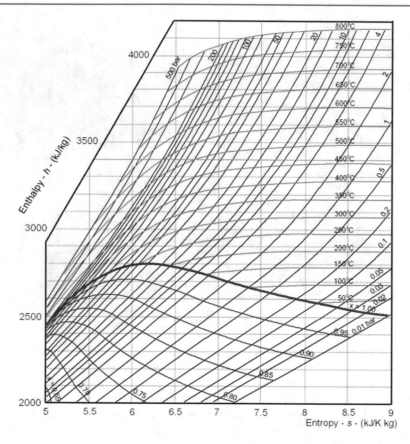

Fig. 4.8 The Mollier diagram for water-steam (http://www.engineeringtoolbox.com) (with permission). The Mollier diagram is useful in determining the values of enthalpy and entropy in various adiabatic flow processes, such as nozzles and turbines

(c) Estimate the dry bulb temperature for an evaporative cooling starting from the state point reached in part (b).

Solution:

(a) Find the intersection (A) of the dry bulb temperature of 74°F and wet bulb temperature of 67°F on the psychrometric chart as shown in the figure below. This intersection (A) is a 'state point' for the air.

We read relative humidity as 70% and dew point temperature as approximately 63.2°F (follow horizontal line, moving left, toward the curved upper boundary of saturation temperatures). The enthalpy is 31.7 Btu/lb dry air.

(b) The heating process moves horizontally to the right along a line of constant humidity ratio (B), and we read the enthalpy as approximately 34.2 Btu/lb dry air and the relative humidity as 50%.

The heat necessary: $q = (34.2–31.7)$ Btu/lb dry air = 2.5 Btu/lb dry air.

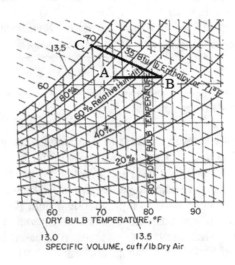

(c) During the evaporative cooling heat contained in the air evaporates water. The process moves upward along the line of constant enthalpy or constant web bulb temperature. Air temperature (dry bulb) drops to 70°F while water content (humidity) rises to the saturation point (C) with a relative humidity of 100%. Evaporation is often used in hot weather to cool ventilation air.

Example 4.19 Determination of moisture from the Mollier chart
Determine the quality and the pressure when the enthalpy and entropy of a saturated steam approximately are 2300 kJ/kg and 6.65 kJ/kg K, respectively.
Solution:
Consider the Mollier diagram for water-steam in Fig. 4.8.
Under the saturation line, we have two-phase region. In this region we can find the intersection point between the approximate values of enthalpy as 2300 kJ/kg and the entropy as 6.65 kJ/kg K.
At this intersection point, the constant pressure line of 0.5 bar and constant quality of 0.85 can be obtained. The moisture is 15% corresponding to the quality of the steam x = 0.85.
The readings are only approximate values.

Example 4.20 Determination of enthalpy and entropy from the Mollier chart
Determine the enthalpy and entropy of a saturated steam using the Mollier chart
when the pressure is 4 bar and temperature is 200 °C.
Solution:
Consider the Mollier diagram for water-steam in Fig. 4.8.
Find the intersection point which is well above the saturation line on the Mollier
chart when the pressure is 4 bar and temperature is 200 °C.
At this intersection point we can read the approximate value of enthalpy as
2850 kJ/kg and the approximate value of entropy as 7.2 kJ/kg K.

4.11   Heat Transfer

During the transfer of energy in the form of work, some energy involved may be
dissipated as heat. Heat and work are the two mechanisms by which energy can be
transferred. Work performed on a body is, by definition, an energy transfer to the
body that is due to a change to external parameters of the body, such as the volume,
magnetization, and center of mass in a gravitational field [1]. Heat is the energy
transferred to the body in any other way. Heat is transferred by conduction, convection, and radiation. Heat transfer by conduction is based on direct molecular
contact, most commonly in solids, while convection heat transfer is based on
moving particles of liquids and gases. Heat transfer by radiation is based on the
emission of electromagnetic waves from warm bodies to their surroundings.

4.11.1   Conduction Heat Transfer

Rate of heat transfer by conduction is estimated by

$$\dot{q} = -kA\frac{(T_2 - T_1)}{\Delta x} \tag{4.41}$$

where $\dot{q}$ is the rate of heat transfer, J/s (Watt, W) or Btu/h, k is the thermal
conductivity in J/m s °C or Btu/ft s °F, A is the area for heat transfer, $(T_2 - T_1)$ is
the temperature difference between state 2 (cold) and state 1 (hot), and Δx is the
distance. As the heat flows from high to low temperature the value of $(T_2 - T_1)$ is
negative. The values of thermal conductivity for various substances are tabulated in
Table A1. Table 4.4 shows the thermal conductivities and heat capacities of some
gases at 1 atm and some liquids and solids [1].

Table 4.4 Thermal conductivities and heat capacities of some gases (at 1 atm), liquids, and solids [1]

Substance	T (K)	k (W/m K)
Gases		
Hydrogen, H_2	300	0.1779
Oxygen, O_2	300	0.0265
Carbon dioxide, CO_2	300	0.0166
Methane, CH_4	300	0.0342
NO	300	0.0259
Liquids		
Water, H_2O	300	0.6089
	350	0.6622
Ethanol, C_2H_5OH	300	0.1676
Carbon tetrachloride, CCl_4	350	0.0893
Acetone, $(CH_3)_2CO$	300	0.1274
Mercury, Hg	372.2	10.50
Solids		
Lead, Pb	977.2	15.1
Aluminum, Al	373.2	205.9
Copper, Cu	291.2	384.1
Cast iron, Fe	373.2	51.9
Steel	291.2	46.9
Silver	373.2	411.9

4.11.2 Convection Heat Transfer

Rate of heat transfer by convection is estimated by

$$\dot{q} = hA(T_2 - T_1) \tag{4.42}$$

where h is the heat transfer coefficient in J/m^2 s °C or Btu/ft^2 s °F, and *heat flux*, $\dot{q}/A$, is the rate of heat transfer per unit cross-sectional area.

4.11.3 Radiation Heat Transfer

All bodies radiate energy in the form of photons moving in a random direction and frequency. When photons reach another surface, they may either be absorbed, reflected, or transmitted [1]:

- α = absorptance – fraction of incident radiation absorbed
- τ = transmittance – fraction of incident radiation transmitted
- ρ = reflectance – fraction of incident radiation reflected

From energy considerations, the three coefficients must sum to unity: $\alpha + \tau +$ = 1. Real bodies radiate less effectively than black bodies. The measurement of this is the emittance ε defined by

$$\varepsilon = \frac{E}{E_b} \tag{4.43}$$

where E is radiation from the real body, and E_b is radiation from a black body at temperature T. Values of emittance are near unity for rough surfaces such as ceramics or oxidized metals, and roughly 0.02 for polished metals or silvered reflectors. The energy radiated per unit area of a black body is estimated by: $E_b = \sigma T^4$ where σ is the Stefan-Boltzmann constant, $\sigma = 5.67 \times 10^{-8}$ W/m^2 K^4. The level of emittance can be related to absorptance by

$$\frac{E}{E_b} = \varepsilon = \alpha \tag{4.44}$$

The relation $\varepsilon = \alpha$ is known as Kirchhoff's law. It implies that good radiators are good absorbers. Heat transfer between gray and planar surfaces is [1]

$$\dot{q} = \frac{\sigma(T_1^4 - T_2^4)}{\frac{1}{\varepsilon_1} + \frac{1}{\varepsilon_2} - 1} \tag{4.45}$$

Example 4.21 Estimation of radiation heat transfer
Use a thermos bottle to estimate the heat transfer between thermos silver walls. The temperature of silver wall is $T_1 = 100\ °C = 373\ K$ and the surroundings temperature is $T_2 = 20\ °C = 293\ K$. The values of the emittances are $\varepsilon_1 = \varepsilon_2 = 0.02$.

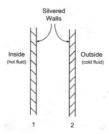

Solution:
The Stefan-Boltzmann constant, $\sigma = 5.67 \times 10^{-8}$ W/m^2 K^4,
$T_1 = 100\ °C = 373\ K$, $T_2 = 20\ °C = 293.15\ K$, and $\varepsilon_1 = \varepsilon_2 = 0.02$.

From Eq. (**4.45**): $\dot{q} = \dfrac{\sigma(T_1^4 - T_2^4)}{\frac{1}{\varepsilon_1} + \frac{1}{\varepsilon_2} - 1} = 6.9\,\text{W/m}^2$

The thermos is a good insulator as the heat transfer is very small from the hot wall to cold wall.

4.12 Exergy

An energy source comes with its amount of energy, and its work potential that is the useful energy or available energy. As heat flows out from a finite energy source with a high temperature to a low energy source, its temperature decreases, the quantity of available energy decreases, and so the quality of energy degrades. Amount and quality of energy in a source can be distinguished by the concept of exergy, *Ex*. Exergy is the maximum amount of work theoretically available by bringing a resource into equilibrium with its surrounding through a reversible process. Exergy is also known as availability. Quality of energy is the ability to produce work under the conditions determined by the *natural environment* called *the reference state*. The temperature and pressure of the natural environment are usually the standard state values, such as 298.15 K and 101.31 kPa. High temperature fluid has a higher potential of heat-to-work conversion than the fluid at a low temperature. That is how the exergy provides a means of defining energy quality [3].

The work done during a process depends on the initial state, final state, and the process path chosen between these initial and final states. The initial state is not a variable in an exergy analysis. In the final state, a process reaches thermodynamic equilibrium with the environment and the system's temperature and pressure become the same with the values of temperature and pressure of the environment. So, the system has no kinetic and potential energy relative to the environment. The work output is maximized when the process takes place between the initial and final states in a reversible manner. In practical processes, this means that work output is maximized when the process proceeds through a path with the lowest possible *entropy production* between the initial and final states.

Kinetic and potential energies can be converted to work completely. Therefore, exergy of the kinetic energy and potential energy of a system are equal to the kinetic energy and potential energy

$$\Delta Ex_{KE} = m\frac{\Delta v^2}{2} \tag{4.46}$$

$$\Delta Ex_{PE} = mg\Delta z \tag{4.47}$$

However, the enthalpy H and internal energy U of a system are not completely available. The exergy change of a closed system during a process is

$$\Delta Ex = \Delta H - T_o \Delta S \tag{4.48}$$

Here the Δ denotes the difference between the initial and final state (reference state), and T_o is the reference temperature. The fluid flow exergy is

$$\dot{Ex} = \dot{m}\left((H - H_o) - T_o(S - S_o) + \frac{\Delta v^2}{2}\left(\frac{\text{kJ/kg}}{1000\,\text{m}^2/\text{s}^2}\right) + g\Delta z\right) \tag{4.49}$$

Exergy change of a fluid flow in an open system is given by

$$\Delta \dot{Ex} = \dot{m}\left(\Delta H - T_o \Delta S + \frac{\Delta v^2}{2}\left(\frac{\text{kJ/kg}}{1000\,\text{m}^2/\text{s}^2}\right) + g\Delta z\right) \tag{4.50}$$

Here, as before, the dot over a symbol indicates a quantity per unit time. If the changes are kinetic and potential energy are neglected then Eq. (4.50) becomes

$$\Delta \dot{Ex} = \dot{m}(\Delta H - T_o \Delta S) \tag{4.51}$$

Heat (especially at low temperature) is much lower in exergy due to its low efficiency in its conversion to other forms of energy. The useful (available) work content of the heat is limited by the Carnot factor, η_{Carnot}, therefore, the available heat $q_{\text{available}}$ becomes

$$Ex_{\text{heat}} = q_{\text{available}} = q\eta_{\text{Carnot}} = q\left(1 - \frac{T_C}{T_H}\right) \tag{4.52}$$

where T_H and T_C are the temperatures for the hot and cold regions, respectively. Equation (4.52) means that a power generator based on heat cannot have efficiency greater than that of a Carnot cycle working between the hot source with T_H and cold source with T_C. It is possible to increase or lower exergy as exergy is not a conserved property [3].

Summary

- *Sign Convention*: If heat flows into a system, it is taken to be positive. If heat goes out of the system, it is taken negative.
- For a closed system, the *internal energy* is essentially defined by

$$\Delta U = q + W$$

where ΔU is the change in internal energy of a system during a process, q is the heat, and W is the mechanical work. The change in internal energy in terms of temperature and volume is

$$dU = C_v dT + \left(\frac{\partial U}{\partial V}\right)_T dV$$

where C_v is the heat capacity at constant volume. The total internal energy U of a system cannot be measured directly and the change in internal energy, $\Delta U = U_{final} - U_{initial}$, is a more useful value.

- *Enthalpy H* of a system is defined by

$$H = U + PV$$

where U is the internal energy, P is the pressure at the boundary of the system and its environment, and V is the volume of the system. The total enthalpy of a system cannot be measured directly; the *enthalpy change* of a system is measured instead.

- *Heat* is part of the total energy flow across a system boundary that is caused by a temperature difference between the system and its surroundings or between two systems. When heat capacity is constant, such as $C_{p,av}$, the amount of heat changed when a substance changed its temperature from T_1 to another temperature T_2 is estimated by

$$q = mC_{p,av}(T_2 - T_1)$$

- When there is no work interaction between a closed system and its surroundings the amount of heat flow is:

$$q = \int_{T_1}^{T_2} C_p dT$$

- *Sensible heat* refers to the amount of heat required to raise the temperature of unit mass of a substance through one degree. Its unit is J/kg K.

$$q = \Delta H = \int_{T_1}^{T_2} C_p dT$$

- *Latent heat* refers to the amount of heat transferred to cause a phase change; *heat of fusion* is required to melt unit mass of a solid into a liquid or to freeze unit mass of the liquid to the solid. *Heat of vaporization* is required to vaporize unit mass of a liquid; *heat of sublimation* is required to sublimate (direct solid–vapor) unit mass of a solid.
- When ΔH_{vap} can be assumed to be constant within the temperature interval of $(T_2 - T_1)$ the *Clausius-Clapeyron* equation is

$$\ln\left(\frac{P_2^{sat}}{P_1^{sat}}\right) = -\frac{\Delta H_{vap}}{R}\left(\frac{1}{T_2} - \frac{1}{T_1}\right)$$

- *Trouton's rule* predicts the heat of vaporization at normal boiling point, ΔH_n, that is at a pressure of 1 standard atm (101.32 kPa).

$$\frac{\Delta H_n}{RT_n} \simeq 10$$

where T_n is the absolute temperature of the normal boiling point and R is the gas constant.

- Standard heat of reaction is expressed by

$$\Delta H_r^o = \sum_i v_i \Delta H_{fi}^o$$

where H_{fi}^o is the *standard heat of formation;* v_i is the stoichiometric coefficient of the reaction.

- Reactions that release heat are called *exothermic reactions* and reactions that require heat are called *endothermic reactions*.
- *Standard enthalpy of vaporization* is the enthalpy required to completely change the state of one mole of substance between liquid and gaseous states under standard conditions.
- Rate of heat transfer by conduction is estimated by

$$\dot{q} = -kA\frac{(T_2 - T_1)}{\Delta x}$$

where $\dot{q}$ is the rate of heat transfer, J/s (Watt, W) or Btu/h, k is the thermal conductivity in J/m s °C or Btu/ft s °F, A is the area for heat transfer, $(T_2 - T_1)$ is the temperature difference between state 2 and state 1, and Δx is the distance.

- Rate of heat transfer by convection is estimated by

$$\dot{q} = hA(T_2 - T_1)$$

where h is the heat transfer coefficient in J/m^2 s °C or Btu/ft² s °F, and *heat flux*, $\dot{q}/A$, is the rate of heat transfer per unit cross-sectional area.

- The fluid-flow exergy is

$$\dot{Ex} = \dot{m}\left((H - H_o) - T_o(S - S_o) + \Delta v^2/2 + g\Delta z\right)$$

- The useful (available) work content of the heat is limited by the Carnot factor, η_{Carnot}; therefore, the available heat $q_{available}$ becomes

$$Ex_{heat} = q_{available} = q\eta_{Carnot} = q(1 - T_C/T_H)$$

where T_H and T_C are the temperatures for the hot and cold regions, respectively.

Problems

4.1 Determine the states of the system containing water under the following conditions by using the steam tables.

T (°C)	P (kPa)	H_{vap} (kJ/kg)	U_{vap} (kJ/kg)	State
175	30	2830.0	2623.6	
45.83	10	2584.8	2438.0	
450	2400	3352.6	3027.1	
100	98	2676.0	2506.5	

4.2 Determine the states of the system containing water under the following conditions by using the steam tables.

T (°F)	P (psia)	H_{vap} (Btu/lb)	U_{vap} (Btu/lb)	State
164	5.99	1134.2	1065.4	
500	20	1286.9	1181.6	
600	1000	1517.4	1360.9	
331.37	105	1188.0	1105.8	

4.3   Determine the states of the system containing water under the following conditions by using the steam tables.

T (°C)	P (kPa)	V (m³/kg)	$H_{sat\ vap}$ (kJ/kg)	State
60	19.92	5.50	2609.7	
110	143.27	1.21	2691.3	
340	8200	0.0280	–	
40	47.36	0.001	–	

4.4   Determine the states of the system containing water under the following conditions by using the steam tables.

T (°C)	P (kPa)	x	H (kJ/kg)	State
120.2	200	0.7	2045.8	
80	47.36	1	2643.8	
150	476	0	632.1	
325	500	–	3116.4	

4.5.   Determine the states of the system containing refrigerant-134a under the following conditions.

T (°F)	P (psia)	H (Btu/lbm)	State
−10	16.67	95.0	
5	23.8	102.47	
30	40.78	20.87	
60	70	110.23	

4.6.   Determine the states of the system containing water under the following conditions by using the steam tables.

T (°C)	P (kPa)	V (m³/kg)	U (kJ/kg)	State
133.5	300	0.50	2196.4	
300	725	0.358	2799.3	
400	6500	0.043	2888.9	
211.1	1950	0.001	900.46	

4.7. Determine the missing property values in the table below for a system that contains water.

P (kPa)	V (m³/kg)	U (kJ/kg)	H (kJ/kg)	x, kg vap/kg total mass	State
9.1		2250			
200				0.85	

4.8. Determine the missing property values in the table below for a system that contains pure refrigerant R-134a.

T (°C)	P (kPa)	V (m³/kg)	U (kJ/kg)	H (kJ/kg)	x
30	836				
−45			204.51		
	300	0.054			
	1124			244.0	
65	1550				

4.9. A 50 kg block of copper (Cu) at 0 °C and a 100 kg block of iron (Fe) at 200 °C are brought into contact in an insulated space. Predict the final equilibrium temperature.

4.10. Estimate the internal energy and enthalpy of 100 lb of water at 220 psia and 500°F.

4.11. A rigid tank contains 500 lb of saturated steam at 230°F. Estimate the internal energy and enthalpy.

4.12. Determine the specific internal energy of 100 L of hydrogen gas at 600 K and 1 atm. The hydrogen gas has a total enthalpy of 19,416 J. Assume that the hydrogen gas is an ideal gas.

4.13. Calculate the change in internal energy when 0.75 m³ of air at 200 kPa is heated from 20 to 250 °C in a constant volume process.

4.14. Calculate the change in internal energy when 15 m³ of air at 110 kPa is heated from 30 to 400 °C in a constant volume process.

4.15. Calculate the change in internal energy when 1.5 m³ of air at 275 kPa is cooled from 150 to 20 °C in a constant volume process.

4.16. Calculate the change in internal energy when 10 m³ of air at 150 kPa is cooled from 300 to 50 °C in a constant volume process.

4.17. Calculate the change in internal energy when 25.1 m³ of carbon dioxide at 250 kPa is heated from 30 to 350 °C in a constant volume process.

4.18. Calculate the change in internal energy when 100 m³ of carbon dioxide at 150 kPa is heated from 0 to 500 °C in a constant volume process.

4.19. Calculate the change in internal energy when 5 m³ of carbon dioxide at 250 kPa is cooled from 270 to 20 °C in a constant volume process.

4.20. Calculate the change in internal energy when 52.8 m^3 of nitrogen at 200 kPa is heated from 15 to 700 °C in a constant volume process.

4.21. Calculate the change in internal energy when 0.92 m^3 of nitrogen at 200 kPa is cooled from 300 to 30 °C in a constant volume process.

4.22. Calculate the change in internal energy when 32.5 ft^3 of nitrogen at 20 psia is heated from 75 to 300°F in a constant volume process.

4.23. Superheated water vapor at 400 °C expands isothermally in a piston-and-cylinder device from 1350 to 770 kPa. Calculate the change in the molar enthalpy and molar internal energy in units of kJ/mol.

4.24. One mole of gas is undergoing an expansion processes, in which it expands from 0.03 to 0.06 m^3 at constant pressure of 2 atm. Estimate the change in internal energy if $C_{v,\text{av}}$ = 12.5 J/mol K.

4.25. A one mole of gas at 300 K expands from 0.03 to 0.06 m^3 at constant temperature. Estimate the change in internal energy of the gas.

4.26. A large rigid tank contains 70 kg of water at 80 °C. If 14 kg of water is in the liquid state estimate the pressure and enthalpy of the water in the tank.

4.27. A large rigid tank contains 40 lb of water at 200°F. If 2.8 lb of the water is in the liquid state estimate the internal energy and enthalpy of the water in the tank.

4.28. A rigid tank is filled with 0.5 lb of refrigerant-134a at T = 79.2°F. If 0.06 lb of the refrigerant is in the liquid state estimate the internal energy and enthalpy of the refrigerant.

4.29. Heat capacity of toluene in Btu/lbmol °F is given by $C_p = 20.869 + 5.293 \times 10^{-2}T$. Estimate the heat capacity in cal/mol K and estimate the heat necessary if 400 g toluene is heated from 17 to 48 °C.

4.30. Determine the enthalpy change and internal energy change of ammonia in J/mol, as it is heated from 30 to 150 °C, using the ideal gas heat capacity given by the following equation: $C_p^{ig} = R(3.578 + 3.02 \times 10^{-3}T)$ T in K (Table B1)

4.31. Determine the enthalpy change and internal energy change of air in kJ/kg, as it is heated from 300 to 750 K, using the following equation for C_p. $C_p^{ig} = R(3.355 + 0.575 \times 10^{-3}T)$ T in K (Table B1)

4.32. An ideal gases mixture of carbon dioxide (CO_2), nitrogen (N_2), and oxygen (O_2) at 200 kPa and 300 °C contains 10 mol% CO_2, 60 mol% N_2, and 30 mol% O_2. The mixture is heated to 600 °C. Calculate the change in the molar internal energy of the mixture.

4.33. 12 kg of benzene is heated from a subcooled liquid at P_1 = 100 kPa and T_1 = 300 K to a superheated vapor at T_2 = 450 K and P_2 = 100 kPa. The heating process is at constant pressure. Calculate the total enthalpy, ΔH, in kJ/kg. Assume the n-heptane vapor behaves as an ideal gas at a constant heat capacity of 85.29 J/mol K. Use the Antoine and Clausius-Clapeyron equations to estimate the heat of vaporization of n-heptane at T_1.

4.34. Water is heated from a subcooled liquid at $P_1 = 200$ kPa and $T_1 = 300$ K to a superheated vapor at $P_2 = 2400$ kPa and $T_2 = 573$ K. Calculate the molar ΔH for each step in the path in kJ/kg.

4.35. Water is cooled from a superheated vapor at $P_1 = 250$ psia and $T_1 = 460°$F to a subcooled liquid at $P_2 = 15$ psia and $T_2 = 140°$F. Calculate the molar ΔH for each step in the path in Btu/lb.

4.36. In a steady-state flow heating process, 50 kmol of methanol is heated from subcooled liquid at 300 K and 3 bar to superheated state at 500 kPa. Estimate the heat flow rate for the process. The saturation temperature for methanol at 3 bar is 368.0 K.

4.37. A one mole of gas at 300 K expands from 0.03 to 0.06 m^3 at adiabatic conditions. Estimate the change in internal energy of the gas. $C_{v.}$ $_{av} = 12.5$ J/mol K

4.38. A piston-and-cylinder device contains 0.3 kg refrigerant R-134a vapor at 12 °C. This vapor is expanded in an internally reversible, adiabatic process until the temperature is −3 °C. Determine the work for this process. The constant heat capacity of refrigerant-134a is $C_{v,av} = 1.7$ kJ/kg °C.

4.39. A piston-and-cylinder device contains 0.5 kg nitrogen at 30 °C. The nitrogen is compressed in an internally reversible, adiabatic process until the temperature is 42 °C. Determine the work for this process. The ratio of constant heat capacities $\gamma = C_p/C_v = 1.4$ for the nitrogen.

4.40. Twenty kg of hydrogen are heated from 25 to 275 °C. Estimate the change in enthalpy in kJ.

4.41. Fifty kg of hydrogen are heated from −10 to 100 °C. Estimate the change in enthalpy in kJ.

4.42. Twenty-five kg of hydrogen are cooled from 20 to −50 °C. Estimate the change in enthalpy ΔH in Btu.

4.43. Ten kg of carbon dioxide is heated from 30 to 300 °C in a steady-state flow process at a pressure sufficiently low that the carbon dioxide may be considered as an ideal gas. (a) Estimate the value of the heat required, (b) Estimate the average heat capacity in J/mol K for the process.

4.44. The molar heat capacity of propane gas in ideal-gas state is given as a function of temperature in Kelvin: $C_p^{ig}/R = 1.213 + 28.785 \times 10^{-3}T - 8.824T \times 10^{-6}$ (Table B1)
Convert this equation to an equation for C_p for temperature in °C.

4.45 The molar heat capacity of methane gas in ideal-gas state is given as a function of temperature in Kelvin: $C_p^{ig}/R = 1.702 + 9.082 \times 10^{-3}T - 2.164T \times 10^{-6}$ (from Table B1)
Convert this equation to an equation for C_p in J/g °C.

4.46 The molar heat capacity of methane gas in ideal-gas state is given as s function of temperature in Kelvin: $C_p^{ig}/R = 1.702 + 9.082 \times 10^{-3}T - 2.164T \times 10^{-6}$ (from Table B1)
Convert this equation to an equation for C_p in Btu/lb °F.

4.47   Ten kg of water at 25 °C in a closed vessel of 0.5 m³ in volume is heated to 120 °C and 1000 kPa. Estimate the change of enthalpy and the final quality of the steam in the vessel.

4.48   Twenty two kg of water at 25 °C in a closed vessel of 2.6 m³ in volume is heated to 120 °C and 1000 kPa. Estimate the change of enthalpy and the quality of the steam in the vessel.

4.49   Estimate the enthalpy of a 10 lb steam at 120 psia with a quality of 27%.

4.50   Calculate the change in enthalpy when 150 lb of nitrogen at 25 psia is heated from 50 to 700°F in a constant pressure process. The heat capacity in Btu/lbmol °F is

$$C_p^{ig} = 6.895 + 0.7624 \times 10^{-3}T - 0.7009 \times 10^{-7}T^2$$

4.51   Calculate the change in enthalpy when 200 lb of nitrogen at 20 psia is cooled from 500 to 100°F in a constant pressure process. The heat capacity in Btu/lbmol °F is

$$C_p^{ig} = 6.895 + 0.7624 \times 10^{-3}T - 0.7009 \times 10^{-7}T^2$$

4.52   Calculate the change in enthalpy when 500 lb of oxygen at 20 psia is heated from 60 to 450°F in a constant pressure process. The heat capacity in Btu/lbmol °F is

$$C_p^{ig} = 6.895 + 0.7624 \times 10^{-3}T - 0.7009 \times 10^{-7}T^2$$

4.53   Calculate the change in enthalpy when 265 ft³ of carbon dioxide at 20 psia is heated from 80 to 360°F in a constant pressure process. The heat capacity in Btu/lbmol °F is

$$C_p^{ig} = 8.448 + 5.757 \times 10^{-3}T - 21.59 \times 10^{-7}T^2$$

4.54   Calculate the change in enthalpy when 1150 ft³ of carbon monoxide at 20 psia is cooled from 250 to 75°F in a constant pressure process. The heat capacity in Btu/lbmol °F is

$$C_p^{ig} = 6.865 + 0.8024 \times 10^{-3}T - 0.7367 \times 10^{-7}T^2$$

4.55   Calculate the change in enthalpy when 500 lb of carbon monoxide at 20 psia is heated from 50 to 290°F in a constant pressure process. The heat capacity in Btu/lbmol °F is

$$C_p^{ig} = 6.865 + 0.8024 \times 10^{-3}T - 0.7367 \times 10^{-7}T^2$$

4.56 Ten kg of carbon dioxide at 50 °C and 1 atm is heated to 200 °C at constant pressure. Carbon dioxide is assumed to be an ideal gas. Estimate the changes in internal energy and enthalpy.

4.57 A 50 ft^3 rigid tank is filled with saturated vapor at 150 psia. The steam is cooled as the heat is transferred from the tank and the final pressure of the steam drops to 20 psia. Estimate: (a) the mass of the steam that has condensed, (b) the final temperature of the steam, and (c) the amount of heat transferred from the steam.

4.58 A piston-and-cylinder device contains R-134a which is initially at −10 °C and 200 kPa. The R-134a undergoes an isochoric heating to 80 °C. Estimate the work in kJ/kg.

4.59 The primary energy source for cells is the aerobic oxidation of sugar called glucose ($C_6H_{12}O_6$). Estimate the standard heat of reaction for the oxidation of glucose:

$$C_6H_{12}O_6(s) + 6O_2(g) \rightarrow 6CO_2(g) + 6H_2O(g)$$

4.60 Synthesis gas is a mixture of H_2 and CO. One of the uses of synthesis gas is to produce methanol from the following reaction: $CO(g) + 2H_2(g) \rightarrow CH_3OH(g)$
Estimate the standard heat of reaction.

4.61 Determine the standard heat of a reaction at 298.15 K for the reactions below:
(a) $N_2(g) + 3H_2(g) \rightarrow 2NH_3(g)$
(b) $CaCO_3(s) \rightarrow CaO(s) + CO_2(g)$

4.62 Estimate the heat of reaction for the combustion of propane:

$$C_3H_8(g) + 5O_2(g) \rightarrow 3CO_2(g) + 4H_2O(g)$$

4.63 Estimate the heat of reaction for the combustion of ethane:

$$C_2H_6 + \frac{7}{2}O_2 \rightarrow 2CO_2(g) + 3H_2O(g)$$

4.64 Estimate the heat of reaction for the combustion of methane:

$$CH_4 + 2O_2 \rightarrow CO_2(g) + 2H_2O(g)$$

4.65 Nitric acid (HNO_3) is produced by the oxidation of ammonia. Estimate the standard heat of reaction: $NH_3 + 2O_2 \rightarrow HNO_3 + H_2O$

4.66 Estimate the maximum temperature that can be reached by the combustion of methane with air at constant pressure. Both the methane and air enter the combustion chamber at 25 °C and 1 atm. The combustion reaction is: $CH_4(g) + 2O_2(g) \rightarrow CO_2(g) + 2H_2O(g)$

4.67   Find the heat of reaction and determine whether the process is exothermic or endothermic for the following chemical reaction: $16H_2S(g) + 8SO_2(g)$ $16H_2O(l) + 3S_8(s)$.

4.68   Heating value of a fuel vary if the water produced is in the vapor state or liquid state. The higher heating value (HHV) of a fuel is its standard heat of combustion when the water product is liquid; the lower heating value when the water product is vapor. Estimate the HHV and the LHV for natural gas represented by pure methane: $CH_4(g) + 2O_2(g) \rightarrow CO_2(g) + 2H_2O(g)$.

4.69   Estimate the HHV and the LHV for heating oil represented by pure n-decane. As liquid n-decane heat of formation is -249.7 kJ/mol: $C_{10}H_{22}(l) + 15.5O_2(g) \rightarrow 10CO_2(g) + 11H_2O(g)$.

4.70   What is the standard heat of combustion at 25 °C of $C_6H_{12}(g)$ when the reaction products are $CO_2(g)$ and $H_2O(g)$?
$C_6H_{12}(g) + 9O_2 \rightarrow 6CO_2(g) + 6H_2O(g)$

4.71   What is the standard heat of combustion at 25 °C of $6CH_3OH$ when the reaction products are $CO_2(g)$ and $H_2O(g)$? The reaction is: $6CH_3OH$ $(g) + 9O_2 \rightarrow 6CO_2(g) + 12H_2O(g)$.

4.72   Estimate the standard heat of combustion of n-pentane gas at 25 °C when the combustion products are $CO_2(g)$ and $H_2O(l)$: C_5H_{12} (g) $+ 8O_2$ (g) $5CO_2(g) + 6H_2O(l)$.

4.73   Estimate the standard heat of reaction of following reactions at 25 °C.
(a) $4NH_3(g) + 5O_2(g) = 4NO(g) + 6H_2O(g)$
(b) $C_2H_4(g) + 1/2O_2(g) = 2(CH_2)O(g)$
(c) $CH_3OH(g) + 1/2O_2(g) = HCHO(g) + H_2O(g)$
(d) $C_2H_5OH(l) + O_2(g) = CH3COOH(l) + H_2(g)$

4.74   A heating oil with an average chemical composition of $C_{10}H_{18}$ is burned with oxygen completely in a bomb calorimeter. The heat released is measured as 43,900 J/g at 25 °C. The combustion on the calorimeter takes place at constant volume and produces liquid water. Estimate the standard heat of reaction of the oil at 25 °C when the products are (g) H_2O and CO_2 (g).

4.75   Derive the following isentropic relation for ideal gases with constant specific heats.

$$\frac{T_2}{T_1} = \left(\frac{P_2}{P_1}\right)^{(\gamma-1)/\gamma}$$

4.76   A piston and cylinder device with a free-floating piston has an initial volume of 0.1 m³ and contains 0.6 kg of steam at 400 kPa. Heat is transferred into the steam until the temperature reaches 300 °C while the pressure remains constant. Determine the heat and work transfer.

4.77   1 kg Superheated steam in a piston-cylinder system is initially at 200 °C and 1300 kPa. The steam is cooled at constant pressure until half of the mass condenses. Estimate the final temperature and the volume change.

4.78 1.21 lb Superheated refrigerant-134a is initially at 150°F and 140 psia. The refrigerant is cooled to 5°F at constant pressure until it exists as compressed liquid. Estimate the changes in internal energy and total volume.

4.79 Refrigerant-134a with a flow rate of 0.12 kg/s is cooled to 40 °C at constant pressure in a condenser by using cooling water. Initially, the refrigerant is at 1200 kPa and 50 °C. Estimate the heat removed by the cooling water.

4.80 Air with a flow rate of 500 ft^3/s is heated from 100 to 910°F at approximately atmospheric pressure for a combustion process. What is the rate of heat transfer required?

4.81 An oven wall is made of 6-inch thick forebrick. If the interior of the wall is at 1550°F and the exterior of the wall is at 350°F, estimate the heat loss per hour. The thermal conductivity of the firebrick is 0.81 Btu/h °F ft and assumed remain constant.

4.82 Estimate the thickness (Δx) needed for a cork insulation with $k = 0.04$ J/m s K when the heat transfer is 13.8 W/m^2 between the walls with temperatures $T_1 = 100$ °C and $T_2 = 20$ °C.

4.83 A resistance heater within a well-insulated tank is used to heat 0.8 kg of water. The heater passes a current of 0.2 A (A) for 30 min from a 120-V source. The water in the tank is originally at 25 °C. Estimate the temperature of the water after 30 min of heating.

4.84 A resistance heater within a well-insulated tank is used to heat 4 kg of water. The heater passes a current of 0.2 A (A) from a 120-V source. The water in the tank is originally at 20 °C. Estimate the time necessary for the temperature of the water to reach 45 °C.

4.85 A resistance heater within a tank is used to heat water. The heater passes a current of 0.4 A (A) for 2 h from a 120-V source. A heat loss of 8.7 kW occurs from the tank to surroundings. Estimate the amount of heat transfer to the water after 2 h of heating.

4.86 A resistance heater within a tank is used to heat 150 kg water. The heater passes a current of 0.6 A (A) for 8 h from a 120-V source. A heat loss of 38.2 kW occurs from the tank to surroundings. The water in the tank is originally at 5 °C. Estimate the amount of heat transfer to the water after 8 h of heating.

4.87 A resistance heater within a well-insulated room is used to heat the inside air. The heater passes a current of 2 A (A) from a 120-V source. Estimate the amount of heat transfer to the air after 1 h of heating.

4.88 A resistance heater within a well-insulated room is used to heat the inside air. The heater passes a current of 2 A (A) for 3 h from a 120-V source. Estimate the amount of heat transfer to the air after 3 h of heating.

4.89 A resistance heater within a well-insulated room is used to heat the inside air. The heater passes a current of 0.8 A (A) for 2 h from a 120-V source. The heat loss from the room is 29.0 kJ. Estimate the amount of heat transfer to the air after 2 h of heating.

4.90 A wall of an oven is made of 10 in thick brick. If the interior wall of the oven is at 2000°F and the exterior of the wall is at 70°F estimate the heat loss per hour. The average value of the thermal conductivity of the brick is 0.81 Btu/h ft °F.

4.91 A vertical plate receives solar radiation of 1000 W/m². The plate is insulated at the back and loses heat by convection heat transfer through the air flow. The heat transfer coefficient is 35 W/m² K and the air temperature is 22 °C. The absorptivity of the surface α is 0.6. Estimate the surface temperature of the plate.

4.92 An insulated frame wall of a house has an average thermal conductivity of 0.0318 Btu/f ft R. The wall is 6 in. thick and it has an area of 160 ft². The inside air temperature is 70°F and the heat transfer coefficient for convection between the inside air and the wall is 1.5 Btu/h ft²°R. On the outside of the wall, the convection heat transfer coefficient is 6 Btu/h ft²°R and the air temperature is −10°F. Ignoring radiation, determine the rate of heat transfer through the wall at steady-state in Btu/h.

4.93 A small rigid cylindrical vessel contains 1 kg of humid air at 80 °C under a total pressure of 1 bar. The humid air inside the vessel contains 0.150 kg of water vapor. Determine the relative humidity of the air in the container.

4.94 A small rigid cylindrical vessel contains 1 kg of humid air at 75 °C under a total pressure of 1 bar. The humid air inside the vessel contains 0.10 kg of water vapor. Determine the relative humidity of the air in the container.

4.95 A water heating system delivers 230 kW with a temperature difference of 20 ° C into circulating water. Estimate the volumetric mass flow rate of the water.

4.96 Heat is added at a rate of 100 J/s to a material whose initial temperature is 20 ° C. At this temperature, the material is a solid with a specific heat of 2.4 J/g °C. The melting point of this material is 36 °C, latent heat of fusion is 340 J/g, and the specific heat at liquid phase is 4.1 J/g °C. If the mass of material is 0.5 kg, how long will it take to reach a final temperature of 40 °C?

4.97 A vessel contains ethanol and water in equilibrium. If the composition of the liquid is 35% ethanol, what is the water concentration in the vapor phase and what percent of liquid is present at these conditions?

4.98 A vessel contains ethanol and water in equilibrium. What is the change in mixing enthalpy if the system starts at 90% ethanol vapor and ends at 1% ethanol liquid?

References

1. Bird RB, Stewart WE, Lightfoot EN (2002) Transport phenomena, 2nd edn. Wiley, New York
2. Himmelblau DM, Riggs JB (2012) Basic principles and calculations in chemical engineering, 8th edn. Prentice Hall, Upper Saddle River
3. Demirel Y (2013) Thermodynamics analysis. The Arabian J Sci Eng 38:221–249

Energy Balances

5

Introduction and Learning Objectives Global conservation law of energy states that the total energy of a system always remains constant. Mass and energy are conserved in a process, while exergy is not conserved. In energy production systems, energy and mass conservations are solved together. By accounting the balance equations, energy interactions have been analyzed in turbines, compressors, nozzles, pumps, valves, mixers, and heat exchangers. Besides these, the Carnot cyclic process is also discussed briefly.

The learning objectives of this chapter are to understand and use:

- Mass, energy, and exergy balance equations,
- Balance equations for analyzing the fluid-flow processes,
- Heat exchangers.

5.1 Balance Equations

According to energy conservation law, the total inflow of energy into a system must be equal to the total outflow of energy from the system. The flow of energy can be in various forms, such as work, heat, electricity, light, and other forms. The total energy of a system can be subdivided and classified into various ways. For example, it is sometimes convenient to distinguish potential energy (which is a function of coordinates) from kinetic energy (which is a function of coordinate time derivatives). It may also be convenient to distinguish gravitational energy, electric energy, and thermal energy from each other.

Figure 5.1 shows an open system with mass and energy interactions with its surroundings. Especially for open systems involving fluid-flow processes, energy balance alone may not fully describe the processes. Therefore, other balance equations of mass, entropy, and exergy may also be required. These balance equations are the main tools of analyses of energy interactions of various processes

© Springer Nature Switzerland AG 2021
Y. Demirel, *Energy*, Green Energy and Technology,
https://doi.org/10.1007/978-3-030-56164-2_5

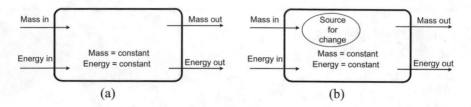

Fig. 5.1 **a** An open steady-state flow process with no internal source of changing mass and energy, **b** an open steady-state flow process with an internal source for change in mass and energy, entropy, and exergy. For example, in an internal chemical reaction in a system, a substance may be consumed, and heat of reaction may be released. So, the chemical reaction acts as source for internal change apart from the boundary interactions of the system with its surroundings

in closed and open systems. Mass and energy are conserved in a process. However, entropy and exergy are not conserved and the balance equations of them account a *source term* (see Fig. 5.1b) [1]. The source term mainly refers to internal change of mass, energy, entropy, and exergy in a system. For example, in an internal chemical reaction in a system, a substance may be consumed, and heat of reaction may be released. So, the chemical reaction acts as a source for internal change apart from the boundary interactions of the system with its surroundings. The conservation equations do not have the source terms and hence are different from the balance equations.

5.2   Mass Balance

In a closed system, mass is constant as the boundary does not permit the transfer of matter between the system and its surroundings. However, mass and energy interactions take place in an open system as seen in Fig. 5.1. The amount of mass flowing through a cross section per unit time is the mass flow rate $\dot{m}$. Here, the dot over the symbol for mass m indicates the amount of mass per unit time. The mass flow rate of a fluid flowing in a pipe is proportional to the cross-sectional area A of the pipe, the density ρ, and the average velocity v of the fluid

$$\dot{m} = vA\rho \tag{5.1}$$

In Eq. (5.1), the fluid flow is assumed to be in a one-dimensional flow and the velocity is assumed to be constant across the cross section. When there is no chemical reaction within a system, *mass balance* for an unsteady-state flow process is

$$\frac{dm}{dt} + \sum_i (\dot{m}_i)_{out} - \sum_i (\dot{m}_i)_{in} = 0 \tag{5.2}$$

As Eq. (5.2) shows that the change of mass with time equals the difference between the total amount of incoming masses and the total amount of outgoing masses. This is known as the *conservation of mass*. This form of mass conservation equation is also called the *continuity equation* [2]. The conservation of mass principle is a fundamental principle in nature. At steady-state flow processes, the accumulation term (dm/dt) becomes zero

$$\sum_i (\dot{m}_i)_{\text{out}} - \sum_i (\dot{m}_i)_{\text{in}} = 0 \qquad (5.3)$$

for example, in a pump, the amount of water entering the pump is equal to the amount of water leaving it at a steady-state pumping process.

Many engineering devices, such as pumps, nozzles, compressors, and turbines, involve a single fluid flow. If we denote the flow at the inlet state by the subscript 1 and the flow at the exit state by the subscript 2, as seen in for a pump in Fig. 5.2a, for a single fluid-flow system mass balance is

$$\dot{m}_1 = \dot{m}_2 \quad \text{or} \quad \rho_1 v_1 A_1 = \rho_2 v_2 A_2 \qquad (5.4)$$

For a mixer chamber shown in Fig. 5.2b, with two inlet flow and one outlet flow, mass balance becomes

$$\dot{m}_1 + \dot{m}_2 = \dot{m}_3 \qquad (5.5)$$

Using the mole flow rate $\dot{n}$, the steady-state balance equation is

$$\sum_i (\dot{n}_i)_{\text{out}} - \sum_i (\dot{n}_i)_{\text{in}} = 0 \qquad (5.6)$$

The volume of the fluid flowing in a pipe or a duct per unit time is known as the *volumetric flow rate* $\dot{Q}$ in m^3/s or ft^3/h

$$\dot{Q} = vA \qquad (5.7)$$

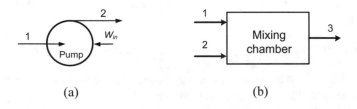

(a)   (b)

Fig. 5.2 Input and output flows in a pump and mixing chamber

where v is the average velocity of the flow and A is the cross-sectional area of a pipe or duct. The mass and volume flow rates are related by

$$\dot{m} = \rho \dot{Q} \tag{5.8}$$

For *incompressible fluids* such as liquids, the density remains constant. The conservation of mass for incompressible fluid flows is

$$\sum_i (\dot{Q}_i)_{\text{out}} - \sum_i (\dot{Q}_i)_{\text{in}} = 0 \tag{5.9}$$

The conservation of mass relations can be simplified for a single incompressible fluid flow by

$$\dot{Q}_1 = \dot{Q}_2 \quad \text{or} \quad v_1 A_1 = v_2 A_2 \tag{5.10}$$

Liquids are mostly incompressible fluids.

5.3   Energy Balance

In open and closed systems, energy can cross the boundary in the form of heat q and work W. If the energy exchange is because of temperature difference between a system and its surroundings, this energy exchange appears as heat; otherwise, it appears as work. When a force acts on a system through a distance, then energy is transferred as work. The heat and work interactions change the internal energy of a closed system ΔU

$$\Delta U = q + W \tag{5.11}$$

According to the modern energy sign convention recommended by the International Union of Pure and Applied Chemistry, heat and work are considered positive for their transfers into the system from their surroundings. Heat and work are negative when they are transferred out of the system to the surroundings. Total internal energy is an extensive property as its value depends on the quantity of material in a system.

Finite exchanges of energy between a system and its surroundings obey the energy balance

$$\Delta(\text{Energy of the system}) + \Delta(\text{Energy of the surroundings}) = 0 \tag{5.12}$$

Figure 5.1 shows the mass and energy interactions between a system and its surroundings. As the energy is conserved, the net rate of change of energy within the system is equal to the net rate of energy exchanged between the system and its surroundings at steady state. Streams in various numbers may flow into and out of the system. These streams may have energy associated with them in the forms of internal, kinetic, and potential energies. All these forms of energy contribute to the energy change of the system [1].

5.3.1 Unsteady-State Flow Systems

When there is no chemical reaction taking place in a system, the general energy balance for an unsteady-state open system is

$$\frac{d(mU)}{dt} + \dot{m}\left(\Delta H + \frac{1}{2}\Delta v^2\left(\frac{kJ/kg}{1000\,m^2/s^2}\right) + \Delta zg\right) = \dot{q} + \dot{W}_s \qquad (5.13)$$

where U is the internal energy, H is the enthalpy, v is the average velocity, z is the elevation above a datum level, and g is the acceleration of gravity. The work $\dot{W}_s$ denotes the work other than the pressure work (PV) and called the shaft work. Each unit of mass flow transports energy at a rate $\dot{m}[\Delta H + (1/2)\Delta v^2\left(\frac{kJ/kg}{1000\,m^2/s^2}\right) + \Delta zg]$.

5.3.2 Steady-State Flow Systems

For *steady-state flow processes*, the accumulation term is zero (d(mU)/dt = 0) and the general energy balance at steady-state fluid-flow process becomes

$$\dot{m}\left(\Delta H + \frac{\Delta v^2}{2}\left(\frac{kJ/kg}{1000\,m^2/s^2}\right) + g\Delta z\right) = \dot{q} + \dot{W}_s \qquad (5.14)$$

For the English engineering systems of units, we have

$$\dot{m}\left(\Delta H + \frac{\Delta v^2}{2g_c}\left(\frac{kJ/kg}{1000\,m^2/s^2}\right) + \frac{g}{g_c}\Delta z\right) = \dot{q} + \dot{W}_s \qquad (5.15)$$

If the kinetic and potential energy changes are small enough, then Eq. (5.15) reduces to

$$\sum_i (\dot{m}_i H_i)_{out} - \sum_i (\dot{m}_i H_i)_{in} = \dot{q} + \dot{W}_s \qquad (5.16)$$

Equation (5.15) is widely applicable to many thermal engineering systems. For a single fluid-flow process, Eq. (5.16) becomes

$$\dot{m}\Delta H = \dot{q} + \dot{W}_s \qquad (5.17)$$

When the energy is expressed per number of moles n, such as kJ/k mole or Btu/lb mole, the energy balance given in Eq. (5.16) becomes

$$\sum_i (\dot{n}_i H_i)_{out} - \sum_i (\dot{n}_i H_i)_{in} = \dot{q} + \dot{W}_s \qquad (5.18)$$

Consider an adiabatic, steady-state, and one-dimensional flow of a compressible fluid in a duct or in a pipe. If there is no shaft work and the changes in potential and kinetic energies are negligible, then Eq. (5.17) reduces to

$$\Delta H = 0 \tag{5.19}$$

All energy types in Eqs. (5.11)–(5.19) require the same energy units. In the SI system, the unit for energy is the joule. Other units are the Btu, ft lb$_f$, and calorie.

Mass and energy are the conserved properties, and they cannot be created or destroyed. However, mass and energy can be converted to each other according to the well-known equation proposed by Einstein

$$E = mc^2 \tag{5.20}$$

where c is the speed of light. Equation (5.20) suggests that the mass of a system is related to its energy. However, for all energy interactions taking place in practice, except the nuclear reactions, the change of mass is extremely small, hence negligible.

Example 5.1 Energy balance in a closed system
A 60 m^3 room is heated by an electrical heater. The room air originally is at 12 °C and 100 kPa. The room loses heat at a rate of 0.2 kJ/s. The air has a constant heat capacity of $C_{v,av} = 0.72$ kJ/kg °C. If the electrical heater supplies energy at a rate of 3 kW, estimate the time necessary for the room temperature to reach 22 °C.

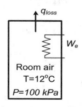

Solution:
Assumptions: The air is ideal gas. This is a closed system of air in the room.
$C_{v,av} = 0.72$ kJ/kg °C, $R = 8.314$ Pa m^3/mol K, $MW_{air} = 29$ kg/kmol
$P_1 = 100$ kPa, $T_1 = 12$ °C (285 K), $V_{room} = 60$ m^3, $T_2 = 22$ °C (295 K)
Electrical energy supplied: $W_e = 3$ kW, $q_{loss} = 0.2$ kW,
$q_{net} = (3 - 0.2)$ kW $= 2.8$ kW
The amount of mass of air in the room:

$$m_{air} = MW(\text{moles of air}) = MW \frac{P_1 V_{room}}{RT_1} = 73.4 \text{ kg}$$

Energy balance: $q_{net}\Delta t = (m\Delta U)_{air} = m_{air}C_{v,av}(T_2 - T_1)$
$(2.8 \text{ kW})(\Delta t) = (73.4 \text{ kg})(0.72 \text{ kJ/kg °C})(22 - 12) \text{ °C}$
$\Delta t = 189$ s $= $ **3.1 min**

5.4 Exergy Balance

The exergy change ΔEx of a closed system is [1, 4]

$$\Delta Ex = \Delta H - T_o \Delta S \tag{5.21}$$

Here, the Δ denotes the difference between the initial and final states (final state–initial state), and T_o is the surrounding (dead state) temperature. Exergy change of a fluid flow in an open system is

$$\Delta Ex = \Delta H - T_o \Delta S + \frac{\Delta v^2}{2} \left(\frac{\text{kJ/kg}}{1000\,\text{m}^2/\text{s}^2} \right) + g\Delta z \tag{5.22}$$

Exergy is an extensive property and a thermodynamic potential. Exergy is not conserved and always lost during a process. The exergy balance states that the exergy change during a process in a system is equal to the difference between the net exergy transfer and the exergy lost during a process within the system because of irreversibilities. The exergy balance is

$$\Delta \dot{Ex} = \dot{Ex}_{in} - \dot{Ex}_{out} - \dot{Ex}_{loss} \tag{5.23}$$

$$\text{Change in exergy} = \text{exergy in} - \text{exergy out} - \text{exergy loss} \tag{5.24}$$

Exergy loss is positive in an irreversible process and vanishes in a reversible process. The rate of loss exergy $\dot{Ex}_{loss}$ represents the overall imperfections and is directly proportional to the rate of entropy production due to irreversibilities in a process. The rate of exergy loss is

$$\dot{Ex}_{loss} = T_o \dot{S}_{prod} \tag{5.25}$$

This equation shows that the rate of exergy transferred into the control volume must exceed the rate of exergy transferred out, and the difference is the exergy lost due to irreversibilities [4].

Example 5.2 Exergy loss estimations
In a mixer, we mix a saturated steam (stream 1) at 110 °C with a superheated steam (stream 2) at 1000 kPa and 300 °C. The saturated steam enters the mixer at a flow rate 1.5 kg/s. The product mixture (stream 3) from the mixer is at 350 kPa and 240 °C. The mixer operates at adiabatic conditions. Determine the rate of energy dissipation if the surroundings are at 290 K.

Solution:
Assume that the kinetic and potential energy effects are negligible, this is a steady adiabatic process, and there are no work interactions.
The properties of steam from the steam tables:

Stream 1: Saturated steam: $T_1 = 110\ °C$, $H_1 = 2691.3\ kJ/kg$, $S_1 = 7.2388\ kJ/kg\ K$
Stream 2: Superheated steam: $P_2 = 1000\ kPa$, $T_2 = 300\ °C$, $H_2 = 3052.1\ kJ/kg$
$S_2 = 7.1251\ kJ/kg\ K$
Stream 3: Superheated steam: $P_3 = 350\ kPa$, $T_3 = 240\ °C$, $H_3 = 2945.7\ kJ/kg$,
$S_3 = 7.4045\ kJ/kg\ K$
The mass, energy, and entropy balances for the mixer at steady state are:
Mass balance: $\dot{m}_{out} = \dot{m}_{in} \rightarrow \dot{m}_1 + \dot{m}_2 = \dot{m}_3$
Energy balance: $\dot{E}_{out} = \dot{E}_{in} \rightarrow \dot{m}_1 H_1 + \dot{m}_2 H_2 = \dot{m}_3 H_3$
Entropy balance: $\dot{S}_{in} - \dot{S}_{out} + \dot{S}_{prod} = 0 \rightarrow \dot{m}_1 S_1 + \dot{m}_2 S_2 - \dot{m}_3 S_3 + \dot{S}_{prod} = 0$
Combining the mass and energy balances, we estimate the flow rate of the super-
heated steam $\dot{m}_2$:
$$\dot{m}_1 H_1 + \dot{m}_2 H_2 = (\dot{m}_1 + \dot{m}_2) H_3$$
$$\dot{m}_2 = \frac{\dot{m}_1 (H_1 - H_3)}{H_3 - H_2} = \frac{1.5\ kg/s(2691.3 - 2945.7)\ kJ/kg}{(2945.7 - 3052.1)\ kJ/kg} = 3.58\ kg/s$$
The mass flow rate of the warm water $\dot{m}_3$: $\dot{m}_3 = \dot{m}_1 + \dot{m}_2 = 1.5 + 3.58 = 5.08\ kg/s$.
The rate of entropy production for this adiabatic mixing process:
$$\dot{S}_{prod} = \dot{m}_3 S_3 - \dot{m}_1 S_1 - \dot{m}_2 S_2$$
$$\dot{S}_{prod} = 5.08\ kg/s(7.4045) - 1.5\ kg/s(7.2388) - 3.58\ kg/s(7.1251) = 1.25\ kJ/s\ K$$
The energy dissipated because of mixing:
$$\dot{E}_{loss} = T_o \dot{S}_{prod} = 290\ K(1.25\ kJ/s\ K) = \mathbf{362.5\ kW}$$

5.5 Fluid-Flow Processes

Many thermal engineering devices and processes operate under the steady-state
conditions for a long period of time. For example, the processes using pumps,
boilers, turbines, and compressors operate at steady state until the system is shut
down for maintenance, and they are analyzed as steady-state flow processes.
Steady-flow energy equation can be applied to [2, 3]:

- Steam turbine
- Compressor
- Pump
- Nozzle
- Boiler
- Condenser
- Throttling
- Adiabatic mixing.

Some examples of unsteady-flow processes are filling of a liquefied petroleum gas (LPG) cylinder in a bottling plant and discharging LPG to the supply line to the gas oven.

5.5.1 Turbines, Compressors, and Pumps

Figure 5.3 shows a single steady-fluid-flow turbine, compressor, and pump. In a steam power plant, the hot and pressurized fluid passes through a turbine and the work is done against the blades, which are attached to the shaft. As a result, the turbine drives the electric generator and produces work. The work is transferred to the surroundings, and its sign is negative. Heat transfer effects in a turbine are usually negligible [1–3].

Compressors, pumps, and fans increase the pressure of fluids; therefore, a work from the surroundings must be supplied. Pumps use incompressible fluids that are liquids, while compressors and fans use compressible fluids of gases. Fans only slightly increase the pressure of gases. Potential energy changes are usually negligible in all these devices. Kinetic energy and heat effects are usually negligible in turbines.

For a pump, we may neglect the kinetic and potential energies and assume that there is no considerable heat exchange with the surroundings. For turbines, compressors, and pumps, we have the following similar relations for energy balance equation

$$\dot{m}\Delta H = \dot{m}(H_2 - H_1) = \dot{q} + \dot{W} \tag{5.26}$$

For a turbine, the work is transferred to the surroundings and will have a negative sign. However, the work is transferred from the surroundings for compressors and pumps, and the sign of work for them is positive.

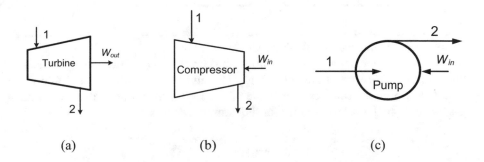

Fig. 5.3 Schematics of **a** turbine, **b** compressor, and **c** pump

Example 5.3 Turbine calculations

A superheated steam (stream 1) expands in a turbine from 5000 kPa and 325 °C to
150 kPa and 200 °C. The steam flow rate is 10.5 kg/s. If the turbine generates
1.1 MW of power, determine the heat loss to the surroundings.

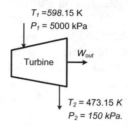

Solution:

Assume that the kinetic and potential energy effects are negligible; this is a steady
process.

The properties of steam from the steam tables:

Stream 1: Superheated steam: $P_1 = 5000$ kPa, $T_1 = 325$ °C (598.15 K),
$H_1 = 3001.8$ kJ/kg

Stream 2: Superheated steam: $P_2 = 150$ kPa, $T_2 = 200$ °C (473.15 K),
$H_2 = 2872.9$ kJ/kg

$\dot{W}_{out} = -1100$ kW $= -1100$ kJ/s

The mass and energy balances for the turbine are:

Mass balance: $\dot{m}_{out} = \dot{m}_{in}$

Energy balance: $\dot{E}_{out} = \dot{E}_{in} \rightarrow \dot{m}_1 (H_2 - H_1) = \dot{q}_{out} + \dot{W}_{out}$

Heat loss from the energy balance

$\dot{q}_{out} = -\dot{W}_{out} + \dot{m}_1 (H_2 - H_1) = 1100\,\text{kJ/s} + (10.5\,\text{kg/s})(2872.9 - 3001.8)\,\text{kJ/kg}$
$= \mathbf{-253.45}$ **kJ/s**

The sign is negative as the heat is lost from the system.

Example 5.4 Compressor calculations

Air enters a compressor at 100 kPa, 300 K, and a velocity of 2 m/s through a feed
line with a cross-sectional area of 1.0 m². The effluent is at 500 kPa and 400 K.
Heat is lost from the compressor at a rate of 5.2 kW. If the air behaves as an ideal
gas, what is the power requirement of the compressor in kW.

$P_1 = 100$ kPa
1 $T_1 = 300$ K

Ws

Compressor

$T_2 = 400$ K
2 $P_2 = 500$ kPa

Solution:
Assume: The compressor is at steady state, air is ideal gas, and the change in the potential and kinetic energy of the fluid from the inlet to the outlet is negligible.
$R = 8.314$ Pa m^3/k mol K, $MW = 29$ kg/k mol; $v_1 = 2$ m/s, $A_1 = 1.0$ m^2
$P_1 = 100$ kPa, $T_1 = 300$ K, $P_2 = 500$ kPa, $T_2 = 400$ K, $q_{loss} = -5.2$ kW
For ideal gas: $V_1 = \dfrac{RT_1}{P_1} = 0.025$ m^3/mol,

the specific volume $(1/\rho)_{air} = \dfrac{V_1}{MW_{air}} = 0.86$ m^3/kg

The mass flow rate of air: $\dot{m} = \dfrac{v_1 A_1}{(1/\rho)} = 1.72$ kg/s

From Table D.1 for ideal gas of air: $H_2 = 400.9$ kJ/kg, $H_1 = 300.2$ kJ/kg
Energy balance: $\dot{m}(H_2 - H_1) - \dot{q} = \dot{W}$
$(1.72 \text{ kg/s})((400.9 - 300.2) \text{ kJ/kg}) + 5.2 \text{ kW} = \dot{W} = \mathbf{178.4 \text{ kW}}$
This is the work needed by the compressor.

Example 5.5 Calculation of shaft work during air compression
An air compressor compresses air from 100 to 600 kPa, whereupon the internal energy of air has increased by 80 kJ/kg. The air flow rate is 9000 kg/h, and specific volumes of air at the inlet and outlet are 0.95 m^3/kg and 0.19 m^3/kg, respectively. The jacket cooling water removes 100 kW of heat generated due to the compression. Estimate the rate of the shaft work input to the compressor.

Solution:
Assume that the air is ideal gas and kinetic and potential energy changes are negligible.
From the first law of thermodynamics: $\Delta U_1 + P_1 V_1 + q = U_2 + P_2 V_2 + W_c$
$\dot{m} = 9000$ kg/h $= (9000 \text{ kg/h}) \times (\text{h}/3600 \text{ s}) = 2.5$ kg/s
$2.5 \text{ kg/s}[(-90 \text{ kJ/kg}) + (100 \times 0.95 - 600 \times 0.19) \text{ kPa m}^3/\text{kg}] - 100 \text{ kJ/s} = W_c$
$2.5 \text{ kg/s}[(-90 \text{ kJ/kg}) + (-19) \text{ kPa m}^3/\text{kg}] - 100 \text{ kJ/s} = W_c$
The shaft work to the compressor is $\mathbf{-372.5 \text{ kW} = W_c}$

Example 5.6 Pump power calculation
A pump increases the pressure in liquid water from 100 kPa at 25 °C to 4000 kPa. Estimate the minimum horsepower motor required to drive the pump for a flow rate of 0.01 m^3/s.

Solution:
Assume that the pump operates adiabatically and nearly isothermally, changes in potential and kinetic energy are negligible, and water is incompressible.
$P_1 = 100$ kPa, $T_1 = 298.15$ K, $P_2 = 4000$ kPa, $Q = 0.01$ m^3/s
Energy balance: $\dot{q} + \dot{W}_{pump} = \dot{m}\Delta H$
For an adiabatic operation ($\dot{q} = 0$): $\dot{W}_{pump} = \dot{m}\Delta H$

Since the water is incompressible, the internal energy is only the function of temperature.

$\Delta U = 0$ as the system is isothermal.

$\Delta H = \Delta U + \Delta(PV) = V\Delta P$

The specific volume of saturated liquid water at 25 °C: $V_1 = 0.001003$ m^3/kg (steam tables).

The mass flow rate: $\dot{m} = \dfrac{Q}{V_1} = 9.97$ kg/s

$\dot{W}_{pump} = \dot{m}V_1(P_2 - P_1) = (9.97$ kg/s$)(0.001003$ m^3/kg$)(4000 - 100)$ kPa
$= \mathbf{39 \ kJ/s = 39 \ kW = 52.3 \ hP}$ (1 hp = 745.7 W).

Example 5.7 Pump work calculations

A centrifugal pump delivers 1.8 m^3/h water to an overhead tank 20 m above the eye of the pump impeller. Estimate the work required by the pump, assuming a steady frictionless flow. Take density of water = 1000 kg/m^3.

Solution:
Assume that the pump operates adiabatically and nearly isothermally, changes in potential and kinetic energy are negligible, and water is incompressible.

Data: $Q = 1.8$ m^3/h and density of water: $\rho = 1000$ m^3/h

Energy balance: $\Delta\left(H + \left(\dfrac{\text{kJ/kg}}{1000 \, \text{m}^2/\text{s}^2}\right)\frac{1}{2}v^2 + zg\right)\dot{m} = \dot{q} + \dot{W}_s$

For an adiabatic operation ($\dot{q} = 0$) with no kinetic and potential energy effects:

$\dot{W}_{pump} = \dot{m}g\Delta z = \left(\dfrac{1.8 \, \text{m}^3/\text{h}}{3600 \, \text{s/h}}\right)(1000) \, \text{kg/m}^3(9.81 \, \text{m/s})(30 \, \text{m}) = 147.15 \, \text{Nm/s}$

$= \mathbf{147.15 \ W}$

5.5.2 Nozzles and Diffusers

Nozzles and diffusers are widely used in manufacturing sector including in jet engines. A converging nozzle increases the velocity of a fluid, while the pressure of the fluid is reduced. A diffuser (diverging duct) increases the pressure of a fluid, while the velocity of the fluid is reduced. In nozzles and diffusers, the rate of heat interactions between the system and its surroundings is usually very small, and process may be analyzed as adiabatic ($q = 0$). Nozzles and diffusers generally involve no work, and changes in potential energy are negligible; therefore, the analyses of such processes may require kinetic energy changes [2, 3]. Figure 5.4a shows schematics of nozzles and diffusers.

For a converging nozzle, the velocity increases; then, we consider the kinetic energy and usually neglect the potential energy. There is no work effect, and energy balance becomes

$$\dot{m}\left(\Delta H + \frac{\Delta v^2}{2}\left(\frac{\text{kJ/kg}}{1000 \, \text{m}^2/\text{s}^2}\right)\right) = \dot{q} \qquad (5.27)$$

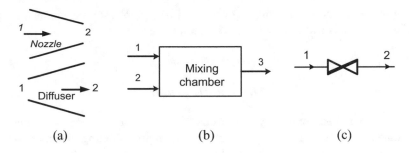

(a)   (b)   (c)

Fig. 5.4 Steady-flow processes: **a** nozzle and diffuser, **b** mixing chamber, **c** throttle valve

---

Example 5.8 Nozzle calculations
In a converging nozzle, steam flow is accelerated from 20 to 200 m/s. The steam enters the nozzle at 3000 kPa and 275 °C and exits at 1700 kPa. The nozzle inlet cross-sectional area is 0.005 m². The nozzle loses heat at a rate of 76 kJ/s. Estimate the steam mass flow rate and exit temperature.

$T_1 = 275°C$
$P_1 = 3000\ kPa$ Steam $P_2 = 1700\ kPa$
$v = 20\ m/s$ Flow $v = 200\ m/s$

Solution:
Assumptions: This is a steady-state flow.
Inlet: $P_1 = 3000$ kPa, $T_1 = 275$ °C, $v_1 = 20$ m/s, $A_1 = 0.005$ m², $\dot{q}_{loss} = -76$ kJ/s
$H_1 = 2928.2$ kJ/kg, $V_1 = 76.078$ cm³/g $= 0.0761$ m³/kg (steam tables)
Outlet: $P_2 = 1700$ kPa, $v_2 = 200$ m/s

The steam mass flow rate: $\dot{m}_s = \dfrac{1}{V_1}A_1v_1 = \mathbf{1.314\ kg/s}$

Energy balance: $\dot{m}_s\left[(H_2 - H_1) + \left(\dfrac{v_2^2}{2} - \dfrac{v_1^2}{2}\right)\left(\dfrac{1\,kJ/kg}{1000\,m^2/s^2}\right)\right] = \dot{q}_{loss}$

Estimate the enthalpy at the exit H_2:

$$(1.314\,kg/s)\left[(H_2 - 2928.2) + \left(\dfrac{200^2}{2} - \dfrac{20^2}{2}\right)\left(\dfrac{1\,kJ/kg}{1000\,m^2/s^2}\right)\right] = -76\,kJ/s$$

$H_2 = 2850.56$ kJ/kg, T_2 (at H_2 and P_2) = **225** °C (approximate)

5.5.3   Mixing Chambers

In thermal engineering applications, mixing two streams in a chamber is very common. For example, mixing of a hot stream with a cold stream produces a warm stream. Conservation of mass and energy principles require that the sum of incoming mass flow rates and enthalpy rates are equal to the mass flow rates and enthalpy rates of the sum of outgoing streams. For example, for the mixing chamber in Fig. 5.4b, mass and energy balances become

$$\dot{m}_1 + \dot{m}_2 = \dot{m}_3 \qquad (5.28)$$

$$\dot{m}_1 H_1 + \dot{m}_2 H_2 = \dot{m}_3 H_3 \qquad (5.29)$$

Mixing chambers are usually well insulated; hence, they are generally adiabatic. The kinetic and potential energy effects are negligible. Mixing chambers do not involve any kind of work production [1].

Example 5.9 Mixing chamber calculations

A hot air at 900 K is mixed with cold air at 300 K in a mixing chamber operating under adiabatic conditions. Flow rate of the cold air is 2.4 kg/s. The outlet stream is warm air at 700 K. Estimate the flow rate of hot air.

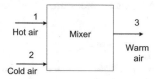

Solution:

Assumptions: This is a steady-state flow process. Kinetic and potential energy changes are negligible. The heat loss from the mixer is negligible. The air streams are ideal gases.

For an ideal gas, enthalpy depends only on the temperature.

Hot air in: $\dot{m}_1$ = unknown; T_1 = 900 K; H_1 = 933.0 kJ/kg (Appendix D Table D1)

Cold air in: $\dot{m}_2$ = 2.4 kg/s; T_1 = 300 K; H_2 = 300.2 kJ/kg (Appendix D Table D1)

Outlet stream: Warm air at 700 K and H_3 = 713.3 kJ/kg (Appendix D Table D1)

Energy and mass balances are: $\dot{m}_1 H_1 + \dot{m}_2 H_2 = \dot{m}_3 H_3$, $\dot{m}_1 + \dot{m}_2 = \dot{m}_3$

Combining and organizing these equations yield the hot air mass flow rate:

$$\dot{m}_1 = \dot{m}_2 \frac{H_3 - H_2}{H_1 - H_3} = (2.4\,\text{kg/s}) \left(\frac{713.3 - 300.2}{933.0 - 713.3} \right) = \textbf{4.51 kg/s}$$

5.5.4 Throttling Valve

In a throttling valve (see Fig. 5.4c), a fluid flows through a restriction, such as an orifice, a partly closed valve, capillary tube, or a porous structure with negligible kinetic energy and potential energy changes. Throttling valve causes a considerable drop in pressure of fluid without involving any work. Throttling valves usually operate under adiabatic conditions [3]. When there is no or negligible heat transfer, then the energy balance equation for the single steady-flow throttling becomes

$$H_1 = H_2 \tag{5.30}$$

Since the values of enthalpy at the inlet and at the outlet are the same, throttling valve is called as an *isenthalpic process*

$$\text{Internal energy} + \text{Pressure energy} = \text{Constant} \tag{5.31}$$

$$(U + PV)_1 = (U + PV)_2 = \text{Constant} \tag{5.32}$$

A throttling process does not change the temperature of an ideal gas since the enthalpy of an ideal gas depends only on temperature. For most real gases, however, pressure drop at constant enthalpy results in temperature decrease. If a wet steam is throttled to a considerably low pressure, the liquid evaporates, and the steam may become superheated. Throttling of a saturated liquid may cause vaporization (or flashing) of some of the liquid, which produces saturated vapor and liquid mixture. For throttling valves with large exposed surface areas such as capillary tubes, heat transfer may be significant.

Example 5.10 Throttling process calculations

(a) A steam at 600 kPa and 200 °C is throttled to 101.32 kPa. Estimate the temperature of the steam after throttling.
(b) A wet steam at 1200 kPa and with a quality of 0.96 is throttled to 101.32 kPa. Estimate the enthalpy and the temperature of the throttled steam.

Solution:
Assume: No heat transfer occurs, and kinetic and potential energy changes are negligible in the throttling process.

Data from steam tables:

(a) Inlet: $P_1 = 600$ kPa, and $T_1 = 200$ °C, H_1 (600 kPa, 200 °C) = 2849.7 kJ/kg

Outlet: $P_2 = 101.32$ kPa
For adiabatic throttling: $H_1 = H_2$
From steam tables: T_2 (101.32 kPa, 2849.7 kJ/kg) = **187.1 °C** (after an interpolation)
The temperature drop ($\Delta T = 2.9$ °C) is small.

(b) Throttling a wet steam to a lower pressure may cause the liquid to evaporate and the vapor to become superheated.

Inlet: The wet steam is saturated at: $P_1 = 1200$ kPa, $T_{sat} = 187.9$ °C, $x = 0.96$ (steam tables).
$H_{1sat\ vap} = 2782.7$ kJ/kg; $H_{1sat\ liq} = 798.4$ kJ/kg
$H_1 = (1 - x)H_{1sat\ liq} + xH_{1sat\ vap} = \textbf{2683.5 kJ/kg}$
Outlet: $P_2 = 101.32$ kPa
T_2 (101.32 kPa, 2683.5 kJ/kg) = **103.6 °C**.
Since $T_2 > T_{2\ sat}$ at 101.32 kPa, the throttled steam is superheated.
The temperature drop ($\Delta T = 84.3$ °C) is considerable because of evaporation of liquid.

Example 5.11 Throttling of a refrigerant
Refrigerant-134a in a refrigerator enters a throttling valve as saturated liquid at 85.8 psia and is throttled to a pressure of 18.8 psia. Estimate the quality of the refrigerant and temperature drop at the exit.

Solution:
Inlet conditions: Saturated liquid (steam tables):
$P_1 = 85.8$ psia and $T_1 = 70$ °F, $H_1 = 33.9$ Btu/lb
Exit conditions: Saturated mixture of liquid and vapor at (steam tables):
$P_{2,sat} = 18.8$ psia and $T_{2,sat} = -5$ °F, $H_{2sat\ liq} = 10.1$ Btu/lb,
$H_{2sat\ vap} = 101.0$ Btu/lb
Throttling process: $H_1 = H_2 = 33.9 = (1 - x) H_{2sat\ liq} + x H_{2sat\ vap}$
Quality of the saturated mixture: $x = \dfrac{m_{vap}}{m_{total}} = \dfrac{33.9 - 10.1}{101 - 10.1} = \textbf{0.25}$
After the throttling process, 25.0% of the saturated liquid is evaporated.
Temperature drop: $\Delta T = T_2 - T_1 = (-5 - 70)$ °F = **−75 °F**.

5.5.5   Heat Exchangers

In shell and tube heat exchangers, hot and cold fluids exchange heat without mixing. Figure 5.5 shows a typical countercurrent shell and tube heat exchanger where hot and cold streams flow in opposite to each other; in the tube side, a cold stream 4 enters and leaves the tube as hot stream 3, while hot stream 1 enters the shell side and leaves as cold stream 2. Heat exchangers may have various designs, such as double pipe and shell and tube, and are widely used in thermal engineering systems [2]. For example, for the heat exchanger shown in Fig. 5.5 mass and energy balances become

$$\dot{m}_1 = \dot{m}_2 \quad \text{and} \quad \dot{m}_3 = \dot{m}_4 \qquad (5.33)$$

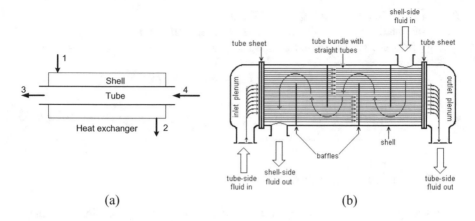

Fig. 5.5 a Schematic of a typical heat exchanger where hot and cold fluid flows do not mix with each other, **b** schematic of a countercurrent shell and tube heat exchanger. Baffles help distribute the shell side fluid around the tube bundle

$$\dot{m}_1(H_2 - H_1) = \dot{m}_3(H_4 - H_3) \tag{5.34}$$

Heat exchangers involve no work interactions and usually operate with negligible kinetic and potential energy changes for hot and cold fluid flows.

Example 5.12 Heat exchanger calculations
A steam is cooled in a condenser using cooling water. The steam enters the condenser at 20 kPa and a quality of 0.9, and exits at 20 kPa as saturated liquid. The steam flow rate is 3 kg/s. The cooling water enters the condenser at 18 °C and exits at 40 °C. Average heat capacity of the cooling water is $C_{p,av} = 4.2$ kJ/kg °C. Estimate the mass flow rate of the cooling water.

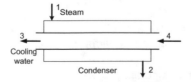

Solution:
Assumptions: This is a steady-state system, and kinetic energy and potential energies are negligible.
Steam flow rate: $\dot{m}_s = 3$ kg/s
Cooling water: $T_4 = 18$ °C, $T_3 = 40$ °C, $C_{p,av} = 4.2$ kJ/kg °C
Steam is saturated liquid and vapor mixture: $P_1 = 20$ kPa, $x_1 = 0.90$
$H_{1\text{sat liq}} = 251.4$ kJ/kg, $H_{1\text{sat vap}} = 2609.7$ kJ/kg (steam tables)

$H_1 = (1 - x_1)H_{1\text{sat liq}} + xH_{1\text{sat vap}} = 2373.9$ kJ/kg

Exit: Saturated liquid at $P_2 = 20$ kPa, $H_2 = 251.4$ kJ/kg (steam tables)

Energy balance: Heat removed by the cooling water = heat lost by the steam

$$\dot{m}_w C_{p,\text{av}}(T_3 - T_4) = \dot{m}_s(H_1 - H_2)$$

$$\dot{m}_w = \dot{m}_s \frac{(H_1 - H_2)}{C_{p,\text{av}}(T_4 - T_3)} = (3 \text{ kg/s})\left(\frac{2373.9 - 251.4}{(4.2 \text{ kJ/kg} \,^\circ\text{C})(40 - 18)\,^\circ\text{C}}\right) = \mathbf{69.24 \text{ kg/s}}$$

5.5.6 Pipe and Duct Flows

The transport of fluids in pipes and ducts has wide applications in many heating/cooling systems including heating, ventilation, and air conditioning (HVAC) systems. Fluids mainly flow through pipes and ducts at steady-state flow conditions. Heat interactions of pipe and duct flows with their surroundings may be considerable, especially for long pipes and ducts. For example, a pipe with a hot fluid flowing inside must be insulated to reduce the heat loss to the surroundings.

The velocities of fluid flows in pipes and ducts are usually low, and the kinetic energy changes may be negligible. This is especially true when the cross-sectional area remains constant and the heat effects are small. Kinetic energy changes, however, may be considerable when the cross-sectional area changes for gases flowing in a duct. The potential energy changes may also be considerable when a fluid undergoes a large elevation change as it flows in a pipe or duct [2, 3].

5.6 Energy Balance in a Cyclic Process

Most of the processes described above are part of some important thermal cyclic process, such as Carnot cycle, heat pump, or refrigeration cycle. These thermal cyclic processes are used for power production as well as for heating, ventilation, and air-conditioning purposes [3, 5].

Fossil fuel-fired power plants mainly consume coal, natural gas, or oil. Figure 5.6 describes the main processes within the cycle. The main processes are:

- Process 4–1: The water is pumped to the boiler.
- Processes 1–2: The high-pressure liquid enters the boiler where it is heated at a constant pressure by an external heat source q_H to become a superheated vapor.
- Processes 2–3: The superheated vapor expands through the turbine and generates power. This decreases the temperature and pressure of the vapor, and some condensation may occur.
- Processes 3–4: The wet vapor then enters the condenser where the condensation of heat is discharged q_C to start a new cycle with the liquid water.

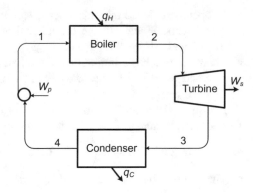

Fig. 5.6 Typical processes in a heat engine cycle to produce work

The process analysis yields the following energy balance equations. Pump power needed

$$W_{p,\text{in}} = \dot{m}V_4(P_1 - P_4) \tag{5.35}$$

where $\dot{m}$ is the steam unit mass flow rate, V_4 is the specific volume at state 4 that is the saturated liquid water, and P is the pressures. Enthalpy rate at state 1, H_1, is

$$H_1 = H_4 + W_{p,\text{in}} \tag{5.36}$$

The heat input q_{in} to the pressurized water in the boiler is

$$q_{\text{in}} = \dot{m}(H_2 - H_1) \tag{5.37}$$

The discharged heat rate q_{out} at the condenser is

$$q_{\text{out}} = \dot{m}(H_4 - H_3) \tag{5.38}$$

The net value of the work, W_{net}, produced is

$$W_{\text{net}} = (q_{\text{in}} - q_{\text{out}}) \tag{5.39}$$

Summary

- According to energy conservation law, the total inflow of energy into a system must be equal to the total outflow of energy from the system. Heat and work are considered positive for their transfers into the system from their surroundings; they are negative when they are transferred out of the system to the surroundings. The flow of energy can be in various forms, such as work, heat, electricity, and light.
- The mass flow rate $\dot{m}$, in terms of the cross-sectional area A, the density ρ, and the average velocity v of the fluid: $\dot{m} = vA\rho$; volumetric flow rate $\dot{Q}$ in m^3/s or ft^3/h: $\dot{Q} = vA$, and $\dot{m} = \rho\dot{Q}$.

- In open and closed systems, energy can cross the boundary in the form of heat q and work W and change the internal energy of a closed system ΔU: $\Delta U = q + W$ and $\Delta H = \Delta U + P\Delta V$.
- There are two basic limitations of the first law: It does not differentiate between heat and work, and it does not permit us to know the direction of energy transfer (heat or work).
- Exergy is the available work potential of the energy processed by a process.
- The exergy change ΔEx of a closed system: $\Delta Ex = \Delta H - T_o \Delta S$.
- Open system: $\Delta \dot{Ex} = \dot{m}\left(\Delta H - T_o \Delta S + \left(\dfrac{\text{kJ/kg}}{1000\,\text{m}^2/\text{s}^2} \right) \Delta v^2/2 + g\Delta z \right)$.
- The exergy balance: $\Delta \dot{Ex} = \dot{Ex}_{\text{in}} - \dot{Ex}_{\text{out}} - \dot{Ex}_{\text{loss}}$. The rate of exergy loss: $\dot{Ex}_{\text{loss}} = T_o \dot{S}_{\text{prod}}$.
- When a system absorbs heat, its internal energy increases, and it can perform an external work (PdV work).
- The net energy transfer by heat, work, and mass is equal to the sum of the change in internal, kinetic, potential energies.
- Enthalpy and internal energy of an ideal gas depend only on the temperature.
- When a system executes a reversible process, the exergy destruction is zero and the reversible work and useful work become identical.

Some equations developed in this chapter:

Closed system $\dot{m} = 0$; $\dot{n} = 0$; $\dot{Q} = 0$	$\frac{dm}{dt} = 0$; $\frac{dn}{dt} = 0$; $\frac{dQ}{dt} = 0$
Open steady-state single fluid flow	$\dot{m}_1 = \dot{m}_2$; $\rho_1 v_1 A_1 = \rho_2 v_2 A_2$
Incompressible fluid flow ρ = constant	$\dot{Q}_1 = \dot{Q}_2$; $v_1 A_1 = v_2 A_2$
Steady-state flow processes	$\dot{m}\left(\Delta H + (1/2), \left(\dfrac{\text{kJ/kg}}{1000\,\text{m}^2/\text{s}^2} \right) \Delta v^2 + \Delta zg \right) = \dot{q} + \dot{W}_s$
With negligible kinetic and potential energies	$\sum_i (\dot{m}_i H_i)_{\text{out}} - \sum_i (\dot{m}_i H_i)_{\text{in}} = \dot{q} + \dot{W}_s$
Steady-state single flow	$\dot{m}\Delta H = \dot{q} + \dot{W}_s$
Isenthalpic process	$\dot{m}\Delta H = 0$
Adiabatic process $\dot{q} = 0$	$\dot{m}\Delta H = \dot{W}_s$
Processes with no work interactions	$\dot{m}\Delta H = \dot{q}_s$
Closed system $\dot{m} = 0$	$\Delta U = q + W$

Problems

5.1 Develop a simple energy balance formulation for a steady-state flow operation: (a) pump; (b) gas turbine, (c) gas compressor, (d) nozzle, (e) throttle valve.

5.2 A copper ingot of 1 kg is heat treated at 400 °C and then quenched in a 50 L oil bath initially at 15 °C. Assuming no heat exchange with the surroundings, determine the temperature of the ingot and the oil when they reach thermal equilibrium using the data: $C_{v,copper} = 0.386$ kJ/kg K, $C_{v,oil} = 1.8$ kJ/kg K, $\rho_{copper} = 8900$ kg/m^3, $\rho_{oil} = 910$ kg/m^3.

5.3   A well-insulated 30-m^3 tank is used to store exhaust steam. The tank contains 0.01 m^3 of liquid water at 30 °C in equilibrium with the water vapor. Estimate the internal energy of the tank.

5.4   A superheated steam at a rate of 0.6 lb/s flows through a heater. The steam is at 100 psia and 380 °F. If an electrical heater supplies 37.7 Btu into the steam at constant pressure, estimate the final temperature of the steam.

5.5   A tank of 4 ft^3 contains superheated steam of 1 lb and 120 psia. The tank is heated at constant pressure. The amount of heat added is 200 Btu. Estimate the temperature of the steam after the heating process.

5.6   A 60 m^3 room is heated by an electrical heater. The room air originally is at 12 °C and 100 kPa. The room loses heat at a rate of 0.2 kJ/s. If the electrical heater supplies 3 kW, estimate the time necessary for the room temperature to reach 22 °C.

5.7   A 1-kW heater is used to heat a room with dimensions (3 m) (4.5 m) (4 m). Heat loss from the room is 0.15 kW, and the pressure is always atmospheric. The air in the room may be assumed to be an ideal gas. The air has a constant heat capacity of $C_{v,av} = 0.72$ kJ/kg °C. Initially, the room temperature is 290 K. Determine the temperature in the room after 10 min.

5.8   A tank filled with 10 lb of air is heated. The air originally is at 40 psia and 100 °F. After heating, the air pressure becomes 60 psia. Estimate the amount of heat transferred. Heat capacity for the air is 0.171 Btu/lb °F and remains constant during the heating process.

5.9   A 2.0 m^3 tank filled with hydrogen (H$_2$) at 200 kPa and 500 K. The hydrogen is cooled until its temperature drops to 300 K. Estimate the amount of heat removed from the tank. Heat capacity for the hydrogen is $C_{v,av} = 10.4$ kJ/kg °C and remains constant during the heating process.

5.10   An electrical resistance heater is used to heat a room from 17 to 22 °C within 30 min. The room has the dimensions of 4 × 4 × 5 m^3. The air in the room is at atmospheric pressure. The heat loss from the room is negligible. Estimate the power needed for this temperature change.

5.11   A room with dimensions of (5 m) (5 m) (6 m) is heated with a radiator supplying 8000 kJ/h. Heat loss from the room is 800 kJ/h. The air is originally at 15 °C. Estimate the time necessary to heat the room to 24 °C.

5.12   A 2-kW heater is used to heat a room with dimensions (3 m) (4.5 m) (4 m). Heat loss from the room is negligible, and the pressure is always atmospheric. The air in the room may be assumed to be an ideal gas, and the heat capacity is 29 J/mol K. Initially, the room temperature is 290 K. Determine the rate of temperature increase in the room.

5.13   A resistance heater within a well-insulated tank is used to heat 20 kg of water. The heater passes a current of 0.4 A from a 120-V source. The water in the tank is originally at 10 °C. Estimate the time necessary for the temperature of the water to reach 60 °C.

5.14 A resistance heater within a tank is used to heat 5 kg of water. The heater passes a current of 0.8 A for 2 h from a 120-V source. A heat loss of 15.6 kW occurs from the tank to surroundings. The water in the tank is originally at 30 °C. Estimate the temperature of the water after 2 h of heating.

5.15 A resistance heater within a tank is used to heat 150 kg of water. The heater passes a current of 0.6 A for 8 h from a 120-V source. A heat loss of 38.2 kW occurs from the tank to surroundings. The water in the tank is originally at 5 °C. Estimate the temperature of the water after 8 h of heating.

5.16 A resistance heater within a well-insulated 120 m^3 room is used to heat the inside air. The heater passes a current of 2 A from a 120-V source. The air originally is at 15 °C and 100 kPa. Estimate the time necessary for the inside air temperature to reach 21 °C.

5.17 A resistance heater within a well-insulated 85 m^3 room is used to heat the inside air. The heater passes a current of 2 A for 3 h from a 120-V source. The air originally is at 20 °C and 100 kPa. Estimate the temperature of the air after 3 h of heating.

5.18 A resistance heater within a well-insulated 105 m^3 room is used to heat the inside air. The heater passes a current of 0.8 A for 2 h from a 120-V source. The air originally is at 17 °C and 100 kPa. The heat loss from the room is 29.0 kJ. Estimate the temperature of the air after 2 h of heating.

5.19 A condenser in a power plant is used to condense the discharged steam from the turbine by using cooling water. The temperature rise of the cooling water is 12 °C. The discharged steam enters the condenser at 10 kPa with a quality of 0.95 and exits as saturated liquid water at constant pressure. If the steam flow rate is 5 kg/s, estimate the flow rate of cooling water.

5.20 A condenser in a power plant is used to condense the discharged steam from the turbine by using cooling water at a flow rate of 160 kg/s. Inlet temperature of the cooling water is 15 °C. The discharged steam enters the condenser at 10 kPa with a quality of 0.95 and exits as saturated liquid water at constant pressure. If the discharged steam flow rate is 6.5 kg/s, estimate the exit temperature of the cooling water.

5.21 A 20 lb refrigerant-134a flowing in a coil is cooled by an air flow. The refrigerant enters the coil at 140 °F and 140 psia and leaves as saturated vapor at 140 psia. The air enters the heat exchanger at 14.7 psia and 480 R. Volumetric flow rate of the air is 150 ft^3/s. Estimate the exit temperature of the air.

5.22 A 5 lb refrigerant-134a flowing in a coil is cooled by an air flow. The refrigerant enters the coil at 200 °F and 100 psia and leaves as saturated vapor at 100 psia. The air enters the heat exchanger at 14.7 psia and 480 R. Volumetric flow rate of the air is 250 ft^3/s. Estimate the exit temperature of the air.

5.23 Refrigerant-134a is cooled in a condenser using cooling water. The refrigerant enters the condenser at 1000 kPa and 90 °C and exits at 1000 kPa and 30 °C. The cooling water at a flow rate of 2.0 kg/s enters the condenser at 20 °C and exits at 27 °C. Estimate the mass flow rate of the refrigerant.

5.24 A steam is cooled in a condenser using cooling water. The steam enters the condenser at 20 kPa and a quality of 0.9 and exits at 20 kPa as saturated liquid. The steam flow rate is 3 kg/s. The cooling water enters the condenser at 18 °C and exits at 26 °C. Estimate the mass flow rate of the cooling water.

5.25 A compressed liquid water at 20 °C and a superheated steam at 150 °C and 400 kPa are fed into a mixing chamber operating under adiabatic conditions. Flow rate of the compressed liquid water is 12 kg/s. The outlet stream is saturated liquid water at 400 kPa. Estimate the flow rate of superheated steam.

5.26 A superheated steam at a flow rate of 50 kg/s and at 150 °C, 200 kPa is mixed with another superheated steam at 350 °C and 200 kPa in a mixing chamber operating under adiabatic conditions. The outlet stream is a superheated steam at 225 °C and 200 kPa. Estimate the flow rate of superheated steam at 350 °C and 200 kPa.

5.27 A hot water at 80 °C is mixed with cold water at 20 °C in a mixing chamber operating under adiabatic conditions. Flow rate of the cold water is 15.0 kg/s. The outlet stream is warm water at 45 °C and 100 kPa. Estimate the flow rate of hot water.

5.28 A hot air at 900 R is mixed with cold air at 520 R in a mixing chamber operating under adiabatic conditions. Flow rate of the cold air is 24 lb mol/s. The outlet stream is warm air at 700 R. Estimate the flow rate of hot air.

5.29 A superheated steam at 300 °C and 300 kPa is mixed with liquid water with a flow rate of 4 kg/s and at 20 °C in a mixing chamber operating under adiabatic conditions. The outlet stream is at 60 °C and 300 kPa. Estimate the flow rate of superheated steam.

5.30 An open feed water heater in a steam power plant operates at steady state with liquid entering at $T_1 = 40$ °C and $P_1 = 7$ bar. Water vapor at $T_2 = 200$ °C and $P_2 = 7$ bar enters in a second feed stream. The effluent is saturated liquid water at $P_3 = 7$ bar. Estimate the ratio of the mass flow rates of the two feed streams.

5.31 Initially, the mixing tank shown below has 100 kg of water at 25 °C. Later, two other water inlet streams 1 and 2 add water and outlet stream 3 discharges water. The water in the tank is well mixed, and the temperature remains uniform and equal to the temperature of outlet stream 3. Stream 1 has a flow rate of 20 kg/h and is at 60 °C, while stream 2 has a flow rate of 15 kg/h and is at 40 °C. The outlet stream has a flow rate of 35 kg/h. Determine the time-dependent temperature of the water in the mixing tank.

5.32   In a mixer, we mix a saturated steam (stream 1) at 110 °C with a super-heated steam (stream 2) at 1000 kPa and 300 °C. The saturated steam enters the mixer at a flow rate 1.5 kg/s. The product mixture (stream 3) from the mixer is at 350 kPa and 240 °C. The mixer loses heat at a rate 2 kW. Determine the rate of energy dissipation if the surroundings are at 300 K.

5.33   In a mixer, we mix liquid water (1) at 1 atm and 25 °C with a superheated steam (2) at 325 kPa and 200 °C. The liquid water enters the mixer at a flow rate of 70 kg/h. The product mixture (3) from the mixer is at 1 atm and 55 ° C. The mixer loses heat at a rate of 3000 kJ/h. Estimate the flow rate of superheated steam and the rate of entropy production.

5.34   In a steady-state mixing process, 50.25 kg/s of saturated steam (stream 1) at 501.15 K is mixed with 7.363 kg/s of saturated steam (stream 2) at 401.15 K. The mixer is well insulated and adiabatic. Determine the energy dissipation (work loss) if the surroundings are at 298.15 K.

5.35   In a steady-state mixing process, 50.0 kg/s of saturated steam (stream 1) at 501.15 K is mixed with 17.0 kg/s of saturated steam (stream 2) at 423.15 K. The product steam (stream 3) is at 473.15 K. Determine the rate of heat loss.

5.36   In a steady-state mixing process, 15 kmol/s of air (stream 1) at 550 K and 2 atm is mixed with 40 kmol/s of air (stream 2) at 350 K and 1 atm. The product (stream 3) is at 300 K and 1 atm. Determine the heat loss.

5.37   A steam turbine produces 3.80 MW. The steam at the inlet is at 1000 psia and 900 °F. The steam at the exits is saturated at 12 psia. Estimate the heat loss if the steam flow rate is 12.8 lb/s.

5.38   A steam turbine produces 3.80 MW. The steam at the inlet is at 900 psia and 900 °F. The steam at the exits is saturated at 5 psia. Estimate the heat loss if the steam flow rate is 16.2 lb/s.

5.39   A steam turbine consumes 4000 lb/h steam at 540 psia and 800 °F. The exhausted steam is at 165 psia. The turbine operation is adiabatic.

  (a) Determine the exit temperature of the steam and the work produced by the turbine.
  (b) Determine the entropy production if the turbine efficiency is 80%.

5.40   A superheated steam (stream 1) expands in a turbine from 5000 kPa and 325 °C to 150 kPa and 200 °C. The steam flow rate is 10.5 kg/s. If the turbine generates 1.1 MW of power, determine the heat loss to the surroundings.

5.41   A superheated steam (stream 1) expands in a turbine from 5000 kPa and 325 °C to 150 kPa and 200 °C. The steam flow rate is 10.5 kg/s. If the turbine generates 1.1 MW of power, determine the heat loss to the surroundings.

5.42   Steam expands in a turbine from 6600 kPa and 300 °C to a saturated vapor at 1 atm. The steam flow rate is 9.55 kg/s. If the turbines' power output is 1 MW, estimate the rate of heat loss.

5.43   Steam expands adiabatically in a turbine from 850 psia and 600 °F to a wet vapor at 12 psia with a quality of 0.9. Estimate the steam flow rate if the turbines' power output is 2 MW.

5.44   Steam expands adiabatically in a turbine from 800 psia and 650 °F to a wet vapor at 10 psia with a quality of 0.95. Estimate the steam flow rate if the turbines' power output is 2 MW.

5.45   Air enters an insulated compressor operating at steady state at 1.0 bar, 300 K with a mass flow rate of 3.6 kg/s and exits at 2.76 bar. Kinetic and potential energy effects are negligible. (a) Determine the minimum theoretical power input required, in kW, and the corresponding exit temperature, in K. (b) If the actual exit temperature is 420 K, determine the power input, in kW, and the isentropic compressor efficiency.

5.46   A compressor increases the pressure of carbon dioxide from 100 to 600 kPa. The inlet temperature is 300 K, and the outlet temperature is 400 K. The mass flow rate of carbon dioxide is 0.01 k mol/s. The power required by the compressor is 55 kW. The temperature of the surroundings is 290 K. Determine the minimum amount of work required.

5.47   Hydrogen gas is compressed from an initial state at 100 kPa, 300 K and 5 m^3 to 300 kPa and 370 K. The compression process is polytropic $(PV^\gamma = constant)$. The average heat capacity of hydrogen is $C_{p,av} = 29.1$ J/mol K. Estimate the total entropy change of the system.

5.48   A compressor receives air at 15 psia and 80 °F with a flow rate of 1.2 lb/s. The air exits at 40 psia and 300 °F. At the inlet, the air velocity is low but increases to 250 ft/s at the outlet of the compressor. Estimate the power input of the compressor if it is cooled at a rate of 200 Btu/s.

5.49   In an adiabatic compression operation, air is compressed from 20 °C and 101.32 kPa to 520 kPa. The air flow rate is 22 mol/s. Estimate the work required under isentropic operation.

5.50   Refrigerant-134a with a flow rate of 0.1 lb/s enters a compressor as saturated vapor at $T_1 = 5$ °F ($P_{sat} = 23.8$ psia) and exits a superheated vapor at 60 °F and 40 psia. Estimate the energy transferred during the compression process.

5.51   Air enters a compressor at 100 kPa, 300 K, and a velocity of 6 m/s through a feed line with a cross-sectional area of 3 m^2. The effluent is at 500 kPa and 400 K and has a velocity of 2 m/s. Heat is lost from the compressor at a rate of 160 kW. If the air behaves as an ideal gas, what is the power requirement of the compressor in kW.

5.52   Carbon dioxide at a flow rate of 1.5 kg/s enters a compressor at 300 K and 100 kPa and exits at 480 K. Estimate the volumetric flow rate of CO_2 and the power for compression.

5.53   Propane with a flow rate of 5 lb/s enters a compressor as a saturated vapor at 30 °F and 66.5 psia. The propane leaves the compressor as superheated vapor at 140 psia and 100 °F. Estimate the heat transfer to the propane in the steady-state flow process.

5.54 In a two-stage continuous compression process, methane (stream 1) enters the first compressor at 300 K and 1 bar. The methane (stream 2) leaves the second compressor at 300 K and 60 bar. The flow rate of methane is 0.5 kg/s. The total power input is 400 kW. The intercooler between the compressors uses cooling water. The surroundings are at 295 K. Estimate the minimum work required and the heat removed in the intercooler.

5.55 In a two-stage continuous compression process, methane (stream 1) enters the first compressor at 300 K and 1 bar. The methane (stream 2) leaves the second compressor at 350 K and 80 bar. The flow rate of methane is 0.6 kg/s. The total power input is 450 kW. The intercooler between the compressors uses cooling water. The cooling water enters the cooler of 295 K and leaves at 305 K. The surroundings are at 295 K. Estimate the cooling water rate.

5.56 In an adiabatic compression operation, air is compressed from 20 °C and 101.32 kPa to 520 kPa with an efficiency of 0.7. The air flow rate is 20 mol/s. The air is assumed to be an ideal gas. Estimate the ideal work required.

5.57 Refrigerant-134a enters a compressor operating at steady state as saturated vapor at 2 bar with a volumetric flow rate of 0.019 m^3/s. The refrigerant is compressed to a pressure of 8 bar in an internally reversible process according to $PV^{1.03}$ = constant. Estimate (a) the power required, in kW, and (b) the rate of heat transfer, in kW.

5.58 Air is compressed from 1 bar and 310 K to 8 bar. Estimate the specific work and heat transfer if the air follows a polytropic process path with $\gamma = 1.32$.

5.59 A pump increases the pressure in liquid water from 100 at 25 °C to 2500 kPa. What is the minimum horsepower motor required to drive the pump for a flow rate of 0.005 m^3/s?

5.60 A pump increases the pressure in liquid water from 100 at 25 °C to 3000 kPa. What is the minimum horsepower motor required to drive the pump for a flow rate of 0.25 m^3/s?

5.61 In a converging nozzle, air flow is accelerated from 20 to 60 m/s. The air at the inlet is at 400 kPa, 127 °C. The inlet cross-sectional area of the nozzle is 90 cm^2. At the outlet, the air flow pressure drops to 100 kPa. Estimate the air mass flow rate and exit temperature.

5.62 In a converging nozzle, steam flow is accelerated from 20 to 200 m/s. The steam enters the nozzle at 3000 kPa and 275 °C and exits at 900 kPa. The nozzle inlet cross-sectional area is 0.005 m^2. The nozzle loses heat at a rate of 76 kJ/s. Estimate the steam mass flow rate and exit temperature.

5.63 In a converging nozzle, carbon dioxide (CO_2) flow is accelerated to 400 m/s. The CO_2 enters the nozzle at 1000 kPa and 500 K and exits at 900 kPa. Flow rate of the CO_2 is 5000 kg/h. The nozzle inlet cross-sectional area is 0.005 m^2. Estimate the inlet velocity and exit temperature of the CO_2.

5.64 Steam expands in a nozzle from inlet conditions of 500 °F, 250 psia, and a velocity of 260 ft/s to discharge conditions of 95 psia and a velocity 1500 ft/s. If the flow is at 10 lb/s and the process is at steady state and adiabatic, determine the outlet temperature.

5.65 Air with a flow rate of 1 kg/s enters a nozzle at 400 K and 60 m/s and leaves the nozzle at a velocity 346 m/s. The air inlet pressure is 300 kPa, while the pressure at the outlet is 100 kPa. The rate of heat lost in the nozzle is 2.5 kJ/kg. Determine the enthalpy at the outlet.

5.66 Steam enters a nozzle at 30 psia and 300 °F and exits as a saturated vapor at 300 °F. The steam enters at a velocity of 1467 ft/s and leaves at 75 ft/s. The nozzle has an exit area of 0.5 ft^2. Determine the rate of energy dissipation when the ambient temperature $T_o = 500$ R.

5.67 A steam enters a nozzle at 500 kPa and 220 °C and exits at 400 kPa and 175 °C. The steam enters at a velocity of 200 m/s and leaves at 50 m/s. The nozzle has an exit area of 0.2 m^2. Determine the rate of heat loss to the surroundings.

5.68 A steam enters a nozzle at 4000 kPa and 425 °C with a velocity of 50 m/s. It exits at 286.18 m/s. The nozzle is adiabatic and has an inlet area of 0.005 m^2. Estimate the enthalpy of the outlet stream.

5.69 A steam enters a nozzle at 3200 kPa and 300 °C with a velocity of 20 m/s. It exits at 274.95 m/s. The nozzle is adiabatic and has an inlet area of 0.01 m^2. Estimate the enthalpy at the outlet and the amount of loss of work if the surroundings are at 300 K.

5.70 Steam at 8200 kPa and 500 °C passes through a throttling process so that the pressure is suddenly reduced to 7400 kPa. What is the expected temperature after the throttle?

5.71 5 kg/s of superheated steam at 400 °C and 1100 kPa is throttled to 125 kPa adiabatically through a valve. Determine the outlet temperature and the work loss if the surroundings are at 298.15 K.

5.72 Refrigerant-134a expands adiabatically in an expansion valve. The refrigerant enters at $P_1 = 140$ psia and $T_1 = 140$ °F and exits at 50 psia. The reference conditions are $P_o = 21.2$ psia and $T_o = 0$ °F. Determine the loss of exergy and second law efficiency.

5.73 Superheated steam with a flow rate of 2.0 kg/s is throttled to 1700 kPa. The steam is initially at 4100 kPa and 300 °C. Estimate the lost work potential of the steam during this throttling process.

5.74 Refrigerant-134a is throttled by a valve. The refrigerant enters the valve as a saturated liquid at 30 psia and exits at 15 psia. Estimate the exit temperature and quality of the refrigerant.

5.75 Steam is throttled by a valve adiabatically. The steam enters the valve at 1000 psia and 600 °F. At the exit, the steam is at 850 psia. Estimate the drop in temperature.

5.76 A steam at 900 psia and 700 °F is throttled in a valve to 55 psia at a rate of 20 lb/min at steady state. Determine the entropy production due to expansion of the steam.

References

1. Çengel YA, Boles MA (2014) Thermodynamics: an engineering approach, 8th edn. McGraw-Hill, New York
2. Çengel YA, Turner R, Cimbala J (2011) Fundamentals of thermal-fluid sciences, 4th edn. McGraw-Hill, New York
3. Chattopadhyay P (2015) Engineering thermodynamics, 2nd edn. Oxford University Press, Oxford
4. Demirel Y (2013) Thermodynamics analysis. Arab J Sci Eng 38:221–249
5. Demirel Y, Gerbaud V (2019) Nonequilibrium thermodynamics. Transport and rate processes in physical, chemical and biological systems, 4th edn. Elsevier, Amsterdam

Energy Production

6

Introduction and Learning Objectives: After transforming the primary energy resources, the secondary energy in the form of liquid and gas fuels of petroleum products, electricity, heat, as well as biofuels can be produced. For example, chemical energy found in fossil fuels such as coal or natural gas is used in steam power plants to produce electricity. Electricity is also produced by other means such as the kinetic energy of flowing water, wind, solar photovoltaic, and geothermal power. Hydroelectric power is produced when falling water turns an electromagnet that generates electricity. Nuclear power is produced when neutrons bombard heavy atoms, which release energy in a process called fission. The world's power demand is expected to rise considerably. Cogeneration is the production of more than one useful form of energy (such as process heat and electric power) from the same energy source.

The learning objectives of this chapter are to understand and discuss:

- Production technologies of electricity and fuels,
- Biofuel production, and
- Fuel cell technologies.

6.1 Energy Production

Energy is mainly produced from transforming the primary energy sources to a secondary energy needed the most. Primary energy sources are coal, natural gas, crude oil, nuclear, waste, solar, wind, biomass, hydro, and geothermal. After refining of petroleum oil into its fractions, many useful types of energy including LPG, gasoline, kerosene, diesel, and fuel oil are produced. Therefore, energy production mainly introduces a transformation process between the primary energy sources and the secondary energy sources. There are various forms of energy production in the world, each with its own risks and benefits [1, 13]. *Cogeneration*

Y. Demirel, *Energy*, Green Energy and Technology,
https://doi.org/10.1007/978-3-030-56164-2_6

refers to the production of more than one useful form of energy, for example power and heat from the same energy source.

6.1.1 Electric Power Production

The fundamental principles of electricity production were discovered during the 1820s and early 1830s by the British scientist Michael Faraday. Based on Faraday's work, electricity is produced by the movement of a loop of wire, or disc of copper between the poles of a magnet [3].

A fossil fuel power plant produces electricity by converting fossil fuel energy to mechanical work. The two major power plant systems are based on the steam turbine cycle and the gas turbine cycle mostly using fossil fuels [13]. The steam power cycle relies on the Rankine cycle in which high pressure and high temperature steam produced in a boiler is expanded through a turbine that drives an electric generator. The discharged steam from the turbine gives up its heat of condensation in a condenser to a heat sink such as water from a river or a lake. The condensate is pumped back into the boiler to restart a new cycle. The heat taken up by the cooling water in the condenser is dissipated mostly through cooling towers into the atmosphere (see Fig. 6.1). Most of the electricity comes from the steam turbines that expand superheated steam produced in the boiler using various heat sources, including fossil fuels, nuclear energy, and renewables.

Assuming the nitrogen remains inert, the complete combustion of a fossil fuel using air as the oxygen source may be represented approximately by

$$C_xH_y + \left(x + \frac{y}{4}\right)O_2 + 3.76\left(x + \frac{y}{4}\right)N_2 \rightarrow xCO_2 + \left(\frac{y}{2}\right)H_2O$$
$$- 3.76\left(x + \frac{y}{4}\right)N_2 + \text{Heat} \qquad (6.1)$$

Fig. 6.1 Schematic of a steam power plant

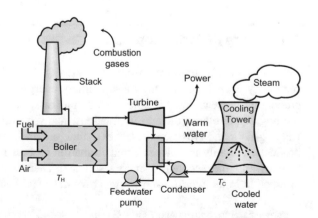

where C_xH_y represents a fossil fuel consisting of carbon (C) and hydrogen (H) only, with stoichiometric coefficients of x and y, respectively, depending on the fuel type. A fossil fuel, such as coal or natural gas, may have other elements such as sulfur and compounds such as minerals besides the carbon and hydrogen. A simple example is the combustion of coal (taken here as consisting of pure carbon): $C + O_2 \rightarrow CO_2$. Coal is prepared by grinding it into powder and mixing it with air which preheats the coal to dry excess moisture content and transports it to the furnace.

Example 6.1 illustrates the analysis of an adiabatic steam turbine. Centralized power production became possible when it was recognized that alternating current power lines can transport electricity at low costs across great distances and power transformers can adjust the power by raising and lowering the voltage.

Renewable energy sources such as solar photovoltaics, wind, biomass, hydro, and geothermal can also provide clean and sustainable electricity. The projections show that renewables will have the highest share compared with coal, natural gas, and nuclear sources in power production [1]. However, renewable energy sources are naturally variable, requiring energy storage or a hybrid system to accommodate daily and seasonal changes. One solution is to produce hydrogen through the electrolysis by splitting of water and to use it in a fuel cell to produce electricity [3].

Example 6.1 Power production by an adiabatic steam turbine
A superheated steam at 4100 kPa and 300 °C expand adiabatically in a steam turbine and exits at 15 kPa with a quality of $x = 0.9$. Velocity of the steam at the inlet is 50 m/s and at the exit is 160 m/s. Elevation at the inlet is 10 m and at the exit is 6 m. Estimate the power produced for the steam flow rate of 1 kg/s.

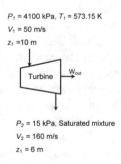

$P_1 = 4100$ kPa, $T_1 = 573.15$ K
$V_1 = 50$ m/s
$z_1 = 10$ m

Turbine W_{out}

$P_2 = 15$ kPa, Saturated mixture
$V_2 = 160$ m/s
$z_1 = 6$ m

Solution:
Assume: Steady-state adiabatic ($q_{loss} = 0$) process.
Steam flow rate: $\dot{m}_s = 1$ kg/s; using steam tables
Inlet conditions: $P_1 = 4100$ kPa and $T_1 = 300$ °C, $v_1 = 50$ m/s, $z_1 = 10$ m, $H_1 = 2958.5$ kJ/kg

Exit conditions: Saturated mixture of liquid and vapor water:

$P_{2,sat} = 15$ kPa, $T_{2,sat} = 54.0\ °C$, $H_{2sat\ liq} = 226.0$ kJ/kg, $H_{2sat\ vap} = 2599.2$ kJ/kg, $x = 0.9$, $v_2 = 160$ m/s, $z_2 = 6$ m

$$H_{2mix} = (1 - x)H_{sat\ liq} + xH_{sat\ vap} = (1 - 0.9)226\,\text{kJ/kg} + (0.9)2599.2\,\text{kJ/kg}$$
$$= 2361.9\,\text{kJ/kg}$$

Energy balance: $\dot{W}_{out} = \dot{m}\left(\Delta H + \dfrac{\Delta v^2}{2}\left(\dfrac{\text{kJ/kg}}{1000\,\text{m}^2/\text{s}^2}\right) + g\Delta z\right)$

$\Delta H = H_2 - H_1 = (2361.9 - 2958.5)\text{kJ/kg} = -596.6\,\text{kJ/kg}$

$\Delta KE = \dfrac{v_2^2 - v_1^2}{2} = \left(\dfrac{160^2 - 50^2}{2}\right)\text{m}^2/\text{s}^2\left(\dfrac{\text{kJ/kg}}{1000\,\text{m}^2/\text{s}^2}\right) = 11.55\,\text{kJ/kg}$

$\Delta PE = g(z_2 - z_1)\left(\dfrac{\text{kJ/kg}}{1000\,\text{m}^2/\text{s}^2}\right) = 9.81(6 - 10)\text{m}^2/\text{s}^2\left(\dfrac{\text{kJ/kg}}{1000\,\text{m}^2/\text{s}^2}\right)$
$$= -0.04\,\text{kJ/kg}$$

$\dot{W}_{out} = \dot{m}\left(\Delta H + \dfrac{\Delta v^2}{2}\left(\dfrac{\text{kJ/kg}}{1000\,\text{m}^2/\text{s}^2}\right) + g\Delta z\right) = (1\ \text{kg/s})(-596.6 + 11.5 - 0.04)\ \text{kJ/kg}$
$$= -585.1\,\text{kW}$$

Contribution of kinetic energy is less than 2%, and the contribution of potential energy is negligible. In practice, changes in kinetic and potential energies are assumed to be negligible.

6.1.2 Distributed Energy Productions

Distributed energy production refers to small power plants or storage systems located close to the point of use. Distributed energy resources usually have higher efficiencies through cogeneration and may reduce emissions of carbon dioxide, because of their use of onsite renewable resources and low greenhouse gas fuels such as natural gas, as seen in Fig. 6.2. Some distributed generation technologies, including photovoltaic and fuel cells, can generate electricity with no emissions or less emissions than that of large-scale *centralized* fossil fuel-fired power plants. Advanced sensors and controls, energy storage and heat exchangers with improved waste heat recovery increase the efficiencies of distributer energy systems. Other advantages include fuel source flexibility, reduced transmission line losses, enhanced power quality and reliability, and more user control. Distributed generation permits consumers who are producing electricity for their own needs, to send their surplus electrical power back into the national power grid [1, 4, 11].

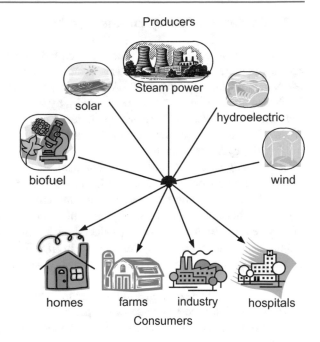

Fig. 6.2 Transition to distributed energy resources and use

6.2   Transmission of Energy

Electricity grid transmits and distributes power from power plants to end user. A combination of sub-stations, transformers, towers, and cables is used to maintain a constant flow of electricity at the required levels of voltage such as 110 or 220 V. Grids also have a predefined carrying capacity or load. New small-scale energy sources may be placed closer to the consumers so that less energy is lost during electricity distribution [11].

Crude oil is carried and distributed through long pipelines between the sources and refineries where it is fractioned into many different types of fuels, such as gasoline, kerosene, diesel, and fuel oil. Pipelines can also distribute the refined fractions of the crude oil to end users. Slurry pipelines are sometimes used to transport coal. Oil pipelines are made of steel or plastic tubes with inner diameter typically from 0.3 to 4 ft (0.1–1.2 m). For natural gas, pipelines are constructed of carbon steel and varying in size from 0.2 to 5 ft (0.06–1.5 m) in diameter, depending on the type of pipeline. The gas is pressurized by compressor stations. Most pipelines are buried at a typical depth of about 3–6 ft (0.9–1.8 m) [1].

6.3 Power-Producing Engine Cycles

The systems used to produce power are called *engines*. Most power-producing engines operate with cyclic processes using a working fluid. Steam power plants use water as the working fluid. Actual engine cycles are complex to analyze. Therefore, it is common to analyze the cycles by assuming that they operate without friction, heat losses, and other complexities. Such a cycle is known as the ideal engine. The analysis of ideal engines yields the major operating parameters controlling the cycle performance.

A *heat engine* converts heat to mechanical energy by bringing a working fluid from a high temperature state T_H to a lower temperature state T_C. Figure 6.3 shows a typical pressure–volume PV and temperature–entropy TS diagrams of ideal engine cycles. On both the PV and TS diagrams, the area that enclosed the process curves of a cycle represents the net work produced during the cycle, which is equivalent to the net heat transfer for the cycle. As seen in Fig. 6.3, the characteristics of cycles on a TS diagram are [5]:

- Heat addition increases pressure and temperature,
- Heat rejection process decreases temperature,
- Isentropic (internally reversible and adiabatic) process takes place at constant entropy.

The area under the heat addition process on a TS diagram measures the total heat input, and the area under the heat rejection process measures the total heat output. The difference between these two areas represents the net heat transfer and the net work output. Therefore, any modifications that improve the net heat transfer rate will also improve the net work output. Example 6.2 illustrates the analysis power output from a steam power plant [2, 4, 10].

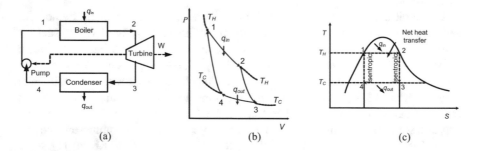

Fig. 6.3 Typical ideal engine cycles: **a** cyclic processes, **b** on a pressure–volume (PV) diagram, **c** on a temperature–entropy (TS) diagram

A heated working fluid at high temperature generates work in the engine while transferring remaining heat to the colder 'sink' until it reaches a low temperature state. The working fluid usually is a liquid or gas. During the operation of an engine some of the thermal energy is converted into work and the remaining energy is lost to a heat sink mainly the general surroundings. Example 6.3 estimates the water mass flow rate for a given power output.

Example 6.2 Steam power production
A steam power production plant uses steam at 8200 kPa and 823.15 K. The turbine discharges the steam at 30 kPa. The turbine and pump operate reversibly and adiabatically. Determine the work produced for every kg steam produced in the boiler.

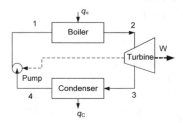

Solution:
Assume that the turbine and pump operate reversibly and adiabatically; no pressure drop across condenser and boiler; only heat transfer occurs at condenser and boiler.
Steam flow rate: $\dot{m}_s = 1$ kg/s
Data from steam tables:
Turbine inlet: $P_2 = 8200$ kPa, $T_2 = 823.15$ K, $H_2 = 3517.8$ kJ/kg, $S_2 = 6.8646$ kJ/kg K
Turbine outlet: $P_3 = 30$ kPa; $T_3 = 342.27$ K, $V_4 = 0.001022$ m^3/kg
$H_{3\text{sat liq}} = 289.30$ kJ/kg $H_{3\text{sat vap}} = 2625.4$ kJ/kg $S_{3\text{sat liq}} = 0.9441$ kJ/kg K, $S_{3\text{sat vap}} = 7.7695$ kJ/kg K
For an ideal operation $S_2 = S_3 = 6.8646$. Since $S_3 < S_{3\text{sat vap}}$ the discharged steam is a mixture of liquid and vapor. Solve for fraction of vapor, x, using entropy balance for the exhaust of the turbine:
$$x_3 = \frac{S_2 - S_{3\text{sat liq}}}{S_{3\text{sat vap}} - S_{3\text{sat liq}}} = \frac{6.8646 - 0.9441}{7.7695 - 0.9441} = 0.867$$
$H_3 = 0.867(2625.4) + (1 - 0.867)(289.3) = 2315.8$ kJ/kg $T_2 = 342.27$ K (saturated)
$H_1 = H_4 + (P_1 - P_4)V_4 = 289.3 + (8200 - 30)(0.001022) = 297.65$ kJ/kg
$q_{\text{out}} = -(H_4 - H_3) = -(289.3 - 2315.8)$ kJ/kg $= 2026.5$ kJ/kg (heat received by the cooling medium)
$q_{\text{in}} = (H_2 - H_1) = (3517.8 - 297.65)$ kJ/kg $= 3220.2$ kJ/kg (heat received by the steam)

Net work output for 1 kg/s steam: $W_{out} = \dot{m}_s(q_{out} - q_{in}) = -1193.7\,\text{kW}$ (work output by the turbine)
Note that the work produced is negative as it is an outgoing work of the system that is the turbine.

Example 6.3 Steam flow rate calculation in a power plant
A steam power plant output is 50 MW. It uses steam (stream 1) at 8200 kPa and 550 °C. The discharged stream (stream 2) is saturated at 75 kPa. If the expansion in the turbine is adiabatic determine the steam flow rate.
Solution:
Assume that kinetic and potential energy are negligible, and the system is at steady state.
(a) Basis: 1 kg/s steam with the properties from the steam tables:
$W = -50\,\text{MW} = -50,000\,\text{kW}$ (turbine is the system)
Turbine   inlet: $H_1 = 3517.8$ kJ/kg, $S_1 = 6.8648$ kJ/kg K   at $T_1 = 550$ °C, $P_1 = 8200$ kPa
Turbine outlet: $S_{2sat\ vap} = 7.3554$ kJ/kg K, $S_{2sat\ liq} = 1.2131$ kJ/kg K at $P_2 = 75$ kPa
$H_{2sat\ vap} = 2663.0$ kJ/kg, $H_{2sat\ liq} = 384.45$ kJ/kg
At isentropic conditions, then we have $S_2 = S_1 < 7.3554$ kJ/kg K
Discharged steam is wet steam: x_s (the quality at isentropic operation):
$$x_s = \frac{6.8646 - 1.2131}{7.4570 - 1.2131} = 0.905$$
The   discharged   steam   enthalpy   at   isentropic   conditions H_{2s} is
$H_{2s} = H_{2sat\ liq}(1 - x_s) + H_{2sat\ vap}x_s = 384.45(1 - 0.905) + 2663(0.905) = 2446.8$ kJ/kg
The steam rate: $\dot{m} = \dfrac{-50000\,\text{kW}}{(H_{2s} - H_1)\,\text{kJ/kg}} = \mathbf{46.7\,kg/s}$

6.3.1 Carnot Cycle

A Carnot *heat engine* performs the conversion of heat to mechanical energy by bringing a working fluid from a high temperature state T_H to a lower temperature state T_C using reversible processes. Figure 6.3 shows a typical pressure–volume *PV* and temperature–entropy *TS* diagrams of ideal engine cycles. On both the *PV* and *TS* diagrams, the area enclosed by the process curves of a cycle represents the net heat transfer to be converted to mechanical energy by the engine.

Carnot cycle consists of four processes shown in Figs. 6.4 and 6.5c:

- Process 1–2: Isothermal heat addition at constant temperature T_H.
- Process 2–3: Isentropic expansion at constant entropy $S_2 = S_3$.
- Process 3–4: Isothermal heat rejection at constant temperature T_C.
- Process 4–1: Isentropic compression at constant entropy $S_4 = S_1$.

From Fig. 6.3c, we can estimate the amounts of added and rejected heats per unit mass flow rate of a working fluid

$$q_{in} = T_H(S_2 - S_1) \tag{6.2}$$

$$q_{out} = T_C(S_4 - S_3) \tag{6.3}$$

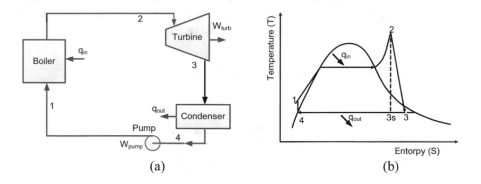

(a) (b)

Fig. 6.4 **a** Schematic of rankine cycle, **b** Rankine cycle processes on a T versus S diagram

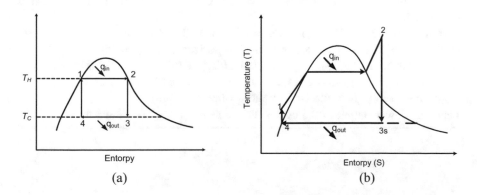

(a) (b)

Fig. 6.5 **a** Carnot cycle, **b** Ideal rankine cycle on a T–S diagram, where on both the cycles pumps and turbines operate at constant entropy (isentropic operation)

Here q_{in} is the heat received by the steam, while q_{out} is the heat rejected into the cooling medium, for example, cooling water. According to the sign convention considered here, heat received by the system is assumed as positive. The net power output becomes

$$W_{out} = q_{out} - q_{in} \tag{6.4}$$

Here, the W_{out} is the work produced by the system that is turbine and hence its sign is negative. Example 6.4 illustrates the analysis of power output from a Carnot cycle [5, 10].

Example 6.4 Power output from a Carnot cycle

A Carnot cycle uses water as the working fluid at a steady-flow process. Heat is transferred from a source at 250 °C and water changes from saturated liquid to saturated vapor. The saturated steam expands in a turbine at 10 kPa, and a heat of 1045 kJ/kg is transferred in a condenser at 10 kPa. Estimate the net power output of the cycle for a flow rate of 10 kg/s of the working fluid.

Solution:

Assumptions: Kinetic and potential energy changes are negligible.

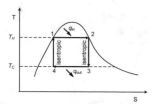

Data from steam tables: $\dot{m} = 10$ kg/s

Turbine inlet: $P^{sat} = 3977.6$ kPa, $T_H = 250$ °C $= 523.15$ K (from the saturated steam table)

Turbine outlet: $T_C = T^{sat}$ (at 10 kPa) $= 45.8$ °C $= 318.8$ K

Heat supplied is equivalent to the heat of vaporization at $T_H = 250$ °C.

$H_{1sat\ liq} = 1085.8$ kJ/kg, $H_{2sat\ vap} = 2800.4$ kJ/kg, at $P_1 = P_2 = 3977.6$ kPa

$q_{in} = H_{2sat\ vap} - H_{2sat\ liq} = 1714.6$ kJ/kg (heat received by the steam)

$q_{out} = 1045$ kJ/kg (heat received by the cooling medium)

Total net power output of the cycle: $\dot{W}_{net} = \dot{m}(q_{out} - q_{in}) = \mathbf{-6696.0\,kW}$

6.3.2 Rankine Cycle

The *Rankine cycle* converts heat into work and generates most of the electric power used throughout the world. Figure 6.4 describes the main processes within the cycle. The Rankine cycle processes are [3, 5]:

- Process 1–2: The high-pressure liquid enters the boiler where it is heated at constant pressure by an external heat source to become a superheated vapor.
- Process 2–3: The vapor expands through the turbine, generating power. This decreases the temperature and pressure of the vapor, and some condensation may occur.
- Process 3–4: The wet vapor discharged from the turbine enters a condenser where it is condensed at a constant pressure.
- Process 4–1: The water is pumped from low to high pressure to start a new cycle.

The Rankine cycle is sometimes referred to as a practical Carnot cycle because in an ideal Rankine cycle the pump and turbine would be *isentropic* and hence maximize the work output shown in Fig. 6.5 as the line 2–3 s where the entropy remains constant. The main difference is that the heat addition (in the boiler) and rejection (in the condenser) are isobaric in the Rankine cycle and isothermal in the Carnot cycle.

6.3.3 Analysis of Ideal Rankine Cycle

The analyses of processes yield the following equations for a Rankine cycle: Pump power needed

$$\dot{W}_{p,\text{in}} = \dot{m} V_4 (P_1 - P_4) \tag{6.5}$$

where $\dot{m}$ is the steam mass flow rate, V_4 is the specific volume at state 4 that is the saturated liquid water, and P_i is the pressure at state i. Enthalpy rate at state 1, H_1, is

$$\dot{m} H_1 = \dot{m} H_4 + \dot{W}_{p,\text{in}} \tag{6.6}$$

For isentropic process $S_1 = S_4$ and $S_2 = S_3$. The quality of the discharged wet steam, x_{3s}, shows the molar or mass fraction of vapor $(S_3 < S_{3\text{sat vap}})$

$$x_{3s} = \frac{(S_3 - S_{3\text{sat liq}})}{(S_{3\text{sat vap}} - S_{3\text{sat liq}})} \tag{6.7}$$

where S_3 is the entropy at state 3, $S_{3\text{sat vap}}$ and $S_{3\text{sat liq}}$ are the entropy of saturated vapor and saturated liquid at state 3, respectively. Enthalpy rate of the wet steam at state 3 is

$$\dot{m}H_3 = \dot{m}[(1 - x_{3s})H_{3\text{sat liq}} + x_{3s}H_{3\text{sat vap}}] \qquad (6.8)$$

where $H_{3\text{sat vap}}$ and $H_{3\text{sat liq}}$ are the saturated vapor and saturated liquid enthalpies at state 3. The rate of heat input, $\dot{q}_{\text{in}}$, to the pressurized water in the boiler is

$$\dot{q}_{\text{in}} = \dot{m}(H_2 - H_1) \qquad (6.9)$$

The discharged heat rate received by a cooling medium (system) $\dot{q}_{\text{out}}$ at the condenser is

$$\dot{q}_{\text{out}} = -\dot{m}\Delta H = -\dot{m}(H_4 - H_3) \qquad (6.10)$$

Net value of the work $\dot{W}_{\text{net}}$ produced by the turbine is

$$\dot{W}_{\text{net}} = (\dot{q}_{\text{out}} - \dot{q}_{\text{in}}) \qquad (6.11)$$

The sign of net work is negative as the output of the system is turbine. Performance of a steam turbine will be limited by the quality of discharged steam. Discharged steam with a low quality may decrease the life of turbine blades and efficiency of the turbine. The easiest way to overcome this problem is by superheating the steam. Example 6.5 illustrates the analysis of an ideal Rankine cycle [2, 5, 10].

Example 6.5 Analysis of a simple ideal Rankine cycle
A steam power plant operates on a simple ideal Rankine cycle shown below. The turbine receives steam at 698.15 K and 4100 kPa, while the discharged steam is at 40 kPa. The mass flow rate of steam is 8.5 kg/s. Determine the net work output.

Solution:
Assume that the kinetic and potential energy changes are negligible. From steam tables:
$\dot{m}_s = 8.5$ kg/s, $P_2 = P_1 = 4100$ kPa; $H_2 = 3272.3$ kJ/kg: $S_2 = 6.8450$ kJ/kg K (steam tables)
Saturated steam properties at $P_3 = P_4 = 40$ kPa, $V_4 = 0.001022$ m³/kg
$H_{3\text{sat vap}} = 2636.9$ kJ/kg; $H_4 = H_{3\text{sat liq}} = 317.65$ kJ/kg;
$S_{3\text{sat vap}} = 7.6709$ kJ/kg K; $S_{3\text{sat liq}} = 1.0261$ kJ/kg K
Basis: mass flow rate of 1 kg/s: $W_{p,\text{in}} = V_1(P_1 - P_4) = (0.001022)(4100 - 40)$
$\left(\dfrac{1\,\text{kJ}}{1\,\text{kPa m}^3}\right) = 4.14\,\text{kJ/kg}$
$H_1 = H_4 + W_{p,\text{in}} = 321.79\,\text{kJ/kg}$
Isentropic process $S_1 = S_4$ and $S_3 = S_2$. The quality of the discharged wet steam ($S_3 < S_{3\text{sat vap}}$):

$x_{3s} = (6.845 - 1.0262)/(7.6709 - 1.0261) = 0.875$
$H_3 = 317.65(1 - 0.875) + 2636.9 \times 0.875 = 2356.6 \, \text{kJ/kg}$
$q_{\text{in}} = H_2 - H_1 = 2950.5 \, \text{kJ/kg}$ (positive heat received by the steam)
$q_{\text{out}} = -\Delta H = -(H_4 - H_3) = 2038.9 \, \text{kJ/kg}$ (positive heat received by the cooling medium)
With a steam flow rate of 8.5 kg/s, we have
$\dot{W}_{\text{net}} = \dot{m}(q_{\text{out}} - q_{\text{in}}) = -7748.6 \, \text{kW} = \textbf{-7.75 MW}$

6.3.4 Brayton Cycle

The gas turbine cycle relies on the *Brayton cycle* using air as the working fluid as shown in Fig. 6.6. An ideal Brayton cycle consists of the following processes:

- *Isentropic process*: Ambient air is drawn into the compressor.
- *Isobaric process*: The compressed air then runs through the combustion chamber where the fuel is burned, and heating that air at a constant pressure process.
- *Isentropic process*: The heated, pressurized air then gives up its energy, expanding through the turbine (or series of turbines).
- *Isobaric process*: Heat rejection to the surroundings.

Some of the electricity produced from the turbine is used to drive the compressor through a shaft arrangement. The gas turbine requires clean fuels such as natural gas or light fuel oil.

Actual Brayton cycle has the processes:

- Compression (adiabatic),
- Heat addition (isobaric),
- Expansion (adiabatic),
- Heat rejection (isobaric).

Fig. 6.6 Schematic of ideal Brayton cycle with processes of adiabatic compression, isobaric heat addition, adiabatic expansion, and isobaric heat rejection

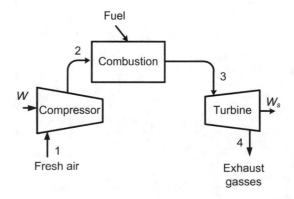

Since neither the compression nor the expansion can be truly isentropic, losses through the compressor and the expander may be considerable. Increasing the compression ratio increases the overall power output of a Brayton system. Inter-cooling the working fluid decreases the amount of work needed for the compression stage overall and increases in fuel consumption of the combustion chamber. To increase the power output for a given compression ratio, air expands through a series of turbines and is then passed through a second combustion chamber before expanding to ambient pressure [2, 10].

6.3.5   Stirling Engine

A Stirling engine operates by cyclic compression and expansion of a working fluid at different temperature levels such that there is a net conversion of heat energy to mechanical work (see Fig. 6.7). The working fluid is mainly air although other gases can also be used. The Stirling engine can use almost any heat source [10]. The Stirling engine is classified as an external combustion engine, as all heat transfers to and from the working fluid take place through the engine wall. The engine cycle consists of compressing gas, heating the gas, expanding the hot gas, and finally cooling the gas before repeating the cycle. Its practical use is largely confined to low-power domestic applications.

6.3.6   Combined Cycles

A combined cycle power plant uses the Brayton cycle of the gas turbine with the Rankine cycle of a heat recovery steam generator. The combined cycle plants are designed in a variety of configurations composed of gas turbines followed by a steam turbine. They generate power by burning natural gas in a gas turbine and use

Fig. 6.7 Schematic of Stirling engine

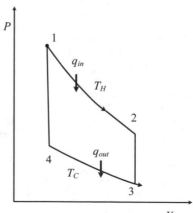

residual heat to generate additional electricity from steam produced by Heat Recovery Steam Generation (HRSG) system [13].

Coal gasification produces a fuel gas that can be used in the gas turbine. By integrating coal gasification with gas turbine and steam cycles, a low pollutant emission can be achieved while using coal. A potential additional advantage of the integrated gasification combined cycle is the capability of capturing carbon dioxide from the fuel gas and making it ready for high-pressure pipeline transportation to a carbon sequestration site [13].

6.4 Improving the Power Production in Steam Power Plants

Improvements on power plant operations may increase the efficiency and power output while reducing the fuel consumption. The modifications for improvements aim at increasing the average temperature at which heat is transferred to the steam in the boiler and decrease the average temperature at which heat is removed in the condenser [2, 4, 5]. Some common modifications in steam power production are summarized in the next sections.

6.4.1 Modification of Operating Conditions of Condensers and Boilers

Superheating the steam to high temperature increases the power output and quality of the discharged steam. Figure 6.8 shows the effects of superheating the steam to higher temperatures and reducing the condenser pressure on the ideal Rankine cycle on a TS diagram. The area enclosed by the process curve "1'-2'-3'-4'" is larger than that of the area for "1-2-3-4" and represents the net work output increase of the

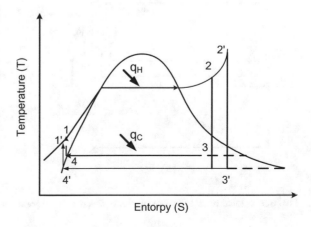

Fig. 6.8 Effects of superheating the steam to higher temperatures and reducing the condenser pressure on an ideal Rankine cycle

cycle under higher boiler temperature and lower condenser pressure. The operating pressures may be as high as 30 MPa (approximately 4500 psia) in modern steam power plants operating at supercritical pressures ($P > 22.09$ MPa). By metallurgical constraints, the temperature of steam is limited. The temperature at the turbine inlet may be as high as 620 °C [5].

6.4.2 Reheating the Steam

Superheating the steam to high temperatures enables the expansion of the steam in various stages instead of a single expansion process. Mainly, reheating increases the power output and the steam quality to protect the material. Figure 6.9 shows a typical reheat Rankine cycle. In an ideal reheat Rankine cycle with two-stage expansion, for example, the steam is expanded to an intermediate pressure isentropically in the high-pressure turbine section and sent to the boiler to be reheated. In the low-pressure turbine section, the reheated steam is expanded to the condenser operating pressure. Example 6.6 illustrates a simple analysis of ideal reheat Rankine cycle.

Example 6.6 Simple reheat Rankine cycle in a steam power plant
A simple ideal reheat Rankine cycle is used in a steam power plant shown in Fig. 6.9. Steam enters the turbine at 9000 kPa and 823.15 K and leaves at 4350 kPa and 698.15 K. The steam is reheated at constant pressure to 823.15 K. The discharged steam from the low-pressure turbine is at 10 kPa. The net power output of the turbine is 40 MW. Determine the mass flow rate of steam.

Solution:
Assume that the kinetic and potential energy changes are negligible, and this is a steady process.
Consider Fig. 6.9.

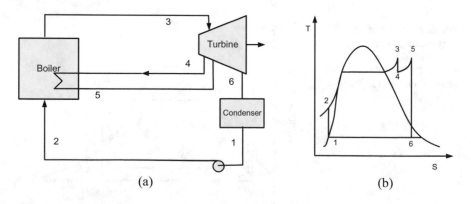

(a)   (b)

Fig. 6.9 **a** Ideal reheat Rankine cycle, **b** T versus S diagram for the reheat Rankine cycle [5]

Data: From steam tables

$V_1 = 0.00101$ m³/kg, $P_3 = 9000$ kPa, $H_3 = 3509.8$ kJ/kg, $S_3 = 6.8143$ kJ/kg K

$P_6 = 10$ kPa

$T_4 = 698.15$ K, $P_4 = 4350$ kPa, $H_4 = 3268.5$ kJ/kg, $S_3 = S_4 = 6.8143$ kJ/kg K

$T_5 = 823.15$ K, $P_5 = 4350$ kPa, $H_5 = 3555.2$ kJ/kg, $S_5 = S_6 = 7.1915$ kJ/kg K

Saturated vapor and liquid data: $H_{6sat\ vap} = H_{1sat\ vap} = 2584.8$ kJ/kg,

$H_{1sat\ liq} = H_{6sat\ liq} = 191.81$ kJ/kg

$S_{6sat\ vap} = 8.1511$ kJ/kg K, $S_{6sat\ liq} = 0.6493$ kJ/kg K

Based on $\dot{m} = 1$ kg/s steam flow rate: $W_{p,in} = V_1(P_2 - P_1)$

$$= 0.00101(9000 - 10)\left(\frac{1\ kJ}{1\ kPa\ m^3}\right) = 9.08\ kJ/kg$$

$H_2 = H_1 + W_{p,in} = 200.9$ kJ/kg

Because this is an isentropic process $S_5 = S_6$ and $S_1 = S_2$. We estimate the quality of the discharged wet steam ($S_6 < S_{6sat\ vap}$) after passing through the turbine:

$x_{6s} = (7.1915 - 0.6493)/(8.1511 - 0.6493) = 0.872$

$x_{6s} = (7.1915 - 0.6493)/(8.1511 - 0.6493) = 0.872$

Turbine outputs: First turbine: $W_{T1} = H_4 - H_3 = -241.3$ kJ/kg

Second turbine: $W_{T2} = H_6 - H_5 = -1276.5$ kJ/kg

The total work done by the turbines: $W_{total} = -1517.8$ kJ/kg

Heat interactions:

$q_{23,in} = H_3 - H_2 = 3308.9$ kJ/kg (heat received by the steam)

$q_{45,in} = H_5 - H_4 = 286.7$ kJ/kg (heat received by the steam during the reheat)

$q_{out} = -(H_1 - H_6) = 2086.9$ kJ/kg (heat is received by the cooling medium)

$q_{in} = q_{23,in} + q_{45.in} = 3595.6$ kJ/kg

$\dot{W}_{out} = \dot{m}(q_{out} - q_{in}) = \dot{m}(2086.9 - 3595.5)$ kJ/kg $= -40000$ kW (work produced by the turbine).

The flow rate of steam is: $\dot{m} = |\dot{W}_{out}|/1508.6 = 26.5$ kg/s

In this ideal reheat Rankine cycle, the expanded steam from the first part of the high-pressure section is reheated in the boiler until it reaches the boiler exit temperature. The reheated steam is expanded through the turbine to the condenser conditions. The reheating decreases the moisture within the discharged steam.

6.4.3 Regeneration

Increasing the boiler water feed temperature by using the expanding steam is possible in a regenerative cycle. Figure 6.12 shows a typical regenerative Rankine cycle. Steam extracted at intermediate pressures from the turbine is used to heat the boiler water feed. The steam at stage 3 leaves the condenser as a saturated liquid at the condenser operating pressure by adjusting the fraction of steam extracted from the turbine. Regeneration helps deaerate the water and control the discharged steam flow rate. Example 6.7 illustrates a simple analysis of ideal regenerative Rankine cycle (Figs. 6.10 and 6.11).

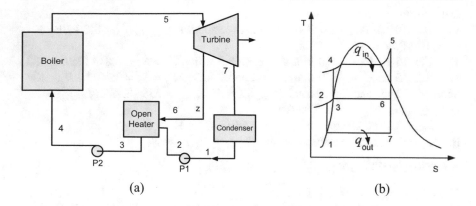

Fig. 6.10 **a** Ideal regenerative Rankine cycle, **b** Temperature T versus entropy S diagram for ideal regenerative Rankine cycle. Here P1 and P2 show the circulation pumps

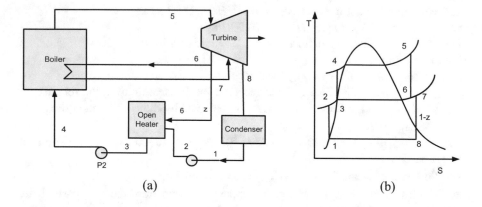

Fig. 6.11 **a** Schematic of ideal reheat-regenerative Rankine cycle, **b** T-S diagram of the cycle

---

Example 6.7 Power output of ideal regenerative Rankine cycle
A steam power plant uses an ideal regenerative Rankine cycle, as shown in Fig. 6.10. Steam enters the high-pressure turbine at 8200 kPa and 773.15 K, and the condenser operates at 20 kPa. The steam is extracted from the turbine at 350 kPa to heat the feed water in an open heater. The water is a saturated liquid after passing through the feed water heater. Determine the net power output of the cycle.

Solution:
Assume that the kinetic and potential energy changes are negligible, and this is a steady-state process.
Consider Fig. 6.10.
Basis: steam flow rate is 1 kg/s (from steam tables):
$V_1 = 0.001017$ m^3/kg, $V_3 = 0.001079$ m^3/kg, $P_1 = P_7 = 20$ kPa,

$T_5 = 773.15$ K, $P_5 = 8200$ kPa, $H_5 = 3396.4$ kJ/kg, $S_5 = 6.7124$ kJ/kg K
$H_{7\text{sat vap}} = H_{1\text{sat vap}} = 2609.9$ kJ/kg, $H_{7\text{sat liq}} = H_{1\text{sat liq}} = 251.45$ kJ/kg,
$S_{7\text{sat vap}} = 7.9094$ kJ/kg K, $S_{7\text{sat liq}} = 0.8321$ kJ/kg K
$P_2 = P_3 = 350$ kPa,
$H_{6\text{sat liq}} = H_{3\text{sat liq}} = 584.27$ kJ/kg, $H_{6\text{satvap}} = H_{3\text{sat vap}} = 2731.50$ kJ/kg
$S_{3\text{sat liq}} = 1.7273$ kJ/kg K, $S_{3\text{sat vap}} = 6.9392$ kJ/kg K
In this ideal regenerative Rankine cycle, the steam extracted from the turbine heats the water from the condenser, and the water is pumped to the boiler. Sometimes, this occurs in several stages. The condensate from the feed water heaters is throttled to the next heater at lower pressure. The condensate of the final heater is flashed into the condenser.

$$W_{p1} = V_1(P_2 - P_1) = 0.001017(350 - 20)\left(\frac{1\,\text{kJ}}{1\,\text{kPa m}^3}\right) = 0.335\,\text{kJ/kg}$$

$$H_2 = H_1 + W_{p1} = 252.78\,\text{kJ/kg}$$

$$W_{p2} = V_3(P_4 - P_3) = 0.001079(8200 - 350)\left(\frac{1\,\text{kJ}}{1\,\text{kPa m}^3}\right) = 8.47\,\text{kJ/kg}$$

$$H_4 = H_3 + W_{p2} = 592.74\,\text{kJ/kg}$$

Because this is an isentropic process, $S_5 = S_6 = S_7$. We estimate the quality of the discharged wet steam at states 6 ($S_5 < S_{6\text{sat vap}}$) and 7 ($S_5 < S_{7\text{sat vap}}$):

$$x_6 = \frac{6.7124 - 1.7273}{6.9392 - 1.7273} = 0.956$$

$$H_6 = 584.27(1 - 0.956) + 2731.50(0.956) = 2638.06\,\text{kJ/kg}$$

$$x_7 = \frac{6.7124 - 0.8321}{7.9094 - 0.8321} = 0.83$$

$$H_7 = 252.45(1 - 0.83) + 2609.9(0.83) = 2211.18\,\text{kJ/kg K}$$

The fraction of steam extracted is estimated from the energy balance:
$$\dot{m}_6 H_6 + \dot{m}_2 H_2 = \dot{m}_3 H_3$$
In terms of the mass fraction: $z = \dot{m}_6/\dot{m}_3$, the energy balance becomes:
$$zH_6 + (1 - z)H_2 = H_3$$
The mass fraction: $z = \dfrac{H_3 - H_2}{H_6 - H_2} = 0.139$

$q_{\text{in}} = (H_5 - H_4) = 2803.66\,\text{kJ/kg}$
$q_{\text{out}} = -(1 - z)(H_1 - H_7) = 1687.32\,\text{kJ/kg}$ (heat received by the cooling medium that is the system)
For a steam flow rate of 1 kg/s: $W_{\text{net}} = \dot{m}(q_{\text{out}} - q_{\text{in}}) = -1116.34\,\text{kW}$
This work is the output of the turbine that is the system.

6.4.4 Reheat-Regenerative Rankine Cycle

In *reheat-regenerative Rankine cycle*, shown in Fig. 6.11, part of the steam extracted from the turbine heats the water from the condenser, and the remaining part is reheated in the boiler. Example 6.7 illustrates a simple analysis of reheat-regenerative Rankine cycle.

Example 6.8 Ideal reheat-regenerative cycle A steam power plant uses an ideal reheat-regenerative Rankine cycle. Steam enters the high-pressure turbine at 9000 kPa and 773.15 K and leaves at 850 kPa. The condenser operates at 10 kPa. Part of the steam is extracted from the turbine at 850 kPa to heat the water in an open heater, where the steam and liquid water from the condenser mix and direct contact heat transfer takes place. The rest of the steam is reheated to 723.15 K and expanded in the low-pressure turbine section to the condenser pressure. The water is a saturated liquid after passing through the water heater and is at the heater pressure. The flow rate of steam is 32.5 kg/s. Determine the power produced.

Solution:
Assume negligible kinetic and potential energy changes, and that this is a steady-state process. The surroundings are at 285 K.
Consider Fig. 6.11. From steam tables:
$P_5 = 9000$ kPa, $H_5 = 3386.8$ kJ/kg, $S_5 = 6.6600$ kJ/kg K, $T_5 = 773.15$ K
$P_1 = P_8 = 10$ kPa, $H_{8\text{sat vap}} = 2584.8$ kJ/kg, $H_{8\text{sat liq}} = 191.83$ kJ/kg,
$V_1 = 0.00101$ m³/kg
$S_{8\text{sat vap}} = 8.1511$ kJ/kg K, $S_{8\text{sat liq}} = 0.6493$ kJ/kg K
$P_3 = 850$ kPa, $H_{3\text{sat liq}} = 732.03$ kJ/kg, $H_{3\text{sat vap}} = 2769.90$ kJ/kg,
$V_3 = 001079$ m³/kg
$P_7 = 850$ kPa, $H_7 = 3372.7$ kJ/kg, $S_7 = 7.696$ kJ/kg K, $T_7 = 723.15$ kPa
$P_6 = 850$ kPa $\rightarrow T_6 = 450.0$ K, $H_6 = 2779.58$ kJ/kg
Work and enthalpy estimations for a unit mass flow rate of steam yield:

$$W_{p1} = V_1(P_2 - P_1) = 0.00101(850 - 10)\left(\frac{1\,\text{kJ}}{1\,\text{kPa}\,\text{m}^3}\right) = 0.848\,\text{kJ/kg}$$

$$H_2 = H_1 + W_{p1} = 192.68\,\text{kJ/kg}$$

$$W_{p2} = V_3(P_4 - P_3) = 0.001079(9000 - 850)\left(\frac{1\,\text{kJ}}{1\,\text{kPa}\,\text{m}^3}\right) = 9.046\,\text{kJ/kg}$$

$$H_4 = H_3 + W_{p2} = 741.07\,\text{kJ/kg}$$

Because this is an isentropic process, $S_5 = S_6$ and $S_7 = S_8$, and we estimate the quality of the discharged wet steam at state 8 ($S_7 < S_{8\text{sat vap}}$):

$$x_8 = \frac{7.696 - 0.6493}{8.1511 - 0.6493} = 0.94$$

$H_8 = 191.83(1 - 0.94) + 2584.8(0.94) = 2439.63$ kJ/kg
The fraction of steam extracted is estimated from the energy balance
$\dot{m}_6 H_6 + \dot{m}_2 H_2 = \dot{m}_3 H_3$.
In terms of the mass fraction: $z = \dot{m}_6/\dot{m}_3$, the energy balance becomes: $z H_6 + (1 - z)H_2 = H_3$.

The mass fraction is: $z = \dfrac{H_3 - H_2}{H_6 - H_2} = \dfrac{732.03 - 192.68}{2779.58 - 192.68} = 0.208$

The turbine work output with $\dot{m} = 1$ kg/s working fluid is:
$\dot{q}_{\text{in}} = \dot{m}[(H_5 - H_4) + (1 - z)(H_7 - H_6)] = 3115.18\,\text{kW}$
$\dot{q}_{\text{reheat}} = \dot{m}z(H_3 - H_6) = 425.89\,\text{kW}$
$\dot{q}_{\text{out}} = -\dot{m}(1 - z)(H_1 - H_8) = 1779.14\,\text{kW}$ (heat received by the cooling medium that is the system)

Total heat input: $\dot{q}_{\text{total in}} = 3115.18 + 425.89 = 3541.1\,\textbf{kW}$

For steam flow rate of 36.5 kg/s workout: $\dot{W}_{\text{net}} = \dot{m}_s(\dot{q}_{\text{out}} - \dot{q}_{\text{in}}) = -\textbf{48765.4\,kW}$
$= -\textbf{48.76\,MW}$

6.5 Geothermal Power Plants

Figure 6.12 shows a schematic of a geothermal power plant. Geothermal power comes from heat energy buried beneath the surface of the Earth. There are three types of geothermal power plants:

- *Dry Steam Power Plants*: The geothermal steam goes directly to a turbine, where it expands and produces power. The expanded steam is injected into the geothermal well.
- *Flash Steam Power Plants*: Geothermal fluids above 360°F (182 °C) can be flashed in a tank at low pressure causing some of the fluid to rapidly vaporize. The vapor then expands in a turbine.
- *Binary-Cycle Power Plants*: Moderate-temperature geothermal fluids between 85 and 170 °C are common. The term 'binary' refers to dual-fluid systems, wherein hot geothermal brine is pumped through a heat exchange network to transfer its energy to a working fluid driving a power train. A hot geothermal fluid and a suitable working fluid with a much lower boiling point than geothermal fluid pass through a heat exchanger. The vaporized working fluid drives the turbines; and no working fluid is emitted into the atmosphere. The working fluids may be isobutene, isopentane, n-pentane, or ammonia.

Fig. 6.12 Schematic of geothermal power plant

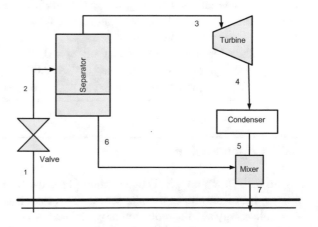

At temperatures below about 200 °C, binary power systems are favored for relative cost-effectiveness. In general, above about 200 °C, flashing geothermal fluids produce steam, and directly driving turbine/generator is preferred. Example 6.9 illustrates a simple analysis of steam power plant using a geothermal source [1, 4, 5].

Example 6.9 A steam power plant using a geothermal energy source
A steam power plant uses a geothermal energy source. The geothermal source is available at 220 °C and 2320 kPa with a flow rate of 200 kg/s. The hot water goes through a valve and a flash drum. Steam from the flash drum enters the turbine at 550 kPa and 428.62 K. The discharged steam from the turbine has a quality of $x_4 = 0.96$. The condenser operates at 10 kPa. The water is a saturated liquid after passing through the condenser. Determine the work output of turbine

Solution:
Assume: The kinetic and potential energy changes are negligible, and this is a steady-state process.
$\dot{m}_1 = 200$ kg/s. From steam tables:
$T_1 = 493.15$ K, $P_1 = 2319.8$ kPa, $H_1 = H_2 = 943.7$ kJ/kg, $S_1 = 2.517$ kJ/kg K
$T_3 = 428.62$ K, $P_3 = 550$ kPa, $H_3 = 2751.7$ kJ/kg $S_3 = 6.787$ kJ/kg K (saturated)
$H_{3sat\,vap} = 2551.7$ kJ/kg, $H_{3sat\,liq} = 655.80$ kJ/kg, $S_{3sat\,vap} = 6.787$ kJ/kg K,
$S_{3sat\,liq} = 1.897$ kJ/kg K, $P_4 = 10$ kPa, $H_{4sat\,vap} = 2584.8$ kJ/kg, $H_{4sat\,liq} = 191.8$ kJ/kg
In this geothermal power plant, the hot water is flashed, and steam is produced and used in the turbine.
The rate of vapor is estimated from the quality at state 2. The fraction of steam after flashing is:
$$x_2 = \frac{943.7 - 655.8}{2751.7 - 655.8} = 0.159$$
$S_2 = (1 - 0.159)1.897 + 0.159(6.787) = 2.6756$
The steam flow rate is: $\dot{m}_3 = x_2(\dot{m}_1) = 0.159(200) = 31.84$ kg/s
From the mass balance around the flash drum, we have $\dot{m}_6 = \dot{m}_1 - \dot{m}_3 = 168.15$ kg/s
The discharged steam has the quality of: $x_4 = 0.96$
$H_4 = (1 - 0.96)H_{4sat\,liq} + (0.96)H_{4sat\,vap} = 2489.08$ kJ/kg
From the flash drum at state 6, we have: $\dot{W}_{net} = \dot{m}_3(H_4 - H_3) = -\mathbf{1993.82\,kW}$

6.6 Cogeneration

Figure 6.13 shows a typical cogeneration plant, which produces electric power and process heat from the same heat source. This may lead to the utilization of more available energy and the reduction of waste heat. A cogeneration plant, for example, may use the waste heat from Brayton engines, typically for hot water production or for space heating. The process heat in industrial plants usually needs steam at

500–700 kPa, and 150–200 °C. The steam expanded in the turbine to the process pressure is used as process heat. Cycles making use of cogeneration may be an integral part of large processes where the energy of the expanded steam from the turbine at intermediate pressure may be fully utilized in producing electricity and process heat simultaneously. The *utilization factor* for a cogeneration plant is the ratio of the energy used in producing power and process heat to the total energy input. Examples 6.10 and 6.11 illustrate simple analyses of energy output in cogeneration plants [3, 4].

$$\text{Cogeneration} = \text{Power} + \text{Heat} \qquad (6.12)$$

Example 6.10 Energy output in a cogeneration plant
A cogeneration plant uses steam at 8200 kPa and 773.15 K (see Fig. 6.13). One-fourth of the steam is extracted at 700 kPa from the turbine for cogeneration. After it is used for process heat, the extracted steam is condensed and mixed with the water output of the condenser. The rest of the steam expands from 8200 kPa to the condenser pressure of 10 kPa. The steam flow rate produced in the boiler is 60 kg/s. Determine the work output and process heat produced.

Solution:
Assume that the kinetic and potential energy changes are negligible, and this is a steady-state process.
Consider Fig. 6.13. From steam tables
$\dot{m} = 60$ kg/s, $z = 0.25$
$P_1 = P_8 = 10$ kPa, $H_{1\text{sat vap}} = 2584.8$ kJ/kg, $H_{1\text{sat liq}} = 191.83$ kJ/kg,
$V_1 = 0.00101$ m³/kg, $S_{1\text{sat vap}} = 8.1511$ kJ/kg K, $S_{1\text{sat liq}} = 0.6493$ kJ/kg K,
$P_3 = P_7 = P_2 = P_4 = 700$ kPa, $H_3 = 697.06$ kJ/kg, $S_3 = 1.9918$ kJ/kg K, $z = 0.25$
$P_6 = 8200$ kPa, $T_6 = 773.15$ K, $H_6 = 3396.4$ kJ/kg, $S_6 = 6.7124$ kJ/kg K

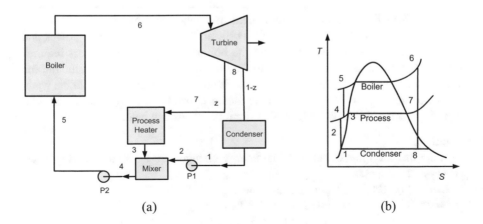

(a) (b)

Fig. 6.13 a Schematic of ideal cogeneration plant, **b** *TS* diagram

In this cogeneration cycle, the steam extracted from the turbine is used as process heat. The liquid condensate from the process heat is combined with the output of the condenser.

Basis: mass flow rate = 1 kg/s

$$W_{p1} = V_1(P_2 - P_1) = 0.00101(700 - 10)\left(\frac{1\,kJ}{1\,kPa\,m^3}\right) = 0.697\,kJ/kg$$

$H_2 = H_1 + W_{p1} = 191.83 + 0.697 = 192.53\,kJ/kg$

From the energy balance around the mixer, we have $\dot{m}_3/\dot{m}_6 = 0.25$

$\dot{m}_6 = \dot{m}_4 = 60\,kg/s$, $\dot{m}_3 = \dot{m}_7 = 15\,kg/s$, $\dot{m}_8 = \dot{m}_1 = 0.75(60) = 45.0\,kg/s$

$\dot{m}_4 H_4 = \dot{m}_2 H_2 + \dot{m}_3 H_3$

$H_4 = [45(192.53) + 15(697.06)]/60 = 318.66\,kJ/kg$

$T_4 = 349.15\,K$, $V_4 = 0.001027\,kg/m^3$

$$W_{p2} = V_4(P_5 - P_4) = 0.001027(8200 - 700)\left(\frac{1\,kJ}{1\,kPa\,m^3}\right) = 7.70\,kJ/kg$$

$H_5 = H_4 + W_{p2} = 326.36\,kJ/kg$

Isentropic processes: $S_6 = S_7 = S_8 = 6.7124$ and $P_7 = 700$ kPa, $H_7 = 2765.68$ kJ/kg.

We estimate the quality of the discharged wet steam at state 8:

$$x_8 = \frac{6.7124 - 0.6493}{8.1511 - 0.6493} = 0.808$$

$H_8 = 191.83(1 - 0.808) + 2584.80(0.808) = 2125.87\,kJ/kg$

The energy balance yields the fraction of steam extracted

$\dot{W}_{total} = \dot{m}_6(H_7 - H_6) + \dot{m}_8(H_8 - H_7) = -66634.44\,kW$

$\sum \dot{W}_{pi} = \dot{m}_1 W_{p1} + \dot{m}_4 W_{p2} = 493.51\,kW$

The net work output: $\dot{W}_{net} = \dot{W}_{total} - \sum \dot{W}_{pi} = -66140.93$ kW (work output of the turbine)

$\dot{q}_{process} = \dot{m}_7(H_3 - H_7) = \mathbf{-31029.3\,kW}$ (heat discharged from the steam)

Example 6.11 Estimation of process heat in a cogeneration plant

A cogeneration plant uses steam at 900 psia and 1000°F to produce power and process heat. The steam flow rate from the boiler is 16 lb/s. The process requires steam at 70 psia at a rate of 3.2 lb/s supplied by the expanding steam in the turbine with a value of $z = 0.2$. The extracted steam is condensed and mixed with the water output of the condenser. The remaining steam expands from 70 psia to the condenser pressure of 3.2 psia. Determine the rate of process heat.

Solution:

Assume that the kinetic and potential energy changes are negligible, and this is a steady-state process.

Consider Fig. 6.13, and from steam tables:

$\dot{m}_6 = 16$ lb/s, $z = \dot{m}_3/\dot{m}_6 = 0.2$, $\dot{m}_3 = 3.2$ lb/s

$P_1 = P_8 = 3.2$ psia, $H_{1\,sat\,vap} = 1123.6$ Btu/lb, $H_{1\,sat\,liq} = 111.95$ Btu/lb

$V_1 = 0.01631$ ft^3/lb, $S_{1sat\ vap} = 0.2051$ Btu/lb R, $S_{1sat\ liq} = 1.8810$ Btu/lb R
$P_3 = 70$ psia, $H_3 = 272.74$ Btu/lb, (Saturated)
$P_6 = 900$ psia, $H_6 = 1508.5$ Btu/lb, $S_6 = 1.6662$ Btu/lb R, $T_6 = 1000°F$
In this cogeneration cycle, the steam extracted from the turbine is used in process heat. The liquid condensate from the process heat is combined with the output of the condenser.
$\dot{m}_2 = \dot{m}_1 = 12.8$ lb/s, $\dot{m}_4 = 16$ lb/s
Because these are isentropic processes, $S_6 = S_{7s} = S_{8s} = 1.6662$ Btu/lb R
$P_7 = 70$ psia, $H_{7s} = 1211.75$ Btu/lb.
Process heat: $\dot{q}_{process} = \dot{m}_3(H_3 - H_7) = \mathbf{-3004.8\,Btu/s}$

6.7 Chemical-Looping Combustion Combined Cycle

Chemical-looping combustion with inherent carbon capture may be an emerging power generation technology. Chemical-looping combustion uses an oxygen carrier that transfers oxygen from the air to the fuel preventing direct contact between them. The oxygen carrier can be alternately oxidized and reduced through the circulation between air and fuel reactor, as seen in Fig. 6.14. Well-designed and operated systems offer scalable, diverse, economical, and environmentally sustainable energy production pathways with inherited carbon capture. The fuel may be coal, natural gas, and biomass. The product gas contains mainly CO_2 and water

Fig. 6.14 Schematic of chemical-looping combustion combined cycle for coal-fired power production [6] (GT: Gas turbine, HRSH: Heat recovery steam generator, ST: Steam turbine, CaSO$_4$ is the oxygen carrier)

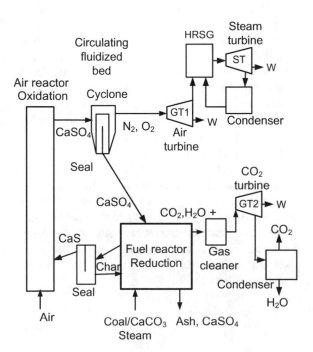

undiluted with nitrogen, and without the production of nitrogen oxides (NO_x) as the high temperatures associated with the use of flame is avoided. The oxidation of the oxygen carrier is strongly exothermic and hence can be used to heat air flow to high temperatures (1000–1200 °C) and can drive a gas turbine [6].

6.8 Nuclear Power Plants

Nuclear power is a method in which steam is produced by heating water through a process called *nuclear fission*. Nuclear fission is a nuclear reaction in which the nucleus of an atom splits into smaller parts, often producing free neutrons and photons in the form of *gamma rays* (γ). *Gamma rays* are electromagnetic radiation of frequencies above 10^{19} Hz, and therefore have energies above 100 keV and wavelength less than 10 pm. In a nuclear power plant, a reactor contains a core of nuclear fuel, primarily uranium. When atoms of uranium fuel are hit by neutrons, they fission (split) releasing heat and more neutrons. Under controlled conditions, these neutrons can strike and split more uranium atoms [9, 14].

Typical fission releases about 200 MeV of energy (1 eV is 96.485 kJ/mol). In contrast, most chemical oxidation reactions, such as burning coal, release at most a few eV. In a nuclear reactor, the energy is converted to heat as the particles and gamma rays collide with the atoms that make up the reactor and its working fluid, usually water or occasionally heavy water [9, 14].

Fusion power, on the other hand, is generated by nuclear fusion reactions where two light atomic nuclei fuse together to form a heavier nucleus and in doing so, releases a large amount of energy. If two light nuclei fuse, they will generally form a single nucleus with a slightly smaller mass than the sum of their original masses. The difference in mass is released as energy according to Albert Einstein's mass–energy equivalence formula $E = mc^2$. The dividing line between 'light' and 'heavy' is iron-56. Above this atomic mass, energy will generally be released by nuclear fission reactions. Most fusion reactions combine isotopes of hydrogen that are protium, deuterium, or tritium to form isotopes of helium ^{3}He or ^{4}He. Most fusion power plants involve using the fusion reactions to create heat, which is then used to operate a steam turbine, which drives generators to produce electricity. This is similar to most coal, oil, and gas-fired power production plants [9, 14].

Nuclear fusion requires precisely controlled temperature, pressure, and magnetic field parameters to generate net energy. If the reactors were damaged, these parameters would be disrupted and the heat generation in the reactor would rapidly cease. In contrast, the fission products in a fission reactor continue to generate heat for several hours or even days after reactor shut-down, meaning that melting of fuel rods is possible even after the reactor has been stopped due to continued accumulation of heat [1, 9, 14].

6.9 Hydropower Plants

Hydropower is a process in which the force of flowing water is used to spin a turbine connected to a generator to produce electricity. Most hydroelectric power comes from the potential energy of dammed water driving a turbine and generator. The power extracted from the water depends on the volume and the amount of potential energy in water is proportional to the head, which is the difference in height between the source and the water's outflow. A simple formula for approximating electric power production at a hydroelectric plant is:

$$\dot{W} = \dot{Q}\rho g \Delta z = \dot{m} g \Delta z \qquad (6.13)$$

where $\dot{W}$ is the power, ρ is the density of water (~ 1000 kg/m^3), Δz is the height, Q is the volumetric flow rate, and g is the acceleration due to gravity. Annual electric energy production depends on the available water supply. In some installations the water flow rate can vary by a factor of 10:1 over the course of a year. Example 6.12 illustrates the analysis of hydroelectric power output.

Hydropower eliminates the use of fossil fuels and hence carbon dioxide emission. Hydroelectric plants also tend to have longer economic lives (50 years or longer) than fuel-fired power production. The sale of hydroelectricity may cover the construction costs after 5 to 8 years of full operation [1]. Operating labor cost is also usually low, as plants are automated. The hydroelectric capacity is either the actual annual energy production or by installed capacity. A hydroelectric plant rarely operates at its full capacity over a full year. The ratio of annual average power output to installed capacity is the *capacity factor* for a hydroelectric power plant. There are large, small, and micro hydropower plant operations, which are summarized below [11]:

- *Large hydroelectric power stations* have outputs from over a few hundred megawatts to more than 10 GW. Many large hydroelectric projects supply public electricity networks. The construction of these large hydroelectric facilities and the changes it makes to the environment are monitored by specialized organizations, such as the International Hydropower Association.
- *Small hydroelectric power plants* have a capacity of up to 10 MW. Some are created to serve specific industrial plants, such as for aluminum electrolytic plants, for which substantial amounts of electricity needed.
- Micro hydroelectric power installations typically produce up to 100 KW of power. These installations can provide power to an isolated home or small community. Sometimes, micro hydro systems may complement photovoltaic solar energy systems because water flow and available hydro power is highest in the winter when solar energy is at a minimum.

Example 6.12 Hydroelectric power output
A hydroelectric plant operates by water falling from a 200 ft height. The turbine in
the plant converts potential energy into electrical energy, which is lost by about 5%
through the power transmission, so the available power is 95%. If the mass flow
rate of the water is 396 lb/s, estimate the power output of the hydro plant.

Solution:

Equation: $\dot{W} = \dfrac{\dot{m}g\Delta z}{g_c}$; Data: $\Delta z = 200$ ft, water flow rate = 396 lb/s; transmission
loss = 5%

$$\dot{W} = -\frac{(396\,\text{lb/s})(32.2\,\text{ft}^2/\text{s})(200\,\text{ft})}{32.2\,\text{ft\,lb}\;/\text{lb}_f\text{s}^2}\left(\frac{1.055\,\text{kW}}{778\,\text{lb}_f\,\text{ft/s}}\right) = -107.4\,\text{kW}$$

(with the sign convention)
With the transmission loss of 5% of the available power:
-107.4 kW $(1 - 0.05) = \mathbf{-102.0\ kW}$.

6.10 Wind Power Plants

The earth is unevenly heated by the sun and the differential heating drives a global
atmospheric convection system leading to the wind. *Wind power* is the conversion
of wind energy into electricity by using wind turbines. A wind turbine is a device
for converting the kinetic energy in wind into the mechanical energy of a rotating
shaft. The generator is usually connected to the turbine shaft through gears that turn
the generator at a different speed than the turbine shaft. Power electronic controls
and converts the electricity into the correct frequency and voltage to feed into the
power grid at 60 or 50 Hertz [9, 11].

The power produced by a wind turbine is proportional to the kinetic energy of
the wind captured by the wind turbine. The kinetic energy of the wind is equal to
the product of the kinetic energy of air per unit mass and the mass flow rate of air
through the blade span area:

Wind power = (efficiency) (kinetic energy) (mass flow rate of air)

$$\dot{W}_{\text{wind}} = \eta_{\text{wind}}\frac{v^2}{2}(\rho A v) = \eta_{\text{wind}}\frac{v^2}{2}\rho\frac{\pi D^2}{4}v \tag{6.14}$$

After rearrangement, we have

$$\dot{W}_{\text{wind}} = \eta_{\text{wind}}\rho\frac{\pi v^3 D^2}{8} = (\text{constant})v^3 D^2 \tag{6.15}$$

$$\text{constant} = \frac{\eta_{\text{wind}}\rho\pi}{8} \tag{6.16}$$

where ρ is the density of air, v is the velocity of air, D is the diameter of the blades of the wind turbine, and η_{wind} is the efficiency of the wind turbine. Therefore, the power produced by the wind turbine is proportional to the cube of the wind velocity and the square of the blade span diameter. The strength of wind varies, and an average value for a given location does not alone indicate the amount of energy a wind turbine could produce there. To assess the frequency of wind speeds at a location, a probability distribution function is often fit to the observed data.

However, it has been assumed that the entire kinetic energy of the wind is available for power generation. Betz's law states that the power output of a wind turbine is maximum when the wind is slowed down to one-third of its initial velocity. The actual efficiency of wind power is

$$\eta_{\text{windturbine}} = \frac{\text{Shaft power out of turbine into gear box}}{\text{Wind power into turbine blades}} \quad (6.17)$$

This efficiency ranges between 20 and 40% and becomes the part of the constant to be used in Eq. (6.16) [1].

Large-scale wind farms are connected to the electric power transmission network, while the smaller facilities are used to provide electricity to isolated locations. Wind energy, as an alternative to fossil fuels, is plentiful, renewable, widely distributed, clean, and produces no greenhouse gas emissions during operation. Wind power is nondispatchable, and for economic operation, all the available output must be taken when it is available. Problems of variability are addressed by grid energy storage, batteries, pumped-storage hydroelectricity, and energy demand management. Wind power has projected long useful life of the equipment, which may be more than 20 years, but a high capital cost. The estimated average cost per unit includes the cost of construction of the turbine and transmission facilities [1, 11].

In a wind farm, individual turbines are interconnected with a medium voltage (often 34.5 kV), power collection system, and communications network. At a sub-station, this medium-voltage electric current is increased in voltage with a transformer for connection to the high-voltage electric power transmission system. Wind turbine generators need extensive modeling of the dynamic electromechanical characteristics to ensure stable behavior. The ratio between annual average power and installed capacity rating of a wind-power production is the *capacity factor*. Typical capacity factors for wind power change between 20 and 40%. Example 6.13 illustrates the windmill calculations.

Wind energy penetration refers to the fraction of energy produced by wind compared with the total available production capacity. There is not a generally accepted maximum level of wind penetration. An interconnected electricity grid usually includes reserve production and transmission capacity to allow for equipment failures; this reserve capacity can also serve to regulate for the varying power production by wind plants. At present, a few grid systems have penetration of wind energy above 5%: Despite the power forecasting methods used, predictability of wind plant output remains low for short-term operation. Pumped-storage hydroelectricity or other forms of grid energy storage can store energy developed by high-wind periods and release it when needed [11].

Aside from the availability of wind itself, other factors include the availability of transmission lines, value of energy, cost of land acquisition, land use considerations, and environmental impact of construction and operations. *Wind power density* is a calculation of the effective power of the wind at a location. A map showing the distribution of wind power density is a first step in identifying possible locations for wind turbines.

Small-scale wind power has the capacity to produce up to 50 kW of electrical power. Isolated communities may use small-scale wind turbines to displace fossil fuel consumption [11].

Example 6.13 Windmill power estimations
A farm of windmills supplies a power output of 1 MW for a community. Each windmill has the blades with 10 m in diameter. At the location of the windmills, the average velocity of the wind is 11 m/s and the average temperature is 20 °C. Estimate the minimum number of windmills to be installed.

Solution:
Air is ideal gas and the pressure is atmospheric.
Inlet: $v = 11$ m/s, $R = 8.314$ kPa m^3/kmol K, $T = 293$ K, $D = 10$ m, $MW_{air} = 29$ kg/kmol
Power output = 1 MW

Density of air $\rho = (MW)\dfrac{P}{RT} = \dfrac{(29\,\text{kg/kmol})101.3\,\text{kPa}}{(8.314\,\text{kPa m}^3/\text{kmol K})(293\,\text{K})}$

Air mass flow rate: $\dot{m} = \rho A v = \rho \pi \dfrac{D^2}{4} v = 1036.2$ kg/s

Power from each windmill: $KE = -\dot{m}\dfrac{v^2}{2} = 1036.2\,\text{kg/s}\dfrac{(11\,\text{m/s})^2}{2}\left(\dfrac{\text{kJ/kg}}{1000\,\text{m}^2/\text{s}^2}\right)$
$= -62.7$ kW
The minimum number of windmills to be installed:
1000 kW/|62.7| kW = **16 windmills**

Example 6.14 Estimation of power available from a wind turbine
A wind turbine with 10 m rotor diameter is to be installed on a hilltop where the wind blows steadily at an average speed of 15 m/s. Determine the maximum power that can be generated by the turbine operated with a 35% efficiency.

Solution:
Assume that the air is ideal gas at 300 K and 1 atm (101.3 kPa).
Inlet: $v = 15$ m/s, $R = 8.314$ kPa m^3/kmol K, $T = 300$ K, $D = 10$ m, $MW_{air} = 29$ kg/kmol

Density of air $\rho = (MW)\dfrac{P}{RT} = \dfrac{(29\,\text{kg/kmol})101.3\,\text{kPa}}{(8.314\,\text{kPa}\,\text{m}^3/\text{kmol K})(300\,\text{K})} = 1.18\,\text{m}^3/\text{kg}$

Air mass flow rate: $\dot{m} = \rho A v = \rho\pi\dfrac{D^2}{4}v = 1389.45\,\text{kg/s}$

Power from a wind turbine with a 35% efficiency:

$KE = -\eta\dot{m}\dfrac{v^2}{2} = (0.35)1389.45\,\text{kg/s}\dfrac{(10\,\text{m/s})^2}{2}\left(\dfrac{\text{kJ/kg}}{1000\,\text{m}^2/\text{s}^2}\right) = -24.31\,\text{kW}$

6.11   Solar Power Plants

Solar power is derived from the energy of sunlight. Average insolation changes from 150 to 300 W/m^2 or 3.5 to 7.0 kWh/m^2 day. There are two main types of technologies for converting solar energy to electricity: photovoltaic and solar thermal electric (see Fig. 6.15).

- *Photovoltaic conversion* produces electricity directly from sunlight in a solar cell. There are many types of photovoltaic cells, such as thin film, monocrystalline silicon, polycrystalline silicon, and amorphous cells, as well as multiple types of concentrating solar power. Photovoltaic power is initially used in small and medium-sized applications, from the calculator powered by a single solar cell to off-grid homes powered by photovoltaic modules.
- *Solar thermal electric production* based on concentrating solar power systems use lenses or mirrors and tracking systems to focus a large area of sunlight into a small beam. The concentrated solar energy heat is used to produce steam to drive turbines and produce electricity. A parabolic trough consists of a linear parabolic

Fig. 6.15 Concentrating solar power technologies offer utility-scale power production [11]

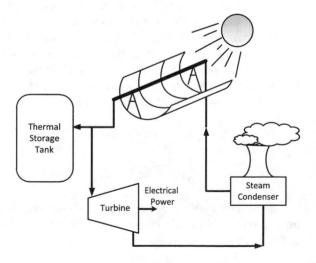

reflector that concentrates light onto a receiver positioned along the reflector's focal line. The receiver is a tube positioned right above the middle of the parabolic mirror and is filled with a working fluid. The reflector is made to follow the sun during the daylight hours by tracking along a single axis.

A Stirling solar dish consists of a stand-alone parabolic reflector that concentrates light onto a receiver positioned at the reflector's focal point. The reflector tracks the sun along two axes and captures the sun's energy through a parabolic dish solar concentrator. The concentrated energy drives a Stirling engine, which spins a generator producing electricity. The advantages of Stirling solar over photovoltaic cells are higher efficiency of converting sunlight into electricity and longer lifetime.

Mirrors, called heliostats, track the sun throughout the day and reflect sunlight to the receiver. Early designs used water heated to steam by the sun's energy to drive turbines. New systems use liquid salt because of its thermal characteristics. The salt is usually a mixture of 60% sodium nitrate and 40% potassium nitrate. The mixture melts at 220 °C/428°F. Cold salt is pumped from a holding tank through the receiver where the focused sunlight heats it to over 1000°F. The hot salt passes through a heat exchanger that makes steam to drive turbines. Power towers are more cost-effective, offer higher efficiency, and better energy storage capability among concentrated solar power technologies [8, 11].

6.12 Fuel Cells

A *fuel cell* oxidizes a fuel, such as hydrogen or methane, electrochemically to produce electric power. It consists of two electrodes separated by an electrolyte. The fuel and oxygen are continuously fed into the cell and the products of reaction are withdrawn continuously. The fuel contacts with the anode, fuel electrode. Oxygen, usually in air, contacts with the cathode, oxygen electrode. Half-cell reactions take place at each electrode. The sum of the half-cell reactions is the overall reaction. The type of electrolyte characterizes the type of fuel cell. Schematic fuel cell using hydrogen as fuel is illustrated in Fig. 6.16. When the electrolyte is acidic, the half-cell reactions occurring at the hydrogen electrode (anode) and at the oxygen electrode (cathode) are

$$H_2 \rightarrow 2H^+ + 2e^- \, (\text{anode}) \tag{6.18}$$

$$\frac{1}{2}O_2 + 2e^- + 2H^+ \rightarrow H_2O(g) \, (\text{cathode}) \tag{6.19}$$

The electrons with negative charge (e^-) are released at the anode. These electrons produce an electric current which is used by the reaction occurring at the cathode. The electric current is carried out by an external circuit. The cation H^+

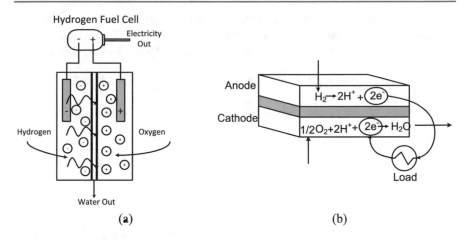

Fig. 6.16 **a** Schematic of a hydrogen fuel cell. The waste product is water *Source* **b** half-cell reactions for a hydrogen/oxygen fuel cell with acidic electrolyte

migrates from anode to cathode through the electrolyte. The sum of the half-cell reaction is the overall reaction taking place at the fuel cell

$$H_2 + \frac{1}{2}O_2 \rightarrow H_2O(g) \tag{6.20}$$

A thin solid polymer known as *proton exchange membrane* serves as an acid electrolyte in the hydrogen/oxygen fuel cell. Each side of the membrane is bonded to a porous carbon electrode impregnated with platinum which serves as a catalyst. The porous electrode provides a very large interface area for the reaction and facilitates the diffusion of hydrogen and oxygen into the cell and the water vapor out of the cell. Cells can be connected in series to make a compact unit with a required level of energy output and operate at a temperature near 60 °C [9, 12].

For each mole of hydrogen consumed, 2 mol of electrons pass to the external circuit. Therefore, the electrical energy (work) is the product of the charge transferred and the voltage V of the cell

$$W_e = -2FV = \Delta G \tag{6.21}$$

where F is the Faraday's constant ($F = 96485$ C/mol) and ΔG is the Gibbs energy. The electric work of reversible and isothermal fuel cell is

$$W_e = \Delta H - q = \Delta G \tag{6.22}$$

Consider a hydrogen/oxygen fuel cell operating at 20 °C and 1 bar with pure hydrogen and oxygen as reactants and water vapor as product. The overall reaction is the standard formation reaction for water and from Appendix C, Table C1

$$\Delta H = \Delta H^o_{f,H_2O} = -241,818\,\text{J/mol and } \Delta G = \Delta G^o_{f,H_2O} = -228,572\,\text{J/mol}$$

Therefore, for the hydrogen/oxygen fuel cell, the electric work and the voltage are

$$W_e = -228,572\,\text{J/mol and } V = \frac{-\Delta G}{2F} = 1.184\,\text{V}$$

Using air instead of pure oxygen in a reversible and isothermal fuel cell, we have

$$W_e = -226,638\,\text{J/mol and } V = 1.174\,\text{V (hydrogen/air fuel cell)}$$

In practice, the operating voltage of hydrogen/oxygen fuel cell is around 0.6-0.7 volts, because of internal irreversibilities which reduce the electric work produced and increase the heat transfer to the surroundings [3].

Fuel cells are very efficient, but expensive to build. Small fuel cells can power electric cars. Large fuel cells can provide electricity in remote places with no power lines. Because of the high cost to build fuel cells, large hydrogen power plants may still not be feasible. However, fuel cells are being used in some places as a source of emergency power, from hospitals to wilderness locations. Portable fuel cells are being manufactured to provide longer power for laptop computers, cell phones, and military applications [10].

6.12.1   Direct-Methanol Fuel Cells

Direct-methanol fuel cells are a subcategory of proton-exchange fuel cells in which methanol is used as the fuel. Methanol is toxic, flammable, and remains in liquid from −97.0 to 64.7 °C at atmospheric pressure. Methanol is fed as a weak solution (usually around 1 M, i.e. about 3% in mass). At the anode, methanol and water are adsorbed on a catalyst usually made of platinum and ruthenium particles and lose protons until carbon dioxide is formed. Direct methanol fuel cells use a methanol solution to carry the reactant into the cell; common operating temperatures are in the range 50–120 °C. Water is consumed at the anode and is produced at the cathode. Protons (H^+) are transported across the proton exchange membrane often made from Nafion to the cathode where they react with oxygen to produce water. Electrons are transported through an external circuit from anode to cathode, providing power to connected devices. The half-reactions take place at the anode and the cathode, and the overall reactions of the cell are

$$CH_3OH + H_2O \rightarrow CO_2 + 6H^+ + 6e^- \text{(Anode)} \tag{6.23}$$

$$3/2O_2 + 6H^+ + 6e^- \rightarrow 3H_2O (Cathode) \tag{6.24}$$

$$CH_3OH + 3/2O_2 \rightarrow CO_2 + 2H_2O + electrical\ energy (Overall\ reaction) \tag{6.25}$$

The only waste products with these types of fuel cells are carbon dioxide and water. Currently, platinum is used as the catalyst for both half-reactions. Efficiency is presently quite low for these cells, so they are targeted especially to portable application, where energy and power density are more important than efficiency. These cells need improvements for the loss of methanol and the management of carbon dioxide created at the anode [9, 10, 12].

6.12.2 Microbial Fuel Cell

A *microbial fuel cell* converts chemical energy, available in a bio substrate, directly into electricity. To achieve this, bacteria are used as a catalyst to convert a variety of organic compounds into CO_2, water, and energy. The microorganisms use the produced energy to grow and to maintain their metabolism. A microbial fuel cell can harvest a part of this microbial energy in the form of electricity. A microbial fuel cell consists of an anode, a cathode, a proton exchange membrane, and an electrical circuit, as seen in Fig. 6.17.

The bacteria live in the anode and convert not only bio substrate such as glucose, acetate but also waste water into CO_2, protons, and electrons

$$C_{12}H_{22}O_{11} + 13H_2O12CO_2 + 48H^+ + 48e^- \tag{6.26}$$

Due to the ability of bacteria to transfer electrons to an insoluble electron acceptor, microbial fuel cell collects the electrons originating from the microbial

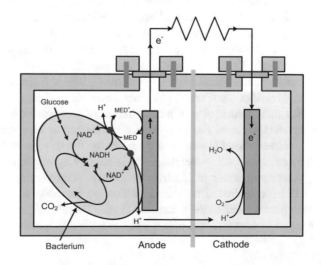

Fig. 6.17 A general layout of a microbial fuel cell in which in the anodic compartment the bacteria can bring about oxidative conversions, while in the cathodic compartment chemical and microbial reductive processes can occur [7]

metabolism. The electrons then flow through an electrical circuit to the cathode. The potential difference (Volt) between the anode and the cathode, together with the flow of electrons (Ampere) results in the generation of electrical power. The protons flow through the proton exchange membrane to the cathode. At the cathode the electrons, oxygen, and protons combine to form only water. The two electrodes are at different potentials (about 0.5 V). Microbial fuel cells use inorganic mediators to tap into the electron transport chain of cells and channel electrons produced. Some possible mediators include natural red, methylene blue, or thionine. The mediator exits the cell laden with electrons such that it shuttles to an electrode where it deposits them. This electrode becomes the anode negatively charged electrode. After releasing the electrons, the mediator returns to its original oxidized state ready to repeat the process. Therefore, the microbial activity is strongly dependent on the redox potential of the anode [7].

6.13 Bioenergy Production

Bioenergy is renewable energy derived from biomass feedstock in a biorefinery platform that can convert various types of biomass into multiproducts, including biofuels and bioproducts. Biofuels derived from energy crops are among the most rapidly growing transportation liquid fuels, including ethanol and biodiesel [7].

Bioethanol Production: Corn and sugarcane are the main energy crops for producing ethanol with the byproducts of dried distillers' grain solubles and baggasse. The most common process uses yeast to ferment the glucose into ethanol and carbon dioxide. The resulting solution is distilled to get fuel-grade ethanol to be blended with conventional gasoline. Lignocellulosic biomass can also be used as feedstock for ethanol production by a more complicated process than conventional ethanol made from fermentation. This process can decouple increasing ethanol production from energy crops from food supply chain [7].

Biodiesel and Green Diesel Production: *Biodiesel* is produced from transesterification of vegetable oils or animal fats with methanol catalytically. One hundred units of feedstock produce approximately 10 units of glycerin and 90 units of biodiesel. The base catalyzed transesterification occurs at low temperature and pressure, and yields high conversion ($\sim 98\%$). Animal and plant fats and oils are triglycerides that are esters containing three free fatty acids and the glycerol. Residual methanol is typically removed through distillation and recycled in the reactor. Glycerol needs to be purified further to be a valuable byproduct [7, 9].

Green diesel is produced by removing the oxygen and saturation of triglycerides by catalytic hydrogeneration. The necessary hydrogen may be produced using wind power and water in electrolytic process. Triglycerides acids contain long, linear aliphatic hydrocarbon chains, which are partially unsaturated and have a carbon number similar to the petroleum diesel. Therefore, green diesel has a heating value equal to conventional diesel and can be used directly in vehicles [7].

Biodiesel has around 11% oxygen, whereas petroleum-based diesel and green diesel have no oxygen. Petroleum diesel has around 10 ppm sulfur and biodiesel and green diesel have less than 1 ppm sulfur. Green diesel yield depends on both feedstock type and the level of hydro isomerization required to achieve product cloud point specification [7, 9].

Energy from Solid Waste: Municipal solid waste contains biomass like paper, cardboard, food scraps, grass clippings, leaves, wood, and others. Municipal solid waste can be burned in special waste-to-energy plants, which produce heat to make steam to heat buildings or to produce electricity. Such plants help reduce the amount of solid waste. Solid waste incinerators can burn the solid waste without electricity production [7].

6.14   Other Energy Production Opportunities

There are other energy production systems and processes either under development or operation with limited scale [3]. Some of these energy production opportunities are:

Solid-state generation (without moving parts) is of particular interest in portable applications. This area is largely dominated by thermoelectric devices, though thermionic and thermophotovoltaic systems have been developed as well.

- *Piezoelectric devices* are used for power generation from mechanical strain, particularly in power harvesting.
- *Betavoltaics* are another type of solid-state power generator which produces electricity from radioactive decay.
- *Fluid-based magnetohydrodynamic* power generation has been studied as a method for extracting electrical power from nuclear reactors and from more conventional fuel combustion systems.
- *Osmotic power* finally is another possibility at places where salt and sweet water merges in rivers and seas.
- Electromagnetic Energy Harvesting.
- Harvesting Circuits.
- Thermoelectrics.
- Microbatteries.

Summary

- Energy production mainly involves converting one form of energy into another form that is needed the most. For example, the sources of chemical energy from fossil fuels and nuclear resources are used to produce approximately 90% of the world's electricity.

- The systems used to produce a net power output are called *engines*. A *heat engine*, for example, converts heat to mechanical energy by bringing a working fluid from a high temperature state T_H to a lower temperature state T_C.
- A Carnot *heat engine* performs the conversion of heat to mechanical energy by bringing a working fluid from a high temperature state T_H to a lower temperature state T_C.
- The *Rankine cycle* converts heat into work and generates about 80% of all electric power used throughout the world.
- Pump power:

$$\dot{W}_{p,\text{in}} = \dot{m}V_4(P_1 - P_4); \dot{q}_{\text{in}} = \dot{m}(H_2 - H_1)$$
$$\dot{q}_{\text{out}} = \dot{m}(H_4 - H_3); \dot{W}_{\text{net}} = (\dot{q}_{\text{in}} - \dot{q}_{\text{out}})$$

- The gas turbine cycle relies on the *Brayton cycle* using air as the working fluid.
- A *Stirling engine* is a heat engine operating by cyclic compression and expansion of the working fluid at different temperature levels.
- A *combined cycle power plant* uses the Brayton cycle of the gas turbine with the Rankine cycle of a heat recovery steam generator.
- Improvements on power plant operations may increase efficiency and power output while reducing the fuel consumption.
- *Reheating the steam* increases the power output and the steam quality to protect the material.
- *Regeneration* uses. Steam extracted at intermediate pressures from the turbine is used to heat the boiler water feed.
- In *reheat-regenerative Rankine cycle* a part of steam extracted from the turbine heats the water from the condenser, and the remaining part is reheated in the boiler.
- *Geothermal power* comes from heat energy buried beneath the surface of the earth.
- *Cogeneration plant* produces electric power and process heat from the same heat source.
- *Nuclear power* is a method in which steam is produced by heating water through a process called *nuclear fission*.
- *Hydropower* is a process in which the force of flowing water is used to spin a turbine connected to a generator to produce electricity. Most hydroelectric power comes from the potential energy of dammed water driving a turbine and generator: $\dot{W} = \dot{Q}\rho g \Delta z = \dot{m} g \Delta z$, where $\dot{W}$ is the power, ρ is the density of water (~ 1000 kg/m^3), Δz is the height, Q is the volumetric flow rate, and g is the acceleration due to gravity.
- *Wind power* is the conversion of wind energy into electricity by using wind turbines. A wind turbine is a device for converting the kinetic energy in wind into the mechanical energy of a rotating shaft:

$$\dot{W}_{\text{wind}} = \eta_{\text{wind}}\rho \frac{\pi v^3 D^2}{8} = (\text{constant})v^3 D^2,$$

$$\text{Constant} = \frac{\eta_{\text{wind}}\rho\pi}{8}$$

where ρ is the density of air, v is the velocity of air, D is the diameter of the blades of the wind turbine, and η_{wind} is the efficiency of the wind turbine.

- *Solar power* is derived from the energy of sunlight. *Photovoltaic conversion* produces electricity directly from sunlight in a solar cell.
- A *fuel cell* oxidizes a fuel, such as hydrogen or methane, electrochemically to produce electric power. The fuel and oxygen are continuously fed into the cell and the products of reaction are withdrawn continuously. For each mole of hydrogen consumed, 2 mol of electrons pass to the external circuit. Therefore, the electrical energy (work) is the product of the charge transferred and the voltage V of the cell: $W_e = -2FV = \Delta G$, where F is the Faraday's constant ($F = 96485$ coulombs/mol) and ΔG is the Gibbs energy. The electric work: $W_e = \Delta H - q = \Delta G$.
- *Direct-methanol fuel cells* are a subcategory of proton-exchange fuel cells in which methanol is used as the fuel.
- A *microbial fuel cell* converts chemical energy, available in a bio substrate, directly into electricity.
- *Bioenergy* is renewable energy derived from biological sources, to be used for heat, electricity, or vehicle fuel. Biofuel derived from plant materials is among the most rapidly growing renewable energy technologies.
- *Bioethanol* is a clear, colorless alcohol fuel made from the sugars found in grains, such as corn, sorghum, and barley, sugarcane, and sugar beets. Therefore, ethanol is a renewable fuel.
- *Biodiesel production* involves transesterification of feedstock of vegetable oils or animal fats catalytically with a short-chain aliphatic alcohol typically methanol.
- *Coal-fired power plants* are required to meet standards that limit the amounts of some of the substances that they release into the air.

Problems

6.1   In a steam-power plant, steam at 200 psia and 600°F enters a turbine and exits at 5 psia and 200°F. The steam enters the turbine through a 2.5-inch-diameter pipe with a velocity of 11 ft/s and exits through a 9-inch-diameter pipe. Estimate the power produced by the turbine.

6.2   A turbine operates at adiabatic and steady-state conditions. At the inlet, a steam at 600 °C and 1100 kPa enters the turbine. The steam flow rate is 3 kg/s. The inlet tube diameter is 10 cm. After expanding in the turbine, the steam exits through a pipe with diameter of 25 cm. At the exit the steam is at 300 °C and 110 kPa. Estimate the work produced by the turbine.

6.3   A turbine operates at adiabatic and steady state conditions. At the inlet, a steam at 550 °C and 1500 kPa enters the turbine. The steam flow rate is 4 kg/s. The inlet tube diameter is 12 cm. After expanding in the turbine, the steam exits through a pipe with diameter of 28 cm. At the exit, the steam is saturated at 110 kPa (at 102.3 °C). Estimate the work produced by the turbine.

6.4   A turbine operates at adiabatic and steady-state conditions. At the inlet, a steam at 600 °C and 4000 kPa enters the turbine. The steam flow rate is 10 kg/s. The inlet tube diameter is 10 cm. After expanding in the turbine, the steam exits through a pipe with diameter of 30 cm. At the exit, the steam is at 150 kPa and 120 °C. Estimate the work produced by the turbine.

6.5   A turbine operates at adiabatic and steady-state conditions. At the inlet, a steam at 600 °C and 8000 kPa enters the turbine. The steam flow rate is 15 kg/s. The inlet tube diameter is 9 cm. After expanding in the turbine, the steam exits through a pipe with diameter of 35 cm. At the exit, the steam is saturated at 100 kPa (99.63 °C). Estimate the work produced by the turbine.

6.6   A superheated steam at 4100 kPa and 300 °C expand adiabatically in a steam turbine and exits at 15 kPa with a quality of $x = 0.87$. Velocity of the steam at the inlet is 50 m/s and at the exit is 160 m/s. Elevation at the inlet is 10 m and at the exit 6 m. Estimate the power produced for the steam flow rate of 1 kg/s.

6.7.   A turbine serves as an energy source for a small electrical generator. The turbine operates at adiabatic and steady-state conditions. At the inlet, a steam at 600 °C and 1100 kPa enters the turbine. The steam flow rate is 3 kg/s. At the exit, the steam is at 300 °C and 110 kPa. Estimate the work produced by the turbine.

6.8   A steam at 8000 kPa and 400 °C expands in a turbine. At the exit, the steam is at 20 kPa with a quality of 0.9. Estimate the power output if the steam flow rate is 11.5 kg/s.

6.9   A steam at 8400 kPa and 400°C expands in a turbine. At the exit, the steam is at 15 kPa with a quality of 0.92. Estimate the power output if the steam flow rate is 11.5 kg/s.

6.10 A steam expands in a turbine. The steam enters the turbine at 9000 kPa and 450 °C and exits at 10 kPa with a quality of 0.95. If the turbine produces a power of 6.5 MW estimate the steam flow rate.

6.11 A hot exhaust gas is heating a boiler to produce superheated steam at 100 psia and 400°F. In the meantime, the exhaust gas is cooled from 2500 to 350°F. Saturated liquid water (stream 1) at 14.7 psia enters the boiler with a flow rate of 200 lb/h. Superheated steam (stream 2) is used in a turbine and discharged as saturated steam (stream 3) at 14.7 psia. Determine the molar flow rate of the exhaust gas needed and the maximum work produced. Assume that the surroundings are at 70°F, and the heat capacity of the flue gas is
$C_p = 7.606 + 0.0006077T$, where T is in Rankine and C_p is in Btu/(lbmol R).

6.12   A steam expands in a turbine. The steam enters the turbine at 1000 psia and 800°F and exits as a saturated vapor at 5 psia. The turbine produces a power of 5 MW. If the steam flow rate is 20 lb/s, estimate the heat loss from the turbine.

6.13   A turbine serves as an energy source for a small electrical generator. The turbine operates at adiabatic and steady-state conditions. At the inlet a steam at 550 °C and 1500 kPa enters the turbine. The steam flow rate is 4 kg/s. At the exit, the steam is saturated at 110 kPa (at 102.3 °C). Estimate the work produced by the turbine.

6.14   A turbine serves as an energy source for a small electrical generator. The turbine operates at adiabatic and steady-state conditions. At the inlet a steam at 600 °C and 4000 kPa enters the turbine. The steam flow rate is 10 kg/s. At the exit, the steam is at 150 kPa and 120 °C. Estimate the work produced by the turbine.

6.15   A turbine serves as an energy source for a small electrical generator. The turbine operates at adiabatic and steady state conditions. At the inlet a steam at 600 °C and 8000 kPa enters the turbine. The steam flow rate is 15 kg/s. At the exit, the steam is saturated at 100 kPa (99.63 °C). Estimate the work produced by the turbine.

6.16   Steam at 8200 kPa and 823.15 K (state 1) is being expanded to 30 kPa in a continuous operation. Determine the final temperature (state 2), entropy produced, and work produced per kg of steam for an isothermal expansion through a turbine.

6.17   Steam enters an adiabatic turbine at 5000 kPa and 450 °C and leaves as a saturated vapor at 140 kPa. Determine the work output per kg of steam flowing through the turbine if the process is reversible and changes in kinetic and potential energies are negligible.

6.18   Steam at 8200 kPa and 823.15 K (state 1) is being expanded to 30 kPa in a continuous operation. Determine the final temperature (state 2), entropy produced, and work produced per kg of steam for an adiabatic expansion through a turbine.

6.19   A steady-flow adiabatic turbine receives steam at 650 K and 8200 kPa and discharges it at 373.15 K and 101.32 kPa. If the flow rate of the steam is 12 kg/s, determine (a) the maximum work and (b) the work loss if the surroundings are at 298.15 K.

6.20   A turbine discharges steam from 6 MPa and 400°C to saturated vapor at 360.15 K while producing 480 kJ/kg of shaft work. The temperature of surroundings is 300 K. Determine maximum possible production of power in kW.

6.21   A Carnot cycle uses water as the working fluid at a steady-flow process. Heat is transferred from a source at 400 °C and water changes from saturated liquid to saturated vapor. The saturated steam expands in a turbine at 10 kPa, and a heat of 1150 kJ/kg is transferred in a condenser at 10 kPa. Estimate the net power output of the cycle for a flow rate of 10 kg/s of the working fluid.

6.22   A Carnot cycle uses water as the working fluid at a steady-flow process. Heat is transferred from a source at 400 °C and water changes from saturated liquid to saturated vapor. The saturated steam expands in a turbine at 30 kPa, and a heat of 1150 kJ/kg is transferred in a condenser at 30 kPa. Estimate the net power output of the cycle for a flow rate of 14.5 kg/s of the working fluid.

6.23   A steam power production plant uses steams at 8200 kPa and 823.15 K. The turbine discharges the steam at 30 kPa. The turbine and pump operate reversibly and adiabatically. Determine the work produced for every kg steam produced in the boiler.

6.24   A steam power plant operates on a simple ideal Rankine cycle shown below. The turbine receives steam at 698.15 K and 4400 kPa, while the discharged steam is at 15 kPa. The mass flow rate of steam is 12.0 kg/s. Determine the net work output.

6.25   A reheat Rankine cycle is used in a steam power plant. Steam enters the high-pressure turbine at 9000 kPa and 823.15 K and leaves at 4350 kPa. The steam is reheated at constant pressure to 823.15 K. The steam enters the low-pressure turbine at 4350 kPa and 823.15 K. The discharged steam from the low-pressure turbine is at 10 kPa. The steam flow rate is 24.6 kg/s. Determine the net power output.

6.26   A reheat Rankine cycle is used in a steam power plant. Steam enters the high-pressure turbine at 10,000 kPa and 823.15 K and leaves at 4350 kPa. The steam is reheated at constant pressure to 823.15 K. The steam enters the low-pressure turbine at 4350 kPa and 823.15 K. The discharged steam from the low-pressure turbine is at 15 kPa. The steam flow rate is 38.2 kg/s. Determine the net power output.

6.27   A simple ideal reheat Rankine cycle is used in a steam power plant shown below. Steam enters the turbine at 9200 kPa and 823.15 K and leaves at 4350 kPa and 698.15 K. The steam is reheated at constant pressure to 823.15 K. The discharged steam from the low-pressure turbine is at 15 kPa. The net power output of the turbine is 75 MW. Determine the mass flow rate of steam.

6.28   A steam power plant uses an actual regenerative Rankine cycle. Steam enters the high-pressure turbine at 11,000 kPa and 773.15 K, and the condenser operates at 10 kPa. The steam is extracted from the turbine at 475 kPa to heat the water in an open heater. The water is a saturated liquid after passing through the water heater. The steam flow rate is 65 kg/s. Determine the work output.

6.29   A steam power plant uses an actual regenerative Rankine cycle. Steam enters the high-pressure turbine at 10,000 kPa and 773.15 K, and the condenser operates at 30 kPa. The steam is extracted from the turbine at 475 kPa to heat the water in an open heater. The water is a saturated liquid after passing through the water heater. The steam flow rate is 45.6 kg/s. Determine the work output.

6.30 A steam power plant uses an ideal regenerative Rankine cycle shown below. Steam enters the high-pressure turbine at 8400 kPa and 773.15 K, and the condenser operates at 10 kPa. The steam is extracted from the turbine at 400 kPa to heat the feed water in an open heater. The water is a saturated liquid after passing through the feed water heater. Determine the net power output of the cycle.

6.31 A steam power plant uses an actual reheat-regenerative Rankine cycle. Steam enters the high-pressure turbine at 11,000 kPa and 773.15 K, and the condenser operates at 10 kPa. The steam is extracted from the turbine at 2000 kPa to heat the water in an open heater. The steam is extracted at 475 kPa for process heat. The water is a saturated liquid after passing through the water heater. Determine the work output for a flow rate of steam of 66.0 kg/s.

6.32 A steam power plant uses an ideal reheat-regenerative Rankine cycle. Steam enters the high-pressure turbine at 9400 kPa and 773.15 K and leaves at 850 kPa. The condenser operates at 15 kPa. Part of the steam is extracted from the turbine at 850 kPa to heat the water in an open heater, where the steam and liquid water from the condenser mix and direct contact heat transfer takes place. The rest of the steam is reheated to 723.15 K and expanded in the low-pressure turbine section to the condenser pressure. The water is a saturated liquid after passing through the water heater and is at the heater pressure. The flow rate of steam is 20 kg/s. Determine the power produced.

6.33 A steam power plant uses an actual reheat regenerative Rankine cycle. Steam enters the high-pressure turbine at 10,800 kPa and 773.15 K, and the condenser operates at 15 kPa. The steam is extracted from the turbine at 2000 kPa to heat the water in an open heater. The steam is extracted at 475 kPa for process heat. The water is a saturated liquid after leaving the water heater. The steam flow rate is 30.8 kg/s. Determine the power produced.

6.34 A steam power plant uses a geothermal energy source. The geothermal source is available at 220 °C and 2320 kPa with a flow rate of 180 kg/s. The hot water goes through a valve and a flash drum. Steam from the flash drum enters the turbine at 550 kPa and 428.62 K. The discharged steam from the turbine has a quality of $x_4 = 0.95$. The condenser operates at 40 kPa. The water is a saturated liquid after passing through the condenser. Determine the net work output.

6.35 A steam power plant uses a geothermal energy source. The geothermal source is available at 220 °C and 2320 kPa with a flow rate of 50 kg/s. The hot water goes through a valve and a flash drum. Steam from the flash drum enters the turbine at 550 kPa and 428.62 K. The discharged steam from the turbine has a quality of $x_4 = 0.90$. The condenser operates at 15 kPa. The water is a saturated liquid after passing through the condenser. Determine the net work output.

6.36 A cogeneration plant uses steam at 5500 kPa and 748.15 K to produce power and process heat. The amount of process heat required is 10,000 kW. Twenty percent of the steam produced in the boiler is extracted at 475 kPa from the turbine for cogeneration. The extracted steam is condensed and mixed with the water output of the condenser. The remaining steam expands from 5500 kPa to the condenser conditions. The condenser operates at 10 kPa. Determine the net work output.

6.37 A cogeneration plant uses steam at 8400 kPa and 773.15 K. One-fourth of the steam is extracted at 600 kPa from the turbine for cogeneration. After it is used for process heat, the extracted steam is condensed and mixed with the water output of the condenser. The rest of the steam expands from 600 kPa to the condenser pressure of 10 kPa. The steam flow rate produced in the boiler is 60 kg/s. Determine the work output.

6.38 A cogeneration plant uses steam at 900 psia and 1000°F to produce power and process heat. The steam flow rate from the boiler is 40 lb/s. The process requires steam at 70 psia at a rate of 5.5 lb/s supplied by the expanding steam in the turbine. The extracted steam is condensed and mixed with the water output of the condenser. The remaining steam expands from 70 psia to the condenser pressure of 3.2 psia. Determine the work output.

6.39 One kmol of air which is initially at 1 atm, −13 °C, performs a power cycle consisting of three internally reversible processes in series. Step 1–2: Adiabatic compression to 5 atm. Step 2–3: Isothermal expansion to 1 atm. Step 3–1: Constant-pressure compression. Determine the net work output.

6.40 A steam power plant output is 62 MW. It uses steam (stream 1) at 8200 kPa and 550 °C. The discharged stream (stream 2) is saturated at 15 kPa. If the expansion in the turbine is adiabatic, and the surroundings are at 298.15 K, determine the steam flow rate.

6.41 A steam power plant output is 55 MW. It uses steam (stream 1) at 8400 kPa and 500 °C. The discharged stream (stream 2) is saturated at 30 kPa. If the expansion in the turbine is adiabatic, and the surroundings are at 298.15 K, determine the steam flow rate.

6.42 In a hydropower plant, a hydro turbine operates with a head of 33 m of water. The inlet and outlet conduits are 1.70 m in diameter. If the outlet velocity of water is 4.6 m/s estimate the power produced by the turbine.

6.43 In a hydropower plant, a hydro turbine operates with a head of 46 m of water. The inlet and outlet conduits are 1.80 m in diameter. If the outlet velocity of water is 5.5 m/s estimate the power produced by the turbine.

6.44 Consider a hydropower plant reservoir with an energy storage capacity of 1.5×10^6 kWh. This energy is to be stored at an average elevation of 40 m relative to the ground level. Estimate the minimum amount of water that must be pumped back to the reservoir.

6.45   Consider a hydropower plant reservoir with an energy storage capacity of 2.0×10^6 kWh. This energy is to be stored at an average elevation of 160 m relative to the ground level. Estimate the minimum amount of water that must be pumped back to the reservoir.

6.46   A farm of windmills supplies a power output of 2 MW for a community. Each windmill has the blades with 10 m in diameter. At the location of the windmills, the average velocity of the wind is 11 m/s and the average temperature is 20 °C. Estimate the minimum number of windmills to be installed.

6.47   A farm of windmills supplies a power output of 3 MW for a community. Each windmill has the blades with 11 m in diameter. At the location of the windmills, the average velocity of the wind is 14 m/s and the average temperature is 20 °C. Estimate the minimum number of windmills to be installed.

6.48   A farm of windmills supplies a power output of 4.2 MW for a community. Each windmill has the blades with 10 m in diameter. At the location of the windmills, the average velocity of the wind is 15 m/s and the average temperature is 20 °C. Estimate the minimum number of windmills to be installed.

References

1. IEA (2019) World energy outlook, International Energy Agency
2. Çengel YA, Boles MA (2014) Thermodynamics: an engineering approach, 8th edn. McGraw-Hill, New York
3. Chattopadhyay P (2015) Engineering thermodynamics, 2nd edn. Oxford Univ Press, Oxford
4. Clark JA (2014) Combined Heat and Power (CHP): integration with industrial processes. In: Anwar S (ed) Encyclopedia of energy engineering and technology, 2nd edn. CRC Press, Boca Raton
5. Demirel Y, Gerbaud V (2019) Nonequilibrium thermodynamics transport and rate processes in physical, chemical and biological systems, 4th edn. Elsevier, Amsterdam
6. Demirel Y, Matzen M, Winters C, Gao X (2015) Capturing and using CO_2 as feedstock with chemical-looping and hydrothermal technologies. Int J Energy Res 39:1011–1047
7. Demirel Y (2018) Biofuels. Comprehensive energy systems. In: Dincer I (ed) vol 1, Part B. Elsevier, Amsterdam, pp 875–908. (https://www.sciencedirect.com/science/article/pii/B9780128095973001255)
8. Dincer I, Ratlamwala T (2013) Development of novel renewable energy-based hydrogen production systems: a comparative study. Energy Convers Manage 7:77–87
9. Jaffe RL, Taylor W (2018) The physics of energy. Cambridge Univeristy Press, Cambridge
10. Moran MJ, Shapiro HN, Boettner DD, Bailey MB (2014) Fundamentals of engineering thermodynamics, 8th edn. Wiley, New York
11. Rabiee MM (2014) Modular smart power system. In: Anwar S (ed) Encyclopedia of energy engineering and technology, 2nd edn. CRC Press, Boca Raton
12. Vielstich W et al (eds) (2009) Handbook of fuel cells: advances in electrocatalysis, materials, diagnostics and durability, vol 6. Wiley, New York

13. Wang X, Demirel Y (2018) Feasibility of power and methanol production by an entrained-flow coal gasification system. Energy Fuels 32:7595–7610. https://doi.org/10.1021/acs.energyfuels.7b03958
14. Wood MB (2014) Nuclear energy: technology. In: Anwar S (ed) Encyclopedia of energy engineering and technology, 2nd edn. CRC Press, Boca Raton

Energy Conversion

7

Introduction and Learning Objectives: Energy can be converted from one form to another in a device or in a system. For example, batteries convert chemical energy to electrical energy and operate mobile electronic equipment. A dam converts gravitational potential energy to kinetic energy of moving water used on the blades of a turbine, which ultimately converts the kinetic energy to electric energy through an electric generator. Heat engines are devices converting heat into work. With each energy conversion, part of the energy becomes dispersed and less useful and hence unavailable for further use. This chapter discusses energy conversion and thermal efficiency of turbines, compressors, and heat engines. Heat engines include Carnot, Rankine, Brayton, Otto, and Diesel cycles.

The learning objectives of this chapter are to understand:

- Energy conversion,
- Thermal efficiencies during these energy conversion processes,
- Improving the efficiency of energy conversions, and
- Energy conversion in biological systems.

7.1 Energy Conversion

In most processes, energy is constantly changing from one form to another. This is called an *energy conversion*. Examples include the living systems converting the solar energy to chemical energy by synthesizing carbohydrates from water and carbon dioxide through the photosynthesis. The mechanical energy of a waterfall can also be converted to electromagnetic energy in a generator. An internal combustion engine converts the chemical energy in gasoline into heat, which is then transformed into the kinetic energy that moves a vehicle. A solar photovoltaic (PV) cell converts solar radiation into electrical energy. The energy that enters a

© Springer Nature Switzerland AG 2021
Y. Demirel, *Energy*, Green Energy and Technology,
https://doi.org/10.1007/978-3-030-56164-2_7

conversion device or a process is turned into other forms of energy, so an equal quantity of energy before and after the conversion is conserved.

Energy is most usable where it is most concentrated as in highly structured chemical bonds in gasoline and sugar. All other forms of energy may be completely converted to heat, but the conversion of heat to other forms of energy cannot be complete. Due to inefficiencies such as friction, heat loss, and other factors, thermal efficiencies of energy conversion are typically much less than 100%. For example, only 35–40% of the heat can be converted to electricity in a steam power plant and a typical gasoline automobile engine operates at around with 25% efficiency. With each energy conversion, a part of the energy is lost, usually in less useful and dispersed form, as illustrated in Fig. 7.1. This shows that there are limitations to the efficiency for energy conversion. Only a part of the energy may be converted to useful work and the remainder of the energy must be reserved to be transferred to a thermal reservoir at a lower temperature [2]. There are many different processes and devices that convert energy from one form to another. Table 7.1 shows a short list of such processes and devices.

When electric current flows in a circuit, it can transfer energy to do work. Devices convert electrical energy into many useful forms, such as heat (electric heaters), light (light bulbs), motion (electric motors), sound (loudspeaker), and information technological processes (computers). Electric energy is one of the most useful forms of output energy, which can be produced by various mechanical and/or chemical devices. There are several fundamental methods of directly transforming other forms of energy into electrical energy [2, 4]:

- *Static electricity* is produced from the physical separation and transport of charge. Electrons are mechanically separated and transported to increase their electric potential and imbalance of positive and negative charges leads to static electricity. For example, lightning is a natural example of static discharge. Also, low conductivity fluids in pipes can build up static electricity.

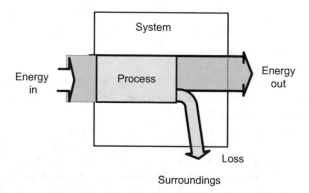

Fig. 7.1 Schematic of the energy usage and conversion in a process. Output energy at a new form is always lower than input energy. The total energy input is recovered in various other forms and hence the energy is conserved

Table 7.1 Some of the processes and devices converting energy from one form to another [2, 4, 7]

Process and device	Energy input	Useful energy output
Steam engine	Heat	Mechanical energy
Photosynthesis	Solar energy	Chemical energy
Hydroelectric dams	Gravitational potential energy	Electric energy
Windmills	Mechanical energy	Electric energy
Electric generator	Mechanical energy	Electric energy
Diesel or petrol engine	Chemical energy	Mechanical energy
Electric motor	Electric energy	Mechanical energy
Fuel cells	Chemical energy	Electric energy
Battery	Chemical energy	Electric energy
Electric bulb	Electric energy	Heat and light
Resistance heater	Electric energy	Heat
Ocean thermal power	Heat	Electric energy
Bioluminescence	Chemical energy	Light energy
Nerve impulse	Chemical energy	Electrical energy
Muscular activity	Chemical energy	Mechanical energy
Geothermal power	Heat	Electric energy
Wave power	Mechanical energy	Electric energy
Friction	Kinetic energy	Heat
Thermoelectric	Heat	Electric energy
Piezoelectrics	Strain	Electric energy

- *Electromagnetic induction*, where an electrical generator, dynamo, or alternator transforms kinetic energy into electricity. Almost all commercial electrical generation is done using electromagnetic induction, in which mechanical energy forces an electrical generator to rotate. There are many different methods of developing the mechanical energy, including heat engines, hydro, wind, and tidal power.
- *Electrochemistry* is the direct transformation of chemical energy into electricity, as in a battery, fuel cell, or nerve impulse.
- *Photoelectric effect* is the transformation of light into electrical energy, as in solar cells.
- *Thermoelectric effect* is the direct conversion of temperature differences to electricity, as in thermocouples, thermopiles, and thermionic converters.
- *Piezoelectric effect* is the electricity from the mechanical strain of electrically anisotropic molecules or crystals.
- *Nuclear transformation* is the creation and acceleration of charged particles. The direct conversion of nuclear energy to electricity by beta decay is used only on a small scale.

7.2 Series of Energy Conversions

It takes a whole series of energy conversions in various processes before a useful form of energy becomes available. For example, when you use your computer, the following energy conversion processes are involved:

- Chemical energy stored in coal is released as heat when the coal is burned.
- The heat is used to produce steam which is converted into mechanical energy in a turbine.
- The generator converts mechanical energy into electric energy that travels through the power lines into your home.
- From the power outlet at home, the computer receives that electric energy.

In a conventional internal combustion engine, these energy transformations are involved:

- Potential energy in the fuel is converted to kinetic energy of expanding gas after combustion.
- Kinetic energy of expanding gas is converted to piston movement and hence to rotary crankshaft movement.
- Rotary crankshaft movement is passed into the transmission assembly to drive the wheels of a car.

7.3 Conversion of Chemical Energy of Fuel to Heat

Chemical energy of a fuel is converted to heat during a combustion or oxidation reaction. Combustion reactions are exothermic and release heat. For example, standard heat of combustion of propane (C_3H_8) can be estimated by using the standard heats of formation tabulated in Table C1 in Appendix C. The combustion reaction of 1 mol propane is:

$$C_3H_8(g) \ + \ 5O_2(g) \rightarrow 3CO_2(g) \ + \ 4H_2O(g) + \text{Heat} \qquad (7.1)$$

When the products and reactants are at their standard states (ideal gas state at 1 bar and 25 °C) the standard heat of reaction is

$$\Delta H_r^o = \sum_i v_i \Delta H_{fi}^o = v_{CO_2}\Delta H_{fCO_2}^o + v_{H_2O}\Delta H_{fH_2O}^o + v_{C_3H_8}\Delta H_{fC_3H_8}^o \qquad (7.2)$$

where v_i is the stoichiometric coefficient of substance i, which is positive for a product and negative for a reactant. For example, for the reaction above: $v_{CO_2} = 3, v_{H_2O} = 4, v_{C_3H_8} = -1$. Standard heat of formation for stable elements such as oxygen, used in the combustion of propane, is zero.

7.3.1   Heating Value of Fuels

The *heating value* of a fuel is the amount of heat released during the combustion in kJ/kg, Btu/m^3, or kcal/kg. The heating values for fuels are expressed as the *higher heating value* (HHV), *lower heating value* (LHV), or *gross heating value* (GHV). Higher heating value is determined by bringing all the products of combustion back to the original pre-combustion temperature and condensing any water vapor produced. Therefore, the value of HHV contains the heat of condensation of the water produced. The combustion process of a fuel consisting of carbon and hydrogen can be approximately represented by

$$[C+H](fuel) + [O_2 + N_2](Air) \longrightarrow CO_2 + H_2O(liquid) + N_2 + Heat(HHV) \quad (7.3)$$

$$[C+H](fuel) + [O_2 + N_2](Air) \longrightarrow CO_2 + H_2O(vapor) + N_2 + Heat(LHV) \quad (7.4)$$

where C = Carbon, H = Hydrogen, O = Oxygen, and N = Nitrogen. Gross heating value considers the heat used to vaporize the water during the combustion reaction as well as the water existing within the fuel before it has been burned. This value is especially important for fuels like wood or coal, which contains some amount of water prior to burning. Higher and lower heating values of some common fuels, and energy density of some fuels are tabulated in Chap. 2 Tables 2.7, 2.8, and 2.9.

Lower heating value (or *net calorific value*) is determined by subtracting the heat of vaporization of the water vapor from the higher heating value. A common method of relating higher heating value to lower heating value is:

$$LHV = HHV - \left(\Delta H_{vap}\right)\left(n_{H_2O,out}/n_{fuel,in}\right)\left(MW_{H_2O,out}/MW_{fuel,in}\right)$$
or $\qquad\qquad\qquad\qquad\qquad\qquad\qquad\qquad\qquad\qquad\qquad\qquad\qquad$ (7.5)
$$LHV = HHV - \left(\Delta H_{vap}\right)\left(m_{H_2O,out}/m_{fuel,in}\right)$$

where ΔH_{vap} is the heat of vaporization of water (in kJ/kg or Btu/lb), $n_{H_2O,out}$ is the moles of water vaporized and $n_{fuel,in}$ is the number of moles of fuel combusted, MW_{H_2O} is the molecular weight of water, and MW_{fuel} is the molecular weight of fuel. Example 7.1 illustrates the estimation of lower heating value from higher heating value, while Example 7.2 shows the estimation of heating values from the standard heat of combustion.

Example 7.1 Estimation of lower heating value from higher heating value
Higher heating value of natural gas is measured as 23,875 Btu/lb around room temperature (70 °F). Convert this higher heating value to lower heating value. Heat of vaporization of water: $\Delta H_{vap} = 1055$ Btu/lb (70 °F).

Solution:
Assume that the natural gas is represented by methane CH_4.
The combustion of methane: $CH_4 + 2O_2 \rightarrow CO_2 + 2H_2O$
Higher heating value: HHV(methane) = 23,875 Btu/lb,

From Table A1:

$MW_{H_2O} = (H_2) + 1/2(O_2) = 2 + 16 = 18$ lb/lbmol

$MW_{CH_4} = (C) + 4(H) = 12 + 4 = 16$ lb/lbmol

$n_{CH_4} = 1$ lbmol and $n_{H_2O} = 2$ lbmol and $n_{H_2O}/n_{CH_4} = 2/1$

Heat of vaporization: $\Delta H_{vap} = 1055$ Btu/lb (70 °F)

Lower heating value:

$LHV = HHV - (\Delta H_{vap})(n_{H_2O,out}/n_{fuel,in})(MW_{H_2O,out}/MW_{fuel,in})$

$LHV = 23{,}875$ Btu/lb $- (1055$ Btu/lb$)(2/1)(18/16) = \mathbf{21{,}500}$ **Btu/lb**

Example 7.2 Estimating the heating values from the standard heat of combustion

The combustion reaction of 1 mol of propane is:

$$C_3H_8(g) + 5O_2(g) \rightarrow 3CO_2(g) + 4H_2O(g)$$

Estimate the higher and lower heating values of 1 kg of propane at room temperature. Heat of vaporization is: $\Delta H_{vap} = 2442$ kJ/kg (25 °C).

Solution:

When the products and reactants are at their standard states (ideal gas state at 1 bar and 25 °C) the standard heat of reaction is expressed by

$$\Delta H_r^o = \sum_i v_i \Delta H_{fi}^o = v_{CO_2}\Delta H_{fCO_2}^o + v_{H_2O}\Delta H_{fH_2O}^o + v_{C_3H_8}\Delta H_{fC_3H_8}^o$$

Stoichiometric coefficients are: $v_{CO_2} = 3 n_{H_2O} = 4, n_{C3H8} = -1$.

Using the standard heats of formation form Table C1, the heat of combustion of one mole of propane (C_3H_8) is estimated by

$$\Delta H_{r298}^o = 3(-393.51)\text{kJ/mol} + 4(-241.818)\,\text{kJ/mol}$$
$$- (-104.7)\text{kJ/mol} = -2043.1\,\text{kJ/mol}$$

Here the standard heat of reaction for oxygen is zero, as it is a naturally existing molecule in the environment.

Molecular weight of propane: $MW = 44$ g/mol.

Heat released after the combustion of 1 kg of propane is the lower heating value since the water product is at vapor state:

$$\Delta H_{r298}^o = LHV = (2043.1 \text{ kJ/mol})(\text{mol}/0.044 \text{ kg}) = 46434.0 \text{ kJ/kg}$$

The higher heating value:

$MW_{H_2O} = (H_2) + 1/2(O_2) = 2 + 16 = 18$ g/mol

$MW_{C_3H_8} = 3(C) + 8(H) = 36 + 8 = 44$ g/mol

$n_{C_3H_8} = 1$ mol and $n_{H_2O} = 4$ mol

$\Delta H_{vap} = 2442$ kJ/kg (25 °C)

Higher heating value becomes:

$\text{HHV} = \text{LHV} + (\Delta H_{vap})\left(n_{H_2O,out}/n_{C_3H_8,in}\right)\left(MW_{H_2O,out}/MW_{C_3H_8,in}\right)$ (from Eq. 7.5)

$\text{HHV} = 46{,}430$ kJ/kg $+ (2442$ kJ/kg$)(4/1)(18/44) = \mathbf{50{,}426\ kJ/kg}$

7.4 Thermal Efficiency of Energy Conversions

Thermal efficiency is a measure of the amount of thermal energy that can be converted to another useful form. For an energy conversion device like a boiler or furnace, the thermal efficiency η_{th} is the ratio of the amount of useful energy to the energy that went into the conversion process

$$\eta_{th} = \frac{q_{useful}}{q_{in}} \tag{7.6}$$

No energy conversion process is 100% efficient. For example, most incandescent light bulbs are only 5–10% efficient because most of the electric energy is lost as heat to the surroundings. An electric resistance heater has a thermal conversion efficiency close to 100%. When fuels are used, thermal efficiency is the ratio of useful energy extracted to the total chemical energy in the fuels. The definition of heating value (HHV or LHV) as input energy affects the value of efficiency.

The heat gained by the steam produced from the boiler-feed water can be estimated as the summation of sensible heat increase plus latent heat of vaporization

$$\dot{q} = \dot{m}(C_{p,av}\Delta T + \Delta H_{vap}) \tag{7.7}$$

where $\dot{q}$ is the total heat gained, $\dot{m}$ is the mass flow rate, $C_{p,av}$ is the average specific heat capacity, ΔT is the temperature difference between inlet and outlet of the water in the boiler, and ΔH_{vap} is the heat of vaporization. Vaporization process needs heat, while the condensation process releases heat as shown below

$$H_2O \text{ (liquid)} + \Delta H_{vap} \rightarrow H_2O \text{ (vapor)} \quad \text{Vaporization} \tag{7.8}$$

$$H_2O \text{ (vapor)} \rightarrow H_2O \text{ (liquid)} + \Delta H_{cond} \quad \text{Condensation} \tag{7.9}$$

The heat of vaporization is equal to the heat of condensation at the same temperature and pressure

$$\Delta H_{vap} = -\Delta H_{cond}(\text{at the same temperature and pressure}) \tag{7.10}$$

The flue gas from boilers contain the water in vapor form and the actual amount of heat available to the boiler is the lower heating value. An accurate control of the air supply is essential to the boiler efficiency. Too much air cools the furnace and carries away useful heat, while the combustion will be incomplete with less than required oxygen in the air, and unburned fuel will be carried over and smoke produced. *Net calorific value* of a fuel excludes the energy in the water vapor discharged to the stack in the combustion process. For heating systems their peak steady-state thermal efficiency is often stated as, for example, this furnace is 90% efficient. However, a more detailed measure of seasonal energy effectiveness is the Annual Fuel Utilization Efficiency (AFUE), which is discussed in Sect. 9.4.

7.5 Ideal Fluid-Flow Energy Conversions

For fluid-flow systems, enthalpy rather than the internal energy is used since the fluid-flow work (PV) is considered within the enthalpy. Figure 7.2 shows an open steady-state flow system.

The general relation between the heat and work for a fluid flow is expressed by:

$$\dot{m}\left(\Delta H + \frac{\Delta v^2}{2}\left(\frac{\text{kJ/kg}}{1000\,\text{m}^2/\text{s}^2}\right) + g\Delta z\right) = \dot{q} + \dot{W}_s \qquad (7.11)$$

Heat transfer and work transfer should be distinguished based on the entropy transfer:

- Energy interaction that is accompanied by entropy transfer is heat transfer.
- Energy interaction that is not accompanied by entropy transfer is work.

For a process producing work, such a turbine as shown in Fig. 7.3a, the *ideal work* is the maximum possible work (isentropic) produced. However, for a process requiring work, such as a compressor shown in Fig. 7.3b, the ideal work is the minimum amount of required work (isentropic). These limiting values of work occur when the change of state in the process is accomplished reversibly [2, 4]. For

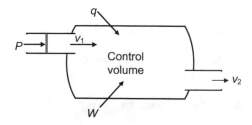

Fig. 7.2 Fluid flow with heat and work interactions between the control volume and surroundings: $\dot{m}\left(\Delta H + \Delta v^2/2\left(\frac{\text{kJ/kg}}{1000\,\text{m}^2/\text{s}^2}\right) + g\Delta z\right) = \dot{q} + \dot{W}_s$ at constant pressure

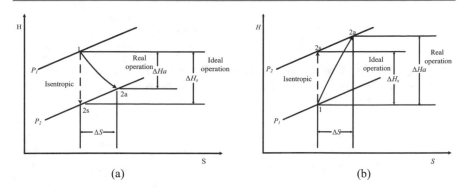

Fig. 7.3 Fluid flow work (**a**) in a turbine ($P_1 > P_2$), (**b**) in a compressor ($P_1 < P_2$); ideal operation is isentropic in a turbine and produces maximum work; ideal operation is isentropic in a compressor and requires minimum work

such processes, friction, heat loss, and other losses are negligible. When the changes in kinetic and potential energies are negligible, Eq. (7.11) becomes

$$\dot{m}\Delta H = \dot{q} + \dot{W}_s \tag{7.12}$$

For a uniform surrounding temperature T_o and using the definition of heat flow as

$$\dot{q} = T_o \Delta (\dot{m}S) \tag{7.13}$$

ideal work becomes

$$W_{\text{ideal}} = W_{\text{rev}} = \Delta(\dot{m}H) - T_o \Delta(\dot{m}S) \tag{7.14}$$

A reversible process, however, is hypothetical and used mainly for determination of the ideal work limit and comparing it with the actual work. When the ideal work is produced (Fig. 7.3a), we have the *adiabatic or isentropic efficiency* defined by

$$\eta_{Ts} = \frac{W_{\text{prod}}}{W_{\text{ideal}}} = \frac{H_{2a} - H_1}{H_{2s} - H_1} = \frac{\Delta H_a}{\Delta H_s} \text{(work produced)} \tag{7.15}$$

With the ideal work required (Fig. 7.3b), such as in a compressor, the *isentropic efficiency* η_s is

$$\eta_{Cs} = \frac{W_{\text{ideal}}}{W_{\text{req}}} = \frac{H_{2s} - H_1}{H_{2a} - H_1} = \frac{\Delta H_s}{\Delta H_a} \text{(work required)} \tag{7.16}$$

Reversible work $\dot{W}_{\text{rev}}$ in terms of the values of exergy [5, 6] for a process between the specified initial and final states is

$$\dot{W}_{\text{rev}} = \dot{m}(Ex_1 - Ex_2) + \left(1 + \frac{T_o}{T}\right)\dot{q} \qquad (7.17)$$

where T_o is the surroundings temperature and T is the system temperature. Example 7.3 illustrates the maximum expansion work calculations, while Example 7.4 calculates the isentropic efficiency.

Example 7.3 Maximum work (ideal work) calculations
One mole of air expands from an initial state of 500 K and 10 atm to the ambient conditions of the surroundings at 300 K and 1 atm. An average heat capacity of the air is $C_{p,\text{av}} = 29.5$ J/mol K. Estimate the maximum work.

Solution:
Assume: Air is an ideal gas system and the change of state is completely reversible.

$T_1 = 500\,\text{K}, \quad P_1 = 10\,\text{atm}, \quad T_2 = 300\,\text{K}, \quad P_2 = 1\,\text{atm},$
$C_{p,\text{av}} = 29.5\,\text{J/mol\,K}, \quad R = 8.314\,\text{J/mol\,K}$

For an ideal gas, enthalpy is independent of pressure and its change is:

$$\Delta H = \int_1^2 C_{p,\text{av}}dT = C_{p,\text{av}}(T_2 - T_1) = -5900 \text{ J/mol}$$

Change in entropy: $\Delta S = C_{p,\text{av}} \ln\left(\frac{T_2}{T_1}\right) - R \ln\left(\frac{P_2}{P_1}\right) = 34.21 \text{ J/mol K}$

Maximum work is the ideal work: $W_{\text{ideal}} = \Delta H - T_o \Delta S = \mathbf{-16163\,J/mol} = \mathbf{-557.9\,J/g}$

(with $MW_{\text{air}} = 28.97$ g/mol)
The negative sign indicates that this expansion work is transferred to the surroundings.

Example 7.4 Isentropic turbine efficiency
An adiabatic turbine is used to produce electricity by expanding a superheated steam at 4100 kPa and 350 °C. The steam leaves the turbine at 40 kPa and 100 °C. The steam mass flow rate is 8 kg/s. Determine the isentropic efficiency of the turbine.

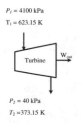
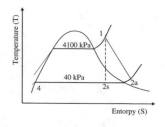
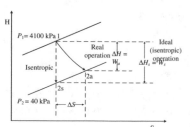

Solution:
Assume: Steady-state adiabatic operation. The changes in kinetic and potential energies are negligible.
Inlet conditions: superheated steam (steam tables)

$$P_1 = 4100\,\text{kPa}, \quad T_1 = 623.15\,\text{K}, \quad H_1 = 3092.8\,\text{kJ/kg}, \quad S_1 = 6.5727\,\text{kJ/kg K}$$

Exit conditions: saturated steam (steam tables)
$P_2 = 40$ kPa, $T_2 = 373.15$ K, $H_{2a} = 2683.8$ kJ/kg, $S_{2\text{sat vap}} = 7.6709$ kJ/kg K, $S_{2\text{sat liq}} = 1.2026$ kJ/kg K
$H_{2\text{sat vap}} = 2636.9$ kJ/kg, $H_{2\text{sat liq}} = 317.6$ kJ/kg at 40 kPa (steam tables)
For the isentropic operation $S_1 = S_2 = 6.5727$ kJ/kg K
Since $S_{2\text{sat liq}} < S_2 = S_{2\text{sat vap}}$ the steam at the exit is saturated liquid–vapor mixture, and the quality of that mixture x_{2s} is

$$x_{2s} = \frac{S_2 - S_{2\text{satliq}}}{S_{2\text{satvap}} - S_{2\text{satliq}}} = \frac{6.5727 - 1.2026}{7.6709 - 1.2026} = 0.83$$

$H_{2s} = (1 - x_{2s})H_{2\text{satliq}} + x_{2s}H_{2\text{satvap}} = 2243.1$ kJ/kg   (isentropic)   and
$H_{2a} = 2683.8$ kJ/kg (real operation)
$H_1 = 3092.8$ kJ/kg

Isentropic efficiency becomes: $\eta_{Ts} = \dfrac{H_{2a} - H_1}{H_{2s} - H_1} = \mathbf{0.48}$ (**or 48%**)

7.6   Lost Work

The difference between the ideal work and the actual work occurs because of irreversibilities within the selected path between the initial and final states for a process. The extent of irreversibility is equivalent to exergy lost and is a measure of lost work potential [6] defined as a difference between the actual and ideal work and related to the rate of entropy production. Actual work is

$$\dot{W}_{\text{act}} = \left[\dot{m} \left(\Delta H + \frac{\Delta v^2}{2} \left(\frac{\text{kJ/kg}}{1000\,\text{m}^2/\text{s}^2} \right) + \Delta gz \right) \right] - \dot{q} \tag{7.18}$$

For surrounding temperature T_o, ideal work is

$$\dot{W}_{\text{ideal}} = \left[\dot{m} \left(\Delta H + \frac{\Delta v^2}{2} \left(\frac{\text{kJ/kg}}{1000\,\text{m}^2/\text{s}^2} \right) + \Delta gz \right) \right] - T_o \Delta(\dot{m}S) \tag{7.19}$$

Then the lost work is given by

$$-\dot{W}_{\text{lost}} = \dot{W}_{\text{act}} - \dot{W}_{\text{ideal}} = \dot{q} - T_o \Delta(\dot{m}S) \tag{7.20}$$

The lost work occurs because of the rate entropy production $\dot{S}_{prod}$, which is defined as

$$\dot{S}_{prod} = \Delta(\dot{m}S) - \frac{\dot{q}}{T_o} \qquad (7.21)$$

Therefore Eq. (7.20) becomes

$$-\dot{W}_{lost} = T_o\dot{S}_{prod} \qquad (7.22)$$

Since the rate of entropy production is always positive ($\dot{S}_{prod} > 0$) for irreversible processes, then $\dot{W}_{lost} > 0$. Only when a process is completely reversible then $\dot{W}_{lost} = 0$ since $\dot{S}_{prod} = 0$. The rate of entropy production and hence the amount of lost work will increase as the irreversibility increases for a process [2, 6]. This leads to increase in the amount of energy that is unavailable and hence wasted. Example 7.5 illustrates the lost work calculations, while Example 7.6 estimates the minimum power required in a compressor.

Example 7.5 Estimation of lost work
A turbine discharges steam from 6 MPa and 400 °C to saturated vapor at 360.15 K while producing 500 kW of shaft work. The temperature of surroundings is 290 K. Determine maximum possible production of power in kW and the amount of work lost.

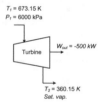

$T_1 = 673.15$ K
$P_1 = 6000$ kPa

$W_{out} = -500$ kW

Turbine

$T_2 = 360.15$ K
Sat. vap.

Solution:
Assume: The turbine operates at steady state. Kinetic and potential energy changes are negligible.
Basis: 1 kg/s steam flow rate
Inlet: superheated steam: $H_1 = 3180.1$ kJ/kg, $S_1 = 6.5462$ kJ/kg K (steam tables)
Outlet: saturated steam: $H_2 = 2655.3$ kJ/kg, $S_2 = 7.5189$ kJ/kg K (steam tables)
$W_{out} = -500$ kW, $T_o = 290$ K
$\Delta H = H_2 - H_1 = (2655.3 - 3180.1)$ kJ/kg $= -524.8$ kJ/kg
The amount of heat transfer: $q = -W + (H_2 - H_1) = -24.8$ kJ/kg
We can determine the entropy production from an entropy balance on the turbine operating at steady state that exchanges heat only with the surroundings:

$$S_{prod} = S_2 - S_1 - \frac{q}{T_o} = 7.5189 - 6.5462 + 24.8 \text{ kJ/kg}/(290)\text{K} = 1.06 \text{ kJ/kg K}$$

Lost work: $W_{lost} = T_o S_{prod} = 307.4$ kJ/kg

Maximum work output: $W_{ideal} = W_{max} = W_{out} - T_o S_{prod} = -500 - 307.4$
$= -807.4\,\text{kJ/kg}$

Example 7.6 Estimation of a minimum power required in a compressor
A compressor receives air at 15 psia and 80 °F with a flow rate of 1.0 lb/s. The air exits at 40 psia and 300 °F. Estimate the minimum power input to the compressor. The surroundings are at 520 R.

Solution:
Assume that the potential energy effects are negligible, and it is a steady process.
Basis: air flow rate = $\dot{m}$ = 1 lb/s. The surroundings are at 520 R
The properties of air from Table D1 after conversions from SI units:
State 1: P_1 = 15 psia, T_1 = 540 R, H_1 = 129.0 Btu/lb, S_1 = 0.6008 Btu/lb R
State 2: P_2 = 40 psia, T_2 = 760 R, H_2 = 182.0 Btu/lb, S_2 = 0.6831 Btu/lb R
Compressor work:
$W_s = \dot{m}(H_2 - H_1) = (1\,\text{lb/s})(182.08 - 129.06)\,\text{Btu/lb} = 53.0\,\text{Btu/s}$
The entropy production: $\dot{S}_{prod} = \dot{m}(S_2 - S_1) = 0.0823$ Btu/s R (Eq. 7.21)
Lost work: $W_{lost} = T_o S_{prod} = 42.8$ Btu/s
Minimum work required:

$$W_{ideal} = W_{min} = \dot{m}\Delta H - T_o S_{prod} = (53.0 - 42.8)\text{Btu/s} = \mathbf{10.2\,Btu/s}$$

7.7 Efficiency of Mechanical Energy Conversions

Transfer of mechanical energy is usually accomplished through a rotating shaft. When there is no loss (for example, in the form of friction) mechanical energy can be converted completely from one mechanical form to another. The mechanical energy conversion efficiency is

$$\eta_{mech} = \frac{\text{Energy out}}{\text{Energy in}} \tag{7.23}$$

For example, a mechanical energy conversion efficiency of 95% shows that 5% of the mechanical energy is converted to heat because of friction and other losses [4, 6].

In fluid-flow systems, a pump receives shaft work usually from an electric motor and transfers this shaft work partly to the fluid as mechanical energy and partly to the frictional losses as heat. As a result, the fluid pressure or velocity, or elevation is increased. A turbine, however, converts the mechanical energy of a fluid to shaft work. The energy efficiency for a pump is

$$\eta_{pump} = \frac{\text{Mechanical energy out}}{\text{Mechanical energy in}} = \frac{\dot{E}_{mech\,out}}{\dot{W}_{shaft\,in}} \tag{7.24}$$

Mechanical energy efficiency of a fan is the ratio of kinetic energy of air at the exit to the mechanical power input. If a fan, for example, is using a 50 W motor and producing an air flow velocity of 15 m/s and the air mass flow rate of 0.3 kg/s, then the mechanical energy efficiency of the fan becomes

$$\eta_{fan} = \frac{\dot{E}_{mech\,out}}{\dot{W}_{shaft\,in}} = \frac{(0.3 \text{ kg/s})(15 \text{ m/s})^2/2}{50 \text{ W}} = 0.675$$

Here, the velocity at the inlet is zero, $v_1 = 0$, and the pressure energy and potential energy are zero ($\Delta P = 0$, and $\Delta z = 0$). The energy efficiency for a turbine is

$$\eta_{turb} = \frac{\text{Mechanical energy output}}{\text{Mechanical energy (extracted from fluid) in}} = \frac{\dot{W}_{shaft\,out}}{\dot{E}_{mech\,in}} \tag{7.25}$$

The motor efficiency and the generator efficiency are

$$\eta_{motor} = \frac{\text{Mechanical power output}}{\text{Electrical energy input}} = \frac{\dot{W}_{shaft\,out}}{\dot{W}_{elect\,in}} \quad (\text{motor}) \tag{7.26}$$

$$\eta_{gen} = \frac{\text{Electrical power output}}{\text{Mechanical power input}} = \frac{\dot{W}_{elect\,out}}{\dot{W}_{shaft\,in}} \quad (\text{generator}) \tag{7.27}$$

A hydraulic turbine is combined with its generator in power production cycles, as shown in Fig. 7.4. A pump is also combined with its motor. Therefore, an overall efficiency for turbine-generator and pump-motor systems is

$$\eta_{turb-gen} = \eta_{turb}\eta_{gen} = \frac{\dot{W}_{elect\,out}}{\Delta\dot{E}_{mech\,in}} \quad (\text{turbine} - \text{generator systems}) \tag{7.28}$$

The overall efficiency of a hydraulic turbine-generator shows the fraction of the mechanical energy of the water converted to electrical energy.

$$\eta_{pump-motor} = \eta_{pump}\eta_{motor} = \frac{\Delta\dot{E}_{mech}}{\dot{W}_{elect\,in}} \quad (\text{pump} - \text{motor systems}) \tag{7.29}$$

The overall efficiency of a pump-motor system shows the fraction of the electrical energy converted to mechanical energy of the fluid [1, 2]. Example 7.7 illustrates the estimation of heat loss in an electric motor, while Example 7.8 illustrates the estimation of mechanical efficiency of a pump.

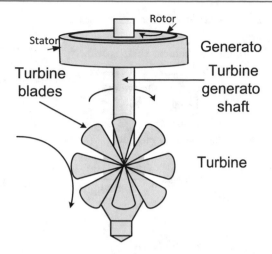

Fig. 7.4 A typical turbine and generator. The overall efficiency of a hydraulic turbine-generator is the ratio of the thermal energy of the water converted to the electrical energy; for a turbine efficiency of 0.8 and a generator efficiency of 0.9, we have: $\eta_{turb-gen} = \eta_{turb}\eta_{gen} = (0.8)(0.95) = 0.76$

Example 7.7 Heat loss in an electric motor

An electric motor attached to a pump draws 10.0 A at 110 V. At steady load the motor delivers 1.32 hp of mechanical energy. Estimate the heat loss from the motor.

Solution:
Assume that the load to motor is steady.
$I = 10.0$ A, $V = 110$ V, $\dot{W}_s = 1.32$ hp $= 983$ W
The electric power received by the motor is used to create a pump work and heat:
Power received: $\dot{W}_e = IV = 1100$ W
Heat loss: $\dot{q}_{loss} = \dot{W}_s - \dot{W}_e = (983-1100) = -117$ W.
Only 983 kW of the 1100 kW is converted to the pump work.

Example 7.8 Mechanical efficiency of a pump

The pump of a water storage tank is powered with a 16-kW electric motor operating with an efficiency of 90%. The water flow rate is 55 l/s. The diameters of the inlet and exit pipes are the same, and the elevation difference between the inlet and outlet is negligible. The absolute pressures at the inlet and outlet are 100 and 300 kPa, respectively. Determine the mechanical efficiency of the pump.

Solution:
Assume: Steady-state one-dimensional and adiabatic flow. Kinetic and potential energy changes are negligible.

$\rho_{water} = 1000 \text{ kg/m}^3 = 1 \text{ kg/l}$; Mass flow rate of water:

$\dot{m}_{water} = \rho \dot{Q} = (1 \text{ kg/l})(55 \text{ l/s}) = 55 \text{ kg/s}$

Electric motor delivers the mechanical shaft work:

$\dot{W}_{pump} = \eta_{motor} \dot{W}_{electric} = (0.9)(16 \text{ kW}) = 14.4 \text{ kW}$

Increase in the mechanical energy of the water:

$$\Delta \dot{E}_{mech} = \dot{m} \left(\frac{P_2 - P_1}{\rho} + \left(\frac{\text{kJ/kg}}{1000 \text{ m}^2/\text{s}^2} \right) \frac{v_2^2 - v_1^2}{2} + g(z_2 - z_1) \right)$$

After neglecting kinetic and potential energies:

$$\Delta \dot{E}_{mech} = \dot{m} \left(\frac{P_2 - P_1}{\rho} \right) = (55 \text{ kg/s}) \left(\frac{(300 - 100) \text{ kPa}}{1000 \text{ kg/m}^3} \right) \left(\frac{\text{kJ}}{\text{kPa m}^3} \right) = 11.0 \text{ kW}$$

The mechanical efficiency of the pump is:

$$\eta_{pump} = \frac{\Delta \dot{E}_{mech}}{\dot{W}_{pump}} = \frac{11 \text{ kW}}{14.4 \text{ kW}} = \mathbf{0.763 \text{ or } 76.3\%}$$

Only 11 kW of the 14.4 kW received by the pump is converted to the pump work. The remaining 3.4 kW is the lost as heat because of friction.

7.8   Conversion of Thermal Energy by Heat Engines

A heat engine is a device that converts heat into work. It does so by operating between two temperature limits T_H and T_L. That is, a heat engine is a device that can operate continuously to produce work receiving heat from a high temperature source T_H and rejecting nonconverted heat to a low temperature sink T_L. Heat engines transform thermal energy q_{in} into mechanical energy or work W_{out}. Some examples of heat engines include the steam engine, heat pump, gasoline (petrol) engine in an automobile, the diesel engine, and gas power cycles. All of these heat engines drive the mechanical motion of the engine by the expansion of heated gases. Some heat engines operate with phase change cycles, such as steam power production by Rankine cycle where liquid water changes to vapor after adding heat. After expansion in the cycle, the vapor condenses into liquid water. Other type of heat engines operates without phase change cycles, such as Brayton cycle where a hot gas is cooled after the expansion in the heat engine. Typical gas power cycles consist of compressing cool gas, heating the gas, expanding the hot gas, and finally cooling the gas before repeating the cycle. Table 7.2 lists some important engines and their cycle processes. Typical thermal efficiency for power plants in the industry is around 33% for coal and oil-fired plants, and up to 50% for combined-cycle gas-fired plants [1–3].

Each process in a cycle is at one of the following states:

- Isothermal (at constant temperature)
- Isobaric (at constant pressure)
- Isochoric (at constant volume)
- Adiabatic (no heat is added or removed).

Table 7.2 Comparison of the processes for some of the heat engines

Cycle	Compression	Heat addition	Expansion	Heat rejection
Rankine	Adiabatic	Isobaric	Adiabatic	Isobaric
Carnot	Isentropic	Isothermal	Isentropic	Isothermal
Ericsson	Isothermal	Isobaric	Isothermal	Isobaric
Stirling	Isothermal	Isochoric	Isothermal	Isochoric
Diesel	Adiabatic	Isobaric	Adiabatic	Isochoric
Otto	Adiabatic	Isochoric	Adiabatic	Isochoric
Brayton	Adiabatic	Isobaric	Adiabatic	Isobaric

Engine types vary as shown in Table 7.3. In the combustion cycles, once the mixture of air and fuel is ignited and burnt, the available energy can be transformed into the work by the engine. In a reciprocating engine, the high-pressure gases inside the cylinders drive the engine's pistons. Once the available energy has been converted to mechanical work, the remaining hot gases are vented out and this allows the pistons to return to their previous positions [1, 2, 7].

Some of the available heat input is not converted into work but is dissipated as waste heat q_{out} into the environment. The thermal efficiency of a heat engine is the percentage of heat energy that is converted into work

$$\eta_{th} = \frac{|W_{out}|}{q_{in}} = 1 - \frac{|q_{out}|}{q_{in}} \tag{7.30}$$

The absolute signs for work out and heat out are for the sign convention that assumes the output of work or heat from a system is negative. The efficiency of even the best heat engines is low; usually below 50% and often far below. So, the energy lost to the environment by heat engines is a major waste of energy resources, although modern cogeneration, combined cycle, and energy recycling schemes are beginning to use this waste heat for other purposes. This inefficiency can be attributed mainly to three causes:

- Thermal efficiency of any heat engine is limited by the Carnot efficiency.
- Specific types of engines may have lower limits on their efficiency due to the inherent irreversibility of the engine cycle they use.
- The nonideal behavior of real engines, such as mechanical friction and losses in the combustion process, may cause further efficiency losses.

Table 7.3 Various types of cycles for heat engines

External combustion cycles	Without phase change: Brayton, Ericsson, Stirling
	With phase change: Rankine, Carnot, Two-phased Stirling
Internal combustion cycles	Without phase change: Diesel, Otto
Refrigeration cycles	Vapor-compression
Heat pump cycles	Vapor-compression

Thermal efficiency of heat engine cycles cannot exceed the limit defined by the Carnot cycle, which states that the overall efficiency is dictated by the difference between the lower and upper operating temperatures of the engine. This limit assumes that the engine is operating in ideal conditions that are reversible with no heat loss. A car engine's real-world fuel economy is usually measured in the units of miles per gallon (or fuel consumption in liters per 100 km). Even when aided with turbochargers and design aids, most engines retain an *average* efficiency of about 30–35% [2, 4, 8]. The efficiencies of rocket engines are better, up to 70%, because they operate at very high temperatures and pressures, and can have very high expansion ratios. For stationery and shaft engines, fuel consumption is measured by calculating the *brake specific fuel consumption* which measures the mass flow rate of fuel consumed divided by the power produced. Example 7.8 illustrates the estimation of fuel consumption of a car.

The engine efficiency alone is only one factor. For a more meaningful comparison, the overall efficiency of the entire energy supply chain from the fuel source to the consumer should be considered. Although the heat wasted by heat engines is usually the largest source of inefficiency, factors such as the energy cost of fuel refining and transportation, and energy loss in electrical transmission lines may offset the advantage of a more efficient heat engine [2, 3]. Engines must be optimized for other goals besides efficiency such as for low emissions, adequate acceleration, fast starting, light weight, and low noise. These requirements may lead to compromises in design that may reduce efficiency. Large stationary electric generating plants have fewer of these competing requirements, so the Rankine cycles are significantly more efficient than vehicle engines. Therefore, replacing internal combustion vehicles with electric vehicles, which run on a battery recharged with electricity generated by burning fuel in a power plant, has the theoretical potential to increase the thermal efficiency of energy use in transportation, thus decreasing the demand for fossil fuels [2, 4].

Real engines have many departures from ideal behavior that waste energy, reducing actual efficiencies far below the theoretical values because of:

- Friction of moving parts.
- Inefficient combustion.
- Heat loss from the combustion chamber.
- Nonideal behavior of the working fluid.
- Inefficient compressors and turbines.
- Imperfect valve timing.

Example 7.9 Carnot limit in heat engines

In a cyclic process, a heat engine takes in energy from a hot reservoir q_h and converts this energy into mechanical work W, discharge some energy q_c to the surroundings and return to its original state.

Solution:
Conservation of energy shows: $q_h = W + q_c$

The entropy added when heat is absorbed, and entropy decreased when energy is discharged are

$$\Delta S_h = \left(\frac{q}{T}\right)_h + \Delta S'_h$$
$$\Delta S_c = \left(\frac{q}{T}\right)_c - \Delta S'_c$$

where S' is the entropy production due to irreversibility and always positive. The net change in entropy is zero in the cyclic process $\Delta S_h = \Delta S_c$ and in the reversible operation, we have

$$\frac{q_h}{q_c} = \frac{T_h}{T_c}$$

Defining the efficiency of a heat engine as: $\eta = \dfrac{W}{q_h}$

And using $q_h = W + q_c$, we have: $\eta = \dfrac{W}{q_h} = \dfrac{q_h - q_c}{q_h} \leq \dfrac{T_h - T_c}{T_h}$

This relation shows the limit on the thermal efficiency of a heat engine formulated by Carnot and known as Carnot limit, which is applied to cyclic operation of heat engines.

$$\eta_C = \frac{T_h - T_c}{T_h} > \frac{W}{q_h}$$

This relation also brings out an important difference between thermal energy associated with entropy which is not the case for mechanical energy. The following table compares the thermal efficiency of various heat engines [3, 4, 8].

Energy source	T_h (K)	T_c (K)	η_C %	η %
Steam turbine nuclear plant	600	300	50	33
Steam turbine coal plant	800	300	60	33
Combined cycle gas turbine	1500	300	83	45
Auto engine (Otto cycle)	2300	300	87	30
Solar energy (photovoltaic module)	6000	300	95	15-20

Example 7.10 Thermal efficiency of a heat engine
Heat is transferred to a heat engine from a furnace at a rate of 90 MW. The waste heat is discharged to the surroundings at a rate of 55 MW. Estimate the net power output and thermal efficiency of the engine if all the other power losses are neglected.

Solution:
Assume: steady-state process with negligible heat losses.
$\dot{q}_{in} = 90\,\text{MW}$, $\dot{q}_{out} = 55\,\text{MW}$ (system is surrounding that receives this heat)
Net power output: $\dot{W}_{net} = \dot{q}_{out} - \dot{q}_{in} = -35\,\text{MW}$ (the system is heat engine)
Thermal efficiency: $\eta_{th} = \dfrac{|\dot{W}_{net}|}{\dot{q}_{in}} = \dfrac{35}{90} = \mathbf{0.388}\,(\textbf{or } \mathbf{38.8\%})$
The heat engine can convert 38.8% of the heat transferred from the furnace.

Example 7.11 Fuel consumption of a car
The overall efficiencies are about 25–28% for gasoline car engines, 34–38% for diesel engines, and 40–60% for large power plants [4]. A car engine with a power output of 120 hp has a thermal efficiency of 24%. Determine the fuel consumption of the car if the fuel has a higher heating value of 20,400 Btu/lb.

Solution:
Assume: the car has a constant power output.
Net heating value $\cong$ higher heating value (0.9) = 18,360 Btu/lb (Approximate)
Car engine power output and efficiency: $\dot{W}_{net} = 120\,\text{hp}$, $\eta_{th} = 0.24$

$$\eta_{th} = \frac{\dot{W}_{net}}{\dot{q}_{in}} \rightarrow \dot{q}_{in} = \frac{\dot{W}_{net}}{\eta_{th}} = \frac{120\,\text{hp}}{0.24}\left(\frac{2545\,\text{Btu/h}}{\text{hp}}\right)$$
$$\dot{q}_{in} = 1,272,500\,\text{Btu/h}$$

Net heating value $\cong$ higher heating value (0.9) = 18,360 Btu/lb
Fuel consumption = $\dot{q}_{in}$/net heating value = (1,272,500 Btu/h)/(18,360 Btu/lb)
= **69.3 lb/h**
Assuming an average density of 0.75 kg/l:
ρ_{gas} = (0.75 kg/l)(2.2 lb/kg)(1/0.264 gal) = 6.25 lb/gal
Fuel consumption in terms of gallon: (69.3 lb/h)/(6.25 lb/gal) = **11.1 gal/h**

7.8.1 Air-Standard Assumptions

Internal combustion cycles of Otto and diesel engines as well as the gas turbines are some well-known examples of engines that operate on gas cycles. In these cycles, the working fluid remains gas for the entire cycle. In the analysis of gas power cycles, the following assumptions known as *air-standard assumptions* are used:

- Working fluid is air and always behaves as an ideal gas.
- All processes in the cycle are internally reversible (isentropic).
- Heat-addition process uses an external heat source.
- Heat-rejection process restores the working fluid to its original state.

It is also assumed that the air has a constant value for the ratio of specific heats determined at room temperature (25 °C or 77 °F). These air-standard assumptions are called the *cold-air-standard assumptions*, which simplify the analysis of gas power cycles without significantly deviating from the actual cycles [2].

7.8.2 Isentropic Processes of Ideal Gases

The entropy change of ideal gas is

$$\Delta S = S_2 - S_1 = \int_1^2 C_p(T) \frac{dT}{T} - R \ln\left(\frac{P_2}{P_1}\right) \tag{7.31}$$

Under isentropic conditions ($\Delta S = 0$) and constant heat capacity $C_{p,\mathrm{av}}$, Eq. (7.31) becomes

$$0 = C_{p,\mathrm{av}} \ln\left(\frac{T_2}{T_1}\right) - R \ln\left(\frac{P_2}{P_1}\right) \tag{7.32}$$

For ideal gas, the heat capacities are related by

$$C_p - C_v = R \tag{7.33}$$

From Eqs. (7.32) and (7.33) the following relations may be derived

$$\left(\frac{T_2}{T_1}\right) = \left(\frac{V_1}{V_2}\right)^{(\gamma-1)} \quad \text{and} \quad \left(\frac{T_2}{T_1}\right) = \left(\frac{P_2}{P_1}\right)^{(\gamma-1)/\gamma} \tag{7.34}$$

where $\gamma = C_p/C_v$.

Under isentropic conditions ($\Delta S = 0$) and variable heat capacities, Eq. (7.31) becomes

$$0 = S_{T2} - S_{T1} - R \ln\left(\frac{P_2}{P_1}\right) \tag{7.35}$$

where S_{T1} and S_{T2} are the values of entropy at temperatures T_1 and T_2, respectively. After rearranging, Eq. (7.35) yields,

$$\frac{P_2}{P_1} = \exp\left(\frac{S_{T2} - S_{T1}}{R}\right) = \frac{\exp(S_{T2}/R)}{\exp(S_{T1}/R)} = \frac{P_{r2}}{P_{r1}} \tag{7.36}$$

where P_{r1} is called the *relative pressure* given by $\exp(S_{T1}/R)$, which is dimensionless quantity. Using the ideal gas relation

$$\frac{P_1 V_1}{T_1} = \frac{P_2 V_2}{T_2} \tag{7.37}$$

the relative specific volume V_r ($V_r = T/P_r$) is derived by

$$\frac{V_2}{V_1} = \frac{T_2/P_{r2}}{T_1/P_{r1}} = \frac{V_{r2}}{V_{r1}} \tag{7.38}$$

The values of P_r and V_r are tabulated for the air against temperature in Appendix D, Table D1. Equations (7.35) to (7.38) account for the variation of specific heats with temperature and are valid only for isentropic processes of ideal gases. Equations (7.36) and (7.38) are useful in the analysis of gas power cycles operating with isentropic processes [2].

7.8.3 Conversion of Mechanical Energy by Electric Generator

In electricity generation, an *electric generator* converts mechanical energy to electrical energy by generally using *electromagnetic induction*. Electromagnetic induction is the production of electric potential (voltage) across a conductor moving through a magnetic field. A generator forces electron in the windings to flow through the external electrical circuit. The source of mechanical energy may be water falling through a turbine, waterwheel, an internal combustion engine, and a wind turbine. The reverse is the conversion of electrical energy into mechanical energy, which is done by an electric motor. Figure 7.4 shows a schematic view of a generator.

The efficiency of a generator is determined by the power of the load circuit and the total power produced by the generator. For most commercial electrical generators, this ratio can be as high as 95% percent. The losses typically arise from the transformer, the copper windings, magnetizing losses in the core, and the rotational friction of the generator [8]. The overall efficiency of a hydraulic turbine-generator is the ratio of the thermal energy of the water converted to the electrical energy and obtained by

$$\eta_{\text{turb-gen}} = \eta_{\text{turb}} \eta_{\text{gen}} \tag{7.39}$$

Michael Faraday discovered the operating principle of electromagnetic generators. The principle, later called Faraday's law, states that an electromotive force is generated in an electrical conductor that encircles a varying magnetic flux. The two main parts of a generator are rotor and stator. Rotor is the rotating part, while stator is the stationary part. The armature of generator windings generates the electric current. The armature can be on either the rotor or the stator. Because power transferred into the field circuit is much less than in the armature circuit, alternating current generators mostly have the field winding on the rotor and the stator as the armature winding.

An electric generator or electric motor uses magnetic field coils. If the field coils are not powered, the rotor in a generator can spin without producing any usable electrical energy. When the generator first starts to turn, a small amount of magnetism present in the iron core provides a magnetic field to get it started, generating a small current in the armature. This flows through the field coils, creating a larger magnetic field which generates a larger armature current. This 'bootstrap' process continues until the magnetic field in the core levels off due to saturation and the generator reaches a steady-state power output. Very large power station generators often utilize a separate smaller generator to excite the field coils of the larger generator [4, 8].

The *dynamo* was the first electrical generator capable of delivering power for industry. The dynamo uses electromagnetic principles to convert mechanical rotation into the direct current. A dynamo consists of a stationary structure, which provides a constant magnetic field, and a set of rotating windings which turn within that field. A *commutator* is a rotary electrical switch in certain types of electric motors or electrical generators that periodically reverses the current direction between the rotor and the external circuit [3].

7.8.4 Carnot Engine Efficiency

A steam power plant is a typical heat engine converting a low-grade energy (heat) to high-grade energy (electricity). In the steam power plant, heat is added to the boiler to produce steam. The steam is then expanded through a steam turbine where the steam's enthalpy drop is converted to shaft work and electricity eventually as a positive work output. The steam is then exhausted into a condenser that serves as a sink where the steam is condensed.

Carnot *heat engine* converts heat to mechanical energy by bringing a working fluid from a high temperature state T_H to a lower temperature state T_C. Figure 7.5 shows a typical pressure–volume PV and temperature–entropy TS diagrams of ideal engine cycles. On both the PV and TS diagrams, the area enclosed by the process curves of a cycle represents the net heat transfer to be converted to mechanical energy by the engine. Therefore, any modifications that improve the net heat transfer rate will also improve the thermal efficiency [2].

A heat source heats the working fluid in the high temperature state. The working fluid usually is a gas or liquid. During the operation of an engine some of the thermal energy is converted into work and the remaining energy is lost to a heat sink, mainly the general surroundings.

Carnot cycle is composed of four totally reversible processes, as shown in Fig. 7.5:

- Process 1–2: Isothermal heat addition at constant temperature T_H
- Process 2–3: Isentropic expansion at constant entropy $S_2 = S_3$

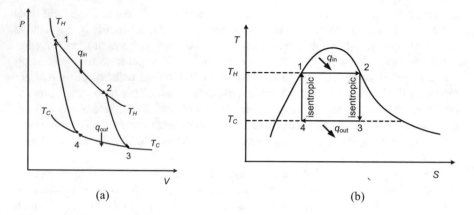

Fig. 7.5 Typical ideal engine cycles **a** on a pressure–volume P–V diagram, **b** on a temperature–entropy T–S diagram

- Process 3–4: isothermal heat rejection at constant temperature T_C
- Process 4–1: isentropic compression at constant entropy $S_4 = S_1$

Thermal efficiency of the Carnot engine is calculated by

$$\eta_{th} = 1 - \frac{|q_{out}|}{q_{in}} \tag{7.40}$$

From Fig. 7.5b, we can estimate the amounts of added q_{in} and rejected heat q_{out} values as

$$q_{in} = T_H(S_2 - S_1) \tag{7.41}$$

$$q_{out} = T_C(S_4 - S_3) \tag{7.42}$$

Using Eqs. (7.41) and (7.42) in Eq. (7.40), we have

$$\eta_{th} = 1 - \frac{T_C(S_4 - S_3)}{T_H(S_2 - S_1)} = 1 - \frac{T_C}{T_H} \tag{7.43}$$

Since the power cycles are isentropic, we have $S_2 = S_3$ and $S_4 = S_1$. Equation (7.43) shows that heat engines efficiency is limited by Carnot's efficiency which is equal to the temperature difference between the hot (T_H) and cold (T_C) divided by the temperature at the hot end, all expressed in absolute temperatures. The Carnot cycle is a standard against which the actual or ideal cycle can be compared. The Carnot cycle has the highest thermal efficiency of all heat engines operating between the same heat source temperature T_H and the same sink temperature T_C.

The cold side of any heat engine is close to the ambient temperature of the environment of around 300 K, such as a lake, a river, or the surrounding air. Therefore, most efforts to improve the thermodynamic efficiencies of various heat engines focus on increasing the temperature of the hot source within material limits. The highest value of T_H is limited by the maximum temperature that the components of the heat engine, such as pistons or turbine blades, can withstand.

The thermal efficiency of Carnot engine is independent of the type of the working fluid. For example, for a hot source temperature of $T_H = 1200$ K and the cold source temperature of $T_C = 800$ K, the maximum possible efficiency becomes:

$$\eta_{Carnot} = 1 - \frac{800 \text{ K}}{1200 \text{ K}} = 0.33 \text{ or } 33\%$$

Practical engines have efficiencies far below the Carnot limit. For example, the average automobile engine is less than 35% efficient. As Carnot's theorem only applies to heat engines, devices that convert the fuel's energy directly into work without burning it, such as fuel cells, can exceed the Carnot efficiency. Actual cycles involve friction, pressure drops as well as the heat losses in various processes in the cycle; hence they operate at lower thermal efficiencies [4, 8].

7.8.5   Endoreversible Heat Engine Efficiency

A Carnot engine must operate at an infinitesimal temperature gradient, and therefore the Carnot efficiency assumes the infinitesimal limit. Endoreversible engine gives an upper bound on the energy that can be derived from a real process that is lower than that predicted by the Carnot cycle, and accommodates the heat loss occurring as heat is transferred. This model predicts how well real-world heat engines can perform by

$$\eta_{max} = 1 - \sqrt{\frac{T_C}{T_H}} \tag{7.44}$$

Table 7.4 compares the efficiencies of engines operating on the Carnot cycle, endoreversible cycle, and actual cycle. As can be seen, the values of observed efficiencies of actual operations are close to those obtained from the endoreversible cycles.

Table 7.4 Efficiencies of engines operating on the Carnot cycle, endoreversible cycle, and actual cycle [1, 2]

Power plant	T_C (°C)	T_H (°C)	η (Carnot)	η (Endoreversible)	η (Observed)
Coal-fired power plant	25	565	0.64	0.40	0.36
Nuclear power plant	25	300	0.48	0.28	0.30
Geothermal power plant	80	250	0.33	0.18	0.16

7.8.6 Rankine Engine Efficiency

The Rankine cycle is used in steam turbine power plants. A turbine converts the kinetic energy of a moving fluid to mechanical energy. The steam is forced against a series of blades mounted on a shaft connected to the generator. The generator, in turn, converts its mechanical energy to electrical energy based on the relationship between magnetism and electricity. The most of the world's electric power is produced with this cycle. Since the cycle's working fluid changes from liquid to vapor and back to liquid water during the cycle, their efficiencies depend on the properties of water. The thermal efficiency of modern steam turbine plants with reheat cycles and combined cycle plants is much higher than that of conventional cycles [6, 11].

Figure 7.6 describes the main processes within the Rankine cycle. The main processes of the Rankine cycle are:

- Process 4–1: The water is pumped from low to high pressure.
- Process 1–2: The high-pressure liquid enters the boiler where it is heated at constant pressure by an external heat source to become a vapor.

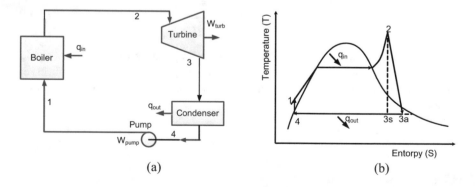

Fig. 7.6 **a** Schematic of Rankine cycle, **b** Rankine cycle processes on a T versus S diagram; Process 2–3 s shows isentropic expansion, while process 2–3a shows a real expansion process where entropy is not constant and the amount of waste heat increases

- Process 2–3: The vapor expands through the turbine, generating power. This decreases the temperature and pressure of the vapor, and some condensation may occur.
- Process 3–4: The wet vapor then enters the condenser where it is condensed at a constant pressure to become a saturated liquid.

The Rankine cycle is sometimes referred to as a practical Carnot cycle (see Fig. 7.5) because in an ideal Rankine cycle the pump and turbine would be *isentropic* and produce no entropy and hence maximize the work output. The main difference is that heat addition (in the boiler) and rejection (in the condenser) are isobaric in the Rankine cycle and isothermal in the Carnot cycle [6]. Example 7.12 analyzes the power output of steam power plant.

Example 7.12 Steam turbine efficiency and power output
An adiabatic turbine is used to produce electricity by expanding a superheated steam at 4100 kPa and 350 °C. The steam leaves the turbine at 40 kPa. The steam mass flow rate is 8 kg/s. If the isentropic efficiency is 0.75, determine the actual power output of the turbine.

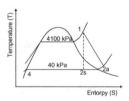

Solution:
Assume: Steady-state adiabatic operation. The changes in kinetic and potential energies are negligible.
Data from steam tables:
Inlet   conditions: $P_1 = 4100$ kPa, $T_1 = 623.15$ K, $H_1 = 3092.8$ kJ/kg, $S_1 = 6.5727$ kJ/kg K
Exit conditions:
$P_2 = 40$ kPa, $S_{2\text{sat vap}} = 7.6709$ kJ/kg K, $S_{2\text{sat liq}} = 1.0261$ kJ/kg K
$H_{2\text{sat vap}} = 2636.9$ kJ/kg, $H_{2\text{sat liq}} = 317.6$ kJ/kg at 40 kPa
For the isentropic operation $S_1 = S_2 = 6.5727$ kJ/kg K
Since $S_1 < S_{2\text{sat vap}}$ the steam at the exit is saturated liquid–vapor mixture
Quality of the saturated mixture: $x_{2s} = \dfrac{S_2 - S_{2\text{sat liq}}}{S_{2\text{sat vap}} - S_{2\text{sat liq}}} = \dfrac{6.5727 - 1.0261}{7.6709 - 1.0261} = 0.83$

$$H_{2s} = (1 - x_{2s})H_{2\text{sat liq}} + x_{2s}H = 2243.1 \, \text{kJ/kg}_{2\text{sat vap}} = 2243.1 \, \text{kJ/kg}$$

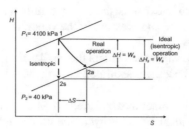

Using isentropic efficiency of 0.75, we can estimate the actual enthalpy H_{2a},

$$\eta_t = \frac{H_1 - H_{2a}}{H_1 - H_{2s}} = 0.75 \rightarrow H_{2a} = 2455.5 \text{ kJ/kg}$$

Actual power output which is not isentropic is $\dot{W}_a$ (see figure above)

$$\dot{m}\Delta H = \dot{W}_a = \dot{m}(H_{2a} - H_1) = (8 \text{ kg/s})(2455.5 - 3092.8) = -5098.4 \text{ kW}$$
$$= -5.098 \text{ MW}$$

Maximum power output: $\dot{W}_s$ (see figure above) is only achievable by isentropic expansion.

$$\dot{m}\Delta H = \dot{W}_s = \dot{m}(H_{2s} - H_1) = (8 \text{ kg/s})(2243.1 - 3092.8) = -6797.6 \text{ kW}$$
$$= -6.797 \text{ MW}$$

The enthalpy-entropy HS diagram above explains the difference between real and ideal turbine operation. The signs are negative for work output.

Example 7.13 Estimation of thermal efficiency of a Rankine cycle
A Rankine cycle shown below uses natural gas to produce 0.12 MW power. The combustion heat supplied to a boiler produces steam at 10,000 kPa and 798.15 K, which is sent to a turbine. The turbine efficiency is 0.7. The discharged steam from the pump efficiency is 0.75. Determine:

(a) The thermal efficiency of an ideal Rankine cycle.
(b) The thermal efficiency of an actual Rankine cycle.

Solution:
Assume that kinetic and potential energy changes are negligible, and the system is at steady state. Consider Fig. 7.6.

(a) The basis is 1 kg/s steam.

Turbine in: $H_2 = 3437.7 \text{ kJ/kg}$, $S_2 = 6.6797 \text{ kJ/kg K}$, at $T_2 = 798.15 \text{ K}$,
$P_2 = 10,000 \text{ kPa}$ (steam tables)

Turbine out: $H_{4\text{sat liq}} = H_{3\text{sat liq}} = 289.3$ kJ/kg, $H_{4\text{sat vap}} = H_{3\text{sat vap}} = 2625.4$ kJ/kg; at $P_3 = P_4 = 30$ kPa, $S_{3\text{sat liq}} = 0.9441$ kJ/kg K; $S_{3\text{sat vap}} = 7.7695$ kJ/kg K; (steam tables)

$\eta_{\text{turb}} = 0.70$; $\eta_{\text{pump}} = 0.75$; $V = 1020$ cm^3/kg at $T = 342.15$ K

For the ideal Rankine cycle, the operation is isentropic, and $S_2 = S_3$. However, $S_2 < S_{3\text{sat vap}}$, and the discharged steam from the turbine is wet steam.

The quality of the wet steam x_{3s}: $x_{3s} = \dfrac{S_2 - S_{3\text{sat liq}}}{S_{3\text{sat vap}} - S_{3\text{sat liq}}} = 0.84$

The enthalpy of the wet steam H_{3s}:

$$H_{3s} = (1 - x_{3s})H_{3\text{satliq}} + x_{3s}H_{3\text{satliq}} = 2252.4 \text{ kJ/kg}$$

Ideal operation: $W_s = \Delta H_s = H_{3s} - H_2 = -1185.3$ kJ/kg (isentropic expansion)

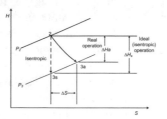

$$P_3 = P_4, P_1 = P_2$$

From the isentropic pump operation, we have: $W_{ps} = V(P_2 - P_3) = 10.2$ kJ/kg

So the enthalpy at point 1: $H_{1s} = \Delta W_{ps} + H_4 = 10.2 + 289.3 = 299.5$ kJ/kg

The heat required in the boiler becomes: $q_{\text{in}} = H_2 - H_{1s} = 3138.2$ kJ/kg

The net work for the ideal Rankine cycle is:

$$W_{\text{net,s}} = W_p + \Delta H_s = (10.2 - 1185.3) \text{ kJ/kg} = -1175.1 \text{ kJ/kg}$$

The efficiency of the ideal Rankine cycle: $\eta_{\text{idealcycle}} = \dfrac{|W_{\text{net,s}}|}{q_{in}} = \mathbf{0.374\, or\, 37.4\%}$

(b)   Actual cycle efficiency: $\dfrac{\Delta H_a}{\Delta H_s} = \eta_{t,\text{actual}}$

With the turbine efficiency of $\eta_t = 0.7$, we have: $\Delta H_a = \eta_t \Delta H_s = -829.7$ kJ/kg

Pump efficiency is: $\dfrac{\Delta W_{ps}}{\Delta W_{pa}} = \eta_{\text{pump}}$; we have: $W_{pa} = \dfrac{W_{ps}}{\eta_{\text{pump}}} = 13.6$ kJ/kg

The net work for the actual cycle is: $W_{\text{net,act}} = W_{pa} + \Delta H_a = -816.1$ kJ/kg

$$H_{1a} = \Delta W_{pa} + H_4 = (13.6 + 289.3) \text{ kJ/kg} = 302.9 \text{ kJ/kg}$$
$$q_{\text{in,act}} = H_2 - H_{1a} = (3437.7 - 302.9) \text{ kJ/kg} = 3134.81 \text{ kJ/kg}$$

Therefore, the efficiency of the actual cycle is: $\eta_{actual} = \dfrac{|W_{net,act}|}{q_{in,act}} = \mathbf{0.260\ or\ 26\%}$

Comparison of the two efficiencies shows that both operations have relatively low efficiencies, although the actual cycle is considerably less efficient than the ideal Rankine cycle (37.4%). The actual cycles involve friction and pressure drop, as well as the heat losses in various processes; therefore, they have lower thermal efficiency than those of ideal cycles.

7.8.7 Brayton Engine Efficiency

The Brayton cycle was first proposed by George Brayton in around 1870. The Brayton cycle is used for gas turbines operating on an open cycle as shown in Fig. 7.7. The Brayton engine consists of three main components: compressor, combustion chamber, and turbine. The air after being compressed in the compressor is heated by burning fuel in the combustion chamber. The heated air expands in the turbine and produces power. The two main applications of gas turbine engines are jet engines and electric power production. Jet engines take a large volume of hot gas from the combustion process and feed it through a nozzle which accelerates the plane to high speed (see Fig. 7.8). Gas turbine cycle engines employ a continuous combustion system where compression, combustion, and expansion occur simultaneously in the engine. The combustion takes place at constant pressure. The fuel must release sufficient heat of combustion to produce necessary power.

Some of the power produced in the turbine of a gas-turbine power plant is used to drive the compressor. The ratio of the compressor work to the turbine work is called the *back-work ratio*. Sometimes, more than one-half of the turbine work may be used by the compressor [2, 3].

The Brayton cycle is analyzed as open system. By neglecting the changes in kinetic and potential energies, the energy balance on a unit-mass basis is

$$\Delta H = (q_{in} - q_{out}) + (W_{in} - W_{out}) \tag{7.45}$$

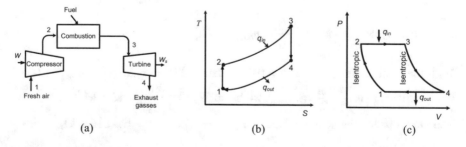

$$\text{(a)} \qquad\qquad\qquad\qquad \text{(b)} \qquad\qquad\qquad\qquad \text{(c)}$$

Fig. 7.7 **a** Schematic of open Brayton cycle, **b** ideal Brayton cycle on a *T–S* diagram with processes of isentropic adiabatic compression, isobaric heat addition, isentropic adiabatic expansion, and isobaric heat rejection, **c** ideal Brayton cycle on a *P–V* diagram

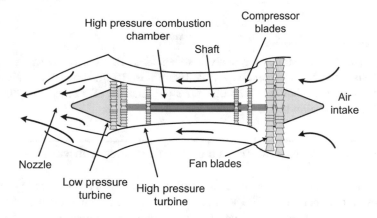

Fig. 7.8 Schematic of a jet engine using gas power cycle

Assuming constant heat capacity, thermal efficiency can be derived using the heat added q_{in} and heat rejected q_{out}

$$\eta_{Brayton} = 1 - \left(\frac{|q_{out}|}{q_{in}}\right) = 1 - \left(\frac{C_{p,av}(T_4 - T_1)}{C_{p,av}(T_3 - T_2)}\right) \tag{7.46}$$

Upon rearrangement, Eq. (7.46) reduces to

$$\eta_{Brayton} = 1 - \left(\frac{T_1}{T_2}\right)\left(\frac{T_4/T_1 - 1}{T_3/T_2 - 1}\right) \tag{7.47}$$

Figure 7.7 shows that processes 1–2 and 3–4 are isentropic, and $P_2 = P_3$ and $P_4 = P_1$, thus the previously derived equations for isentropic process are expressed by

$$\left(\frac{T_2}{T_1}\right) = \left(\frac{P_2}{P_1}\right)^{(\gamma-1)/\gamma} \quad \text{and} \quad \left(\frac{T_3}{T_4}\right) = \left(\frac{P_3}{P_4}\right)^{(\gamma-1)/\gamma}$$

Therefore, the thermal efficiency is

$$\eta_{Brayton} = 1 - \left(\frac{1}{r_p^{(\gamma-1)/\gamma}}\right) \quad \text{(for constant heat capacity)} \tag{7.48}$$

where r_p is the compression ratio (P_2/P_1) and $\gamma = C_p/C_v$. Equation (7.48) shows that the efficiency of Brayton cycle depends on the pressure ratio of the gas turbine and the ratio of specific heats of the working fluid. The typical values of pressure ratio change between 5 and 20 [1]. As seen from Fig. 7.7, part of the work produced is used to drive the compressor. The back-work ratio r_{bw} shows the part of the produced energy is diverted to the compressor

$$r_{bw} = \frac{W_{comp.in}}{W_{turb.out}} \tag{7.49}$$

Example 7.14 illustrates the analysis of an ideal Brayton cycle when the heat capacity is temperature-dependent, while Example 7.15 illustrates the analysis of a real Brayton cycle. Example 7.16 illustrates the analysis of an ideal Brayton cycle with constant specific heats.

Example 7.14 Simple ideal Brayton cycle calculations with variable specific heats

A power plant is operating on an ideal Brayton cycle with a pressure ratio of $r_p = 9$. The fresh air temperature is 295 K at the compressor inlet and 1300 K at the end of the compressor (inlet of the turbine). Using the standard-air assumptions, determine the thermal efficiency of the cycle.

Solution:
Assume that the cycle is at steady-state flow and the changes in kinetic and potential energy are negligible. Heat capacity of air is temperature-dependent, and the air is an ideal gas.
Consider Fig. 7.7.
Basis: 1 kg air. Using the data from the Appendix: Table D1
Process 1–2 isentropic compression at $T_1 = 295$ K, $H_1 = 295.17$ kJ/kg; $P_{r1} = 1.3068$ (P_r is the relative pressure defined in Eq. 7.36).

$$\frac{P_{r2}}{P_{r1}} = \frac{P_2}{P_1} = P_r \rightarrow P_{r2} = (9)(1.3068) = 11.76$$

Approximate values from Table D1 for the compressor exit at $P_{r2} = 11.76$: $T_2 = 550$ K, $H_2 = 555.74$ kJ/kg
Process 3–4 isentropic expansion in the turbine is seen on the T–S diagram in Fig. 7.7.
$T_3 = 1300$ K, $H_3 = 1395.97$ kJ/kg; $P_{r3} = 330.9$ (from Table D1)

$$\frac{P_{r4}}{P_{r3}} = \frac{P_4}{P_3} \rightarrow P_{r4} = \left(\frac{1}{9}\right)(330.9) = 36.76$$

Approximate values from Table D1 at the exit of turbine $P_{r4} = 36.76$: $T_4 = 745$ K and $H_4 = 761.87$ kJ/kg
The work input to the compressor for 1 kg/s mass flow rate:
$\dot{W}_{comp.in} = \dot{m}(H_2 - H_1) = (555.74 - 295.17)$ kW $= 260.6$ kW   (work received from the compressor)
The work output of the turbine:
$W_{turb.out} = \dot{m}(H_4 - H_3) = (761.87 - 1395.97)$ kW $= -634.1$ kW (work produced from the turbine)

The net work out: $W_{net} = W_{out} - W_{in} = -(634.1 - 260.6) = -373.53$ kJ/kg

The back-work ratio r_{bw} becomes: $r_{bw} = \dfrac{W_{comp.in}}{W_{turb.out}} = \dfrac{260.6}{634.1} = 0.41$ or 41%

This shows that 41% of the turbine output has been used in the compressor.
The amount of heat added: $q_{in} = H_3 - H_2 = 1395.97 - 555.74 = 840.23$ kJ/kg
The amount of heat rejected by the hot gas:
$q_{out} = H_1 - H_4 = 295.17 - 761.87 = -466.7$ kJ/kg
With respect to surrounding that receives the heat $q_{out} = 466/7$ kJ/kg

The thermal efficiency: $\eta_{th} = \dfrac{|W_{net}|}{q_{in}} = 1 - \dfrac{|q_{out}|}{q_{in}} = \mathbf{0.444\ or\ (44.4\%)}$

Example 7.15 Thermal efficiency of an actual Brayton cycle with variable specific heats

A power plant is operating on an ideal Brayton cycle with a pressure ratio of $r_p = 9$. The fresh air temperature is 295 K at the compressor inlet and 1300 K at the end of the compressor and at the inlet of the turbine. Assume the gas-turbine cycle operates with a compressor efficiency of 85% and a turbine efficiency of 85%. Determine the thermal efficiency of the cycle.

Solution:
Assume that the cycle is at steady-state flow and the changes in kinetic and potential energy are negligible. Heat capacity of air is temperature-dependent, and the air is an ideal gas. The standard-air assumptions are applicable:
Basis: 1 kg air.

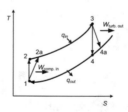

At $T_1 = 295$ K, $H_1 = 295.17$ kJ/kg; $P_{r1} = 1.3068$; (Table D1)
P_r is the relative pressure (Eq. 7.36).

$$\frac{P_{r2}}{P_{r1}} = \frac{P_2}{P_1} = P_r \rightarrow P_{r2} = (9)(1.3068) = 11.76$$

Approximate values from Table D1 for the compressor exit at $P_{r2} = 11.76$:
$T_2 = 550$ K and $H_2 = 555.74$ kJ/kg
Process 3–4 isentropic expansion in the turbine as seen on the T–S diagram above
$T_3 = 1300$ K, $H_3 = 1395.97$ kJ/kg; $P_{r3} = 330.9$ (From Table D1)

$$\frac{P_{r4}}{P_{r3}} = \frac{P_4}{P_3} \rightarrow P_{r4} = \left(\frac{1}{9}\right)(330.9) = 36.76$$

Approximate values from Table D1 at the exit of turbine at 36.76:
$T_4 = 745$ K and $H_4 = 761.87$ kJ/kg
The work input to the compressor for a mass flow rate of 1 kg/s:

$$W_{\text{comp.in}} = \dot{m}(H_2 - H_1) = (555.74 - 295.17) \text{ kW} = 260.57 \text{ kW}$$

The work output of the turbine:

$$W_{\text{turb.out}} = \dot{m}(H_4 - H_3) = (761.87 - 1395.97) \text{ kW} = -634.10 \text{ kW}$$

From the efficiency definitions for compressor and turbine, we have:

$$\eta_C = \frac{W_{Cs}}{W_{Ca}} \rightarrow W_{Ca} = \frac{W_{Cs}}{\eta_C} = \frac{260.6 \text{ kJ/kg}}{0.85} = 306.6 \text{ kW (actual compression work)}$$

(where $W_{\text{comp.in}} = W_{Cs} = 260.6$ kW in Example 7.13)
The work output of the turbine with 85% efficiency:

$$\eta_T = \frac{W_{Ta}}{W_{Ts}} \rightarrow W_{Ta} = \eta_T(W_{Ts}) = 0.85(-634.1 \text{ kW}) = -539.0 \text{ kW}$$

(where $W_{\text{turb.out}} = W_s = -634.1$ kW in ideal operation in Example 7.13)
The net work out: $W_{\text{net}} = W_{\text{out}} - W_{\text{in}} = (-539.0 + 306.6) = -232.4 \text{ kW}$
The back-work ratio r_{bw} becomes: $r_{bw} = \left|\dfrac{W_{\text{comp.in}}}{W_{\text{turb.out}}}\right| = \left|\dfrac{306.6}{539.0}\right| = 0.568$
This shows that the compressor is now consuming 56.8% of the turbine output. The value of back-work ratio increased from 41 to 56.8% because of friction, heat losses, and other nonideal conditions in the cycle.
Enthalpy at the exit of compressor:

$$W_{Ca} = H_{2a} - H_1 \rightarrow H_{2a} = W_{Ca} + H_1 = (306.6 + 295.2)\text{kJ/kg} = 601.8 \text{ kJ/kg}$$

Heat added: $q_{\text{in}} = H_3 - H_{2a} = 1395.97 - 601.8 = 794.2 \text{ kJ/kg}$
The thermal efficiency: $\eta_{th} = \dfrac{W_{\text{net}}}{q_{\text{in}}} = \dfrac{|232.4 \text{ kJ/kg}|}{794.2 \text{ kJ/kg}} = \mathbf{0.292 \text{ (or } 29.2\%)}$

The actual Brayton-gas cycle thermal efficiency drops to 29.9% from 44.4%. Efficiencies of the compressor and turbine affects the performance of the cycle. Therefore, for a better cycle thermal efficiency, significant improvements are necessary for the compressor and turbine operations [6].

	Ideal	Actual
W_{net}, kJ/kg	−373.5	−232.4

(continued)

(continued)

	Ideal	Actual
W_C, kJ/kg	260.6	306.6
W_T, kJ/kg	−634.1	−539.0
q_{in}, kJ/kg	840.2	794.2
q_{out}, kJ/kg	−466.7	−561.8
η_{th} %	44.4	29.2

Example 7.16 Ideal Brayton cycle with constant specific heats A power plant is operating on an ideal Brayton cycle with a pressure ratio of $r_p = 9$. The air temperature is 300 K at the compressor inlet and 1200 K at the end of the compressor. Using the standard-air assumptions and $\gamma = 1.4$, determine the thermal efficiency of the cycle.

Solution:
Assume that the cycle is at steady-state flow and the changes in kinetic and potential energy are negligible. The specific heat capacities are constant, and the air is an ideal gas.
Consider Fig. 7.7.
Basis: 1 kg air and $\gamma = 1.4$. Using the data from the Appendix: Table D1
Process 1–2 isentropic compression: $T_1 = 300$ K, $H_1 = 300.2$ kJ/kg

From Eq. 7.34: $\dfrac{T_2}{T_1} = \left(\dfrac{P_2}{P_1}\right)^{(\gamma-1)/\gamma} \rightarrow T_2 = 300(9)^{0.4/1.4} = 562$ K

Process 3–4 isentropic expansion in the turbine as seen on the T–S diagram above

$T_3 = 1200$ K and $\dfrac{T_4}{T_3} = \left(\dfrac{P_4}{P_3}\right)^{(\gamma-1)/\gamma} \rightarrow T_4 = 1200\left(\dfrac{1}{9}\right)^{0.4/1.4} \quad 640.5$ K

From   Eq. 7.46: $\eta_{th} = 1 - \dfrac{q_{out}}{q_{in}} = 1 - \dfrac{|C_p(T_1 - T_4)|}{C_p(T_3 - T_2)} = 1 - \dfrac{640.5 - 300}{1200 - 562}$

$= \mathbf{0.466\ or\ 46.6\%}$

From Eq. (7.48): $\eta_{Brayton} = 1 - \left(\dfrac{1}{r_p^{(\gamma-1)/\gamma}}\right) = \mathbf{0.463\ or\ 46.3\%}$

The results of efficiency calculations are close to each other; for a constant heat capacity it is easy to use Eq. (7.48) directly to estimate the efficiency using the compression ratio r_p.

7.8.8   Otto Engine Efficiency

The *Otto cycle* is named after Nikolaus A. Otto who manufactured a four-stroke engine in 1876 in Germany. The Otto cycle describes the functioning of a typical ideal cycle for spark-ignition reciprocating piston engine. This cycle is common in automobile internal combustion engines using a chemical fuel with oxygen from the air. The combustion process results in heat, steam, carbon dioxide, and other chemicals at very high temperature. Petroleum fractions, such as diesel fuel, gasoline, and petroleum gases, are the most common fuels. Bioethanol, biodiesel, and hydrogen can also be used as fuels with modified engines.

Internal combustion engines require either spark ignition or compression ignition of the compressed air–fuel mixture [2, 4]. Gasoline engines take in a mixture of air and gasoline and compress it to around 12.8 bar (1.28 MPa), then use a high-voltage electric spark to ignite the mixture. The compression level in diesel engines is usually twice or more than a gasoline engine. Diesel engines will take in air only, and shortly before peak compression, a small quantity of diesel fuel is sprayed into the cylinder via a fuel injector that allows the fuel to instantly ignite. Most diesels also have a battery and charging system. Diesel engines are generally heavier, noisier, and more powerful than gasoline engines. They are also more fuel-efficient in most circumstances and are used in heavy road vehicles, some automobiles, ships, railway locomotives, and light aircraft.

Figure 7.9 shows the Otto engine cycles on *P–V* and *T–S* diagrams. The ideal cycle processes are:

- Process 1–2 is an isentropic compression of the air as the piston moves from bottom dead center to top dead center (*Intake stroke*).
- Process 2–3 is a constant volume heat transfer to the air from an external source while the piston is at top dead center. This process leads to the ignition of the fuel–air mixture (*Compression stroke*).
- Process 3–4 is an isentropic expansion (*Power stroke*).
- Process 4–1 completes the cycle by a constant volume process in which heat is rejected from the air while the piston is at bottom dead center (*Exhaust stroke*).

In the case of a four-stroke Otto cycle, technically there are two additional processes as shown in Fig. 7.9a: one for the exhaust of waste heat and combustion products (by isobaric compression), and one for the intake of cool oxygen-rich air (by isobaric expansion). However, these are often omitted in a simplified analysis.

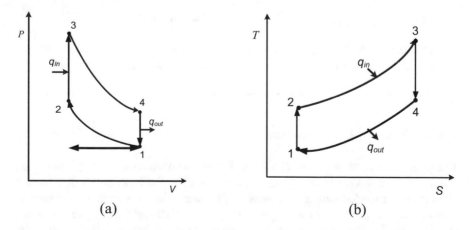

Fig. 7.9 Thermodynamic diagrams of Otto cycle; **a** Otto cycle on a *P–V* diagram, **b** Otto cycle on a *T–S* diagram

Processes 1–2 and 3–4 do work on the system but no heat transfer occurs during adiabatic expansion and compression. Processes 2–3 and 4-1 are isochoric (constant volume); therefore, heat transfer occurs but no work is done as the piston volume does not change.

Idealized P–V diagram in Fig. 7.9a of the Otto cycle shows the combustion heat input q_{in} and the waste exhaust output q_{out}. The power stroke is the top curved line and the bottom is the compression stroke.

The Otto cycle shown in Fig. 7.9 is analyzed as a close system. The energy balance on a unit-mass basis is

$$\Delta U = (q_{in} - q_{out}) + (W_{in} - W_{out}) \tag{7.50}$$

Assuming that the heat capacity is constant, thermal efficiency can be derived by the heat added q_{in} and heat rejected q_{out}

$$\eta_{Otto} = 1 - \left(\frac{|q_{out}|}{q_{in}} \right) = 1 - \left(\frac{|C_{v,av}(T_1 - T_4)|}{C_{v,av}(T_3 - T_2)} \right) \tag{7.51}$$

In an ideal Otto cycle, there is no heat transfer during the processes 1–2 and 3-4 as they are reversible adiabatic processes. Heat is supplied only during the constant volume process 2–3 and heat is rejected only during the constant volume process 4–1. Upon rearrangement:

$$\eta_{Otto} = 1 - \left(\frac{T_1}{T_2} \right) \left(\frac{T_4/T_1 - 1}{T_3/T_2 - 1} \right) \tag{7.52}$$

Since $T_4/T_1 = T_3/T_2$ (see Fig. 7.8), Eq. (7.52) reduces to

$$\eta_{Otto} = 1 - \left(\frac{T_1}{T_2} \right) \tag{7.53}$$

From the isentropic equations of ideal gases, we have

$$C_v \ln \left(\frac{T_2}{T_1} \right) - R \ln \left(\frac{V_2}{V_1} \right) = 0 \tag{7.54}$$

$$\left(\frac{T_2}{T_1} \right) = \left(\frac{V_1}{V_2} \right)^{(\gamma-1)} = r^{(\gamma-1)} \tag{7.55}$$

where r is the compression ratio (V_1/V_2) and $\gamma = C_p/C_v$, and $R = C_p - C_v$ for an ideal gas. Then the thermal efficiency becomes

$$\eta_{Otto} = 1 - \left(\frac{1}{r^{(\gamma-1)}} \right) \quad \text{(constant specific heats)} \tag{7.56}$$

The specific heat ratio of the air–fuel mixture γ varies somewhat with the fuel but is generally close to the air value of 1.4, and when this approximation is used the cycle is called an *air-standard cycle*. However, the real value of γ for the combustion products of the fuel/air mixture is approximately 1.3.

Equation (7.56) shows that the efficiency of the Otto cycle depends upon the compression ratio $r = V_1/V_2$. At higher compression ratio, the efficiency is higher and the temperature in the cylinder is higher. However, the maximum compression ratio is limited approximately to 10:1 for typical automobiles. Usually this does not increase much because of the possibility of auto ignition, which occurs when the temperature of the compressed fuel/air mixture becomes too high before it is ignited by the flame front leading to *engine knocking*. Example 7.17 illustrates the analysis of an ideal Otto engine with temperature-dependent heat capacities, while Example 7.18 illustrates with constant heat capacities.

Knocking: Under high temperature and pressure hydrocarbon fuels break down or combust before ignition. This is called '*knock*' reducing the efficiency of engine and possibly causing damage. To prevent the knock, a compression ratio between 9:1 and 10.5:1 is used in spark ignition engines. For example, with a compression ratio of 10:1 the ideal Otto cycle efficiency of spark ignition engine is around 50%. In real operations, the efficiency may be reduced to 35% or even lower [2].

Example 7.17 Estimation of the efficiency of an Otto engine

Compute the efficiency of an air-standard Otto cycle if the compression ratio is 9 and $\gamma = 1.4$. Compare this with a Carnot cycle operating between the same temperature limits if the lowest temperature of the cycle is 80 °F and the highest temperature is 1000 °F.

Solution:

The efficiency of the Otto cycle is a function only of the compression ratio.

$$\eta_{Otto} = 1 - \left(\frac{1}{r^{(\gamma-1)}} \right) \text{(constant specific heats)} \rightarrow \eta_{Otto} = 1 - \left(\frac{1}{9^{(1.4-1)}} \right)$$
$$= 0.5847 (58.47\%)$$

For the Carnot cycle: $\eta_{Carnot} = 1 - \dfrac{T_C}{T_H} = 1 - \dfrac{(460 + 80)\ R}{(460 + 1000)\ R} = 0.63.01 (63.01\%)$

Example 7.18 Estimation of the work output of an Otto engine

Assuming that the value of $C_v = 0.17$ Btu/lb$_m$ R and the compression ratio of 9, estimate: (a) the heat added and the heat rejected per pound of fluid for the case in which the peak temperature of the cycle is 1200 °F and the lowest temperature of the cycle is 80 °F, (b) the work output, and (c) the efficiency of the Otto cycle based on these values.

Solution:
Assume that $\gamma = 1.4$.

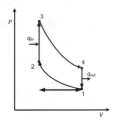

$$\frac{T_3}{T_2} = \left(\frac{V_2}{V_3}\right)^{\gamma-1} = (9)^{0.4} = 2.408 T_3 = 2.408(540) = 1300\,R$$

$$q_{in} = C_v(T_3 - T_2) = 0.17(1660 - 1300) = 61.20\,Btu/lb$$

To determine T_4, therefore: $\dfrac{T_4}{T_3} = \left(\dfrac{V_3}{V_2}\right)^{\gamma-1} = \left(\dfrac{1}{9}\right)^{0.4} = \dfrac{1}{2.408}$ and

$T_4 = (1660/2.408) = 689.36\,R$

$q_{out} = -C_v(T_1 - T_4) = -0.17(540 - 689.36) = 25.39\,Btu/lb$ (heat received by the surrounding)

The net work out from the engine that is the system is:

$$W_{out} = q_{out} - q_{in} = (25.39 - 61.20)Btu/lb = -35.81\,Btu/lb$$

Therefore, the efficiency is: $\eta = \dfrac{|W_{out}|}{q_{in}} = \dfrac{|-38.51|}{61.20} = 0.585(58.5\%)$

Example 7.19 Efficiency calculations of an ideal Otto engine with variable specific heats

An ideal Otto cycle operates with a compression ratio ($r = V_{max}/V_{min}$) of 8.8. Air is at 101.3 kPa and 280 K at the start of compression (state 1). During the constant volume heat addition process, 1000 kJ/kg of heat is transferred into the air from a source at 1900 K. Heat is discharged to the surroundings at 280 K. Determine the thermal efficiency of energy conversion.

Solution:
Assume that the surroundings are at 280 K and the kinetic and potential energy changes are negligible. The specific heats are temperature-dependent.

Heat capacity of air is temperature-dependent, and the air is an ideal gas. The P–V diagram below shows the cycle and the four processes

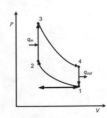

Processes in the cycle:

1-2 Isentropic compression

2-3 Constant volume heat transfer

3-4 Isentropic expansion

4-1 Constant volume heat discharge

Basis: 1 kg air. Using the data from the Appendix D:
q_{in} = 1000 kJ/kg, U_1 = 199.75 kJ/kg, V_{r1} = 783.0 at T = 280 K from Table D1
(V_r is the relative specific volume defined in Eq. 7.38)

$$\frac{V_{r2}}{V_{r1}} = \frac{V_2}{V_1} = \frac{1}{8.8} \rightarrow V_{r2} = \frac{V_{r1}}{r} = \frac{783}{8.8} = 88.97$$

At the value of V_{r2} = 88.97, the air properties from Table D1:

$$U_2 = 465.5 \, kJ/kg \text{ and } T_2 = 640 \, K$$

From isentropic compression of air, we estimate
$$P_2 = P_1 \left(\frac{T_2}{T_1}\right)\left(\frac{V_1}{V_2}\right) = 101.3 \left(\frac{640}{280}\right) 8.8 = 2037 \text{ kPa (ideal gas equation)}$$
The   heat   transferred   in   the   path   2–3: q_{in} = 1000 kJ/kg = $U_3 - U_2$
$\rightarrow U_3$ = 1465.5 kJ/kg
At U_3 = 1465.5 kJ kg, we estimate T_3 and V_{r3} by interpolation using the data
below:

T (K)	U (kJ/kg)	V_r
1750	1439.8	4.328
1800	1487.2	3.994
1777	**1465.5**	**4.147**

We estimate the pressure at state 3:
$$P_3 = P_2 \left(\frac{T_3}{T_2}\right)\left(\frac{V_2}{V_3}\right) = 2037 \left(\frac{1777}{640}\right)(1) = 5656 \text{ kPa}$$

Internal energy of air at state 4:
$$\left(\frac{V_{r4}}{V_{r3}}\right) = \left(\frac{V_4}{V_3}\right) = 8.8 \rightarrow V_{r4} = (8.8)(4.147) = 36.5$$
At V_{r4} = 36.5, approximate values from Table D1 are: U_4 = 658 kJ/kg and
T_4 = 880 K
The process 4-1 is a constant heat discharge. We estimate the discharged heat q_{out}
q_{out} = $-(U_1 - U_4) \rightarrow q_{out}$ = 458.2 kJ/kg (heat received by the surroundings)

$$P_4 = P_3 \left(\frac{T_4}{T_3}\right)\left(\frac{V_3}{V_4}\right) = 5656 \left(\frac{880}{1777}\right)\left(\frac{1}{8.8}\right) = 318.3 \text{ kPa}$$

The net work output: W_{net} = $q_{out} - q_{in}$ = (458.2 − 1000)kJ/kg = −541.8 kJ/kg

So the thermal efficiency: $\eta_{th} = \dfrac{|W_{net}|}{q_{in}} = \mathbf{0.542\ or\ 54.2\%}$

Although all the processes are internally reversible, the heat transfer and discharge take place at finite temperature difference and are irreversible.

Example 7.20 Efficiency calculations of an ideal Otto cycle with constant specific heats

An ideal Otto cycle operates with a compression ratio (V_{max}/V_{min}) of 8. Air is at 101.3 kPa and 300 K at the start of compression (state 1). During the constant volume heat addition process, 730 kJ/kg of heat is transferred into the air from a source at 1900 K. Heat is discharged to the surroundings at 300 K. Determine the thermal efficiency of energy conversion. The average specific heats are $C_{p,av}$ = 1.00 kJ/kg K and $C_{v,av}$ = 0.717 kJ/kg K.

Solution:

Assume: The air-standard assumptions are applicable. The changes in kinetic and potential energy s are negligible. The specific heats are constant.

The P–V diagram below shows the cycle and the four states

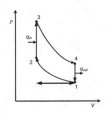

Processes in the cycle:

1-2 Isentropic compression

2-3 Constant volume heat transfer

3-4 Isentropic expansion

4-1 Constant volume heat discharge

Basis: 1 kg air.

q_{in} = 730 kJ/kg; Using the data from Table D1 at T_o = 300 K: U_1 = 214.1 kJ/kg. The average specific heats: $C_{p,av}$ = 1.0 kJ/kg K and $C_{v,av}$= 0.717 kJ/kg K and γ = 1/0.717 = 1.4.

From Eq. 7.34: $\dfrac{T_2}{T_1} = \left(\dfrac{V_1}{V_2}\right)^{\gamma-1} \rightarrow T_2 = \dfrac{V_1}{V_2}T_1 = 8(300\ \text{K})^{0.4} = 689\ \text{K}$

Estimate T_3 from the heat transferred at constant volume in process 2–3:

$$q_{in} = U_3 - U_2 = C_{v,av}(T_3 - T_2) \rightarrow T_3 = 1703\ \text{K}$$

Process 3–4 isentropic expansion: $T_4 = T_3\left(\dfrac{V_3}{V_4}\right)^{\gamma-1} = 1703.0\left(\tfrac{1}{8}\right)^{0.4} = 741.3\ \text{K}$

Process   4–1: $q_{out} = -(U_1 - U_4) = -C_{v,av}(T_1 - T_4) = 316.4$ kJ/kg   (heat discharged)

$$q_{in} = 730 \text{kJ/kg}$$

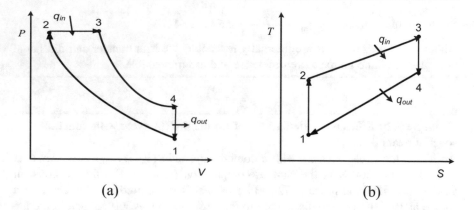

Fig. 7.10 Ideal diesel cycle **a** on P–V diagram, **b** on T–S diagram; the cycle follows the numbers 1–4 in clockwise direction

Net heat transfer is equal to the net work output:
$$W_{\text{net}} = (q_{\text{out}} - q_{\text{in}}) = (316.4 - 730.0) = -413.6\,\text{kJ/kg}$$
So, the thermal efficiency is $\eta_{\text{th}} = \dfrac{|W_{\text{net}}|}{q_{\text{in}}} = \mathbf{0.566\ or\ 56.6\%}$

7.8.9 Diesel Engine Efficiency

Figure 7.10 shows an ideal diesel engine cycle on the P–V and T–S diagrams. The compression ignited engine is first proposed by Rudolph Diesel in 1890. Most diesel engines use a similar cycle to spark-ignited gasoline engine but with a compression heating ignition system, rather than needing a separate ignition system. In combustion-ignited engine, the air is compressed to a temperature that is above the autoignition temperature of the fuel. Therefore, the combustion starts as the fuel is injected into the hot air at constant pressure. This variation is called the diesel cycle.

The diesel cycle is analyzed as a piston-cylinder of a close system. The amount of heat transferred to the working fluid air at constant pressure on a unit-mass basis is

$$q_{\text{in}} = P_2(V_3 - V_2) + (U_3 - U_2) = H_3 - H_2 = C_{p,\text{av}}(T_3 - T_2) \tag{7.57}$$

The amount of heat rejected at constant volume is

$$q_{\text{out}} = U_1 - U_4 = C_{v,\text{av}}(T_1 - T_4) \tag{7.58}$$

Alternatively, thermal efficiency can be derived by the heat added q_{in} and heat rejected q_{out}

$$\eta_{\text{Diesel}} = 1 - \left(\frac{|q_{\text{out}}|}{q_{\text{in}}}\right) = 1 - \left(\frac{(T_4 - T_1)}{\gamma(T_3 - T_2)}\right) = 1 - \left(\frac{T_1(T_4/T_1 - 1)}{\gamma T_2(T_3/T_2 - 1)}\right) \tag{7.59}$$

where $\gamma = C_p/C_v$. In an ideal diesel cycle, the cutoff ratio r_c is defined as the ratio of the cylinder volume after and before the combustion

$$r_c = \frac{V_3}{V_2} \qquad (7.60)$$

After using the cutoff ratio r_c and compression ratio r with the isentropic ideal-gas relations, given in Eq. (7.33), the thermal efficiency of energy conversion becomes

$$\eta_{\text{Diesel}} = 1 - \frac{1}{r^{(\gamma-1)}}\left(\frac{r_c^{\gamma} - 1}{\gamma(r_c - 1)}\right) \quad \text{(With constant specific heats)} \qquad (7.61)$$

Equation (7.61) shows that as the cutoff ratio r_c decreases, the efficiency of the diesel cycle increases. For the limiting ratio of $r_c = 1$, the value of the bracket in Eq. (7.61) becomes unity, and the efficiencies of the Otto and diesel cycles become identical. Thermal efficiencies of diesel engines vary from 35 to 40% [2, 4]. Example 7.18 illustrates the analysis of diesel engine with constant heat capacity, while Example 7.19 with temperature-dependent heat capacity.

Example 7.21 Thermal efficiency of an ideal diesel engine with the constant specific heats
An ideal diesel cycle has an air-compression ratio of 20 and a cutoff ratio of 2. At the beginning of the compression, the fluid pressure, temperature, and volume are 14.7 psia, 70 °F, and 120 in^3, respectively. The average specific heats of air at room temperature are $C_{p,av} = 0.24$ Btu/lb R and $C_{v,av} = 0.171$ Btu/lb R. Estimate the thermal efficiency with the cold-air-standard assumptions.

Solution:
Assume: the cold-air-standard assumptions are applicable. Air is ideal gas. The changes in kinetic and potential energies are negligible. Consider Fig. 7.10.
$P_1 = 14.7$ psia, $T_1 = 70$ °F, and $V_1 = 120$ in^3, $C_{p,av} = 0.24$ Btu/lb R and, $C_{v,av} = 0.171$ Btu/lb R,
$R = 10.73$ psia ft^3/lbmol R, $MW = 29$ lb/lbmol, $\gamma = C_{p,av}/C_{v,av} = 1.4$, $r = 20$ and $r_c = 2$.
The air mass: $m_{\text{air}} = MW\frac{PV}{RT} = 0.0052$ lb
The volumes for each process:
$V_1 = V_4 = 120$ in^3, $V_2 = V_1/r = 6$ in^3, $V_3 = V_2\, r_c = 12$ in^3

Process   1–2:   Isentropic   compression: $\left(\dfrac{T_2}{T_1}\right) = \left(\dfrac{V_1}{V_2}\right)^{(\gamma-1)} \rightarrow T_2 = 1756.6\,\text{R}$

(Eq. 7.34)

Process 2–3: Heat addition at constant pressure for ideal gas: $\left(\dfrac{T_3}{T_2}\right) = \left(\dfrac{V_3}{V_2}\right)$
$\rightarrow T_3 = 3513\,\text{R}$

$$q_{\text{in}} = m(H_3 - H_2) = mC_{p,av}(T_3 - T_2) = 2.19\,\text{Btu}$$

Process 3–4: Isentropic expansion: $\left(\dfrac{T_4}{T_3}\right) = \left(\dfrac{V_3}{V_4}\right)^{(\gamma-1)} \rightarrow T_4 = 1398\,\text{R}$

Process 4–1: Heat rejection at constant volume for ideal gas

$$q_{\text{out}} = -m(H_1 - H_4) = mC_{v,\text{av}}(T_4 - T_1)$$
$$= 0.772\,\text{Btu (heat received by the surroundings)}$$

The net work output is equal to the difference between heat input (q_{in}) and waste heat (q_{out})

$$W_{\text{out}} = q_{\text{out}} - q_{\text{in}} = (0.772 - 2.19) = -1.418\,\text{Btu}$$

Thermal efficiency: $\eta_{\text{th}} = \dfrac{|W_{\text{net}}|}{q_{\text{in}}} = \mathbf{0.647\,(\text{or } 64.7\%)}$

Thermal efficiency can also be estimated from Eq. (7.61):

$$\eta_{\text{Diesel}} = 1 - \frac{1}{r^{(\gamma-1)}}\left(\frac{r_c^\gamma - 1}{\gamma(r_c - 1)}\right) = \mathbf{0.647\ (\text{or } 64.7\%)}$$

Example 7.22 Thermal efficiency of an ideal diesel engine with variable specific heats

An ideal diesel cycle has an air-compression ratio of 18 and a cutoff ratio of 2. At the beginning of the compression, the fluid pressure and temperature are 100 kPa, 300 K, respectively. Utilizing the cold-air-standard assumptions, determine the thermal efficiency.

Solution:

Assume: the cold-air-standard assumptions are applicable. Air is ideal gas. The changes in kinetic and potential energies are negligible. The specific heats depend on temperature.

Consider Fig. 7.10: $P_1 = 100$ kPa, $T_1 = 300$ K, $r = 18$ and $r_c = V_3/V_2 = 2$.

At 300 K, $U_1 = 214.1$ kJ/kg, $V_{r1} = 621.2$ (V_r is the relative specific volume) (Table D1)

From Eq. 7.38: $\dfrac{V_{r2}}{V_{r1}} = \dfrac{V_2}{V_1} = \dfrac{1}{18} \rightarrow V_{r2} = \dfrac{V_{r1}}{r} = \dfrac{621.2}{18} = 34.5$

At this value of $V_{r2} = 34.5$, approximate values are: $H_2 = 932.9$ kJ/(kg K), $T_2 = 900$ K (Table D1)

The heat transferred at constant pressure in process 2-3:

$\dfrac{V_3}{T_3} = \dfrac{V_2}{T_2} = \rightarrow T_3 = \dfrac{V_3}{V_2}T_2 = 2T_2 = 1800$ K

$H_3 = 2003.3$ kJ/kg $V_{r3} = 4.0$ (Table D1)

$$q_{\text{in}} = H_3 - H_2 = 1070.4\,\text{kJ/kg}$$

Processes 3–4: Next, we need to estimate the internal energy of air at state 4:

$$\left(\frac{V_{r4}}{V_{r3}}\right) = \left(\frac{V_4}{V_3}\right) = \left(\frac{V_4}{2V_2}\right) = \left(\frac{18}{2}\right) \rightarrow V_{r4} = (9)(4.0) = 36$$

At $V_{r4} = 36$, $U_4 = 659.7$ kJ/kg (Table D1)

The process 4–1 is heat discharge at constant volume. We estimate the discharged heat q_{out}

$q_{out} = -(U_1 - U_4) = 659.7 - 214.1 = 445.6\,\text{kJ/kg}$ (heat received by the surroundings)

The net heat transfer is equal to the net work output: $W_{net} = q_{out} - q_{in} = 445.6 - 1070.4 = -624.8\,\text{kJ/kg}$

So, the thermal efficiency is $\eta_{th} = \dfrac{|W_{net}|}{q_{in}} = \mathbf{0.583\ or\ 58.3\%}$

7.8.10 Ericsson and Stirling Engine Efficiency

The Ericsson engine is an '*external combustion engine*', as is externally heated. The four processes of the ideal Ericsson cycle are:

- Process 1–2: Isothermal compression. The compressed air flows into a storage tank at constant pressure.
- Process 2–3: Isobaric heat addition. The compressed air flows through the regenerator and picks up heat.
- Process 3–4: Isothermal expansion. The cylinder expansion-space is heated externally, and the gas undergoes isothermal expansion.
- Process 4–1: Isobaric heat removal. Before the air is released as exhaust, it is passed back through the regenerator, thus cooling the gas at a low constant pressure, and heating the regenerator for the next cycle.

The *Stirling engine* operates by cyclic compression and expansion of air or other gas at different temperature levels and converts the heat of hot gas into mechanical work. The Stirling engine has high efficiency compared to steam engines and can use almost any heat source [2, 4]. Stirling cycle has four totally reversible processes:

- Process 1–2: Isothermal heat addition from external source
- Process 2–3: Internal heat transfer from working fluid to regenerator at constant volume
- Process 3–4: Isothermal heat rejection
- Process 4–1: Internal heat transfer from regenerator back to working fluid at constant volume.

Heat addition during process 1–2 at T_H and heat rejection process 3–4 at T_C are both isothermal. For an isothermal process heat transfer is

$$q = T\Delta S \tag{7.62}$$

The change in entropy of an ideal gas at isothermal conditions is

$$\Delta S_{1-2} = -R \ln \frac{P_2}{P_1} \qquad (7.63)$$

Using Eqs. (7.62) and (7.63) the heat input and output are estimated by

$$q_{in} = T_H(S_2 - S_1) = T_H\left(-R \ln \frac{P_2}{P_1}\right) = RT_H \ln \frac{P_1}{P_2} \qquad (7.64)$$

$$q_{out} = T_C(S_4 - S_3) = T_C\left(-R \ln \frac{P_4}{P_3}\right) = RT_C \ln \frac{P_3}{P_4} \qquad (7.65)$$

The thermal efficiency of the Ericsson cycle becomes

$$\eta_{Ericsson} = 1 - \frac{|q_{out}|}{q_{in}} = 1 - \frac{RT_C \ln(P_4/P_3)}{RT_H \ln(P_1/P_2)} = 1 - \frac{T_C}{T_H} \qquad (7.66)$$

Since $P_1 = P_4$ and $P_3 = P_2$.

Both the ideal Ericsson and Stirling cycles are external combustion engines. Both the cycles are totally reversible as is the Carnot cycle and have the same thermal efficiency between the same temperature limits when using an ideal gas as the working fluid

$$\eta_{Ericsson} = \eta_{Stirling} = \eta_{Carnot} = 1 - \frac{T_C}{T_H} \qquad (7.67)$$

Both the cycles utilize regeneration to improve efficiency. Between the compressor and the expander, heat is transferred to a thermal energy storage device called the regenerator during one part of the cycle and is transferred back to the working fluid during another part of the cycle. Equation (7.67) is valid for both the closed and steady-state flow cycles of the engines.

7.8.11   Atkinson Engine Efficiency

The *Atkinson engine* is an internal combustion engine and used in some modern hybrid electric applications. Figure 7.11 shows the cyclic processes in the Atkinson engine. In the engine, the expansion ratio can differ from the compression ratio and the engine can achieve greater thermal efficiency than a traditional piston engine [1, 2]. Expansion ratios are obtained from the ratio of the combustion chamber volumes when the piston is at bottom dead center and top dead center.

In Atkinson cycle, the intake valve allows a reverse flow of intake air. The goal of the modern Atkinson cycle is to allow the pressure in the combustion chamber at the end of the power stroke to be equal to atmospheric pressure; when this occurs,

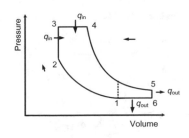

The ideal Atkinson cycle consists of following operations:
1-2 Isentropic or reversible adiabatic compression
2-3 Isochoric heating
3-4 Isobaric heating
4-5 Isentropic expansion
5-6 Isochoric cooling
6-1 Isobaric cooling

Fig. 7.11 Ideal Atkinson gas cycle

all the available energy has been utilized from the combustion process. For any given portion of air, the greater expansion ratio allows more energy to be converted from heat to useful mechanical energy, hence the engine is more efficient. The disadvantage of the four-stroke Atkinson cycle is the reduced power output as a smaller portion of the compression stroke is used to compress the intake air. Atkinson cycles with a supercharger to make up for the loss of power density are known as *Miller engines* [2, 4]. The power from the Atkinson engine can be supplemented by an electric motor. This forms the basis of an Atkinson cycle-based hybrid electric drivetrain.

7.9 Improving the Efficiency of Heat Engines

The followings may be some options toward increasing the efficiency of heat engines [2, 4]:

- Increased hot-side temperature is the approach used in modern combined-cycle gas turbines. However, the construction of materials and environmental concerns regarding NO_x production limit the maximum temperature on the heat engines.
- Lowering the output temperature by using mixed chemical working fluids, such as using a 70:30 mix of ammonia and water may also increase the efficiency. This mixture allows the cycle to generate useful power at considerably lower temperatures.
- Use of supercritical fluids, such as CO_2, as working fluids may increase the efficiency.
- Exploit the chemical properties of the working fluid, such as nitrogen dioxide (NO_2), which has a natural dimer as di-nitrogen tetra oxide (N_2O_4). At low temperature, the N_2O_4 is compressed and then heated. The increasing temperature causes each N_2O_4 to break apart into two NO_2 molecules. This lowers the molecular weight of the working fluid, which drastically increases the efficiency of the cycle. Once the NO_2 has expanded through the turbine, it is cooled and recombined into N_2O_4. This is then fed back to the compressor for another cycle.

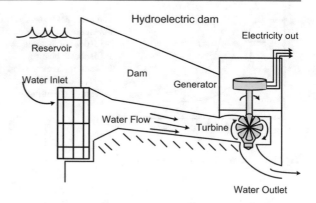

Fig. 7.12 Cross-section of a conventional hydroelectric dam

- New fuels, such as hydrogen, may have positive impact on the performance of the engines since the energy density of hydrogen is considerably higher than that of electric batteries.

7.10 Hydroelectricity

Hydroelectricity is the production of electrical power using the gravitational force of falling or flowing water. Figure 7.12 shows a schematic of a conventional hydroelectric dam, in which hydroelectric power comes from the potential energy of dammed water driving a water turbine and generator. The power extracted from the water depends on the volume and on the difference in height between the source and the water's outflow. This height difference is called the head. The amount of potential energy in water is proportional to the head. A large pipe called the penstock delivers water to the turbine. Example 7.23 illustrates the efficiency calculations for a hydraulic turbine.

Pumped-storage method produces electricity to supply high peak demands by moving water between reservoirs at different elevations. At times of low electrical demand, excess generation capacity is used to pump water into the higher reservoir. When there is higher demand, water is released back into the lower reservoir through a turbine. Compared to wind farms, hydroelectricity power plants have a more predictable load factor. If the project has a storage reservoir, it can generate power when needed. Hydroelectric plants can be easily regulated to follow variations in power demand [4, 7]. Example 7.23 illustrates the pumped energy in a hydropower plant.

Example 7.23 Efficiency of a hydraulic turbine
Electricity is produced by a hydraulic turbine installed near a large lake. Average depth of the water in the lake is 45 m. The water mass flow rate is 600 kg/s. The produced electric power is 220 kW. The generator efficiency is 95%. Determine the overall mechanical efficiency of the turbine-generator and the shaft work transferred from the turbine to the generator.

Solution:

Assume: The mechanical energy of water at the turbine exit is small and negligible. The density of the water is 1000 kg/m^3.

$\dot{m}$ = 600 kg/s, η_{gen} = 0.95, $\dot{W}_{out}$ = -220 kW

$$W_{fluid} = W_{in} = PE = \dot{m}gz = (600\,\text{kg/s})(9.81\,\text{m/s}^2)(45\,\text{m})(\text{kJ/kg} = 1000\,\text{m}^2/\text{s}^2)$$
$$= 264.9\,\text{kW}$$

The overall mechanical efficiency of the turbine-generator:

$$\eta_{overall} = \eta_{turb}\eta_{gen} = \frac{|\text{energy out}|}{\text{energy in}} = \frac{|220\,\text{kW}|}{264.9\,\text{kW}} = \mathbf{0.83}\ (\eta_{gen} = 0.95)\ \text{and}\ \eta_{turb}$$
$$= \frac{\eta_{overall}}{\eta_{gen}} = 0.87$$

The shaft power transferred from the turbine:

$W_{turb} = \eta_{turb}W_{fluid} = 0.87(264.9\,\text{kW}) = 230.5\,\text{kW}$

The lake supplies 264.9 kW of mechanical energy to the turbine. Only 87% of the supplied energy is converted to shaft work. This shaft work drives the generator and 220 kW is produced.

Example 7.24 Pumped energy in a hydropower plant

Consider a hydropower plant reservoir with an energy storage capacity of 1×10^6 kWh. This energy is to be stored at an average elevation of 60 m relative to the ground level. Estimate the minimum amount of water that has to be pumped back to the reservoir.

Solution:

Assume that the evaporation of the water is negligible.

$$PE = 1 \times 10^6\,\text{kWh},\ Dz = 60\,\text{m}\quad g = 9.8\,\text{m/s}^2$$

Energy of the work potential of the water: $PE = mg\Delta z$

Amount of water:

$$m = \frac{PE}{g\Delta z} = \frac{1 \times 10^6\,\text{kWh}}{9.8\,\text{m/s}^2(60\,\text{m})}\left(\frac{3600\,\text{s}}{\text{hr}}\right)\left(\frac{1000\,\text{m}^2/\text{s}^2}{1\,\text{kJ/kg}}\right) = \mathbf{6.122 \times 10^9 kg}$$

7.11 Wind Electricity

The power produced by a wind turbine is proportional to the kinetic energy of the wind captured by the wind turbine

$$\dot{W}_{wind} = \eta_{wind}\rho \frac{\pi v^3 D^2}{8} \qquad (7.68)$$

where ρ is the density of air, v is the velocity of air, D is the diameter of the blades of the wind turbine, and η_{wind} is the efficiency of the wind turbine. Therefore, the power produced by the wind turbine is proportional to the cube of the wind velocity and the square of the blade span diameter. The strength of wind varies, and an average value for a given location does not alone indicate the amount of energy a wind turbine could produce [1]. Example 7.25 illustrates the efficiency calculations for wind turbine.

Example 7.25 Efficiency of a wind turbine
A wind turbine-generator with a 25-foot-diameter blade produces 0.5 kW of electric power. In the location of the wind turbine, the wind speed is 11 mile per hour. Determine the efficiency of the wind turbine generator.

Solution:
Assume: The wind flow is steady. The wind flow is one-dimensional and incompressible. The frictional effects are negligible.
$\rho_{air} = 0.076$ lb/ft^3, $D = 25$ ft, Actual power production = 0.5 kW
Kinetic energy can be converted to work completely. The power potential of the wind is its kinetic energy.

Average speed of the wind: $v_1 = (11\,\text{mph})\left(\dfrac{1.4667\,\text{ft/s}}{1\,\text{mph}}\right) = 16.13\,\text{ft/s}$

The mass flow rate of air: $\dot{m} = \rho v_1 A = \rho v_1 \dfrac{\pi D^2}{4} = 601.4$ lb/s

$$\dot{W}_{max} = \dot{m}\frac{v_1^2}{2g_c} = (601.4\,\text{lb/s})\frac{(16.13\,\text{ft/s})^2}{2}\left(\frac{1\,\text{lb}_f}{32.2\,\text{lb}_m\,\text{fts}^2}\right)\left(\frac{1\,\text{kW}}{737.56\,\text{lb}_f\,\text{ft/s}}\right)$$
$$= 3.29\,\text{kW}$$

This is the available energy to the wind turbine. The turbine-generator efficiency is

$$\eta_{windturb} = \frac{\dot{W}_{act}}{\dot{W}_{max}} = \frac{0.5\,\text{kW}}{3.29\,\text{kW}} = \mathbf{0.152\ or\ 15.2\%}$$

Only 15.2% of the incoming kinetic energy is converted to electric power. The remaining part leaves the wind turbine as outgoing kinetic energy.

7.12 Geothermal Electricity

Geothermal electricity refers to the energy conversion of geothermal energy to electric energy. Technologies in use include dry steam power plants, flash steam power plants, and binary cycle power plants. Geothermal power is sustainable because the heat extraction is small compared with the earth's heat content.

Geothermal electric plants have until recently been built exclusively where high temperature geothermal resources are available near the surface. The development of binary cycle power plants and improvements in drilling and extraction technology may enable enhanced geothermal systems over a much greater geographical range [2].

The thermal efficiency of geothermal electric plants is low, around 10–23%, because geothermal fluids are at a low temperature compared with steam from boilers. This low temperature limits the efficiency of heat engines in extracting useful energy during the generation of electricity. Exhaust heat is wasted, unless it can be used directly, for example, in greenhouses, timber mills, and district heating. In order to produce more energy than the pumps consume, electricity generation requires high temperature geothermal fields and specialized heat cycles [1, 2].

7.13 Ocean Thermal Energy Conversion

Ocean thermal energy conversion uses the difference between cooler deep and warmer shallow or surface ocean waters to run a heat engine and produce useful work, usually in the form of electricity (see Fig. 7.13). Warm surface seawater is pumped through a heat exchanger to vaporize the fluid. The expanding vapor turns the turbo-generator. Cold water, pumped through a second heat exchanger, condenses the vapor into a liquid, which is then recycled through the system. In the tropics, the temperature difference between surface and deep water is a modest 20–25 °C. Ocean thermal energy conversion systems is still considered an emerging technology with a thermal efficiency of 1–3%, which is well below the theoretical maximum for this temperature difference of between 6 and 7% [9]. The most used heat cycle for ocean thermal energy conversion systems is the Rankine cycle using a low-pressure turbine system. Closed-cycle engines use working fluids such as ammonia or refrigerant such as R-134a. Open-cycle engines use vapor from the seawater itself as the working fluid.

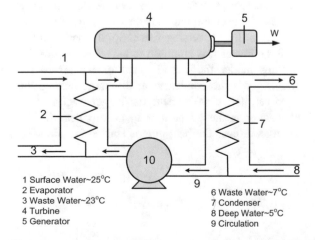

Fig. 7.13 Diagram of a closed cycle ocean thermal energy conversion (OTEC) plant

1 Surface Water~25°C
2 Evaporator
3 Waste Water~23°C
4 Turbine
5 Generator

6 Waste Water~7°C
7 Condenser
8 Deep Water~5°C
9 Circulation

7.14 Thermoelectric Effect

Thermoelectric effect involves energy conversions between heat and electricity as a temperature difference creates an electric potential or electric potential creates a heat flow leading to a temperature difference. There are two thermoelectric effects:

- The *Seebeck effect* refers to conversion of heat to electricity.
- The *Peltier effect* refers to conversion of electricity to heat.

Measuring temperature by *thermocouples* operating between a hot and a cold junction is based on the Seebeck effect. A commonly used thermoelectric material in such applications is Bismuth telluride (Bi_2Te_3). The thermoelectric efficiency approaches to the Carnot limit.

Thermoelectric materials can be used as refrigerators, called 'thermoelectric coolers', or 'Peltier coolers', although it is far less common than vapor-compression refrigeration. Compared to a vapor-compression refrigerator, the main advantages of a Peltier cooler are its lack of moving parts or circulating liquid, and its small size and flexible shape. Another advantage is that Peltier coolers do not require refrigerant liquids, which can have harmful environmental effects. The main *disadvantage* of Peltier coolers is that they cannot simultaneously have low cost and high-power efficiency. Advances in thermoelectric materials may lead to the creation of Peltier coolers that are both cheap and efficient [6, 8].

7.15 Efficiency of Heat Pumps and Refrigerators

Heat pumps, refrigerators, and air conditioners use an outside work to move heat from a colder to a warmer region, so their function is the opposite of a heat engine. Since they are heat engines, these devices are also limited by the Carnot efficiency. A heat pump can be used for heating or cooling as part of *heating, ventilation, and air conditioning* applications. The heat pump can heat and when necessary it uses the basic refrigeration cycle to cool. To do that, a heat pump can change which coil is the condenser and which is the evaporator by controlling the flow direction of the refrigerant by a reversing valve. So, the heat can be pumped in either direction. In cooler climates it is common to have heat pumps that are designed only to heat. In heating mode, the outside heat exchanger is the evaporator and the indoor exchanger is the condenser to discharge heat to the inside air. In cooling mode, however, the outside heat exchanger becomes the condenser and the indoor exchanger is the evaporator to absorb heat from the inside air. The following sections discuss the heat pumps and refrigerators [1, 4].

7.15.1   Heat Pumps

A heat pump is a reversed heat engine that receives heat from a low temperature reservoir and rejects it to a high temperature reservoir by consuming an external work [1]. The efficiency of a heat pump cycle is usually called the *coefficient of performance*. The external work supplied for obtaining that desired effect is to supply heat to the hot body. A heat pump uses a fluid which absorbs heat as it vaporizes and releases the heat when it condenses. Figure 7.14 shows a typical heat pump drawing heat from the ambient air. A heat pump requires external work to extract heat q_C (q_{in}) from the outside air (cold region) and deliver heat q_H (q_{out}) to the inside air (hot region). The most common heat pump is a phase-change heat pump. During the cycle of such a heat pump, the compressor pumps a gas through the condenser where it gets cooled down and finally condenses into liquid phase after releasing heat of condensation q_{out}. Then the liquid flows through an expansion valve where its pressure and temperature both drop considerably. Further it flows through the evaporator where it warms up and evaporates to gaseous phase again by extracting heat of vaporization q_{in} from the surroundings.

   In heat pumps, the work energy W_{in} provided mainly in the form of electricity is converted into heat, and the sum of this energy and the heat energy that is moved from the cold reservoir (q_C) is equal to the total heat energy added to the hot reservoir q_H

$$\dot{q}_H = \dot{q}_C + \dot{W}_{in} \tag{7.69}$$

   Thermal efficiency of heat pumps is measured by the coefficient of performance COP_{HP}

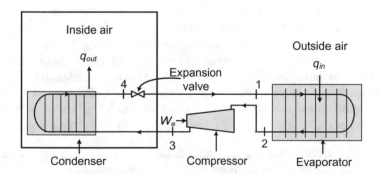

Fig. 7.14 A typical heat pump for heating a room; compressor pumps a gas from a low temperature region to a high temperature region. Within the evaporator, heat of vaporization is absorbed from the outside surroundings. Compressor increases the pressure and temperature of the vapor and pumps it through the cycle. In the condenser, the vapor condenses and releases heat of condensation to the inside surroundings. The expansion valve causes a flash of the liquid to boil because of sudden drop in pressure; complete vaporization of the liquid takes place in the evaporator

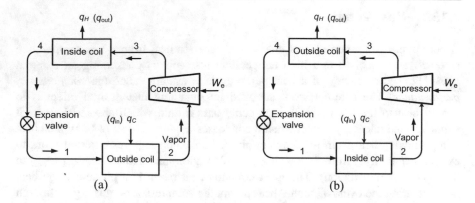

Fig. 7.15 A heat pump can be used: **a** to heat a house in the winter, and **b** to cool in the summer. A reversing valve reverses the direction of the fluid flow so the inside coil in the summer is used as outside coil in the winter

$$COP_{HP} = \frac{\dot{q}_H}{\dot{W}_{in}} \qquad (7.70)$$

The amount of heat they move can be greater than the input work. Therefore, heat pumps can be a more efficient way of heating than simply converting the input work into heat, as in an electric heater or furnace. The limiting value of the Carnot efficiency for the heat pump is

$$COP_{HP} \leq \frac{T_H}{T_H - T_C} \quad \text{(heating)} \qquad (7.71)$$

Equation (7.71) shows that the COP decreases with increasing temperature difference between the hot and cold regions.

Figure 7.15 compares the flow directions of the working fluid when the heat pump is used for heating and for cooling. A reversing valve reverses the direction of the fluid flow so the inside coil in the summer is used as the outside coil in the winter. This means that the working fluid is evaporated in the inside coil extracting heat from the warm air and is condensed discharging heat to warm outside surroundings in the summer [1, 3]. Example 7.26 illustrates the analysis of a heat pump.

Example 7.26 Effect of the COP on heat pump performance
A heat pump provides 60 MJ/h to a house. If the compressor requires an electrical energy input of 5 kW, calculate the COP. If electricity costs $0.08 per kWh and the heat pump operates 100 h per month, how much money does the homeowner save by using the heat pump instead of an electrical resistance heater?

Solution:
The heat pump operates at steady state.

COP for a heat pump with a heat supply of 60 MJ/h = 16.66 kW: and W_{HP} = 5 kW

$$COP_{HP} = \frac{q_{out}}{W_{HP}} = 3.33$$

An electrical resistance heater converts all the electrical work supplied W_e into heat q_H. Therefore, in order to get 16.66 kW into your home, you must buy 16.66 kW of electrical power.
Cost of resistance heater:
Power = 16.66 kW

$$Cost(\$/month) = Electricity(\$0.08/kWh)Power(16.66\,kW)Time(100\,h/month)$$
$$= 133.3\$/month$$

Cost of heat pump with a power of 5 kW:

$$Cost(\$/month) = Electricity(\$0.08/kWh)Power(5.0\,kW)Time(100\,h/month)$$
$$= 40.0\$/month$$

Therefore, monthly saving is (133.3–40.0) $/month = **93.3 $/month**.
Electrical resistance heaters are not very popular, especially in cold climates. The thermal efficiency of a heat pump drops significantly as the outside temperature falls. When the outside temperature drops far enough that the $COP_{HP} \sim 1$, it becomes more practical to use the resistance heater.

7.15.2 Refrigerators

A refrigerator is similar to a heat pump that extracts heat from the cold body/space and deliver it to high temperature body/surroundings using an external work. Figure 7.16 shows a typical refrigeration cycle. A refrigerator is a heat engine in which work is required to extract energy from the freezing compartment and discharge that heat to the room through a condenser at the back of the refrigerator. This leads to further cooling of the cold region. One of the common refrigerants is 1,1,1,2-tetrafluoroethane (CF_3CH_2F) known as R-134a. R-134a has a boiling point temperature of –26.2 °C (–15 °F) and a latent heat of 216.8 kJ/kg at 1.013 bar. It is compatible with most existing refrigeration equipment. R-134a has no harmful influence on the ozone layer of the earth's atmosphere. It is noncorrosive and nonflammable. R-134a is used for medium-temperature applications, such as air conditioning and commercial refrigeration. Another refrigerant is pentafluoroethane (C_2HF_5) known as R-125, which is used in low- and medium-temperature applications. With a boiling point of -55.3 °F at atmospheric pressure, R-125 is nontoxic, nonflammable, and noncorrosive [1, 2, 7].

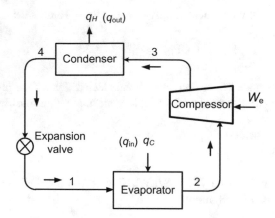

Fig. 7.16 A typical refrigeration cycle; the evaporator is the freezing compartment of the fridge, where the refrigerant absorbs heat to be vaporized. The condenser is usually at the back of the fridge, where the heat of condensation is discharged to outside hot surroundings

Thermal efficiency of refrigerators and air conditioners is called the coefficient of performance COP_R:

$$COP_R = \frac{\dot{q}_C}{\dot{W}_{in}} \tag{7.72}$$

The limiting value of the Carnot efficiency for the refrigeration processes is

$$COP_R \leq \frac{T_C}{T_H - T_C} \quad \text{(cooling)} \tag{7.73}$$

Equation (7.73) shows that the COP_R decreases with increasing temperature difference between the hot and cold regions. Example 7.27 illustrates the analysis of a refrigerator. When the desired effect is cooling the heat resulting from the input work is just an unwanted byproduct. In everyday usage the efficiency of air conditioners is often rated by the *Seasonal Energy Efficiency Ratio* (SEER), which is discussed in detail in Sect. 9.4. Example 7.28 illustrates the estimation of the heat rejected in refrigeration cycle, while Example 7.29 illustrates the coefficient of performance estimation [1, 4].

Example 7.27 Analysis of a refrigeration cycle
In a refrigeration cycle, the superheated R-134a (state 2) enters a compressor at 263.15 K and 0.18 MPa. The R-134a (state 3) leaves the compressor at 313.15 K and 0.6 MPa, and enters a condenser, where it is cooled by cooling water. The R-134a (state 4) leaves the condenser at 293.15 K and 0.57 MPa as saturated liquid

and enters a throttling valve. The partially vaporized R-134a (state 1) leaves the valve at 0.293 MPa. The cycle of R-134a is completed when it passes through an evaporator to absorb heat from the matter to be refrigerated. The flow rate of R-134a is 0.2 kg/s. The total power input is 60 kW. The surroundings are at 290 K. Determine the coefficient of performance and the exergy loss of the cycle.

Solution:
Assume that kinetic and potential energy changes are negligible, and the system is at steady state.
Consider Fig. 7.16.
From Tables E1 to E2 the data for R-134a:
$H_2 = 242.06$ kJ/kg, $S_2 = 0.9362$ kJ/(kg K), at $T_2 = 263.15$ K; $P_2 = 0.18$ MPa
(superheated vapor)
$H_3 = 278.09$ kJ/kg, $S_3 = 0.9719$ kJ/(kg K), at $T_3 = 313.15$ K, $P_3 = 0.6$ MPa
(superheated vapor)
$H_4 = 77.26$ kJ/kg, at $T_4 = 293.15$ K, $P_4 = 0.571$ MPa (saturated liquid)
$P_1 = 0.293$ MPa, $T_1 = 273.15$ K, $W_{in} = 60$ kW, $\dot{m}_r = 0.2$ kg/s
$T_o = 290$ K, $T_{evaporator} = 273$ K, $T_{condenser} = 290$ K
The throttling process where $H_4 = H_1 = 77.26$ kJ/kg (Stage 1) causes partial vaporization of the saturated liquid coming from the condenser.
The vapor part of the mixture ('quality') can be obtained using the enthalpy values at 0.293 MPa
$H_{1,sat\ liq} = 50.02$ kJ/kg, $H_{1,sat\ vap} = 247.23$ kJ/kg, $S_{1,sat\ liq} = 0.1970$ kJ/kg K,
$S_{1,sat\ vap} = 0.9190$ kJ/kg K

$$x_1 = \frac{77.26 - 50.02}{247.23 - 50.02} = 0.138$$

Then the value of entropy:
$S_1 = (1 - 0.138)S_{1sat\ liq} + 0.138 S_{1sat\ vap} = 0.2967 kJ/kg\ K$
For the cycle, the total enthalpy change is zero.
At the compressor, outside energy W_{in} is needed.
At the evaporator, heat transfer, q_{in}, is used to evaporate the refrigerant R-134a.
The heat absorbed within the evaporator is: $\dot{q}_{in} = \dot{m}_r(H_2 - H_1) = 32.96\,kW$
$= 32.96$ kW
The energy balance indicates that the total energy $(W_{in} + q_{in})$ is removed:
$\dot{W}_{in} + \dot{q}_{in} = \dot{q}_{out} = 92.96$ kW $\dot{W}_{ideal,in} = \dot{m}_r(H_3 - H_2) = 7.20$ kW
Coefficient of performance (COP) of the refrigerator:
$$COP = \frac{\dot{q}_{in}}{\dot{W}_{ideal,in}} = \frac{H_2 - H_1}{H_3 - H_2} = \mathbf{4.57}$$
The total work (exergy) loss:
$$\dot{Ex}_{total} = W_{in} + \left(1 - \frac{T_0}{T_{evaporater}}\right)\dot{q}_{in} - \left(1 - \frac{T_0}{T_{condenser}}\right)(-\dot{q}_{out}) = \mathbf{58.94\,kW}$$

Exergy analysis identifies the performance of individual processes. Finding ways to improve the thermodynamic performance of individual steps is equally important.

Example 7.28 Heat rejection by a refrigerator

Food compartment of a refrigerator is maintained at 4 °C by removing heat from it at a rate of 350 kJ/min. If the required power input of the refrigerator is 1.8 kW, determine (a) the coefficient of performance (COP) of the refrigerator, (b) the rate of heat discharged to the surroundings.

Solution:
Assume: Steady-state operation.

$$\dot{q}_c = 350\,\text{kJ/min}, \dot{W}_{\text{net,in}} = 1.8\,\text{kW},$$

$$\text{COP}_R = \frac{\dot{q}_C}{\dot{W}_{\text{net,in}}} = \frac{350\,\text{kJ/min}}{1.8\,\text{kW}}\left(\frac{\text{kW}}{60\,\text{kJ/min}}\right) = \mathbf{3.2}$$

This means that 3.2 kJ of heat is removed from the refrigerator for each kJ of energy supplied.

(b) Energy balance

$$\dot{q}_H = \dot{q}_C + \dot{W}_{\text{net,in}} = 350\text{k J/min} + 1.8\text{kW}\left(\frac{60\,\text{kJ/min}}{kW}\right) = \mathbf{458\,kJ/min}$$

Both the heat removed from the refrigerator space and the unused energy supplied to the refrigerator as electrical work are discharged to surrounding air.

Example 7.29 Coefficient of performance of a vapor-compression refrigeration cycle

An ideal vapor-compression refrigeration cycle uses refrigerant-134a. The compressor inlet and outlet pressures are 120 and 900 kPa. The mass flow rate of refrigerant is 0.04 kg/s. Determine the coefficient of performance.

Solution:
Assume: Steady-state operation. The changes in kinetic and potential energies are negligible.

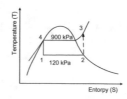

Solution:

Assume: Steady-state adiabatic operation. The changes in kinetic and potential energies are negligible.

The refrigerant mass flow rate = 0.04 kg/s

$P_2 = 120$ kPa, $H_2 = 233.86$ kJ/kg, $S_2 = 0.9354$ kJ/kg K

$P_3 = 900$ kPa, for the isentropic compression: $S_3 = S_2 = 0.9354$ kJ/kg K, $\rightarrow$, $H_3 = 276.7$ kJ/kg,

$P_4 = 900$ kPa, $H_{4sat\ liq} = 99.56$ kJ/kg; $H_{1sat\ liq} = H_{4sat\ liq}$ (throttling)

Power input to the compressor:

$\dot{W}_{in} = \dot{m}(H_3 - H_2) = (0.04$ kg/s$)(276.7 - 233.86) = 1.71$ kW

Heat removed from the refrigerator:

$\dot{q}_C = \dot{m}(H_2 - H_1) = (0.04$ kg/s$)$ $(233.86{-}99.56)$ kJ/kg $= 5.37$ kW

Heat discharged from the refrigerator:

$\dot{q}_H = \dot{m}(H_3 - H_4) = (0.04$ kg/s$)$ $(276.7{-}99.56)$ kJ/kg $= 7.08$ kW

$$\mathrm{COP}_R = \frac{\dot{q}_C}{\dot{W}_{in}} = \frac{5.37\ \mathrm{kW}}{1.71\ \mathrm{kW}} = 3.14 = 3.14$$

This refrigerator is capable of removing 3.14 units of energy from the refrigerated space for each unit of electric energy it consumes.

7.16   Efficiency of Fuel Cells

A *fuel cell* is an electrochemical cell that converts chemical energy of a fuel into electric energy by the reaction between a fuel supply and an oxidizing agent. A hydrogen fuel cell uses hydrogen as fuel and oxygen (usually from air) as oxidant. Other fuels include hydrocarbons and alcohols. A typical fuel cell produces a voltage from 0.6 to 0.7 V at full rated load. To deliver the desired amount of energy, the fuel cells can be combined in series and parallel circuits, where series circuits yield higher voltage, and parallel circuits allow a higher current to be supplied. Such a design is called a *fuel cell stack*. The cell surface area can be increased to allow stronger current from each cell [10].

The efficiency of a fuel cell depends on the amount of power drawn from it. Drawing more power means drawing more current and hence increasing the losses in the fuel cell. Most losses appear as a voltage drop in the cell, so the efficiency of a cell is almost proportional to its voltage. For this reason, it is common to show graphs of voltage versus current (polarization curves) for fuel cells. A typical cell running at 0.7 V has an efficiency of about 50%, meaning that 50% of the energy content of the hydrogen is converted into electrical energy and the remaining 50% will be converted into heat. Fuel cells are not heat engines and so their efficiency is

Table 7.5 Types of fuel cell using polymer membrane as electrolyte [4.10]

Fuel cell	Power output	T (°C)	Cell efficiency (%)
Proton exchange Membrane fuel cell	100 W–500 kW	150–120 (Nafion)	50–70
Direct methanol fuel cell	100 mW–1 kW	90–120	20–30
Microbial fuel cell	Low	<40	Low

not limited by the Carnot cycle efficiency. Consequently, they can have very high efficiencies in converting chemical energy to electrical energy, especially when they are operated at low power density, and using pure hydrogen and oxygen as reactants. Fuel cell vehicles running on compressed hydrogen may have a power-plant-to-wheel efficiency of 22% if the hydrogen is stored as high-pressure gas [10]. Table 7.5 compares the efficiency of several fuel cells.

The overall efficiency (electricity to hydrogen and back to electricity) of such plants (known as *round-trip efficiency*) is between 30 and 50%, depending on conditions. While a much cheaper lead-acid battery might return about 90%, the electrolyzer/fuel cell system can store indefinite quantities of hydrogen, and is therefore better suited for long-term storage [10].

7.17 Energy Conversions in Biological Systems

Energy production by way of biochemical reactions, its conversion, and storage is the basis of *bioenergetics* of living systems [6, 9]. Outside energy is supplied with the intake of food or with solar radiation. Living systems convert part of this energy to produce electrons and protons. Flow of protons leads to the production of adenosine triphosphate (ATP). The hydrolysis of ATP is coupled to synthesizing protein molecules, transporting ions and substrates, producing mechanical work, and other metabolic activity.

All living systems must convert energy to the chemical energy in the form of energy-rich chemical compounds such as ATP. The two biochemical cyclic processes for such conversions are the oxidative phosphorylation in animals and the photosynthesis in plants. These cycles are discussed briefly in the following sections.

7.17.1 Energy Conversion by Oxidative Phosphorylation

In oxidative phosphorylation, the electrons are removed from food molecules in electron transport chain. A series of proteins in the membranes of mitochondria use the energy released from passing electrons from reduced molecules like NADH onto oxygen to pump protons across the membrane in mitochondria. The flow of protons causes the rotation of stalk subunit of a large protein called the ATPase, which helps synthesize the adenosine triphosphate (ATP) from adenosine diphosphate (ADP) and inorganic phosphorus (Pi)

$$ADP + Pi = ATP \qquad (7.74)$$

ATP is an energy-rich compound having three phosphate groups attached to a nucleoside of adenine called adenosine. Of the three phosphate groups, the terminal one has a weak linkage. This phosphate group can break spontaneously whenever ATP forms a complex with an enzyme. The breaking up of this bond releases chemical energy causing an immediate shift in the bond energy giving rise to ADP. The energy of ATP is used for all the activity of living cells, such as transport of ions and molecules, synthesis of new proteins and other substances, and the growth and development. ATP therefore acts as 'energy currency of the cell' and is used to transfer chemical energy between different biochemical reaction cycles [6].

7.17.2   Energy from Photosynthesis

Photosynthesis is the synthesis of carbohydrates from sunlight, water, and carbon dioxide (CO_2). The capture of solar energy is similar in principle to oxidative phosphorylation, as the proton motive force then drives ATP synthesis. The electrons needed to drive this electron transport chain come from light-gathering proteins called photosynthetic reaction centers. In plants, cyanobacteria and algae, oxygenic photosynthesis splits water, with oxygen produced as a waste product. This process uses the ATP and NADPH produced by the photosynthetic reaction centers. This carbon-fixation reaction is carried out by the enzyme RUBisCO [6].

7.17.3   Metabolism

Metabolism is the set of biochemical reactions that occur in living organisms to maintain life. Metabolism is usually divided into two categories: catabolism and anabolism. Catabolism breaks down organic matter, for example to harvest energy in cellular respiration. Anabolism uses energy to construct components of cells such as proteins and nucleic acids. Adenosine triphosphate (ATP) is used to transfer chemical energy between different biochemical reaction cycles. ATP in cells is continuously regenerated and acts as a bridge between catabolism and anabolism, with catabolic reactions generating ATP and anabolic reactions consuming it [6, 9].

7.17.4   Biological Fuels

Biological fuels can be categorized into three groups: carbohydrates (CH), representing a mixture of mono-, di-, and poly-saccharides, fats (F), and proteins (Pr) [14]. Carbohydrates are straight-chain aldehydes or ketones with many hydroxyl groups that can exist as straight chains or rings. Carbohydrates are the most abundant biological molecules, and play numerous roles, such as the storage and transport of energy (starch, glycogen) and structural components such as cellulose in plants [6].

The fuel value is equal to the heat of reaction of combustion (oxidation). Carbohydrates and fats can be completely oxidized while proteins can only be partially oxidized and hence have a lower fuel value. The energy expenditure may be calculated from the energy balance. Assume that (i) carbohydrate (CH), fat (F), and protein (Pr) are the only compounds involved in the oxidation process, (ii) the other compounds are stationary, and (iii) the uptake and elimination of oxygen, carbon dioxide, and nitrogen is instantaneous. Energy balance is

$$\dot{E} = \sum_i (\dot{m}\Delta H_r)_i = (\dot{m}\Delta H_r)_{\text{CH}} + (\dot{m}\Delta H_r)_{\text{F}} + (\dot{m}\Delta H_r)_{\text{Pr}} = \dot{q} + \dot{W} \qquad (7.75)$$

7.17.5   Converting Biomass to Biofuels

Technologies for converting biomass to biofuels include *thermochemical* and *biochemical conversions:*

- *Thermochemical conversion* involves applying heat to break down biomass into chemical intermediates that can be used to make fuel substitutes. Many of these thermal technologies are known for well over a century and are used primarily in transforming coal and natural gas into fuels. Gasification combined with catalytic conversion of syngas (mainly carbon monoxide and hydrogen) is used to produce fuels, for example, hydrogen, liquid fuels, mixed alcohols, or dimethyl ether. Pyrolysis of biomass produces bio-oils that could serve as intermediates in biofuel synthesis.
- *Biochemical conversion* focuses on conversion of organic materials including animal waste, plant residues, and energy crops, such as corn and soybean into fuels and bioproducts by enzymes, bacteria, and microorganisms. For example, anaerobic digestion or fermentation converts organic waste. biogas, biofuel, and bioproducts.

Optimal combinations of both biochemical and thermochemical fuel production may lead to greater energy efficiency in the transformation of the energy of biomass into useable fuels.

Accessing energy in lignocellulosic biomass greatly increases the potential supply of biofuels. Figure 7.17 provides a schematic representation of the conversion process of lignocellulosic feedstock to biofuels. Cellulosic biomass needs pretreatment and enzymatic processes before it can be converted to biofuel by fermentation. The critical technology advancements need of biological catalysts that can cost-effectively break down the carbohydrate polymers in biomass to sugars and microbes that can ferment all of the sugars in biomass to ethanol [5, 9].

Fig. 7.17 Process schematic of biological conversion of lignocellulosic biomass to ethanol

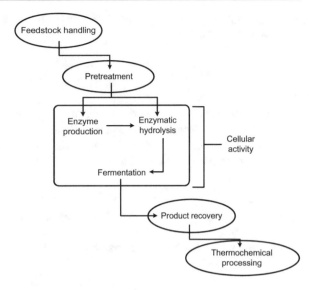

Government incentives for biodiesel have spurred significant growth in the vegetable oil-based biodiesel fuel substitute. Converting vegetable oil to biodiesel is a relatively simple thermochemical process in which the oil with methanol is converted to fatty acid methyl ester known as biodiesel [5].

Summary

- *Energy conversion* examples include the use of electric energy to produce various mechanical and/or chemical energies.
- Higher heating value (HHV) is determined by bringing all the products of combustion back to the original pre-combustion temperature, and in particular condensing any water vapor produced.

$$[C + H](fuel) + [O_2 + N_2](Air) \rightarrow CO_2 + H_2O(liquid) + N_2 + Heat \ (HHV)$$

where C = Carbon, H = Hydrogen, O = Oxygen, and N = Nitrogen. Higher heating value is related to lower heating value: $LHV = HHV - (\Delta H_{vap})$ $(m_{H_2O,out}/m_{fuel,in})$, where ΔH_{vap} is the heat of vaporization of water (in kJ/kg or Btu/lb), MW_{H_2O} and MW_{fuel} are the molecular weight of water and fuel.

- *Thermal efficiency* is a measure of the amount of thermal energy that can be converted to another useful form

$$\eta_{th} = q_{useful}/q_{in}$$

- For *fluid flow systems* the general relation between the heat and work for a fluid flow is expressed by:

$$\dot{m}\left(\Delta H + \frac{\Delta v^2}{2}\left(\frac{\text{kJ/kg}}{1000\,\text{m}^2/\text{s}^2} \right) + g\Delta z \right) = \dot{q} + \dot{W}_s$$

- For a process producing work, such as a turbine, the *ideal work* is the maximum possible work (isentropic) produced. However, for a process requiring work, such as a compressor, the ideal work is the minimum amount of required work (isentropic). Ideal and actual works are

$$W_{\text{ideal}} = \Delta(\dot{m}H) - T_o\Delta(\dot{m}S)$$
$$\dot{W}_{\text{act}} = \Delta\left[\dot{m}\left(H + v^2/2 + gz \right) \right] - \dot{q}$$

- *Lost work* is defined as a difference between the actual and ideal work

$$\dot{W}_{\text{lost}} = \dot{W}_{\text{act}} - \dot{W}_{\text{ideal}}$$

- *Transfer of mechanical energy* is usually accomplished through a rotating shaft. The motor efficiency and the generator efficiency are used in overall efficiency for turbine-generator

$$\eta_{\text{turb}-\text{gen}} = \eta_{\text{turb}}\eta_{\text{gen}} = \frac{\dot{W}_{\text{electout}}}{\Delta\dot{E}_{\text{mechin}}} \text{ (turbine-generator systems)}$$

- *Heat engines* transform thermal energy q_{in} into mechanical energy or work W_{out}. The thermal efficiency of a heat engine is

$$\eta_{\text{th}} = \frac{W_{\text{out}}}{q_{\text{in}}} = 1 - \frac{q_{\text{out}}}{q_{\text{in}}}$$

- Thermal efficiency of the Carnot engine is

$$\eta_{\text{th}} = 1 - \frac{T_C}{T_H}$$

- The *Rankine cycle* is used in steam turbine power plants. A turbine converts the kinetic energy of a moving fluid to mechanical energy.
- The *Brayton cycle* is used for gas turbines operating on an open cycle and consists of three main components: a compressor, a combustion chamber, and a turbine. The thermal efficiency is

$$\eta_{\text{Brayton}} = 1 - \left(r_p^{(\gamma-1)/\gamma} \right)^{-1} \text{(Constant heat capacity)}$$

where r_p is the compression ratio (P_2/P_1) and $\gamma = C_p/C_v$. The *back-work ratio* r_{bw} shows the part of the produced energy is diverted to the compressor

$$r_{\mathrm{bw}} = \frac{W_{\mathrm{comp.in}}}{W_{\mathrm{turb.out}}}$$

- *Otto cycle* describes the functioning of a typical ideal cycle for spark-ignition reciprocating piston engine. The thermal efficiency is

$$\eta_{\mathrm{Otto}} = 1 - \left(\frac{1}{r^{(\gamma-1)}}\right) \quad \text{(Constant specific heats)}$$

- *Diesel cycles* use a similar cycle to spark-ignited gasoline engine, but with a compression heating ignition system. The thermal efficiency of energy conversion is

$$\eta_{\mathrm{Diesel}} = 1 - \frac{1}{r^{(\gamma-1)}} \left(\frac{r_c^{\gamma} - 1}{\gamma(r_c - 1)}\right) \quad \text{(With constant specific heats)}$$

- The Ericsson engine is an '*external combustion engine*' because it is externally heated. A *Stirling engine* operates by cyclic compression and expansion of air or other gas at different temperature levels and converts the heat of hot gas into mechanical work.
- The *Atkinson engine* is an internal combustion engine and used in some hybrid electric applications.
- *Hydroelectricity* is the production of electrical power using the gravitational force of falling or flowing water.
- The power produced by a *wind turbine* is proportional to the kinetic energy of the wind captured by the wind turbine, and estimated by

$$\dot{W}_{\mathrm{wind}} = \eta_{\mathrm{wind}} \rho \frac{\pi v^3 D^2}{8}$$

where ρ is the density of air, v is the velocity of air, D is the diameter of the blades of the wind turbine, and η_{wind} is the efficiency of the wind turbine.
- *Geothermal electricity* refers to the use of geothermal energy to electric energy.
- *Ocean thermal energy conversion* uses the difference between cooler deep and warmer shallow or surface ocean waters to run a heat engine and produce useful work.
- *Thermoelectric* effect involves energy conversions between heat and electricity
- A *heat pump* uses a fluid which absorbs heat as it vaporizes and releases the heat when it condenses. Thermal efficiency of heat pumps is measured by the coefficient of performance $\mathrm{COP_{HP}}$

$$\mathrm{COP_{HP}} = \frac{\dot{q}_H}{\dot{W}_{\mathrm{in}}}$$

- A *refrigerator* is a heat engine in which work is required to extract energy from the freezing compartment and discharge that heat to the room through a condenser.
- A *fuel cell* is an electrochemical cell that converts chemical energy of a fuel into electric energy. Electricity is generated from the reaction between a fuel supply and an oxidizing agent.
- In *oxidative phosphorylation*, the electrons are removed from food molecules in electron transport chain. A series of proteins in the membranes of mitochondria uses the energy released from passing electrons from reduced molecules like NADH onto oxygen to pump protons across the membrane in mitochondria. The flow of protons helps synthesize adenosine triphosphate (ATP) from adenosine diphosphate (ADP) and inorganic phosphorus (Pi)

$$ADP + Pi = ATP$$

ATP is an energy-rich compound having three phosphate groups attached to a nucleoside of adenine called adenosine.
- *Metabolism* is usually divided into two categories: catabolism and anabolism. *Catabolism* breaks down organic matter; for example, to harvest energy in cellular respiration. *Anabolism* uses energy to construct components of cells such as proteins and nucleic acids.
- *Biological fuels* can be categorized into three groups: carbohydrates, fats, and proteins.
Technologies for converting biomass to biofuels are often classified into two main categories: *thermochemical conversion* and *biological conversion*.
- *Thermochemical conversion* involves applying heat to break down biomass into chemical intermediates that can be used to make fuel substitutes.
- *Biochemical conversion* focuses on fermentation of carbohydrates in biomass to ethanol and other chemicals.

Problems

7.1   A pump delivers water from the bottom of a storage tank open to the atmosphere containing water at 75 °F. The water in the storage tank is 10 ft deep and bottom of the tank is 50 ft above the ground. The pump delivers the water at 80 °F and 100 psia to another tank, which is 10 ft above the ground. If the water flow rate is 5000 lb/h and pump efficiency is 80%, estimate the power in hp required.

7.2.   The pump of a water storage tank is powered with an 18-kW electric motor operating with an efficiency of 90%. The water flow rate is 40 L/s. The diameters of the inlet and exit pipes are the same, and the elevation difference between the inlet and outlet is negligible. The absolute pressures at the inlet and outlet are 100 and 400 kPa, respectively. Determine the mechanical efficiency of the pump.

7.3. The pump of a water storage tank is powered with an 18-kW electric motor operating with an efficiency of 85%. The water flow rate is 30 L/s. The diameters of the inlet and exit pipes are the same, and the elevation difference between the inlet and outlet is negligible. The absolute pressures at the inlet and outlet are 100 and 500 kPa, respectively. Determine the mechanical efficiency of the pump.

7.4. A wind turbine generator with a 24-foot-diameter blade produces 0.35 kW of electric power. In the location of the wind turbine, the wind speed is 11 miles per hour. Determine the efficiency of the wind turbine generator.

7.5. A wind turbine generator with a 27-foot-diameter blade produces 0.4 kW of electric power. In the location of the wind turbine, the wind speed is 10 miles per hour. Determine the efficiency of the wind-turbine generator.

7.6. A hydroelectric plant operates by water falling from a 25 m height. The turbine in the plant converts potential energy into electrical energy with an assumed efficiency of 85%. The electricity is lost by about 8% through the power transmission. Estimate the mass flow rate of the water necessary to power a 500 W light bulb.

7.7. A hydroelectric plant is generating 125 kW of power. If their turbines only convert the potential energy form the water with a 75% efficiency, find the height change for the water flow rate of 480 kg/s must fall in order to continue to make 125 kW of power.

7.8. A hydroelectric plant operates by water falling from a 20 m height. The turbine in the plant converts potential energy into electrical energy with an assumed efficiency of 80%. The electricity is lost by about 8% through the power transmission. Estimate the mass flow rate of the water necessary to power a 1500 W light bulb.

7.9. A hydroelectric plant operates by water falling from a 100 ft height. The turbine in the plant converts potential energy into electrical energy with an assumed efficiency of 82%. The electricity is lost by about 6% through the power transmission. Estimate the mass flow rate of the water necessary to power a 3500 W light bulb.

7.10. A hydroelectric plant operates by water falling from a 50 ft height. The turbine in the plant converts potential energy into electrical energy with an assumed efficiency of 85%. The electricity is lost by about 5% through the power transmission, so the available power is 95%. Estimate the mass flow rate of the water necessary to power a 3500 W light bulb.

7.11. A hydroelectric plant operates by water falling from a 75 ft height. The turbine in the plant converts potential energy into electrical energy with an assumed efficiency of 85%. The power is lost by about 5% through the power transmission, so the available power is 95%. If the mass flow rate of the water 396 lb/s, estimate the power output of the hydro plant.

7.12. An electric motor attached to a pump draws 10.2 A at 110 V. At steady load, the motor delivers 1.32 hp of mechanical energy. Estimate the rate of heat transfer from the motor.

7.13. An electric motor attached to a pump draws 12 A at 110 V. At steady load, the motor delivers 1.5 hp of mechanical energy. Estimate the rate of heat transfer from the motor.

7.14. Air enters an insulated compressor operating at steady state at 1.0 bar, 300 K with a mass flow rate of 3.6 kg/s and exits at 2.76 bar. Kinetic and potential energy effects are negligible. (a) Determine the minimum theoretical power input required, in kW, and the corresponding exit temperature, in K. (b) If the actual exit temperature is 420 K, determine the power input, in kW, and the isentropic compressor efficiency.

7.15. Air enters an insulated compressor operating at steady state at 1.0 bar, 300 K with a mass flow rate of 2.5 kg/s and exits at 2.6 bar. Kinetic and potential energy effects are negligible. (a) Determine the minimum theoretical power input required, in kW, and the corresponding exit temperature, in K. (b) If the actual exit temperature is 420 K, determine the power input, in kW, and the isentropic compressor efficiency.

7.16. A compressor increases the pressure of carbon dioxide from 100 to 600 kPa. The inlet temperature is 300 K and the outlet temperature is 400 K. The mass flow rate of carbon dioxide is 0.01 kmol/s (440 g/s). The power required by the compressor is 55 kW. The temperature of the surroundings is 290 K. Determine the minimum amount of work required and the coefficient of performance.

7.17. A compressor increases the pressure of carbon dioxide from 100 to 500 kPa. The inlet temperature is 300 K and the outlet temperature is 400 K. The mass flow rate of carbon dioxide is 0.015 kmol/s (660 g/s). The power required by the compressor is 60 kW. The temperature of the surroundings is 290 K. Determine the minimum amount of work required and the coefficient of performance.

7.18 A compressor receives air at 15 psia and 80 °F with a flow rate of 1.2 lb/s. The air exits at 40 psia and 300 °F. At the inlet, the air velocity is low but increases to 250 ft/s at the outlet of the compressor. Estimate the power input to the compressor if it is cooled at a rate of 200 Btu/s.

7.19 A compressor receives air at 15 psia and 80 °F with a flow rate of 1.5 lb/s. The air exits at 50 psia and 300 °F. At the inlet, the air velocity is low but increases to 250 ft/s at the outlet of the compressor. Estimate the power input to the compressor if it is cooled at a rate of 150 Btu/s.

7.20. In an adiabatic compression operation, air is compressed from 20 °C and 101.32–520 kPa with an efficiency of 0.7. The air flow rate is 22 mol/s (637.3 g/s). Assume that the air remains ideal gas during the compression. The surroundings are at 298.15 K. Determine the thermodynamic efficiency η_{th} and the rate of energy dissipated $\dot{E}_{loss}$.

7.21. In an adiabatic compression operation, air is compressed from 25 °C and 101.32–560 kPa with an efficiency of 0.7. The air flow rate is 20 mol/s (579.4 g/s). Assume that the air remains ideal gas during the compression. The surroundings are at 298.15 K. Determine the thermodynamic efficiency η_{th} and the rate of energy dissipated $\dot{E}_{loss}$.

7.22. The power required to compress 0.05 kg/s of steam from a saturated vapor state at 50 °C to a pressure of 800 kPa at 200 °C is 15 kW. Find the conversion rate of power input to heat loss from the compressor.

7.23. The power required to compress 0.1 kg/s of steam from a saturated vapor state at 80 °C to a pressure of 1000 kPa at 200 °C is 28 kW. Find the conversion rate of power input to heat loss from the compressor.

7.24. A steam turbine consumes 4000 lb/h steam at 540 psia and 800 °F. The exhausted steam is at 165 psia. The turbine operation is adiabatic.

    (a) Determine the exit temperature of the steam and the work produced by the turbine.

    (b) Determine the thermal efficiency.

7.25. A steam turbine consumes 3800 lb/h steam at 540 psia and 800 °F. The exhausted steam is at 180 psia. The turbine operation is adiabatic.

    (a) Determine the exit temperature of the steam and the work produced by the turbine.

    (b) Determine the thermal efficiency.

7.26. A superheated steam (stream 1) expands in a turbine from 5000 kPa and 325 °C to 150 kPa and 200 °C. The steam flow rate is 15.5 kg/s. If the turbine generates 1.1 MW of power, determine the heat loss to the surroundings and thermal efficiency.

7.27. A superheated steam (stream 1) expands in a turbine from 6000 kPa and 325 °C to 150 kPa and 200 °C. The steam flow rate is 16.2 kg/s. If the turbine generates 1.2 MW of power, determine the heat loss to the surroundings and thermal efficiency.

7.28. Steam expands in a turbine from 6600 kPa and 300 °C to a saturated vapor at 1 atm. The steam flow rate is 9.55 kg/s. If the turbine generates a power of 1 MW, determine the thermal efficiency.

7.29. Steam expands in a turbine from 7000 kPa and 300 °C to a saturated vapor at 1 atm. The steam flow rate is 14.55 kg/s. If the turbine generates a power of 1.5 MW, determine the thermal efficiency.

7.30. Steam expands adiabatically in a turbine from 850 psia and 600 °F to a wet vapor at 12 psia with a quality of 0.9. The turbine produces a power output of 1500 Btu/s. Estimate the thermal efficiency for a steam flow rate of 12.8 lb/s.

7.31. Steam expands adiabatically in a turbine from 900 psia and 600 °F to a wet vapor at 10 psia with a quality of 0.9. The turbine produces a power output of 1550 Btu/s. Estimate the thermal efficiency for a steam flow rate of 14.2 lb/s.

7.32. A turbine produces 65,000 kW electricity with an efficiency of 70%. It uses a superheated steam at 8200 kPa and 550 °C. The discharged stream is a saturated mixture at 75 kPa. If the expansion in the turbine is adiabatic, and the surroundings are at 298.15 K, determine the thermodynamic efficiency and the work loss.

7.33. A turbine produces 70 MW electricity with an efficiency of 70%. It uses a superheated steam at 8800 kPa and 550 °C. The discharged stream is a saturated mixture at 15 kPa. If the expansion in the turbine is adiabatic, and the surroundings are at 298.15 K, determine the thermodynamic efficiency and the work loss.

7.34. A Carnot cycle uses water as the working fluid at a steady-flow process. Heat is transferred from a source at 250 °C and water changes from saturated liquid to saturated vapor. The saturated steam expands in a turbine at 10 kPa, and heat is transferred in a condenser at 10 kPa. Estimate the thermal efficiency and net power output of the cycle.

7.35. A Carnot cycle uses water as the working fluid at a steady-flow process. Heat is transferred from a source at 200 °C and water changes from saturated liquid to saturated vapor. The saturated steam expands in a turbine at 20 kPa, and heat is transferred in a condenser at 20 kPa. Estimate the thermal efficiency of the cycle and the amount of heat transferred in the condenser for a flow rate of 5.5 kg/s of the working fluid.

7.36. A Carnot cycle uses water as the working fluid at a steady-flow process. Heat is transferred from a source at 200 °C and water changes from saturated liquid to saturated vapor. The saturated steam expands in a turbine at 10 kPa, and heat is transferred in a condenser at 10 kPa. Estimate the thermal efficiency of the cycle and the amount of heat transferred in the condenser for a flow rate of 7.5 kg/s of the working fluid.

7.37. A Carnot cycle uses water as the working fluid at a steady-flow process. Heat is transferred from a source at 400 °F and water changes from saturated liquid to saturated vapor. The saturated steam expands in a turbine at 5 psia, and heat is transferred in a condenser at 5 psia. Estimate the thermal efficiency of the cycle and the amount of heat transferred in the condenser for a flow rate of 10 lb/s of the working fluid.

7.38. Consider a simple ideal Rankine cycle. If the turbine inlet temperature and the condenser pressure are kept the same, discuss the effects of increasing the boiler pressure on: (a) Turbine power output, (b) heat supplied, (c) thermal efficiency, (d) heat rejected.

7.39. Consider a simple ideal Rankine cycle. If the boiler and the condenser pressures are kept the same, discuss the effects of increasing the temperature of the superheated steam on: (a) Turbine power output, (b) heat supplied, (c) thermal efficiency, (d) heat rejected.

7.40. A steam power production plant burns fuel at 1273.15 K (T_H), and cooling water is available at 290 K (T_C). The steam produced by the boiler is at 8200 kPa and 823.15 K. The condenser produces a saturated liquid at 30 kPa. The turbine and pump operate reversibly and adiabatically. Determine the thermal efficiency of the cycle for the steam flow rate of 1 k/s.

7.41. A steam power plant operates on a simple ideal Rankine cycle. The boiler operates at 3000 kPa and 350 °C. The condenser operates at 30 kPa. The mass flow rate of steam is 22 kg/s. Estimate the thermal efficiency of the cycle and the net power output.

7.42. A steam power plant operates on a simple ideal Rankine cycle shown below. The turbine receives steam at 698.15 K and 4100 kPa, while the discharged steam is at 40 kPa. The mass flow rate of steam is 3.0 kg/s. In the boiler, heat is transferred into the steam from a source at 1500 K (T_H). In the condenser, heat is discharged to the surroundings at 298 K (T_C). The condenser operates at 298 K. Determine the thermal efficiency of the cycle.

7.43. A steam power plant operates on a simple ideal Rankine cycle. The boiler operates at 5000 kPa and 300 °C. The condenser operates at 20 kPa. The mass flow rate of steam is 25 kg/s. Estimate the thermal efficiency of the cycle and the net power output.

7.44. A steam power plant operates on a simple ideal Rankine cycle. The boiler operates at 10,000 kPa and 400 °C. The condenser operates at 30 kPa. The power output of the cycle is 140 MW. Estimate the thermal efficiency of the cycle and the mass flow rate of the steam.

7.45. A steam power plant is operating on the simple ideal Rankine cycle. The steam mass flow rate is 20 kg/s. The steam enters the turbine at 3500 kPa and 400 °C. Discharge pressure of the steam from the turbine is 15 kPa. Determine the thermal efficiency of the cycle.

7.46. A simple ideal Rankine cycle is used in a steam power plant. Steam enters the turbine at 6600 kPa and 798.15 K. The net power output of the turbine is 35 kW. The discharged steam is at 10 kPa. Determine the thermal efficiency.

7.47. A steam power plant operates on a simple ideal Rankine cycle. The boiler operates at 10,000 kPa and 500 °C. The condenser operates at 10 kPa. The power output of the cycle is 175 MW. Turbine operates with an isentropic efficiency of 0.80, while the pump operates with an isentropic efficiency of 0.90. Estimate the thermal efficiency of the cycle and the heat transferred in the condenser.

7.48. A steam power plant operates on a simple ideal Rankine cycle. The boiler operates at 10,000 kPa and 400 °C. The condenser operates at 10 kPa. The power output of the cycle is 145 MW. Turbine operates with an isentropic efficiency of 0.85, while the pump operates with an isentropic efficiency of 0.95. Estimate the thermal efficiency of the cycle and the heat transferred in the condenser.

7.49. A steam power plant operates on a simple ideal Rankine cycle. The boiler operates at 10,000 kPa and 400 °C. The condenser operates at 10 kPa. The mass flow rate of steam is 110 kg/s. Turbine operates with an isentropic efficiency of 0.85, while the pump operates with an isentropic efficiency of 0.95. Estimate the thermal efficiency and the power output of the cycle.

7.50. A steam power plant operates on a simple ideal Rankine cycle. The boiler operates at 8000 kPa and 400 °C. The condenser operates at 20 kPa. The mass flow rate of steam is 80 kg/s. Turbine operates with an isentropic efficiency of 0.85, while the pump operates with an isentropic efficiency of 0.90. Estimate the thermal efficiency and the power output of the cycle.

7.51. A steam power plant shown below uses natural gas to produce 0.1 MW power. The combustion heat supplied to a boiler produces steam at 10,000 kPa and 798.15 K. The turbine efficiency is 0.7. The discharged steam from the turbine is at 30 kPa and is sent to a condenser. The condensed water is pumped to the boiler. The pump efficiency is 0.90. Determine the thermal efficiency of the cycle.

7.52. A steam power plant shown below uses natural gas to produce 0.12 MW power. The combustion heat supplied to a boiler produces steam at 10,000 kPa and 798.15 K. The turbine efficiency is 0.75. The discharged steam from the turbine is at 30 kPa and is sent to a condenser. The condensed water is pumped to the boiler. The pump efficiency is 0.85. Determine the thermal efficiency of an ideal and actual Rankine cycles.

7.53. A steam power plant shown below uses natural gas to produce 0.12 MW power. The combustion heat supplied to a boiler produces steam at 9000 kPa and 798.15 K. The turbine efficiency is 0.8. The discharged steam from the turbine is at 10 kPa and is sent to a condenser. The condensed water is pumped to the boiler. The pump efficiency is 0.9. Determine the thermal efficiency of an ideal and actual Rankine cycles.

7.54. A simple ideal reheat Rankine cycle is used in a steam power plant shown below. Steam enters the turbine at 9000 kPa and 823.15 K and leaves at 4350 kPa and 698.15 K. The steam is reheated at constant pressure to 823.15 K. The discharged steam from the low-pressure turbine is at 10 kPa. The net power output of the turbine is 40 MW. In the boiler, heat is transferred into the steam from a source at 1600 K (T_H). In the condenser, heat is discharged to the surroundings at 298 K (T_C). The condenser operates at 298 K. The turbine efficiency is 0.8 and the pump efficiency is 0.9. Determine the mass flow rate of steam and the thermal efficiency of the cycle.

7.55. A steam power plant uses an ideal regenerative Rankine cycle shown below. Steam enters the high-pressure turbine at 8200 kPa and 773.15 K, and the condenser operates at 20 kPa. The steam is extracted from the turbine at 350 kPa to heat the feed water in an open heater. The water is a saturated liquid after passing through the feed water heater. The work output of the turbine is 50 MW. In the boiler, heat is transferred into the steam from a source at 1600 K (T_H). In the condenser, heat is discharged to the surroundings at 285 K (T_C). Determine the thermal efficiency of the cycle.

7.56. A steam power plant uses an ideal reheat regenerative Rankine cycle shown below. Steam enters the high-pressure turbine at 9000 kPa and 773.15 K and leaves at 850 kPa. The condenser operates at 10 kPa. Part of the steam is extracted from the turbine at 850 kPa to heat the water in an open heater, where the steam and liquid water from the condenser mix and direct contact heat transfer takes place. The rest of the steam is reheated to 723.15 K and expanded in the low-pressure turbine section to the condenser pressure. The water is a saturated liquid after passing through the water heater and is at the heater pressure. The work output of the turbine is 75 MW. In the boiler, heat is transferred into the steam from a source at 1600 K (T_H). In the condenser, heat is discharged to the surroundings at 285 K (T_C). Determine the thermal efficiency of the cycle.

7.57. A steam power plant uses a geothermal energy source. The geothermal source is available at 220 °C and 2320 kPa with a flow rate of 200 kg/s. The hot water goes through a valve and a flash drum. Steam from the flash drum enters the turbine at 550 kPa and 428.62 K. The discharged steam from the turbine has a quality of $x_4 = 0.96$. The condenser operates at 10 kPa. The water is a saturated liquid after passing through the condenser. Determine the thermal efficiency of the cycle.

7.58. A geothermal power production plant produces 7 MW power. Inlet temperature of the hot geothermal liquid source is 150 °C. The flow rate of the hot liquid water is 220 kg/s. The reference state is at 25 °C. Estimate the second law efficiency of the plant.

7.59. A reheat Rankine cycle is used in a steam ower plant. Steam enters the high-pressure turbine at 9000 kPa and 823.15 K and leaves at 4350 kPa. The steam is reheated at constant pressure to 823.15 K. The steam enters the low-pressure turbine at 4350 kPa and 823.15 K. The discharged steam from the low-pressure turbine is at 10 kPa. The net power output of the turbine is 65 MW. The isentropic turbine efficiency is 80%. The pump efficiency is 95%. In the boiler, heat is transferred into the steam from a source at 1600 K. In the condenser, heat is discharged to the surroundings at 298 K. The condenser operates at 298 K. Determine the thermal efficiency.

7.60. A reheat Rankine cycle is used in a steam power plant. Steam enters the high-pressure turbine at 10,000 kPa and 823.15 K and leaves at 4350 kPa. The steam is reheated at constant pressure to 823.15 K. The steam enters the low-pressure turbine at 4350 kPa and 823.15 K. The discharged steam from the low-pressure turbine is at 15 kPa. The net power output of the turbine is 65 MW. The isentropic turbine efficiency is 80%. The pump efficiency is 95%. In the boiler, heat is transferred into the steam from a source at 1600 K. The condenser operates at 298 K. Determine the thermal efficiency.

7.61. A steam power plant uses an actual regenerative Rankine cycle. Steam enters the high-pressure turbine at 11,000 kPa and 773.15 K, and the condenser operates at 10 kPa. The steam is extracted from the turbine at 475 kPa to heat the water in an open heater. The water is a saturated liquid after passing through the water heater. The work output of the turbine is 90 MW. The pump efficiency is 95% and the turbine efficiency is 75%. In the boiler, heat is transferred into the steam from a source at 1700 K. In the condenser, heat is discharged to the surroundings at 285 K. Determine the thermal efficiency.

7.62. A steam power plant uses an actual regenerative Rankine cycle. Steam enters the high-pressure turbine at 10,000 kPa and 773.15 K, and the condenser operates at 15 kPa. The steam is extracted from the turbine at 475 kPa to heat the water in an open heater. The water is a saturated liquid after passing through the water heater. The work output of the turbine is 90 MW. The pump efficiency is 90% and the turbine efficiency is 80%. In the boiler, heat is transferred into the steam from a source at 1700 K. In the condenser, heat is discharged to the surroundings at 285 K. Determine the thermal efficiency.

7.63. A steam power plant uses an actual reheat regenerative Rankine cycle. Steam enters the high-pressure turbine at 11,000 kPa and 773.15 K, and the condenser operates at 10 kPa. The steam is extracted from the turbine at 2000 kPa to heat the water in an open heater. The steam is extracted at 475 kPa for process heat. The water is a saturated liquid after passing through the water heater. The work output of the turbine is 90 MW. The turbine efficiency is 80%. The pumps operate isentropically. In the boiler, heat is transferred into the steam from a source at 1700 K. In the condenser, heat is discharged to the surroundings at 290 K. Determine the thermal efficiency of the plant.

7.64. A steam power plant uses an actual reheat-regenerative Rankine cycle. Steam enters the high-pressure turbine at 10,000 kPa and 773.15 K, and the condenser operates at 15 kPa. The steam is extracted from the turbine at 2000 kPa to heat the water in an open heater. The steam is extracted at 475 kPa for process heat. The water is a saturated liquid after passing through the water heater. The work output of the turbine is 85 MW. The turbine efficiency is 80%. The pumps operate isentropically. In the boiler, heat is transferred into the steam from a source at 1700 K. In the condenser, heat is discharged to the surroundings at 290 K. Determine the thermal efficiency of the plant.

7.65. A steam power plant operates on a regenerative cycle. Steam enters the turbine at 700 psia and 800 °F and expands to 1 psia in the condenser. Part of the steam is extracted at 60 psia. The efficiencies of the turbine and pump are 0.80 and 0.95, respectively. If the mass flow rate of steam is 9.75 lb/s estimate the thermal efficiency of the turbine.

7.66. A steam power plant operates on a regenerative cycle. Steam enters the turbine at 750 psia and 800 °F and expands to 5 psia in the condenser. Part of the steam is extracted at 60 psia. The efficiencies of the turbine and pump are 0.80 and 0.90, respectively. If the mass flow rate of steam is 10.5 lb/s estimate the thermal efficiency of the turbine.

7.67. A cogeneration plant shown below uses steam at 900 psia and 1000 °F to produce power and process heat. The steam flow rate from the boiler is 16 lb/s. The process requires steam at 70 psia at a rate of 3.2 lb/s supplied by the expanding steam in the turbine. The extracted steam is condensed and mixed with the water output of the condenser. The remaining steam expands from 70 psia to the condenser pressure of 3.2 psia. In the boiler, heat is transferred into the steam from a source at 3000 R. In the condenser, heat is discharged to the surroundings at 540 R. If the turbine operates with an efficiency of 80% and the pumps with an efficiency of 85%, determine the thermal efficiency of the cycle.

7.68. A power plant is operating on an ideal Brayton cycle with a pressure ratio of $r_p = 9$. The fresh air temperature is 300 K at the compressor inlet and 1200 K at the end of the compressor and at the inlet of the turbine. Using the standard-air assumptions, determine the thermal efficiency of the cycle.

7.69. A power plant is operating on an ideal Brayton cycle with a pressure ratio of $r_p = 9$. The fresh air temperature is 300 K at the compressor inlet and 1200 K at the end of the compressor and at the inlet of the turbine. Assume the gas-turbine cycle operates with a compressor efficiency of 80% and a turbine efficiency of 80%. Determine the thermal efficiency of the cycle.

7.70. The net work of a power cycle is 8×10^6 Btu/s and the heat transfer to the cold reservoir, q_C, is 12×10^6 Btu/s. The hot source operates at 1400 R and the cold source temperature is 560 R. What is the ratio of the achieved thermal efficiency to the maximum thermal efficiency of the cycle?

7.71. The net work of a power cycle is 25 MW and the heat transfer to the cold reservoir, q_C, is 67 MW. The hot source operates at 573 K and the cold source temperature is 285. What is the ratio of the achieved thermal efficiency to the maximum thermal efficiency of the cycle?

7.72. The net work of a power cycle is 52,000 Btu/s and the heat transfer to the cold reservoir, q_C, is 69,000 Btu/s. The hot source operates at 1450 R and the cold source temperature is 550 R. What is the ratio of the achieved thermal efficiency to the maximum thermal efficiency of the cycle?

7.73. The net work of a power cycle is 75,000 Btu/s and the heat transfer to the cold reservoir, q_C, is 85,000 Btu/s. The hot source operates at 1500 R and the cold source temperature is 530 R. What is the ratio of the achieved thermal efficiency to the maximum thermal efficiency of the cycle?

7.74. An ideal Otto cycle operates with a compression ratio (V_{max}/V_{min}) of 8.9. Air is at 101.3 kPa and 300 K at the start of compression (state 1). The maximum and minimum temperatures in the cycle are 1360 and 300 K. Specific heats depend on the temperature. Determine the thermal efficiency of the cycle and the thermal efficiency of a Carnot cycle working between the same temperature limits.

7.75. An ideal Otto cycle operates with a compression ratio (V_{max}/V_{min}) of 9.2. Air is at 101.3 kPa and 300 K at the start of compression (state 1). During the constant volume heat addition process, 730 kJ/kg of heat is transferred into the air from a source at 1900 K. Heat is discharged to the surroundings at 280 K. Determine the thermal efficiency of energy conversion.

7.76. An ideal Otto cycle operates with a compression ratio (V_{max}/V_{min}) of 9. Air is at 101.3 kPa and 295 K at the start of compression (state 1). During the constant volume heat addition process, 900 kJ/kg of heat is transferred into the air from a source at 1800 K. Heat is discharged to the surroundings at 295 K. Determine the thermal efficiency of energy conversion.

7.77. An ideal Otto cycle operates with a compression ratio (= V_{max}/V_{min}) of 8.5. Air is at 101.3 kPa and 285 K at the start of compression (state 1). During the constant volume heat addition process, 1000 kJ/kg of heat is transferred into the air from a source at 1800 K. Heat is discharged to the surroundings at 280 K. Determine the thermal efficiency of energy conversion.

7.78. An ideal Otto cycle operates with a compression ratio (= V_{max}/V_{min}) of 10. Air is at 101.3 kPa and 295 K at the start of compression (state 1). During the constant volume heat addition process, 1000 kJ/kg of heat is transferred into the air from a source at 1800 K. Heat is discharged to the surroundings at 295 K. Determine the thermal efficiency of the energy conversion.

7.79. An ideal Otto cycle operates with a compression ratio ($r = V_{max}/V_{min}$) of 8. Air is at 101.3 kPa and 300 K at the start of compression (state 1). The maximum and minimum temperatures in the cycle are 1300 and 300 K, respectively. Determine the thermal efficiency of energy conversion and the thermal efficiency of the Carnot engine. The average specific heats are $C_{p,av}$ = 1.00 kJ/kg K and $C_{v,av}$ = 0.717 kJ/kg K.

7.80. An ideal Otto cycle operates with a compression ratio ($r = V_{max}/V_{min}$) of 8.8. Air is at 101.3 kPa and 300 K at the start of compression (state 1). The maximum and minimum temperatures in the cycle are 1400 and 300 K, respectively. Determine the thermal efficiency of energy conversion and the thermal efficiency of the Carnot engine. The average specific heats are $C_{p,av}$ = 1.00 kJ/kg K and $C_{v,av}$ = 0.717 kJ/kg K.

7.81 One kmole of carbon dioxide is initially at 1 atm and −13 °C, and performs a power cycle consisting of three internally reversible processes in series. Step 1–2: Adiabatic compression to 5 atm. Step 2–3: Isothermal expansion to 1 atm. Step 3–1: Constant-pressure compression. Determine the net work, in Btu per lb_m and the thermal efficiency.

7.82 An ideal diesel cycle has an air-compression ratio of 18 and operating with maximum temperature of 2660 R. At the beginning of the compression, the fluid pressure and temperature are 14.7 psia, 540, respectively. The average specific heats of air at room temperature are $C_{p,av}$ = 0.24 Btu/lb R and $C_{v,av}$ = 0.171 Btu/lb R. Utilizing the cold air-standard assumptions, determine the thermal efficiency.

7.83 An ideal diesel cycle has an air-compression ratio of 18 and operating with maximum temperature of 2200 K. At the beginning of the compression, the fluid pressure and temperature are 100 kPa, 290 K, respectively. The average specific heats of air at room temperature are $C_{p,av}$ = 1.0 kJ/kg K and $C_{v,\text{av}} = 0.718$ kJ/kg K. Utilizing the cold air-standard assumptions, determine the thermal efficiency of the cycle.

7.84   An ideal diesel cycle has an air-compression ratio of 16 and a cutoff ratio of 2. At the beginning of the compression, the fluid pressure, temperature, and volume are 100 kPa, 300 K, respectively. The average specific heats of air at room temperature are $C_{p,\text{av}} = 1.005$ kJ/kg K and $C_{v,\text{av}} = 0.7181$ kJ/kg K. Utilizing the cold air-standard assumptions, determine the thermal efficiency of the cycle.

7.85.   An ideal diesel cycle has an air-compression ratio of 16 and a cutoff ratio of 2. At the beginning of the compression, the fluid pressure, temperature, and volume are 100 kPa, 290 K, respectively. The average specific heats of air at room temperature are $C_{p,\text{av}} = 1.005$ kJ/kg K, $C_{v,\text{av}} = 0.7181$ kJ/kg K. Utilizing the cold air-standard assumptions, determine the thermal efficiency of the cycle.

7.86.   A refrigerator using tetrafluoroethane (R-134a) as refrigerant operates with a capacity of 10,000 Btu/h. The refrigerated space is at 15 °F. The evaporator and condenser operate with a 10 °F temperature difference in their heat transfer. Cooling water enters the condenser at 70°F. Therefore, the evaporator is at 5 °F, and the condenser is at 80 °F. Determine the ideal and actual power necessary if the compressor efficiency is 85%.

7.87.   In a refrigeration cycle, the superheated R-134a (state 2) enters a compressor at 200 °F and 90.0 psia. The R-134a (state 3) leaves the compressor at 360 °F and 140 psia, and enters a condenser, where it is cooled by cooling water. The R-134a (state 4) leaves the condenser at 90.5 °F and 120 psia as saturated liquid and enters a throttling valve. The partially vaporized R-134a (state 1) leaves the valve at 100 psia. The cycle of R-134a is completed when it passes through an evaporator to absorb heat from the matter to be refrigerated. The flow rate of R-134a is 0.2 lb/s. The total power input is 85 Btu/s. The cooling water enters the condenser at 80 °F and leaves at 115 °F. The surroundings are at 210 °F. Determine the overall exergy loss.

7.88.   In a tetrafluoroethane (R-134a) refrigeration cycle, the superheated R-134a (state 1) enters a compressor at 253.15 K and 0.14 MPa. The R-134a (state 2) leaves the compressor at 303.15 K and 0.5 MPa, and enters a condenser, where it is cooled by cooling water. The R-134a (state 3) leaves the condenser at 299.87 K and 0.75 MPa and enters a throttling valve. The partially vaporized R-134a (state 4) leaves the valve at 0.205 MPa. The cycle is completed by passing the R-134 through an evaporator to absorb heat from the matter to be refrigerated. The R-134a (state 1) leaves the evaporator as superheated vapor. The flow rate of R-134a is 0.16 kg/s. The total power input is 750 kW. Estimate the coefficient of performance.

7.89.   A refrigerator using tetrafluoroethane (R-134a) as refrigerant operates with a capacity of 250 Btu/s. Cooling water enters the condenser at 70 °F. The evaporator is at 10 °F, and the condenser is at 80 °F. The refrigerated space is at 20 °F. Determine the ideal and actual power necessary if the compressor efficiency is 75%. Assume that the kinetic and potential energy changes are negligible.

7.90. A refrigerator using tetrafluoroethane (R-134a) as refrigerant operates with a capacity of 2500 kW. Cooling water enters the condenser at 280 K. Evaporator is at 271.92 K, and the condenser is at 299.87 K. The refrigerated space is at 280 K. Determine the ideal and actual power necessary if the compressor efficiency is 80%. Assume that the kinetic and potential energy changes are negligible.

7.91. A refrigeration cycle has a COP = 3.0. For the cycle, $q_H = 2000$ kJ. Determine q_C and W_{net}, each in kJ.

7.92. A refrigeration cycle has a COP = 2.5. For the cycle, $q_H = 1500$ kJ. Determine q_C and W_{net}, each in kJ.

7.93. In a pentafluoroethane (R-125) refrigeration cycle, the saturated R-125 (state 1) enters a compressor at 250 K and 3 bar. The R-125 (state 2) leaves the compressor at 320 K and 23.63 bar, and enters a condenser, where it is cooled by cooling water. The R-125 (state 3) leaves the condenser as saturated liquid at 310 K and 18.62 bar and enters a throttling valve. The partially vaporized R-125 (state 4) leaves the valve at 255 K and 3.668 bar. The cycle is completed by passing the R-125 through an evaporator to absorb heat from the matter to be refrigerated. The R-125 leaves the evaporator as saturated vapor. The evaporator temperature is 275.15 K. The flow rate of R-125 is 0.75 kg/s. The total power input is 60 kW. The cooling water enters the condenser at 293.15 K and leaves at 295.15 K. The surroundings are at 298.15 K. Determine the coefficient of performance.

7.94. A refrigerator using tetrafluoroethane (R-134a) as refrigerant operates with a capacity of 10,000 Btu/h. The refrigerated space is at 15 °F. The evaporator and condenser operate with a 10 °F temperature difference in their heat transfer. Cooling water enters the condenser at 70 °F. Therefore, the evaporator is at 5 °F, and the condenser is at 80 °F. Determine the coefficient of performance if the compressor efficiency is 85%.

7.95. In a refrigeration cycle, the superheated R-134a (state 2) enters a compressor at 200 °F and 90.0 psia. The R-134a (state 3) leaves the compressor at 360 °F and 140 psia, and enters a condenser, where it is cooled by cooling water. The R-134a (state 4) leaves the condenser at 90.5 °F and 120 psia as saturated liquid and enters a throttling valve. The partially vaporized R-134a (state 1) leaves the valve at 100 psia. The cycle of R-134a is completed when it passes through an evaporator to absorb heat from the matter to be refrigerated. The flow rate of R-134a is 0.2 lb/s. The total power input is 85 Btu/s. The cooling water enters the condenser at 80 °F and leaves at 115 °F. The surroundings are at 210 °F. Determine the coefficient of performance.

7.96. In a tetrafluoroethane (R-134a) refrigeration cycle, the superheated R-134a (state 1) enters a compressor at 253.15 K and 0.14 MPa. The R-134a (state 2) leaves the compressor at 303.15 K and 0.5 MPa, and enters a condenser, where it is cooled by cooling water. The R-134a (state 3) leaves the condenser at 299.87 K and 0.75 MPa and enters a throttling valve. The partially vaporized R-134a (state 4) leaves the valve at 0.205 MPa. The cycle is completed by passing the R-134 through an evaporator to absorb heat from the matter to be refrigerated. The R-134a (state 1) leaves the evaporator as superheated vapor. The flow rate of R-134a is 0.16 kg/s. The total power input is 750 kW. Estimate the coefficient of performance.

7.97. A refrigerator using tetrafluoroethane (R-134a) as refrigerant operates with a capacity of 250 Btu/s. Cooling water enters the condenser at 70 °F. The evaporator is at 10 °F, and the condenser is at 80 °F. The refrigerated space is at 20 °F. Determine the coefficient of performance if the compressor efficiency is 75%. Assume that the kinetic and potential energy changes are negligible.

7.98. A refrigerator using tetrafluoroethane (R-134a) as refrigerant operates with a capacity of 2500 kW. Cooling water enters the condenser at 280 K. Evaporator is at 271.92 K, and the condenser is at 299.87 K. The refrigerated space is at 280 K. Determine the ideal and actual power necessary if the compressor efficiency is 80%. Assume that the kinetic and potential energy changes are negligible.

References

1. Aye L (2014) Heat Pumps. In: Anwar S (ed) Encyclopedia of energy engineering and technology, 2nd edn. CRC Press, Boca Raton
2. Çengel YA, Boles MA (2014) Thermodynamics: an engineering approach, 8th edn. McGraw-Hill, New York
3. Çengel YA, Turner R, Cimbala J (2011) Fundamentals of thermal-fluid sciences, 4th edn. McGraw-Hill, Ney York
4. Chattopadhyay P (2015) Engineering thermodynamics, 2nd edn. Oxford Univ Press, Oxford
5. Demirel Y (2018) Sugar versus lipid for sustainable biofuels. Int J Energy Res 42:881–884
6. Demirel Y, Gerbaud V (2019) Nonequilibrium thermodynamics. Transport and rate processes in physical, chemical and biological systems, 4th edn. Elsevier, Amsterdam
7. IRENA (2020) Global renewables outlook: energy transformation 2050 (Edition: 2020). International Renewable Energy Agency, Abu Dhabi
8. Jaffe RL, Taylor W (2018) The Physics of Energy. Cambridge University Press, Cambridge
9. Küçük K, Tevatia R, Sorgüven E, Demirel Y, Özilgen M (2015) Bioenergetics of growth and lipid production in Chlamydomonas reinhardtii. Energy 83:503–505
10. Li X, Gholamreza Karimi G (2014) Fuel cells: intermediate and high temperature. In: Anwar S (ed) Encyclopedia of energy engineering and technology, 2nd edn. CRC Press, Boca Raton
11. Wang X, Demirel Y (2018) Feasibility of power and methanol production by an entrained-flow coal gasification system. Energy Fuels 32:7595–7610. https://doi.org/10.1021/acs.energyfuels.7b03958

Energy Storage

8

Introduction and Learning Objectives: Devices or physical media can store some form of energy to perform a useful operation later or/and at a different location. Energy storage reduces the mismatches between the energy production and demand. For example, if it is stored, the solar energy would still be available during the night. Also, the stored energy may be a supplement during the peak demand for energy. Besides, a stored energy can be transported. A battery, for example, makes it possible to use a wristwatch, mobile phone, or a laptop computer. This chapter begins by underlying the importance of energy storage and regulation by water and hydrogen and later discusses thermal, electric, chemical, and mechanical energy storage systems. Solar energy storage by sensible and/or latent heat and for short- and long-term applications is discussed briefly. Some common phase-changing materials and usage of them for the latent heat storage technique are described. Underground thermal energy systems are discussed briefly. Capacitor, hydroelectric, and battery are discussed in storing electricity. Chemical energy storage by biosynthesis is briefly discussed. Later, mechanical energy storage by compressed air, flywheel, hydraulic, and springs is discussed.

The learning objectives of this chapter are to understand:

- Thermal energy storage,
- Phase-changing materials,
- Electrical energy storage,
- Chemical energy storage, and
- Mechanical energy storage.

© Springer Nature Switzerland AG 2021
Y. Demirel, *Energy*, Green Energy and Technology,
https://doi.org/10.1007/978-3-030-56164-2_8

8.1 Energy Storage

Devices or physical media can store some form of energy to perform a useful operation later time. A battery stores readily convertible chemical energy to operate a mobile phone, a hydroelectric dam stores energy in a reservoir as gravitational potential energy, and ice storage tanks store thermal energy to meet the peak demand for cooling. Battery and fossil fuels such as coal and gasoline store chemical energy. Synthesis of some energy-rich biomolecules such as adenosine triphosphate stores and transforms chemical energy in living systems. Sunlight is also captured by plants as *chemical potential energy* when carbon dioxide and water are converted into carbohydrates.

Two important physical media for storage and regulation of thermal energy are water and hydrogen. Table 8.1 lists some physical properties of water and hydrogen, including the heats of formation of water in liquid and vapor states. Their material and physical properties make them unique not only in storing and transforming energy but also in regulation of the temperature of the earth. Hydrogen represents a store of potential energy which can be released by nuclear fusion in the sun in the form of light. Solar energy may be stored as chemical energy by the photosynthesis after it strikes the earth. Also, when water evaporates from oceans and is deposited high above the sea level, it drives turbine/generators to produce electricity, hence sunlight may again be stored as gravitational potential energy [2, 3].

8.1.1 Ecological Regulation by Water

Approximately 80% of Earth's surface is water. Water has a high specific heat capacity of 4.185 kJ/kg °C, which is higher than that of soil and rock (see Table 8.1). This enables the water to store and retain large amount of heat, making the oceans one of Earth's temperature regulators. Therefore, the temperature of water changes more slowly than that of Earth's surface. Besides that, water has large heat of vaporization (2257 kJ/kg at 20 °C) that can be transferred over long distances in vapor form and released after condensing as rain (see Fig. 8.1). Accumulation of the rain at reservoirs stores large amount of potential energy, which drives turbines to produce electricity. The oceans control Earths' energy balance as it absorbs nearly four times more insolation than the Earth.

Table 8.1 Physical properties of water and hydrogen [1]

Compound	MW	ρ (kg/m^3)	$C_{p,av}$ (kJ/kg K)	T_m (°C)	T_b (°C)	ΔH_v (kJ/kg)	ΔH_m (kJ/kg)
Water (H$_2$O)	18.02	1000	4.18	0.0	100.0	2257	333.7
Hydrogen (H$_2$)	2.02	70.7	10.0	−259.2	−252.8	445.7	59.5

MW: molecular weight, T_m: temperature of melting, $C_{p,av}$: average heat capacity, ρ: density
ΔH_m: Heat of fusion, ΔH_{vap}: Heat of vaporization

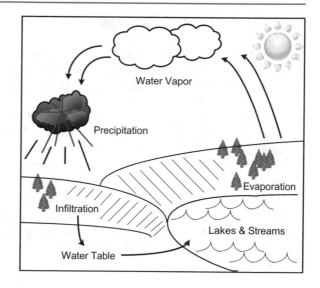

Fig. 8.1 Water circulation on the Earth

The liquid water also helps to control the amount of carbon dioxide in the atmosphere by dissolving it into 'carbonic acid'. Some of the dissolved carbon dioxide would combine with minerals in the water and settle at the ocean floor to form limestone. Water is also essential for all life forms as most animals and plants contain more than 60% water by volume, which helps organisms to regulate their body temperature. Therefore, water regulates the temperature of the Earth as well as the temperature of living systems [1, 7].

8.1.2 Hydrogen

Hydrogen is a versatile storage medium for its conversion to useful work in a fuel cell to produce electricity and can be an electrical power storage medium. Hydrogen can also be used in an internal combustion engine as a clean fuel emitting water only. Hydrogen must first be manufactured by other energy sources and may become a significant energy storage medium for renewable energies, including solar and wind power. Hydrogen can be manufactured by steam reforming of fossil fuels or by electrolysis of water using renewable power. Large quantities of gaseous hydrogen can be stored in underground caverns and depleted oil or gas fields for many years and can function as multipurpose energy storage [10].

8.2   Types of Energy Storage

A method of energy storage may be chosen based on stability, ease of transport, energy capture, and release. Main energy storage technologies involve the store of energy in thermal, electrical, chemical, and mechanical forms: [7, 9]

- *Thermal*: Thermal energy can be stored by sensible and latent heats at temperatures above or below the ambient temperatures. Heat is also stored for short- or long-term applications. Thermal energy from the sun, for example, can be captured by solar collectors and stored in a reservoir for daily or seasonal use. Thermal energy for cooling can be stored in ice.
- *Electrical*: Dams can be used to store hydroelectricity by accumulating large amounts of water at high altitude. The water then turns a turbine and generates electricity. Also, by using pumped-storage hydroelectricity we can store energy by pumping water back into the reservoir. A capacitor can store electric charge as a part of electrical energy production system. Battery can also store electric energy to be used in portable electrical devices.
- *Chemical*: Stable chemical compounds such as fossil fuels store chemical energy. Biological systems can store energy in chemical bonds of energy-rich molecules, such as glucose and adenosine triphosphate (ATP). Other forms of chemical energy storage include hydrogen, synthetic hydrocarbon fuel, and batteries.
- *Mechanical*: Energy can be stored in pressurized gases such as in compressed air, and used to operate vehicles and power tools. Hydraulic accumulator, flywheel, and springs can also store mechanical energy.

8.3   Thermal Energy Storage

The thermal storage medium may be maintained at a temperature above (hotter) or below (colder) than that of the ambient temperature of 25 °C. In hot climates, the primary applications of thermal energy storage are cold storage because of large electricity demand for air conditioning. Thermal energy storage may be planned for short term or seasonal. Some advantages of utilizing thermal energy storage are [2, 3, 7]:

- Reduced energy consumptions and carbon footprint,
- Reduced initial equipment and maintenance costs,
- Reduced pollutant emissions,
- Increased flexibility of operation, efficiency, and effectiveness of equipment utilization,
- Process application in portable and rechargeable way at the required temperature,
- Isothermal and higher storage capacity per unit weight,
- Energy from any source (thermal or electrical) when needed.

Thermal energy is generally stored in the form of sensible heat and latent heat. Figure 8.2 compares the characteristics of heat storage by sensible and latent heats. Temperature keeps increasing as a material absorbs and stores sensible heat until it reaches a temperature where a phase changing occurs. Under their melting points, the solid–liquid phase-changing materials store sensible heat and their temperature rises as they absorb the heat (see Fig. 8.2a). When phase-changing materials reach their melting temperature they absorb large amounts of latent heat at an almost constant temperature until melting is completed. If the outside heat supply continues, then the melted material starts to store sensible heat. A material may freeze at a temperature lower than its actual freezing temperature. This is called the *subcooling*. Also, a material may melt at a temperature higher than its actual melting temperature. This is called the *superheating*. Subcooling and superheating of a phase-changing material create an interval for temperature control (see Fig. 8.2b). This temperature interval between the subcooling and superheating may be reduced by adding nucleating agents to phase-changing material.

Stored heat is transferred to a *heat transfer fluid* such as water or air in a heat exchanger system, as seen in Fig. 8.3. When air is used as heat transfer fluid, the cold air inside the room extracts the heat as it flows through a heat storage system and the warm air out of the storage is fed directly into the room.

Night-time low-cost electricity may be used to store energy as sensible heat or/and latent heat. The stored energy is then used for the cooling/heating needs during the peak demand hours. Thermal energy storage systems can be designed using chilled water, ice, and encapsulated phase-change material. Phase-change material modules installed in a chilled water tank has proven to increase the energy density and improve the performance of water and ice-based thermal energy storage systems. Figure 8.4 shows a typical arrangement of cooling using phase-change material. A phase-change material in the supplementary tank is charged by the chiller during the off-peak hours. During peak demand hours the cooling is

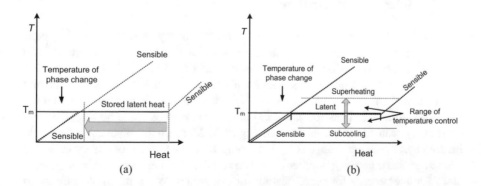

Fig. 8.2 a Comparison of sensible heat with latent heat storage, **b** temperature control during phase-change energy storage; subcooling and superheating of a phase-changing material create a range for temperature control; this range may be reduced by adding nucleating agents to phase-changing material

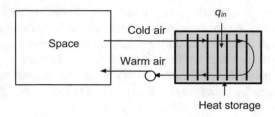

Fig. 8.3 Direct space heating using air as the heat transfer fluid. Cold air inside the room flows through a heat storage system and the warm air is fed directly into the room for heating

Fig. 8.4 Configuration of a thermal energy storage system with phase-change material (PCM)

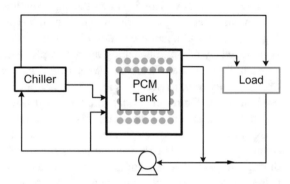

achieved by the stored energy in the phase-change material energy storage tank as the heat exchange fluid passes through phase-change material tank. This stabilizes the load during peak demand hours and shifts the load to the off-peak period [13].

8.3.1   Solar Energy Storage

Due to time-dependence of sun exposure, the efficiency of solar thermal systems relies on the well-integrated thermal energy storage technology. For example, in utilizing solar-water collectors with thermal energy storage, higher efficiency is achieved by bridging the gap between solar heat availability and hot water demand. Solar thermal energy is usually captured by active solar collectors and transferred to insulated storage systems for various applications, such as space heating, domestic or process water heating, as seen in Fig. 8.5. Most of the practical active solar heating systems have storage for a few hours to a day's worth of energy collected. There are also a growing number of seasonal thermal storage systems being used to store summer energy for space heating during winter. Water-based technology for thermal energy storage is practical because of the large heat of fusion of water (see Table 8.1). The original definition of a 'ton' of cooling capacity was the heat to melt one ton of ice every 24 h. This definition has since been replaced by one ton heating ventilation and air conditioning (HVAC) capacity of 12,000 Btu/h [9].

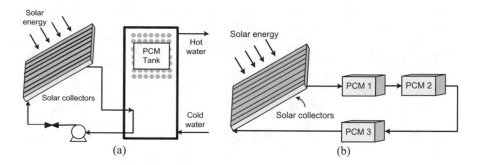

Fig. 8.5 Configuration of solar energy storage in the form of sensible and latent heats: **a** Solar water heater with heat storage and phase-changing material (PCM), **b** Hot and cold energy storage of solar energy using various phase-changing materials working at high and low temperatures; PCM 1 to PCM 3 refer to phase-changing materials with different heat storage characteristics

Many of the early applications utilized solar-water collectors with water storage because of the advantages of increased efficiency and reduced size. For home heating, water has to go through a heat exchanger to release its heat to a room. However, solar heating systems that use air as the transfer medium are being recommended for home usage because of direct heating and less potential for damage from a leak or frozen water. In *direct heating*, the warm air coming from the solar air collector can be fed to a room for heating. Also, air collectors and ducting are usually cheaper and require less maintenance [6, 13].

8.3.2 Sensible Heat Storage

Sensible heat is stored by raising its temperature of a material

$$q = mC_{p,\mathrm{av}}\Delta T = \rho V C_{p,\mathrm{av}}\Delta T \qquad (8.1)$$

where q is the sensible heat stored (J, Btu), V is the volume (m^3, ft^3), ρ is the density (kg/m^3, lb/ft^3), m is the mass (kg, lb), $C_{p,\mathrm{av}}$ is the average specific heat capacity of material (kJ/kg °C, Btu/lb °F), and ΔT is the temperature change (°C, °F). Energy density or volumetric heat capacity is defined by the product of density and heat capacity: $\rho C_{p,\mathrm{av}}$. Table 8.2 shows some materials used for heat storage in the form of sensible heating. Stored heat is usually transferred by a heat transfer fluid such as water or air to the location where to be used as seen in Fig. 8.5. Example 8.1 illustrates the sensible heat storage calculations.

Example 8.1 Sensible heat storage calculations

(a) Estimate sensible heat stored in 2 m^3 water and 2 m^3 granite heated from 20 to 40 °C.

(b) Estimate the heat stored in 1000 lb of water heated from 20 to 30 °F. Assume that average heat capacity is: $C_{p,\mathrm{av}} = 1.0$ Btu/lb °F.

Table 8.2 Some materials used for heat storage in the form of sensible heating, volumetric heat capacity, or energy density is defined by $\rho C_p = $ J/m^3 K [6, 13]

Material	Temperature range (°C)	Density ρ (kg/m^3)	Heat capacity $C_{p,av}$ (J/kg °C)	Energy density $\rho C_{p,av}$ (kJ/m^3 °C)
[a]50% ethylene glycol and 50% water	0–100	1075	3480	3741
Dowtherm A	12–260	867	2200	1907
Therminol 66	−9 to 343	750	2100	1575
Water	0–100	1000	4190	4190
Granite	–	2400	790	1896
Draw salt: 50% NaNO$_3$–50% KNO$_3$	220–540	1733	1550	2686
[a]Molten salt: 50% KNO$_3$– 40% NaNO$_2$–7% NaNO$_3$	142–540	1680	1560	2620

[a]Weight percentages

Solution:

Assume that the values of heat capacity remain constant.

(a) Sensible heat storage by water and granite:

Use Eq. (8.1) with the data for heat capacity $C_{p,av}$ and density ρ from Table 8.2:

Water:

$q = \rho V C_{p,av} \Delta T = \left(1000 \text{ kg/m}^3\right)\left(2 \text{ m}^3\right)(4190 \text{ J/kg °C})(40 - 20) \text{ °C}$
$= 167{,}600 \text{ kJ} = 167{,}600 \text{ kJ}/(3600 \text{ s/h}) = \mathbf{46.5\,kWh}$

Granite:

$q = \rho V C_{p,av} \Delta T = \left(2400 \text{ kg/m}^3\right)\left(2 \text{ m}^3\right)(790 \text{ J/kg °C})(40 - 20) \text{ °C}$
$= 75840 \text{ kJ} = (75840 \text{ kJ})/(3600 \text{ s/h}) = \mathbf{21.06\,kWh}$

(b) Heat stored in 1000 lb of water heated when the average specific heat of water is 1.0 Btu/lb °F:

$q = m C_{p,av} \Delta T = (1000 \text{ lb})(1.0 \text{ Btu/lb °F})(30 - 20) \text{ °F}$
$= 10000 \text{ Btu} = 10{,}550 \text{ kJ} = 10{,}550 \text{ kJ}/(3600 \text{ s/h}) = \mathbf{2.93\,kWh}$

8.3.3 Latent Heat Storage by Phase-Changing Material

Latent heat storage can be achieved through solid–solid, solid–liquid, solid–gas, and liquid–gas phase changes. Liquid–gas transitions have a higher heat of transformation than solid–liquid transitions. However, liquid–gas phase changes are not practical because of the large volumes or high pressures required to store the materials when they are in their gas phase states. Solid–solid phase changes are typically very slow and have a rather low heat of transformation. Therefore, the main phase change used for heat storage is the solid–liquid change. Figure 8.6 shows the change of enthalpy of water from solid to vapor state on enthalpy–temperature diagram. Between −30 and 0 °C the sensible heat of water increases. At 0 °C the water starts to melt by absorbing heat of melting. After the melting is completed, the liquid water is heated until 100 °C [3, 11].

Phase-changing materials used for storing heat are chemical substances that undergo a solid–liquid transition at temperatures within the desired range for heating and cooling purposes. During the transition process, the material absorbs energy as it goes from a solid to a liquid state and releases energy as it goes back to a solid from liquid state. A phase-changing material should possess:

- Melting temperature in the desired operating temperature range,
- High latent heat of fusion per unit volume,
- High specific heat, density, and thermal conductivity,
- Small changes of volume and vapor pressure on phase change at operating temperatures,

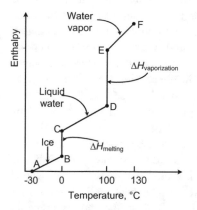

- Process A-B: Increase in sensible heat of ice between −30°C and 0 °C
- Process B-C: Phase change form solid to liquid: melting of ice at 0 °C; heat of melting ($\Delta H_{melting}$) is required.
- Process C-D: Increase in sensible heat of liquid water between 0°C –100 °C.
- Process D-E: Phase change from liquid to vapor at 100 °C; heat of vaporization ($\Delta H_{vaporization}$) is required.
- Process E-F: Increase in sensible heat of water vapor between 100°C and 130 °C.

Fig. 8.6 Change of enthalpy values of water from solid state (ice) to vapor state between −30 and 130 °C. Between −30 and 0 °C the sensible heat of water increases. At 0 °C the water starts to melt by absorbing heat of melting. After the melting process is completed, the liquid water is heated until 100 °C at which the transformation from liquid to vapor starts. After absorbing the heat of vaporization all the liquid water becomes vapor. Between 100 and 130 °C, the water in vapor state increases its sensible heat

- Congruent melting (solid compositions is the same with the composition of the liquid melt),
- Chemical stability and complete reversible freeze/melt cycles,
- Noncorrosive, nontoxic, nonflammable, and nonexplosive,
- Low cost and availability.

Latent heat storage technology reduces temperature fluctuations and offers a higher heat storage capacity per volume/mass. The temperature and the amount of energy stored can be adjusted by selecting a specific phase-changing material. Total heat stored by a solid-to-liquid phase-changing material between initial and final temperatures is

$$q_{stored} = \text{Solid sensible heat} + \text{Latent heat} + \text{Liquid sensible heat}$$
$$q_{stored} = mC_{ps,av}(T_m - T_i) + m\Delta H_m + mC_{pl,av}(T_f - T_m)(T_f > T_m > T_i) \tag{8.2}$$

where m is the mass of phase-changing material, $C_{ps,av}$ and $C_{pl,av}$ are the average heat capacities for solid and liquid phases, respectively, T_m is the temperature of melting, T_i and T_f are the initial and final temperatures, respectively, and ΔH_m is the heat of melting. Stored heat is transferred by a heat transfer fluid such as water or air in a heat exchanger. Consider a phase-changing material, which is melted; when the heat transfer fluid temperature T_f is lower than the melting temperature ($T_f < T_m$), the phase-changing material solidifies and releases its heat of melting to the heat transfer fluid. When the temperature of heat transfer fluid is higher than the melting temperature ($T_f > T_m$), phase-changing material starts melting and stores heat.

Various phase-changing materials are available in temperature ranges from −5 to 190 °C storing 5–14 times more heat per unit volume than conventional storage materials such as water [3, 11]. Tables 8.3 and 8.4 present short lists of some common materials for heat storage. The most commonly used phase-change materials are salt hydrates, fatty acids, esters, and various paraffins (such as octadecane). The chemical composition of salts is varied in the mixture to achieve required phase-change temperature and stability in phase transformations. Special nucleating agents may be added to the mixture to minimize phase-change salt separation and eliminating subcooling. Salt hydrates have high volumetric latent heat storage capacity, sharp melting point, high heat of fusion, and thermal conductivity. Salt hydrates are nontoxic, nonflammable, and economical. They, however, can be corrosive and have a finite life of around 10,000 cycles. All salt-based phase-changing material solutions must be encapsulated to prevent water evaporation or uptake. The packaging material should conduct heat well. It should be durable enough to withstand frequent changes in the storage material's volume as phase changes occur. Packaging must also resist leakage and corrosion. Materials of encapsulation include stainless steel, polypropylene, and polyolefin [6, 13].

Organic phase-changing materials, such as paraffin (C_nH_{2n+2}) and fatty acids ($CH_3(CH_2)_{2n}COOH$) freeze without much subcooling with self-nucleating properties [11]. They are chemically stable with relatively high heat of fusion, safe,

Table 8.3 Comparison of typical storage densities of various materials for energy storage in the form of sensible and latent heats [3, 6, 11, 13]

Method/Material	(kJ/l)	(kJ/kg)	Temperature (°C)
Sensible heat			
Granite	50	17	$\Delta T = 20$
Water	84	84	$\Delta T = 20$
Latent heat of melting			
Water	330	330	0
Lauric acid		178	42–44
Capric acid		153	32
Butyl stearate		140	19
Paraffin	180	200	5–130
Paraffin C18	196	244	28
Salt hydrate	300	200	5–130
Salt	600–1500	300–700	300–800
Latent heat of evaporation			
Water	2452	2450	1 atm and 25 °C

Table 8.4 Properties of some phase-change materials (PCMs) [3, 6, 11, 13]

PCM	T_m (°C)	ΔH_m (kJ/kg)	$C_{pl,av}$ (kJ/kg K)	$C_{ps,av}$ (kJ/kg K)	k_l (W/m K)	k_s (W/m K)	ρ_l (kg/m³)	ρ_s (kg/m³)
Water	0	333	4.19	2.0	0.595	2.2	1000	920
n-Hexadecane	18	235	2.1	1.95	0.156	0.43	765	835
Octadecane $CH_3(CH_2)_{16}CH_3$	28	243	2.2	1.8	0.15	0.42	775	~900
n-Eicosane	36.4	248	2.4	1.92	0.146	0.426	769	910
n-Docosane $CH_3(CH_2)_{20}CH_3$	44.5	196–252	~2.5	~1.9	0.15	0.4	778	920
Paraffins	10–70	125–240	1.7	1.7	0.15	0.25	~700	~900

T_m: Temperature of melting; H_m: Heat of fusion; $C_{pl,av}$: Average heat capacity at liquid state
$C_{ps,av}$: Average heat capacity at solid state; ρ_l: Density at liquid state; ρ_s: Density at solid state

nonreactive, and recyclable. On the other hand, they have low volumetric latent heat storage capacity and low thermal conductivity in their solid state, so requiring high heat transfer rates during the freezing cycle. They can be expensive and flammable at high temperatures. Sometimes mixtures of phase-change materials may be more beneficial. *Thermal-composite* is a term given to combinations of phase-change materials and other (usually solid) structures. A simple example is a copper-mesh immersed in paraffin. Table 8.4 shows a short list of some phase-change materials and their thermo-physical properties. Example 8.2 illustrates the heat storage estimations for home heating.

Example 8.2 Solar Energy Storage by Phase-Changing Material

A typical square two-story home with a roof surface area of 1260 ft^2, and a wall surface area of 2400 ft^2 is to be heated with solar energy storage using a salt hydrate as phase-change material. It presently has an insulation of 6 inches in the roof and 1 inch in the walls. Inside temperature will be held at 70 °F and expected outside low temperature is 10 °F. Average solar radiation is 650 Btu/ft^2. Each solar collector has the total absorber area of 20 ft^2. Estimate the total energy needed from the storage amount of salt hydrate and number of solar air collectors.

Solution:

Assume that the approximate thermal conductivity of the walls and roof is: $k = 0.025$ Btu/h °F ft^2.

Average solar radiation is 650 Btu/ft^2 and the cost of solar air collector is $1.1/ft^2. The salt hydrate costs around $0.15/lb and its heat of melting is 145 Btu/lb.

Heating requirement of building with present insulation:

Heat loss from roof ($\Delta x = 6/12 = 0.5$ ft); $\Delta T = (10 - 70) = -60$ °C

$$q_{loss,roof} = -kA\frac{\Delta T}{\Delta x} = -(0.025 \text{ Btu/h °F ft}^2)(1260 \text{ ft}^2)(-60 \text{ °F/0.5 ft}) = 3780 \text{ Btu/h}$$

Heat loss from roof ($\Delta x = 6/12 = 0.5$ ft); $\Delta T = (10 - 70) = -60$ °C

$$q_{loss,roof} = -kA\frac{\Delta T}{\Delta x} = -(0.025 \text{ Btu/h °F ft}^2)(1260 \text{ ft}^2)(-60 \text{ °F/0.5 ft}) = 3780 \text{ Btu/h}$$

Heat loss from walls ($\Delta x = 1$ in $= 0.083$ ft)

$$q_{loss,wall} = -kA\frac{\Delta T}{\Delta x} = -(0.025 \text{ Btu/h °F ft}^2)$$
$$(2400 \text{ ft}^2)(-60 \text{ °F/0.083 ft}) = 43373 \text{ Btu/h}$$

Total heat loss = Total heating requirement

$$q_{totalloss} = 3780 \text{ Btu/h} + 43373 \text{ Btu/h} = 47153 \text{ Btu/h}$$

Total heat storage material needed :

$m_s = (47153 \text{ Btu/h})(24 \text{ h})/145 \text{ Btu/lb} = \mathbf{7804.0\,lb}$

Size and cost of collector to meet the present heating requirements:

Area of collectors $= (47153 \text{ Btu/h})(24 \text{ h})/(650 \text{ Btu/ft}^2) = \mathbf{1740\,ft^2}$

Number of collectors : $(1740 \text{ ft}^2)/(12 \text{ ft}^2) = \mathbf{87}$

Annual cost of heating may be optimized by compromising between the two opposing effects of capital costs of thermal energy storage system and fuel as shown below.

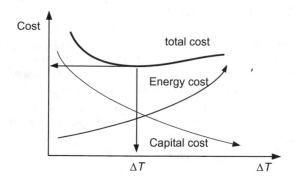

As the temperature difference between hot and cold space increases, the capital cost may decrease while the energy (fuel) cost increases. The total cost should be optimized for a required value for ΔT.

8.3.4 Ice Storage

Ice storage is the thermal energy storage using ice. It is practical because of the large heat of fusion of water. One metric ton of water (one cubic meter) can store 334 MJ or 317,000 Btu, 93 kWh, or 26.4 ton-hours. The original definition of a 'ton' of cooling capacity was the heat to melt one ton of ice every 24 h. One ton heating ventilation and air conditioning (HVAC) capacity is known as 12,000 Btu/h. A small storage unit can hold enough ice to cool a large building for a day or a week, by using, for example, off peak power and other such intermittent energy source.

Cold thermal energy storage may be an economically viable technology. It has become a key component in many successful thermal systems. Cold is a valuable commodity. Generally, a cold energy storage system comprises three basic operations in a full cycle: charging, storing, and discharging. A typical cold storage system consists of a tank containing a fixed quantity of storage fluid and heat-transfer coil through which a heat transfer fluid is circulated. Usually, kinetic and potential energies and pump work are negligible compared to cold input or cold accumulation and hence ignored. Therefore, an energy balance over an entire cycle of the cold storage becomes

$$\text{Cold input} - (\text{Cold loss} + \text{Cold recovered}) = \text{Cold accumulation}$$

Here, cold input is the heat removed from the cold thermal energy storage source by the heat transfer fluid during charging, the cold recovered is the heat removed from the heat transfer fluid by the cold storage medium, the cold loss is the heat gained from the environment to the storage medium during charging, storing, and discharging, and the cold accumulation is the decrease in internal energy of the storage medium during one complete cycle [6, 13].

8.3.5 Molten Salt Technology

Molten salt can retain heat collected by a solar tower to generate electricity. The molten salt is a mixture of 60% sodium nitrate and 40% potassium nitrate called *saltpeter*. It is nonflammable, nontoxic and is used in the chemical and metals industries as a heat transfer fluid. The salt is kept as liquid at 288 °C (550 °F) in an insulated storage tank and is pumped through panels in a solar collector where the focused sun heats it to 566 °C (1051 °F). It is then sent to a well-insulated storage tank. When electricity is needed, the hot salt is pumped to a conventional steam-generator to produce superheated steam for a turbine/generator of a power plant [7, 13].

8.3.6 Seasonal Thermal Energy Storage

A *seasonal thermal storage* retains heat deposited during the hot summer months for use during colder winter weather for heating. For cooling, the winter is used to store cold heat to be used in the summer. The heat is typically captured using solar collectors, although other energy sources are sometimes used separately or in parallel [2, 3]. Seasonal (or 'annualized') thermal storage can be divided into three broad categories:

- Low-temperature systems use the soil adjoining the building as a heat store medium. At depths of about 20 feet (6 m) temperature is naturally 'annualized' at a stable year-round temperature. Two basic techniques can be employed:

(i) In the *passive seasonal heat storage*, solar heat is directly captured by the structure's spaces through windows and other surfaces in summer and then passively transferred by conduction through its floors, walls, and sometimes, roof into adjoining thermally buffered soil. It is then *passively* returned by conduction and radiation as those spaces cool in winter.

(ii) The *active seasonal heat storage* concept involves the capture of heat by solar collectors or geothermal sources and deposited in the earth or other storage masses or mediums. A heat transfer fluid in a coil embedded in the storage medium is used to deposit and recover heat.

- Warm-temperature systems also use soil or other heat storage mediums to store heat, but employ active mechanisms of solar energy collection in summer to store heat and extract in winter. Water circulating in solar collectors transfers the heat to the storage units beneath the insulated foundation of buildings. A ground source heat pump may be used in winter to extract the heat from the storage system. As the heat pump starts with a relatively warm temperature, the coefficient of performance may be high.

- High-temperature systems are essentially an extension of the building's HVAC and water-heating systems. Water or a phase-change material is normally the storage medium.

Advantages of seasonal storage systems are:

- Seasonal storage is a renewable energy utilization system.
- Stored thermal energy would be available in desired amount, time, and location.
- Nontoxic, noncorrosive, and inexpensive storage medium can be selected for a required application.
- Seasonal heat storage needs relatively moderate initial and maintenance costs.
- Size and capacity of heat storage system can be adjusted based on the size of space to be conditioned.
- It can be used as preheater and precooler of existing heating and cooling systems.
- It is possible to use water as heat transfer fluid during charging the storage with water solar heaters, and airflow for discharging mode.

8.3.7 Seasonal Solar Thermal Energy Storage for Greenhouse Heating

Figure 8.7 shows the schematics of a seasonal solar energy storage system using solar air heaters and paraffin as phase changing material for greenhouse heating with data acquisition and control system [2, 3, 12]. The three common processes in a seasonal thermal energy storage for greenhouse heating are:

- Charging: The charging is for capturing solar energy by solar air heaters and feeding the warm air to the storage unit through the coils embedded in the heat storage tank.
- Storing: The storage process is for storing the captured solar energy by sensible and latent heat using a phase-change material in a well-insulated heat storage tank.
- Discharging: The discharge process, on the other hand, is for recovering the stored heat by the cold air flowing through the heat storage tank and delivering the warm air to the greenhouse directly when necessary.

Figure 8.8 shows the charging and discharging operations within the three units of

- Unit 1: Solar air heaters
- Unit 2: Heat storage
- Unit 3: Greenhouse.

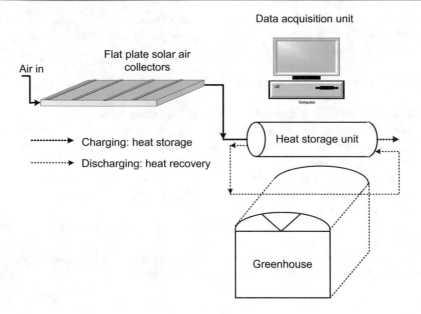

Fig. 8.7 Seasonal heat storage system using paraffin as a phase-change material for heating a greenhouse [2, 12]

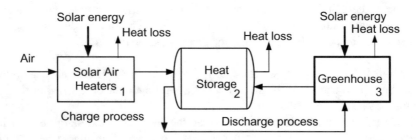

Fig. 8.8 Schematic structure with three components of the seasonal heat storage system: Unit 1 is for capturing solar energy to heat the air and send it to the heat storage, Unit 2 is heat storage by sensible and latent heats of paraffin, and Unit 3 is the greenhouse. The discharge process takes place between the greenhouse and heat storage when the temperature in the greenhouse drops below a set value

Ambient air is heated to around 60–75 °C by the solar radiation using solar air heaters. The warm air flows in spiral coils embedded inside the storage tank and heats the paraffin. When the temperature of greenhouse drops below a set point of around 14 °C, the discharge process is activated so that cold air from the greenhouse flows through the melted paraffin, recovers a part of the stored heat, and delivers it to the greenhouse until raising its temperature to a required level. The data acquisition system controls the activation and deactivation of the discharge process. Energy losses occur in various levels in all the three units [2].

Figure 8.9 shows typical temperature profiles of paraffin and airflow in charging and discharging modes. Here, the paraffin starts with the cold-solid temperature of T_{sc} and ends with the storage-hot temperature of T_{sh}, representing the cold and hot level of temperatures within the storage unit. An average temperature of T_s may be assumed for thermal energy storage calculations. The m_c is the air flow rate for charging and m_d is the air flow rate for discharging, while T_{ci} and T_{co} are the inlet and outlet temperatures for charging (loading) and T_{di} and T_{do} are inlet and outlet temperatures for discharging (recovery) process, respectively.

The latent heat storage system undergoes a temperature difference of $(T_{sh} - T_{sc})$. The solar heat captured through the series of solar collectors is

$$\dot{q}_c = \dot{m}_c C_{p,av}(T_{ci} - T_{co}) \qquad (8.3)$$

where $\dot{m}_c$ is the charging fluid flow rate, T_{ci} and T_{co} are the inlet and outlet temperatures of the air, and $C_{p,av}$ is the average heat capacity of air. The heat captured will be delivered to the heat storage unit by the charging process. The heat stored by the paraffin q_s will be (i) sensible heat until the temperature of the solid paraffin reaches its lower temperature of melting, (ii) latent heat, and (iii) sensible heat until the temperature of the melted paraffin reaches the temperature T_{sh}. The amount of stored heat q_s is

$$q_s = m_s[C_{ps,av}(T_s - T_{sc}) + \Delta H_m + C_{pl,av}(T_{sh} - T_s)] \qquad (8.4)$$

where ΔH_m is the heat of melting, T_{sc} and T_{sh} are the lowest and highest melting points of the paraffin, and $C_{ps,av}$ and $C_{pl,av}$ denote the average heat capacities of

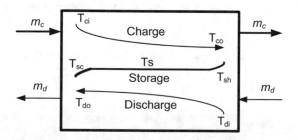

Fig. 8.9 Approximate temperature profiles of airflow and heat storage material; m_c is the air flow rate for charging and m_d is the air flow rate for discharging, T_{ci} and T_{co}: inlet and outlet temperatures for charging (loading); T_{di} and T_{do}: inlet and outlet temperatures for discharging (recovery); T_{sc} and T_{sh}: cold and hot temperatures of paraffin temperatures, respectively

solid and liquid states of the paraffin, respectively. T_s is an average temperature of the paraffin, which may be approximated as $T_s = (T_{sc} + T_{sh})/2$.

The thermal efficiency is the ratio of amount of heat stored to the amount of heat captured at a certain time interval

$$\eta_s = \frac{\text{Actual heat stored}}{\text{Maximum energy gain}} = \frac{\dot{q}_s}{\dot{q}_c} \tag{8.5}$$

The recovered heat through the discharge process is

$$\dot{q}_d = \dot{m}_d C_{p,av}(T_{di} - T_{do}) \tag{8.6}$$

where T_{di} is the temperature of air from the greenhouse, T_{do} is the temperature of the air leaving the heat storage system, and $C_{p,av}$ is the average heat capacity of air in the discharge process. The approximate thermal efficiency of the discharged process is

$$\eta_d = \frac{\dot{q}_d}{\dot{q}_s} \tag{8.7}$$

All the temperatures are time-dependent, and the charging and discharging cycles need to be monitored over the time of operation. Example 8.3 illustrates the latent heat storage calculations [13].

Example 8.3 Latent heat storage calculations
As a phase-change material of 60 kg, octadecane is heated from 20 to 30 °C by a solar energy system, which supplies heat at a rate of 2 kW. Assume that the octadecane is fully melted. Estimate the minimum size of the storage unit and the time necessary for the charging process.
Solution:
Assume that there is no heat loss from the thermal energy storage system.

$T_{sc} = 20°C$, $T_{sh} = 30°C$, $T_s = 28°C$, $m_s = 60.0$ kg, $\dot{q}_{net} = 2$ kW, $\dot{q}_{loss} = 0$ kW

Use data from Table 8.4.

PCM	T_m (°C)	ΔH_m (kJ/kg)	$C_{pl,av}$ (kJ/kg K)	$C_{ps,av}$ (kJ/kg K)	k_l (W/m K)	k_s (W/m K)	ρ_l (kg/m^3)	ρ_s (kg/m^3)
Octadecane CH$_3$(CH$_2$)$_{16}$CH$_3$	28	243	2.2	1.8	0.15	0.42	775	~900

Total heat stored:

$$q_s = m_s[C_{ps,av}(T_s - T_{sc}) + \Delta H_m + C_{pl,av}(T_{sh} - T_s)]$$
$$q_s = 60 \text{ kg}[(1.8 \text{ kJ/kg K})(28{-}20) \text{ °C} + 243 \text{ kJ/kg} + (2.2 \text{ kJ/kg K})$$

$$(30{-}28) \text{ °C}] = 15,708 \text{ kJ} \quad V_{tank} = m/\rho_s = 60 \text{ kg}/(775\text{kg/m}^3) = \mathbf{0.08\,m^3}$$

Energy balance : energy supplied = energy stored + energy lost

Energy lost:

$$\text{Energy supplied} = (\dot{q}_{net})(\Delta t) = q_s = 15708 \text{ kJ}$$
$$\Delta t = \mathbf{7854\,s = 2.2\,h}$$

Contributions of sensible heats:

$$q_s = 60 \text{ kg}[(1.8 \text{ kJ/kg K})(28{-}20) \text{ °C} + (2.2 \text{ kJ/kg K })(30{-}28) \text{ °C}]/15708 \text{ kJ}$$
$$q_s = (864 + 264) \text{ kJ}/15708 \text{ kJ} = 0.055 + 0.017$$

Contributions of sensible heats are 5.5% for solid state and 1.7% for liquid state; therefore, the main contribution toward heat storage comes from latent heat of octadecane. The charging process needs 2.2 h to supply the heat of 15,708 kJ required.

8.3.8 Underground Thermal Energy Storage Systems

Underground thermal energy storage (UTES) can store large amount of low temperature heat for space heating and cooling as well as for preheating and precooling. Common energy sources include winter ambient air, heat-pump reject water, solar energy, and process heat. Underground thermal energy storage may supply all or part of heating and/or cooling requirements of the buildings or processes. A heat pump may be used to decrease or increase the storage temperature for cooling or heating. Underground thermal energy may store energy, which is actively gathered, or store waste, or byproduct energy, which is called the double-effect storage and more likely to be more economical [5, 7]. Underground thermal energy storage encompasses both aquifer thermal energy storage (ATES) and borehole thermal energy storage (BTES) systems. These systems are discussed within the following sections.

8.3.9   Aquifer Thermal Energy Storage

Aquifers are underground, geological formations and can have gravel, sand, or rocks. Aquifer thermal energy storage may be used on a short-term or long-term basis for:

- The sole source of energy for partial storage,
- A temperature useful for direct application,
- Combination with a dehumidification system, such as desiccant cooling.

Cold storage water is usually supplied at 2−5 °C, with a cooling power typically ranging from 200 to 20 MW, and with the stored cooling energy of 29 GWh. Cold storage underground is now a standard design option in several countries such as Sweden. The duration of storage depends on the local climate and the type of building and/or process.

Aquifer thermal energy storage is used extensively for various applications such as heating of greenhouses. In summer, the greenhouse is cooled with ground water, pumped from an aquifer, which is the cold source. This heats the water, which is then stored by the aquifer system. In winter, the warm water is pumped up to supply heat to the greenhouse. The now cooled water is returned to the cold source. The combination of cold and heat storage with heat pumps has an additional benefit for greenhouses, as it may be combined with humidification. In the closed-circuit system, the hot water is stored in one aquifer, while the cold water is stored in another. The water is used to heat or cool the air, which is moved by fans. Such a system can be completely automated [9].

Chemical changes in ground water due to temperature and pressure variations with aquifer thermal energy storage may cause operational and maintenance problem. These problems are avoidable and manageable. Flushing is a recommended practice to maintain the efficiency of well. Potential environmental concerns over the use of earth energy heat pump and UTES are:

- The possible leakage of the heat exchanger fluid into the natural environment,
- Thermally induced biochemical effects on ground water quality,
- Ecological distress due to chemical and thermal pollution,
- External contaminants entering the ground water.

All these and other possible problems are addressed within the guidelines and standards for planning, construction, and operation of underground thermal energy storage.

Ground source heat pump systems can significantly lower the heating and cooling operating costs and can qualify for renewable energy credits under sustainable building rating programs. Approximately, 65% less energy consumption is attainable with existing technologies at reasonable costs, and a heat pump supported underground thermal energy storage can play useful role in achieving these

Table 8.5 Four types of application of UTES and their typical performances [9]

Application	Energy source (%)	[a]Payback period (year)	[b]COP
Direct heating and cooling	90–95	0–2	20–40
Heating and cooling with heat pump	80–87	1–3	5–7
Heat pump supported heating only	60–75	4–8	3–4
Direct cooling only	90–97	0–2	20–60

[a]Payback period is the time in years to recover the cost of investment fully
[b]COP is the coefficient of performance

reductions [5, 9]. Table 8.5 displays the performances of various types of underground thermal energy storage applications.

8.3.10 Borehole Thermal Energy Systems

Borehole thermal energy storage(BTES) applications involve the use of boreholes, typically 5–200 m deep, and operate in closed loop, in which there is no contact between the natural ground water and heat exchanger fluid. Typically, a borehole thermal energy storage includes one or more boreholes containing borehole heat exchangers, such as U-tubes, through which waste energy (hot or cold) is circulated and transferred to the ground for storage. Borehole thermal energy storage is typically applied for combined heating and cooling often supported with heat pumps for a better usage of the low-temperature heat from the storage [5].

In a solar energy integrated application, solar heated water is pumped into a borehole thermal energy storage system consisting of many boreholes and operates with the ground temperature of around 90 °C (194 °F). During the winter, the hot water flows from the borehole thermal energy storage system to the houses through a distribution network. Once inside the house, it flows through a heat exchanger with coil units, over which air is blown. The hot air then heats the house. Each house also has an independent solar thermal system installed on its sloped roof to provide domestic hot water. This system has a 90% solar fraction, meaning 90% of the energy required to heat the air and water within the community is provided by the sun. This results in considerable emission reduction [5].

8.4 Electric Energy Storage

A *capacitor* is a device for storing electric charge and used as parts of electrical systems. Figure 8.10 shows a flat plate model for a capacitor. The forms of practical capacitors vary widely, but all contain at least two conductors separated by a

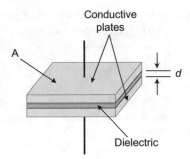

Fig. 8.10 Parallel plate model for a capacitor. The capacitance increases with area and decreases with separation; A is the surface area of the capacitor, and d is the thickness of dielectric

nonconductor. For example, a capacitor consists of metal foils separated by a layer of insulating film. When there is a potential difference (voltage) across the conductors, a static electric field develops across the dielectric, causing positive charge to collect on one plate and negative charge on the other plate. Energy is stored in the electrostatic field. An ideal capacitor is characterized by a single constant value called the capacitance, measured in Farads. Capacitance is the ratio of the electric charge on each conductor to the potential difference between them. Capacitors may be used to produce high intensity releases of energy, such as a camera's flash [7, 8].

The dielectric is just an insulator. Examples of dielectric mediums are glass, air, paper, vacuum, and even a semiconductor depletion region chemically identical to the conductors. A capacitor is assumed to be self-contained and isolated, with no net electric charge and no influence from any external electric field. The conductors thus hold equal and opposite charges on their facing surfaces, and the dielectric develops an electric field. In SI units, a capacitance of one farad means that one coulomb of charge on each conductor causes a voltage of one volt across the device. In practice, the dielectric between the plates passes a small amount of leakage current and has an electric field strength limit, resulting in a breakdown voltage [7, 8].

The capacitor is a reasonably general model for electric fields within electric circuits as shown in Fig. 8.11. An ideal capacitor is fully characterized by a constant capacitance C, defined as the ratio of charge $\pm Q$ on each conductor to the voltage V between them:

$$C = \frac{Q}{V} \tag{8.8}$$

Work must be done by an external influence to transport charge between the conductors. When the external influence is removed, the charge separation persists in the electric field and energy is stored. The energy is released when the charge is allowed to return to its equilibrium position. The work done in establishing the electric field, and hence the amount of energy stored is

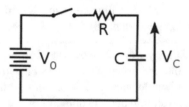

Fig. 8.11 A simple resistor-capacitor circuit demonstrates charging of a capacitor

$$W = \int\limits_{0}^{Q} V \mathrm{d}Q = \int\limits_{0}^{Q} \frac{Q}{C} \mathrm{d}Q = \frac{1}{2}\frac{Q^2}{C} = \frac{1}{2}VQ \qquad (8.9)$$

As the capacitor reaches equilibrium with the source voltage, the voltage across the resistor and the current through the entire circuit decay exponentially. *Discharging* of a charged capacitor demonstrates exponential decay [7, 8].

The capacitance increases with area and decreases with separation. The capacitance is therefore greatest in devices made from materials with a high permittivity, large plate area, and small distance between plates. The maximum energy is a function of dielectric volume, permittivity, and dielectric strength per distance. So, increasing the plate area and decreasing the separation between the plates while maintaining the same volume does not change the amount of energy the capacitor can store [7].

The dual of the capacitor is the inductor, which stores energy in the magnetic field rather than the electric field. Inductors consist of coils of wire for concentrating the magnetic field and collecting the induced voltage. Generated voltage is proportional to the rate of change in current in a circuit and the proportionality coefficient is the inductance L in henry in SI units

$$V = L\frac{\mathrm{d}I}{\mathrm{d}t} \qquad (8.10)$$

8.4.1 Hydroelectric Energy Storage

Pumped-storage hydroelectricity is a type of hydroelectric power generation used by some power plants for *load balancing*. 1 kg mass elevated to 1000 m can store 9.8 kJ of energy. The pumped-storage hydroelectricity stores energy in the form of water, pumped from a lower elevation reservoir to a higher elevation. Low-cost off-peak electric power is used to run the pumps. The stored water is released through the turbines to produce electric power. Although the losses of the pumping process make the plant a net consumer of energy overall, the system increases revenue by selling more electricity during periods of *peak demand*, when electricity prices are the highest. Example 8.4 illustrates the electric energy storage by the pumped energy in a hydropower plant [8].

Example 8.4 Pumped Energy in a Hydropower Plant
One method of meeting the additional electric power demand at peak usage is to pump some water from a source such as a lake back to the reservoir of a hydro-power plant at a higher elevation when the demand or the cost of electricity is low. Consider a hydropower plant reservoir with an energy storage capacity of 0.5×10^6 kWh. This energy is to be stored at an average elevation of 40 m relative to the ground level. Estimate the minimum amount of water has to be pumped back to the reservoir.

Solution:
Assume that the evaporation of the water is negligible.

$PE = 0.5 \times 10^6$ kWh, $\Delta z = 40$ m, $g = 9.8$ m/s^2
Energy of the work potential of the water: $PE = mg\Delta z$

Amount of water: $m = \dfrac{PE}{g\Delta z} = \dfrac{0.5 \times 10^6 \text{ kWh}}{9.8 \text{ m/s}^2 (40 \text{ m})} \left(\dfrac{3600 \text{ s}}{\text{h}}\right) \left(\dfrac{1000 \text{ m}^2/\text{s}^2}{1 \text{ kJ/kg}}\right) = \mathbf{4.59\,10^9 kg}$

8.4.2 Electric Energy Storage in Battery

An electrical *battery* is one or more electrochemical cells that can store electric energy in the form of chemical energy as well as convert stored chemical energy into electrical energy. Alessandro Volta invented the first battery in 1800. Batteries come in various sizes, such as miniature cells and car batteries as shown in Fig. 8.12. There are two types of batteries: *primary batteries*, which are disposable and designed to be used once and discarded, and *secondary batteries*, which are rechargeable and designed to be recharged and used multiple times. Rechargeable batteries can have their chemical reactions reversed by supplying electrical energy to the cell (Fig. 8.12f).

Zinc-carbon batteries and alkaline batteries are the common types of disposable batteries. Generally, these batteries have higher energy densities than rechargeable batteries. Rechargeable batteries, therefore, can store electric energy in the form of

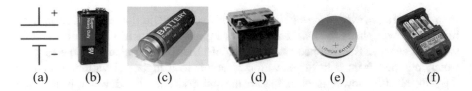

(a)   (b)   (c)   (d)   (e)   (f)

Fig. 8.12 Various types of batteries: **a** the symbol for a battery in a circuit diagram, **b** 9 V battery, **c** 1.5 V battery, **d** rechargeable car battery, **e** lithium battery, and **f** battery charger

chemical energy. For example, they can be used to store electrical energy in battery to be used in electric vehicles. Battery of electric vehicles can be charged from the grid. Because the stored energy is derived from electricity, it is possible to use other forms of alternative energy such as wind, solar, geothermal, nuclear, or hydro-electric in electric vehicles.

The oldest form of rechargeable battery is the lead-acid battery used in cars. Low manufacturing cost and its high surge current levels (450 A) make lead-acid battery use common where a large capacity (over approximately 10 Ah) is required. The anode of the battery is lead metal while the cathode is made up of lead 4 oxide. During the discharge, the chemical reaction that takes place at the anode, negative electrode, is

$$Pb(s) + SO_4^{-2}(aq) \rightarrow PbSO_4(s) + 2e^- \tag{8.11}$$

The chemical reaction that takes place at the cathode, positive electrode, is

$$PbO_2(s) + SO_4^{-2}(aq) + 4H^+(aq) + 2e^- \rightarrow PbSO_4(s) + 2H_2O(l) \tag{8.12}$$

During discharge, sulfuric acid is used up and lead sulfate is deposited on both electrodes. The battery becomes 'dead' when the electrodes are fully covered with lead sulfate. During recharging the chemical reaction that takes place at the anode is

$$PbSO_4(s) + 2H_2O(l) \rightarrow PbO_2(s) + SO_4^{-2}(aq) + 4H^+(aq) + 2e^- \tag{8.13}$$

The chemical reaction that takes place at the cathode is

$$PbSO_4(s) + 2e^- \rightarrow Pb(s) + SO_4^{-2}(aq) \tag{8.14}$$

By passing a current the battery can be recharged. Recharging reverses both reactions and leads to sulfuric acid regeneration and the removal of the lead sulfate from the electrodes. Lead-acid batteries should not be discharged to below 20% of their full capacity, because internal resistance will cause heat and damage when they are recharged [7, 8, 9].

8.4.3 Rechargeable Battery for Electric Car

Electric car uses a bank of onboard batteries, generally lithium ion batteries, to generate power for the motor. The choice of material used for the anode, cathode, and electrolyte determines the performance, safety, and lifespan of batteries. Many electric cars utilize lithium ion batteries, which contains layers of lithium isolated between plates which are made from a combination of cobalt and oxygen atoms [7, 8].

The batteries in electric cars must be charged from an external power source. Charging time depends on the type and capacity of the batteries and the type of outlet they are charging from. A standard 120 V outlet can provide about 1.5 kW per hour of charge. If you plug it into a 240 V circuit, then about 6.6 kW per hour can be charged, drastically cutting charge time. When calculating efficiency this must be taken into account.

Battery life in cars is measured in cycles rather than time or distance traveled. A battery cycle is one charge and one discharge and the number of cycles is usually expressed as a range because how the battery is used and charged can affect its life. There are various kinds of batteries commonly considered and used in electric vehicles as well as for general purposes [8]:

- Nickel-Cadmium (NiCd) batteries are the rechargeable battery used by power tools, portable electronics, and electric toys; all use NiCd batteries because of their long life and low cost. With a 40–60 Wh/kg (watt-hour per kilogram) energy to weight ratio and a life cycle of around 1000–2000 cycles they are inexpensive source of power.
- Nickel-metal hydride (NiMH) batteries have better energy to weight ratio than NiCd at 30–80 Wh/kg and are free of the toxic element of cadmium. Nickel-metal hydride batteries have a life of about 800–1000 cycles. Low-capacity nickel-metal hydride batteries (1700–2000 mAh) can be charged for about 1000 cycles, whereas high-capacity NiMH batteries (above 2500 mAh) can be charged for about 500 cycles.
- Lithium-ion battery powers the laptop or other battery-powered consumer electronics. An efficient 100–160 Wh/kg energy-to-weight ratio makes these the battery of choice in most high-power/low-weight applications. Lithium-ion batteries have a life of approximately 1200 cycles [7].

The standard hybrid vehicle increases fuel efficiency by adding additional battery capacity and charging capability. Plug-in vehicles thereby operate on energy derived from the electrical grid rather than from gasoline. Electric motors are more efficient than internal combustion engines. A computer inside the vehicle determines when it is most efficient for the vehicle to be fueled by electricity or by liquid fuel (either petroleum or biofuels) for hybrid vehicles.

Gasoline engines can be sized slightly smaller and therefore operate in a more efficient range because the electric motor can supplement the engine when needed. Gasoline engines can be shut down during periods when it would be operating inefficiently, particularly idling at stops, but also at slow speeds. Electric vehicle enables regenerative braking whereby some of the vehicle's kinetic energy is converted into electricity by the electric motor acting as a generator and charging the battery.

In plug-in hybrid vehicles, the small battery is replaced with a larger battery pack. Plug-in hybrid vehicles use both electric and gasoline drive and can be plugged into a standard home voltage. The vehicles have onboard controls to use the appropriate power source for maximum efficiency [8].

8.5 Chemical Energy Storage

The free energy available from the electron-transport cycle is used in producing the proton flow through the inner membrane in mitochondria to synthesize adenosine triphosphate (ATP) in living cells. ATP is an energy-rich compound having three phosphate groups attached to a nucleoside of adenine, called adenosine (adenine + pentose sugar). Of the three phosphate groups, the terminal one has a weak bond linkage and can break spontaneously whenever ATP forms a complex with an enzyme. The breaking up of this bond releases chemical energy, causing an immediate shift in the bond energy giving rise to adenosine diphosphate. ATP formation therefore acts as chemical energy storage of the living cell [4].

In plants, ATP is used in biochemical cycles to synthesize carbohydrates such as glucose from carbon dioxide and water. The glucose is stored mainly in the form of starch granules consisting of a large number of glucose units. This polysaccharide is produced by all green plants as chemical energy store. Starch is processed to produce many of the sugars in processed foods.

A *biological battery* generates electricity from sugar in a way that is similar to the processes observed in living organisms. The battery generates electricity through the use of enzymes that break down carbohydrates, which are essentially sugar. A similarly designed sugar drink can power a phone using enzymes to generate electricity from carbohydrates that covers the phone's electrical needs.

8.5.1 Bioenergy Sources

Carbohydrates, fats, and proteins are the major metabolic fuels (sources of energy) of living systems. Chemical energy released by oxidation of these fuels is stored as energy-rich molecules such as adenosine triphosphate in living cells. Energy contents of the major metabolic fuels are:

- Carbohydrates release about 4 kcal/g (17 kJ/g),
- Fats release about 9 kcal/g (37 kJ/g),
- Proteins release about 4 kcal/g (17 kJ/g).

In a human body excess energy is stored mainly by:

- Fat-triacylglycerol (triglyceride): triacylglycerol is the major energy store of the body.
- Glycogen: Energy stored by glycogen is relatively small but it is critical. For example, muscle glycogen is oxidized for muscle contraction.

Metabolism is the set of biochemical reactions that occur in living organisms to maintain life and consists of two major pathways: catabolism and anabolism. In catabolism, organic matter is broken down, for example, to harvest energy. In anabolism, the energy is used to construct components of cells such as proteins and nucleic acids. Daily energy expenditure includes the energy required for the basal metabolic rate and for the physical activity in every day. Basal metabolic rate for a person is approximately 24 kcal/kg (101 kJ/kg) body weight per day [4].

8.5.2 Energy Storage in Biofuels

Various biofuels, or biomass, can be helpful in reducing the use of hydrocarbon fuels. Some chemical processes, such as Fischer-Tropsch synthesis can convert the carbon and hydrogen in coal, natural gas, biomass, and organic waste into short hydrocarbons suitable as replacements for existing hydrocarbon fuels. Many hydrocarbon fuels have the advantage of being immediately usable in existing engine technology and existing fuel distribution infrastructures. A long-term high oil price may make such synthetic liquid fuels economical on a large-scale production, despite some of the energy in the original source is lost in the conversion processes.

Carbon dioxide and hydrogen can be converted into methane and other hydrocarbon fuels [10] with the help of energy from another source, preferably a renewable energy source. As hydrogen and oxygen are produced in the electrolysis of water,

$$2H_2O \rightarrow 2H_2 + O_2 \tag{8.15}$$

hydrogen would then be reacted with carbon dioxide to produce methane and water in the Sabatier reaction

$$CO_2 + 4H_2 \rightarrow CH_4 + 2H_2O \tag{8.16}$$

Produced water would be recycled back to the electrolysis stage, reducing the need for new pure water. In the electrolysis stage oxygen would also be stored for methane combustion in a pure oxygen environment (eliminating nitrogen oxides). In the combustion of methane, carbon dioxide and water are produced.

$$CH_4 + 2O_2 \rightarrow CO_2 + 2H_2O \tag{8.17}$$

Produced carbon dioxide would be recycled back to the Sabatier process. Methane production, storage and adjacent combustion would recycle all the reaction products, creating a cycle.

Methane is the simplest hydrocarbon with the molecular formula CH_4. Methane can be stored more easily than hydrogen and the technologies for transportation, storage and combustion infrastructure are highly developed. Methane would be stored and used to produce electricity later. Besides that, methanol can also be synthesized from carbon dioxide and hydrogen. If the hydrogen is produced by the electrolysis using the electricity from the wind power then the electricity is stored by the production of methane or methanol [10].

8.5.3 Energy Storage in Voltaic Cell

Each voltaic cell consists of two half cells connected in series by a conductive electrolyte containing anions and cations. One half-cell includes electrolyte and the positive electrode called the cathode. Negatively charged ions called the anions migrate to the cathode. The other half-cell includes electrolyte and the negative electrode called anode. The positively charged ions called the cations migrate to anode. In the redox reaction that powers the battery, cations are reduced (electrons are added) at the cathode, while anions are oxidized (electrons are removed) at the anode. The electrodes do not touch each other but are electrically connected by the electrolyte. Figure 8.13 shows the half cells of a copper-zinc battery [7].

In a copper-zinc battery, the tendency for Zn to lose electron is stronger than that for copper. When the two cells are connected by a *salt bridge* and an *electric conductor* as shown in Fig. 8.14 to form a closed circuit for electrons and ions to flow, copper ions (Cu^{2+}) gain electron to become copper metal. Electrons flow through the electric conductors connecting the electrodes and ions flow through the salt bridge. The overall reaction of the cell is

$$Zn + Cu^{2+} = Zn^{2+} + Cu \qquad (8.18)$$

The net electromotive force of the cell is the difference between the reduction potentials of the half-reactions. The electrical driving force of a cell is known as the *terminal voltage* (ΔV) and is measured in volts. An ideal cell has negligible internal resistance, so it would maintain a constant terminal voltage until exhausted. If such a cell maintained 1.5 volts and stored a charge of one coulomb then on complete discharge it would perform 1.5 J of work. A battery is a collection of multiple electrochemical cells; for example, a 1.5 V AAA battery is a single 1.5 V cell, and a 9 V battery has six 1.5 V cells in series (see Fig. 8.12).

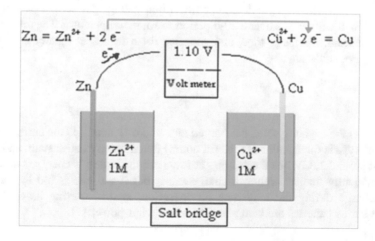

Fig. 8.13 Half cells of a copper-zinc battery and copper-zinc voltaic cells; Zn electrode to loose electron according to the reaction, so it is cathode: $Zn = Zn^{2+} + 2e^-$

Fig. 8.14 Schematic of a dry cell

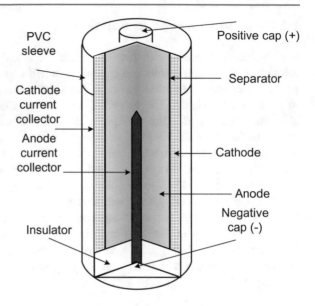

Other portable rechargeable batteries include several 'dry cell' types using a paste in place of liquid electrolyte and are therefore useful in appliances such as mobile phones and laptop computers. Cells of this type include nickel-cadmium (NiCd), nickel-metal hydride (NiMH), and lithium-ion (Li-ion) cells. Lithium-ion has the highest share of the dry cell rechargeable market. Meanwhile, NiMH has replaced NiCd in most applications due to its higher capacity, but NiCd remains in use in power tools, two-way radios, and medical equipment. Figure 8.14 shows the main layers of a dry cell [7, 8].

The battery capacity printed on a battery is usually the product of 20 h multiplied by the maximum constant current that a new battery can supply for 20 h at 68 °F (20 °C), down to a pre-determined terminal voltage per cell. A battery rated at 100 Ah will deliver 5 A over a 20-h period at room temperature. The relationship between current, discharge time, and capacity for a lead acid battery is approximated by Peukert's law:

$$t = \frac{Q_B}{I^k} \qquad (8.19)$$

where Q_B is the capacity when discharged at a rate of 1 amp, I is the current drawn from battery, t is the amount of time (in hours) that a battery can sustain, and k is a constant around 1.3. When discharging at low rate, the battery's energy is delivered more efficiently than at higher discharge rates, but if the rate is too low, it will self-discharge during the long time of operation, again lowering its efficiency. Sodium-sulfur batteries are being used to store wind power [7].

8.6 Mechanical Energy Storage

Energy may also be stored in pressurized gases or alternatively in a vacuum. Compressed air, for example, may be used to operate vehicles and power tools. Large-scale compressed air energy storage facilities may be used to supplement demands on electricity generation by providing energy during peak hours and storing energy during off-peak hours. Such systems save on expensive generating capacity since it only needs to meet average consumption [9].

8.6.1 Compressed Air Energy Storage

Compressed air energy storage (CAES) technology stores low-cost off-peak energy, in the form of compressed air in an underground reservoir. The air is then released during peak load hours and heated with the exhaust heat of a standard combustion turbine. This heated air is converted to energy through expansion turbines to produce electricity. Energy storage systems often use large underground caverns due to very large volume, and thus large quantity of energy can be stored only with a small pressure change. The cavern space can be compressed adiabatically with little temperature change and heat loss. Compressed air energy storage can also be employed on a smaller scale such as exploited by air cars and air-driven locomotives. Compressed air energy storage application using sequestered carbon dioxide is also possible for energy storage [9].

Compression of air generates a lot of heat and the air gets warmer. If no extra heat is added, the air will be much colder after decompression, which requires heat. If the heat generated during compression can be stored and used again during decompression, the efficiency of the storage improves considerably. Compressed air energy storage can be achieved in adiabatic, diabatic, or isothermic processes:

- *Adiabatic storage* retains the heat produced by compression and returns it to the air when the air is expanded to generate power. Heat can be stored in a fluid such as hot oil (up to 300 °C) or molten salt solutions (600 °C).
- *Diabatic storage* dissipates the extra heat with intercoolers (thus approaching isothermal compression) into the atmosphere as waste. Upon removal from storage, the air must be re-heated prior to expansion in the turbine to power a generator which may be accomplished, for example, with a natural gas-fired burner.
- *Isothermal compression* and expansion approaches attempt to maintain operating temperature by constant heat exchange to the environment. They are only practical for low power levels and some heat losses are unavoidable. Compression process is not exactly isothermal and some heat losses will occur.

The ideal gas law for an isothermal process is

$$PV = nRT = \text{constant} \tag{8.20}$$

$$W_{AB} = nRTln(V_B/V_A) = nRTln(P_A/P_B) \tag{8.21}$$

where A and B are the initial and final states of the system and W represents the maximum energy that can be stored or released. In practice, no process is perfectly isothermal and the compressors and motors will have heat losses. Compressed air can transfer power at very high flux rates, which meets the principal acceleration and deceleration objectives of transportation systems, particularly for hybrid vehicles. Examples 8.5 and 8.6 illustrate the simple analysis of mechanical energy storage by compressed air.

Example 8.5 Maximum Air Compressed Energy Storage
Determine the maximum available energy if we compress 1450 kg air from 100 kPa to 1200 kPa at 300 K at isothermal conditions with a heat loss of 24,000 kJ.

Solution:
Assume that air is ideal gas.

$MW = 29$ kg/kmol, $R = 8.314$ kJ/kmol K,

$P_1 = 100$ kPa, $P_2 = 1200$ kPa, $m = 1450$ kg

Number of moles of air $= 1450/28.97 = 50$ kmol

Mechanical energy stored:

$$W_{max} = -nRT \ln\left(\frac{P_1}{P_2}\right)$$

$q_{loss} = 24000$ kJ

Net energy to be stored:

$W_{max} - q_{loss} = (309892.7 - 24000)$ kJ $= \mathbf{285,892.7\,kJ} = \mathbf{285.9\,MJ}$

Example 8.6 Maximum Air Compressed Energy Storage in a Large Cavern
An underground cavern will be used to store energy of a compressed air. If the cavern has a volume of 29,000 m^3, determine the value of stored energy by the compression of air from 100 to 1500 kPa at 300 K and isothermal conditions with a heat loss of 55,000 kJ.

Solution:
Assume that air is ideal gas. Cavern space is under ideal gas conditions and at standard temperature and pressure (25 °C and 1 atm).

$MW = 28.97$ kg/kmol, $R = 8.314$ kJ/kmol K, $P_1 = 100$ kPa, $P_2 = 1500$ kPa
Under ideal gas conditions and at standard temperature and pressure (25 °C and 1 atm)
1 mol of air occupies 22.4 L and 1 kmol occupies 2.24×10^4 L

Volume $= 29000$ m^3 $= 2.9 \times 10^7$ liters

Approximate number of moles of air $= 2.9 \times 10^7$ liters/2.24×10^4 liters $= 1294.6$ kmol

Energy stored:

$$W = -nRT \ln\left(\frac{P_1}{P_2}\right) = -(1294.6 \text{ kmol})(8.314 \text{ kJ/kmol K})(300 \text{ K})\ln(100/1500)$$
$$= 8,744,270.6 \text{ kJ}$$

$q_{loss} = 55000$ kJ
Net energy to be stored $= W_{max} - q_{loss} = \mathbf{8,689,270.6}$ **kJ or 8,689.3 MJ**

8.6.2 Flywheel Energy Storage

Flywheel energy storage works by accelerating a rotor (flywheel) to a very high speed and maintaining the energy in the system as rotational energy. When energy is extracted from the system, the flywheel's rotational speed is reduced as a consequence of the principle of conservation of energy. Most flywheel energy storage systems use electricity to accelerate and decelerate the flywheel, but devices that directly use mechanical energy are also being developed. Advanced flywheel energy storage systems have rotors made of high strength carbon filaments, suspended by magnetic bearings, and spinning at speeds from 20,000 to over 50,000 rpm in a vacuum enclosure [9].

Example 8.7 Energy Storage in a Flywheel
Estimate the energy stored in a flywheel of steel disk with a mass of 5 kg and radius of 0.2 m rotating at 40,000 rpm.
The moment of inertia of the disk of density ρ, radius r, and height z is

$$I = \frac{\pi \rho z r^4}{2} = \frac{mr^2}{2} = \frac{5 \text{ kg } (0.2 \text{ m})^2}{2} = 0.1 \text{ kg m}^2$$

$\omega = 40,000(2\pi/60) = 4200 1/s$
Approximate stored energy is

$$E = \frac{I\omega^2}{2} = \frac{0.1 \text{ kg m}^2 (4200/s)^2}{2} = 0.88 \text{ MJ}$$

8.6.3 Hydraulic Accumulator

A *hydraulic accumulator* is an energy storage device in which a noncompressible hydraulic fluid is held under pressure by an external source. That external source can be a spring, a raised weight, or a compressed gas. An *accumulator* is used in a hydraulic system so that the pump doesn't need to be so large to cope with the extremes of demand. Also, the supply circuit can respond more quickly to any temporary demand and to smooth pulsations. Compressed gas accumulators are the most common type. These are also called hydro-pneumatic accumulators [9].

8.6.4 Springs

A *spring* is an elastic material used to store mechanical energy. When a spring is compressed or stretched, the force it exerts is proportional to its change in length. The *spring constant* of a spring is the change in the force it exerts divided by the change in deflection of the spring. An extension or compression spring has units of force divided by distance, for example lb_f/in or N/m. The stiffness (or rate) of springs in parallel is additive, as is the compliance of springs in series. Depending on the design and required operation, any material can be used to construct a spring, as long as the material has the required combination of rigidity and elasticity [9].

Summary

- Devices or physical media can store some form of energy to perform a useful operation at a later time. Sunlight is also captured by plants as *chemical potential energy*, when carbon dioxide and water are converted into carbohydrates. Two important physical media for storage and regulation of thermal energy are water and hydrogen.
- *Thermal energy storage* technologies balance the energy demand and energy production. For example, the solar energy would be available during the night time or winter season if it is stored previously. Also, a stored and transformed energy can be used somewhere else different from the location where it is captured or produced.
- Thermal energy is generally stored in the form of sensible heat and latent heat. A material may freeze at a temperature lower than its actual freezing temperature. This is called the *subcooling*. Also, a material may melt at a temperature higher than its actual melting temperature. This is called the *superheating*.
- Due to time-dependence of sun exposure, the efficiency of solar thermal systems relies on the well-integrated thermal energy storage technology. For example, in utilizing solar-water collectors with thermal energy storage, higher efficiency is achieved by bridging the gap between solar heat availability and hot water demand.

- Many of the early applications utilized solar-water collectors with water storage because of the advantages of increased efficiency and reduced size. For home heating, water has to go through a heat exchanger to release its heat to a room. However, solar heating systems that use air as the transfer medium are being recommended for home usage because of direct heating, less potential for damage from a leak or frozen water. In *direct heating*, the warm air coming from the solar air collector can be fed to a room for heating.
- Sensible heat is stored by raising its temperature of a material
- Latent heat storage can be achieved through solid–solid, solid–liquid, solid–gas, and liquid–gas phase changes. Phase-changing materials used for storing heat are chemical substances that undergo a solid–liquid transition at temperatures within the desired range for heating and cooling purposes. Latent heat storage technology reduces temperature fluctuations and offers a higher heat storage capacity per volume/mass.
- Various phase-changing materials are available in any required temperature range from $-5\ °C$ up to $190\ °C$ storing 5–14 times more heat per unit volume than conventional storage materials such as water. Organic phase-changing materials, such as paraffin (C_nH_{2n+2}) and fatty acids ($CH_3(CH_2)_{2n}COOH$), freeze without much subcooling with self-nucleating.
- *Ice storage* is the thermal energy storage using ice. It is practical because of the large heat of fusion of water. One metric ton of water (one cubic meter) can store 334 MJ or 317,000 Btu, 93 kWh, or 26.4 ton-hours.
- *Molten salt* can be employed as a heat store to retain heat collected by a solar tower so that it can be used to generate electricity.
- A *seasonal thermal storage* retains heat deposited during the hot summer months for use during colder winter weather for heating. For cooling, the winter is used to store cold heat to be used in the summer. The heat is typically captured using solar collectors, although other energy sources are sometimes used separately or in parallel.
- The three common processes in a seasonal thermal energy storage for greenhouse heating are: charging, storing, and discharging.
- Underground thermal energy storage (UTES) can store large amount of low temperature heat for space heating and cooling as well as for preheating and precooling. Aquifers are underground, geological formations and can have gravel, sand, or rocks. Aquifer thermal energy storage may be used on a short-term or long-term basis.
- *Borehole thermal energy storage* (BTES) applications involve the use of boreholes, typically 5–200 m deep, and operate in closed loop, in which there is no contact between the natural ground water and heat exchanger fluid.
- A *capacitor* is a device for storing electric charge and used as parts of electrical systems. The forms of practical capacitors vary widely, but all contain at least two conductors separated by a nonconductor.
- *Pumped-storage hydroelectricity* is a type of hydroelectric power generation used by some power plants for *load balancing*. 1 kg mass elevated to 1000 m

can store 9.8 kJ of energy. The pumped-storage hydroelectricity stores energy in the form of water, pumped from a lower elevation reservoir to a higher elevation.

- An electrical *battery* is one or more electrochemical cells that can store electric energy in the form of chemical energy as well as convert stored chemical energy into electrical energy.

- Electric car uses a bank of onboard *rechargeable battery*, generally lithium ion batteries, to generate power for the motor. The batteries in electric cars must be charged from an external power source.

- In plug-in hybrid vehicles, the small battery is replaced with a larger battery pack. Plug-in hybrid vehicles use both electric and gasoline drive and can be plugged into a standard home voltage. The vehicles have onboard controls to use the appropriate power source for maximum efficiency. Plug-in hybrid vehicles with modest battery packs could run primarily on the electric motor with stored energy for around 20 miles/day, thereby reducing the gasoline use.

- The free energy available from the electron-transport cycle is used in producing the proton flow through the inner membrane in mitochondria to synthesize adenosine triphosphate (ATP) in living cells.

- A *biological battery* generates electricity from sugar in a way that is similar to the processes observed in living organisms. The battery generates electricity through the use of enzymes that break down carbohydrates, which are essentially sugar. A similarly designed sugar drink can power a phone using enzymes to generate electricity from carbohydrates that covers the phone's electrical needs.

- Bioenergy sources are carbohydrates, fats, and proteins. Metabolism is the set of biochemical reactions that occur in living organisms to maintain life and consists of two major pathways of catabolism and anabolism. In catabolism, organic matter is broken down, for example, to harvest energy. In anabolism, the energy is used to construct components of cells such as proteins and nucleic acids.

- Various biofuels, or biomass, can be helpful in reducing the use of hydrocarbon fuels. Some chemical processes, such as Fischer-Tropsch synthesis, can convert the carbon and hydrogen in coal, natural gas, biomass, and organic waste into short hydrocarbons suitable as hydrocarbon fuels.

- Each voltaic cell consists of two half cells connected in series by a conductive electrolyte containing anions and cations. One half-cell includes electrolyte and the positive electrode called the cathode. Negatively charged ions called the anions migrate to the cathode. The other half-cell includes electrolyte and the negative electrode called anode. The positively charged ions called the cations migrate to anode.

- In *mechanical energy storage*, energy may also be stored in pressurized gases or alternatively in a vacuum. Compressed air, for example, may be used to operate vehicles and power tools.

- *Flywheel energy storage* works by accelerating a rotor (flywheel) to a very high speed and maintaining the energy in the system as rotational energy. When energy is extracted from the system, the flywheel's rotational speed is reduced as a consequence of the principle of conservation of energy.

- A *hydraulic accumulator* is an energy storage device in which a noncompressible hydraulic fluid is held under pressure by an external source. That external source can be a spring, a raised weight, or a compressed gas.
- A *spring* is an elastic material used to store mechanical energy. When a spring is compressed or stretched, the force it exerts is proportional to its change in length.

Problems

8.1. (a) Estimate sensible heat stored in 6 m^3 water and 6 m^3 granite heated from 20 to 40 °C.

(b) Estimate the heat stored in 500 lb of water heated from 20 to 30 °F.

8.2. (a) Estimate sensible heat stored in 1.1 m^3 Dowtherm A and 1.1 m^3 therminol 66 heated from 20 to 240 °C.

(b) Estimate the heat stored in 500 lb of molten salt heated from 150 to 300 °C.

8.3. (a) Estimate sensible heat stored in 2.5 m^3 mixture of 50% ethylene glycol and 50% water and 2.5 m^3 water heated from 20 to 65°C.

(b) Estimate the heat stored in 1500 lb of draw salt heated from 250 to 400 °C.

8.4. A typical square two-story home with a roof surface area of 1260 ft^2 and a wall surface area of 2400 ft^2 is to be heated with solar energy storage using a salt hydrate as phase-change material. It presently has an insulation of 6 inches in the roof and 1 inch in the walls. Inside temperature will be held at 70 °F and expected outside low temperature is 10 °F. Average solar radiation is 300 Btu/ft^2 and the cost of solar air collector is $4.5/ft^2. The salt hydrate costs around $0.5/lb. Estimate the costs of salt hydrate and solar air collectors.

8.5. A large building with a roof surface area of 3200 ft^2 and a wall surface area of 4200 ft^2 is to be heated with solar energy storage using a salt hydrate as phase-change material. It presently has an insulation of 6 inches in the roof and 1 inch in the walls. Inside temperature will be held at 70 °F and expected outside low temperature is 10 °F. Average solar radiation is 350 Btu/ft^2 and the cost of solar air collector is $3.4/ft^2. The salt hydrate costs around $0.4/lb. Estimate the costs of salt hydrate and solar air collectors.

8.6. Energy from a thermal energy storage system is used to heat a room from 15 to 25 °C. The room has the dimensions of (4 × 4 × 5) m^3. The air in the room is at atmospheric pressure. The heat loss from the room is negligible. Estimate the amount of n-hexadecane needed if the heat of melting is the stored energy for this temperature change.

8.7. Energy from a thermal energy storage system is used to heat a room from 12 to 27 °C. The room has the dimensions of (4 × 5 × 5) m^3. The air in the room is at atmospheric pressure. The heat loss from the room is negligible. Estimate the amount of n-octadecane needed if the heat of melting is the stored energy for this temperature change.

8.8.   Energy from a thermal energy storage system is used to heat a room from 14
        to 24 °C. The room has the dimensions of $(5 \times 5 \times 5)$ m^3. The air in the
        room is at atmospheric pressure. The heat loss from the room is negligible.
        Estimate the amount of paraffin needed if the heat of melting is the stored
        energy for this temperature change.

8.9   A room with dimensions of $(5 \times 5 \times 6)$ m^3 is heated with a thermal energy
        storage supplying a heating rate of 80 kJ/h. Heat loss from the room is
        2 kJ/h. The air is originally at 15 °C. Estimate the time necessary to heat the
        room to 24 °C.

8.10   A 0.5-kilowatt solar thermal energy storage system is used to heat a room
        with dimensions 3.0 m $\times$ 4.5 m $\times$ 4.0 m. Heat loss from the room is
        0.15 kW and the pressure is always atmospheric. The air in the room may be
        assumed to be an ideal gas. The air has a constant heat capacity of $C_{v,}$
        $_{av} = 0.72$ kJ/kg °C. Initially the room temperature is 15 °C. Determine the
        temperature in the room after 30 min.

8.11.   A 60 m^3 room is heated by a thermal energy storage system. The room air
        originally is at 12 °C and 100 kPa. The room loses heat at a rate of 0.2 kJ/s.
        If the thermal energy storage system supplies 0.8 kW, estimate the time
        necessary for the room temperature to reach 22 °C.

8.12.   A superheated steam at a rate of 0.6 lb/s flows through a heater. The steam is
        at 100 psia and 380 °F. If a heat storage system supplies 37.7 Btu/s into the
        steam at constant pressure estimate the final temperature of the steam.

8.13.   A superheated steam at a rate of 0.9 lb/s flows through a heater. The steam is
        at 100 psia and 380 °F. If a heat storage system supplies 25.4 Btu/s into the
        steam at constant pressure estimate the final temperature of the steam.

8.14.   A thermal energy storage system is used to heat 20 kg of water in a
        well-insulated tank. The storage system supplies a heat at a rate of 1.5 kW.
        The water in the tank is originally at 10 °C. Estimate the time necessary for
        the temperature of the water to reach 50 °C and the amount of salt hydrate
        with a heat of melting of 200 kJ/kg.

8.15.   A thermal energy storage system is used to heat 120 kg of water in a
        well-insulated tank. The storage system supplies a heat at a rate of 1.5 kW.
        The water in the tank is originally at 15 °C. Estimate the time necessary for
        the temperature of the water to reach 50 °C and the amount of n-hexadecane
        with a heat of melting of 235 kJ/kg (Table 8.4).

8.16.   8.16. A thermal energy storage system is used to heat 150 kg of water. The
        heat is stored by latent heat of paraffin. The storage supplies heat at a rate of
        2.3 kW. A heat loss of 0.35 kW occurs from the tank to surroundings. The
        water in the tank is originally at 20 °C. Estimate the temperature of the water
        after 2 h of heating and amount of paraffin required.

8.17.   A thermal energy storage system is used to heat 50 kg of water. The heat is
        stored by latent heat of paraffin. The storage supplies heat at a rate of 2.0 kW.
        A heat loss of 0.4 kW occurs from the tank to surroundings. The water in the
        tank is originally at 5 °C. Estimate the temperature of the water after 3 h of
        heating and amount of paraffin required.

8.18. As a phase-change material 20 kg octadecane is heated from 20 to 30 °C by a solar energy system, which supplies heat at a rate of 2 kW. Assume that the octadecane is fully melted. Estimate the time necessary for the process and the minimum size of the storage unit.

8.19. As a phase-change material 50 kg n-hexadecane is heated from 12 to 26 °C by a solar energy system, which supplies heat at a rate of 1.5 kW. Assume that the n-hexadecane is fully melted. Estimate the time necessary for the process and the minimum size of the storage unit.

PCM	T_m (°C)	ΔH_m (kJ/kg)	$C_{pl,av}$ (kJ/kg K)	$C_{ps,av}$ (kJ/kg K)	k_l (W/m K)	k_s (W/m K)	ρ_l (kg/m³)	ρ_s (kg/m³)
Water	0	333	4.19	2.0	0.595	2.2	1000	920
n-Hexadecane	18	235	2.1	1.95	0.156	0.43	765	835

8.20 As a phase-change material 120 kg octadecane is heated from 20 to 30 °C by a solar energy system, which supplies heat at a rate of 2 kW. Assume that the octadecane is fully melted. Estimate the time necessary for the process and the minimum size of the storage unit.

8.21 As a phase-change material 60 kg n-eicosane is heated from 20 to 42 °C by a solar energy system, which supplies heat at a rate of 1.8 kW. Assume that the n-eicosane is fully melted. Estimate the time necessary for the process and the minimum size of the storage unit.

8.22 As a phase-change material 100 kg docosane is heated from 20 to 55 °C by a solar energy system, which supplies heat at a rate of 1.2 kW. Assume that the docosane is fully melted. Estimate the time necessary for the process and the minimum size of the storage unit.

8.23 As a phase-change material 30 kg docosane is heated from 20 to 55 °C by a solar energy system, which supplies heat at a rate of 1.1 kW. Assume that the docosane is fully melted. Estimate the time necessary for the process and the minimum size of the storage unit.

8.24 One method of meeting the additional electric power demand at peak usage is to pump some water from a source such as a lake back to the reservoir of a hydropower plant at a higher elevation when the demand or the cost of electricity is low. Consider a hydropower plant reservoir with an energy storage capacity of 2×10^6 kWh. This energy is to be stored at an average elevation of 60 m relative to the ground level. Estimate the minimum amount of water that has to be pumped back to the reservoir.

8.25 One method of meeting the additional electric power demand at peak usage is to pump some water from a source such as a lake back to the reservoir of a hydropower plant at a higher elevation when the demand or the cost of electricity is low. Consider a hydropower plant reservoir with an energy storage capacity of 1.2×10^6 kWh. This energy is to be stored at an average elevation of 52 m relative to the ground level. Estimate the minimum amount of water that has to be pumped back to the reservoir.

8.26   One method of meeting the additional electric power demand at peak usage is to pump some water from a source such as a lake back to the reservoir of a hydropower plant at a higher elevation when the demand or the cost of electricity is low. Consider a hydropower plant reservoir with an energy storage capacity of 3.45×10^6 kWh. This energy is to be stored at an average elevation of 44 m relative to the ground level. Estimate the minimum amount of water that has to be pumped back to the reservoir.

8.27   Determine the maximum available energy from compression of 725 kg air from 100 to 1200 kPa at 300 K at isothermal conditions with a loss of 24,000 kJ.

8.28   Determine the maximum available energy if we compress of 1100 kg air from 100 to 1100 kPa at 290 K at isothermal conditions with a loss of 21,200 kJ.

8.29   Determine the maximum available energy if we compress of 540 kg air from 100 to 1100 kPa at 290 K at isothermal conditions with a loss of 12,000 kJ.

8.30   Determine the maximum available energy if we compress of 1450 kg air compression from 100 to 1200 kPa at 300 K at isothermal conditions with a loss of 24,000 kJ.

8.31   An underground cavern will be used to store energy of a compressed air. If the cavern has a volume of 58,000 m^3 determine the maximum available energy from compression of air from 100 to 1500 kPa at 300 K at isothermal conditions with a loss of 55,000 kJ.

8.32   An underground cavern will be used to store energy of a compressed air. If the cavern has a volume of 90,000 m^3 determine the maximum available energy from compression of air from 100 to 1400 kPa at 300 K at isothermal conditions with a loss of 50,000 kJ.

8.33   An underground cavern will be used to store energy of a compressed air. If the cavern has a volume of 58,000 ft^3 determine the maximum available energy from compression of air from 15 psia to 600 psia at 540 R at isothermal conditions with a loss of 50,000 Btu.

8.34   An underground cavern will be used to store energy of a compressed air. If the cavern has a volume of 150,000 ft^3 determine the maximum available energy from compression of air from 15 psia to 500 psia at 540 R at isothermal conditions with a loss of 100,000 Btu.

8.35   A well-insulated 30-m^3 tank is used to store exhaust steam. The tank contains 0.01 m^3 of liquid water at 30 °C in equilibrium with the water vapor. Determine the amount of wet-exhaust steam, in kg, from a turbine at 1 atm at the end of an adiabatic filling process. The wet steam has the quality of 90%, and the final pressure within the tank is 1 atm. Assume that heat transfer between the liquid water and the steam is negligible.

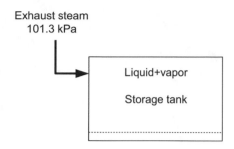

References

1. Çengel YA, Turner R, Cimbala J (2011) Fundamentals of thermal-fluid sciences, 4th edn. McGraw-Hill, Ney York
2. Demirel Y, Ozturk HH (2006) Thermoeconomics of seasonal heat storage system. Int J Energy Res 30:1001–1012
3. Demirel Y (2007) Heat storage by phase changing materials and thermoeconomics. Therm Energy Storage Sustain Energy Consump NATO Sci Series 234:133–151
4. Demirel Y, Gerbaud V (2019) Nonequilibrium thermodynamics: transport and rate processes in physical, chemical and biological systems, 4th edn. Elsevier, Amsterdam
5. Dincer I, Rosen MA (2007) A unique borehole thermal storage system at University of Ontario Institute of Technology. In: Paksoy HO (ed) Thermal energy storage for sustainable energy consumption. NATO Science Series, Springer, Dordrecht
6. Kenisarin M, Mahkamov K (2007) Solar energy storage using phase change materials. Renew Sustain Energy Rev 11:1913–1965
7. Jaffe RL, Taylor W (2018) The physics of energy. Cambridge Univ. Press, Cambridge
8. Loisel R, Mercier A, Gatzen C, Elms N, Petric H (2010) Valuation framework for large scale electricity storage in a case with wind curtailment. Energy Policy 38:7323–7337
9. Marloth R (2014) Energy storage. In: Anwar S (ed) Encyclopedia of energy engineering and technology, 2nd edn. CRC Press, Boca Raton
10. Matzen M, Alhajji M, Demirel Y (2015) Chemical storage of wind energy by renewable methanol production: Feasibility analysis using a multi-criteria decision matrix. Energy 93:343–353
11. Mehling E, Cabeza LF (2007) Phase changing materials and their basic properties. In: Paksoy HO (ed) Thermal energy storage for sustainable energy consumption. NATO Science Series, Springer, Dordrecht
12. Ozturk HH, Demirel Y (2004) Exergy-based performance analysis of packed bed solar air heaters. Int J Energy Res 28:423–432
13. Sharma A, Tyagi VV, Chen CR, Buddhi D (2009) Review on thermal energy storage with phase change materials and applications. Renew Sustain Energy Rev 13:318–345

Energy Conservation

9

Introduction and Learning Objectives Energy conservation can be achieved through increased efficiency in energy and deceased use in total energy. Energy recovery also may be a part of energy conservation through captured and hence reduced waste energy. Energy conservation is an important part of controlling climate change by reducing greenhouse gas emissions. This chapter discusses the possibilities in conserving energy in power production, in compressor work, and in furnaces and boilers. This chapter also explores the energy conservation options in heating and cooling of houses as well as in lighting and driving. Some worked examples show that it is always possible to conserve energy by process improvements and by increasing efficiency of processes.

The learning objectives of this chapter are to understand how to conserve energy in:

- Power production,
- Compressor work,
- Furnaces and boilers,
- Heating and cooling of houses as well as in lighting and driving.

9.1 Energy Conservation and Recovery

Energy conservation mainly refers to reducing energy consumption and increasing efficiency in energy usage. Energy conservation may lead to increased security, financial gain, and environmental protection. For example, electric motors consume considerable amount of electrical energy and operate at efficiencies between 70 and 90%. Therefore, using an electric motor operating with better efficiency will conserve energy throughout its useful life [3, 4].

© Springer Nature Switzerland AG 2021
Y. Demirel, *Energy*, Green Energy and Technology,
https://doi.org/10.1007/978-3-030-56164-2_9

Energy recovery leads to reducing the energy input by reducing the overall waste energy from a system. For example, a waste energy, mainly in the form of sensible or latent heat, from a subsystem may be usable in another part of the same system. Therefore, energy recovery may be a part of energy conservation. There is a large potential for energy recovery in industries and utilities leading to reduced use of fossil fuels and hence greenhouse gas emission. Some examples of energy recovery [6] are:

- Hot water from processes such as power plants and steel mills may be used for heating of homes and offices in the nearby area. Energy conservation through insulation or improved buildings may also help. Low temperature heat recovery would be more effective for a short distance from producer to consumer.
- Regenerative brake is used in electric cars and trains, where the part of kinetic energy is recovered and stored as chemical energy in a battery.
- Active pressure reduction systems where the differential pressure in a pressurized fluid flow is recovered rather than converted to heat in a pressure reduction valve.
- Energy recycling.
- Water heat recycling.
- Heat recovery steam generator (HRSG).
- Heat regenerative cyclone engine.
- Thermal diode.
- Thermoelectric modules.

9.2   Conservation of Energy in Industrial Processes

Most of the industrial processes depend on stable and affordable energy supply to be competitive. Some of these industrial processes produce energy, while others use energy. Energy conservation in industrial sector will reduce wasted energy and GHG emissions [3].

9.2.1   Energy Conservation in Power Production

Section 6.7 discusses some possible modifications in improving the efficiency of power plants. Here, some examples underline the importance of energy conservation of Rankine cycle and Brayton cycle. In gas-turbine engines, the temperature of the exhaust gas T_4 is often higher than the temperature of the gas leaving the compressor T_2 as seen in Fig. 9.1. Therefore, the gas leaving the compressor can be heated in a regenerator by the hot exhaust gases as shown in Fig. 9.1b. Regenerator is a counter-current heat exchanger, which is known as *recuperator*, and recovers waste heat. The thermal efficiency of the Brayton cycle increases as a result of regeneration because the portion of energy of the exhaust gases is used to preheat

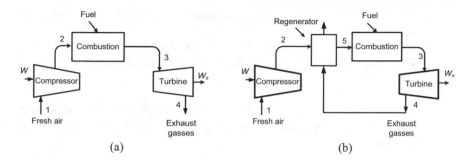

Fig. 9.1 **a** Simple Brayton cycle, **b** Brayton cycle with regeneration; the condition for regeneration is $T_4 > T_2$

the gas entering the combustion chamber. Thus, in turn, regeneration can reduce the fuel input required for the same work output from the cycle. The addition of a regenerator (operating without thermal losses) does not affect the network output of the cycle. A regenerator with higher effectiveness will conserve more fuel. The effectiveness ε of the regenerator operating under adiabatic conditions is

$$\varepsilon = \left(\frac{H_5 - H_2}{H_4 - H_2}\right) \tag{9.1}$$

The enthalpies H_i are shown in Fig. 9.1b. Under the cold-air standard temperature assumptions, the effectiveness ε of the regenerator is

$$\varepsilon \simeq \left(\frac{T_5 - T_2}{T_4 - T_2}\right) \tag{9.2}$$

The regeneration is possible only when $T_4 \gg T_2$. The effectiveness of most regenerators used in practical engine operations is below 0.85 [2, 3].

Under the cold-air-standard temperature assumptions, thermal efficiency of an ideal Brayton cycle with regeneration depends on the ratios of the temperatures and the pressures, and it is estimated by

$$\eta_{th.regen.} = 1 - \left(\frac{T_1}{T_3}\right)(r_p)^{(\gamma-1)/\gamma} \tag{9.3}$$

where r_p is the compression ratio (P_2/P_1) and $\gamma = C_p/C_v$.

Example 9.1 Energy conservation by regeneration in a Brayton cycle
A power plant is operating on an ideal Brayton cycle with a pressure ratio of $r_p = 9$. The fresh air temperature at the compressor inlet is 295 K. The air temperature at the inlet of the turbine is 1300 K. The cycle operates with a compressor efficiency of 80% and a turbine efficiency of 80%. The cycle operates 360 days per year.

(a) Using the standard-air assumptions, determine the thermal efficiency of the cycle.
(b) If the power plant operates with a regenerator with an effectiveness of 0.78, determine the thermal efficiency of the cycle and annual conservation of fuel.

Solution:
Assume that the cycle is at steady-state flow and the changes in kinetic and potential energy are negligible. Heat capacity of the air is temperature dependent, and the air is an ideal gas.

(a) Basis: 1 kg/s air. Data given: $\eta_{turb} = 0.8$, $\eta_{comp} = 0.8$, $r_p = P_2/P_1 = 9$

Process 1–2: isentropic compression

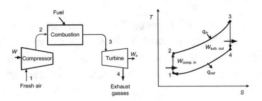

Data from (Appendix D Table D1):
$T_1 = 295$ K, $H_1 = 295.17$ kJ/kg, $P_{r1} = 1.3068$

P_r shows the relative pressure defined in Eq. (7.36): $\dfrac{P_{r2}}{P_{r1}} = \dfrac{P_2}{P_1} = r_p \rightarrow$

$P_{r2} = (9)(1.3068) = 11.76$
Approximations from Table D1 for the compressor exit: $P_{r2} = 11.76$: $T_2 = 550$ K and $H_2 = 555.74$ kJ/kg
Process 3–4: isentropic expansion in the turbine as seen on the TS diagram above
$T_3 = 1300$ K, $H_3 = 1395.97$ kJ/kg, $P_4/P_3 = 1/r_p = 1/9$, $P_{r3} = 330.9$
$\dfrac{P_{r4}}{P_{r3}} = \dfrac{P_4}{P_3} \rightarrow P_{r4} = \left(\dfrac{1}{9}\right)(330.9) = 36.76$
Approximate values from Table D1 at the exit of turbine: at $P_{r4} = 36.76$: $T_4 = 745$ K, $H_4 = 761.87$ kJ/kg
The work input to the compressor:
$$W_{comp.in} = \frac{H_2 - H_1}{\eta_{comp}} = \frac{(335.74 - 295.17)\ \text{kJ/kg}}{0.8} = 325.7\ \text{kJ/kg}$$
The work output of the turbine:
$W_{turb.out} = \eta_{turb}(H_4 - H_3) = 0.8(761.87 - 1395.97) = -507.3$ kJ/kg
The back-work ratio r_{bw}: $r_{bw} = \dfrac{W_{comp.in}}{|W_{turb.out}|} = \dfrac{325.7}{507.3} = 0.64$
This shows that 64% of the turbine output has been used in the compressor.
Work output: $W_{net} = W_{out} - W_{in} = -181.6$ kJ/kg
Enthalpy at the compressor outlet:
$H_{2a} = H_1 + W_{comp.in} = (295.17 + 325.70)\text{kJ/kg} = 620.87$ kJ/kg

The amount of heat added: $q_{in} = H_3 - H_{2a} = 1395.97 - 620.87 = 775.1$ kJ/kg

The thermal efficiency: $\eta_{th} = \dfrac{|W_{net}|}{q_{in}} = \dfrac{181.6}{775.1} = $ **0.234 or 23.4%**

The temperature of the exhaust air, T_4, and the actual enthalpy are estimated from the energy balance: $W_{turbout} = H_{4a} - H_3$

$H_{4a} = 1395.97 - 507.3 = 888.67$ kJ/kg $\rightarrow T_4 = 860$ K (From Table D1)

Advances in the compressor and turbine designs with minimal losses increase the efficiency of these components. In turn, a significant increase in the thermal efficiency of the cycle is possible.

(b) $T_4 = 860$ K and $T_2 = 550$ K, since $T_4 > T_2$ regeneration is possible.

Regeneration with effectiveness of $\varepsilon = 0.78$:

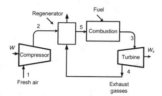

$\varepsilon = \left(\dfrac{H_5 - H_2}{H_4 - H_2}\right) = \dfrac{H_5 - 620.9}{888.67 - 620.9} = 0.78 \rightarrow H_5 = 829.80$ kJ/kg

$q_{in} = H_3 - H_5 = 1395.97 - 829.8 = 566.2$ kJ/kg

This represents a conservation of $840.2 - 566.2 = 274.0$ kJ/kg from the fuel required.

The thermal efficiency: $\eta_{th} = \dfrac{|W_{net}|}{q_{in}} = \dfrac{181.6}{566.2} = $ **0.32 or 32%**

Days of operation 360, hours of operation per year $= 360(24) = 8640$ h/year

Conserved fuel: (274 kJ/kg)(1 kg/s) (8640 h/year) = 2,367,360 kWh/year

After the regeneration, the thermal efficiency has increased from 23 to 32% in the actual Brayton cycle operation. The addition of a regenerator (operating without thermal losses) does not affect the net work output of the cycle.

9.2.2 Conservation of Energy by Process Improvements

Some of the possible modifications in operation of steam power plants to increase the efficiency are [1, 2, 4]:

- *Increasing the efficiency of a Rankine cycle by reducing the condenser pressure.*
  Example 9.2 illustrates the estimation of efficiency of a Rankine cycle operating at two different discharge pressures. The thermal efficiency increased from 27.6 to 33.4% by reducing the condenser pressure from 78.5 to 15.0 kPa. However, the quality of the discharged steam decreased from 0.9 to 0.84, which is not desirable for the blades of the turbine. Savings are considerable since the power

output increased. Example 9.3 compares the actual thermal efficiency with the maximum thermal efficiency obtained from fully reversible Carnot cycle.

- *Increasing the efficiency of a Rankine cycle by increasing the boiler pressure.* Example 9.4 illustrates the estimation of thermal efficiency of a Rankine cycle operating at a higher pressure. The thermal efficiency increased from 0.276 to 0.326 by increasing the boiler pressure from 3500 to 9800 kPa. However, the quality of the discharged steam decreased from 0.9 to 0.80, which is not desirable for the blades of the turbine.
- *Increasing the efficiency of a Rankine cycle by increasing the boiler temperature.* Example 9.5 illustrates the estimation of thermal efficiency of a Rankine cycle operating at a higher temperature. The thermal efficiency increased from 0.276 to 0.298 by increasing the boiler temperature from 400 to 525 °C. The quality of the discharged steam increased from 0.9 to 0.96, which is desirable for the protection of the turbine blades. Example 9.6 compares the estimated thermal efficiency with the maximum thermal efficiency obtained from the full reversible Carnot cycle.

Example 9.2 Increasing the efficiency of a Rankine cycle by reducing the condenser pressure

A steam power plant is operating on the simple ideal Rankine cycle. The steam mass flow rate is 20 kg/s. The steam enters the turbine at 3500 kPa and 400 °C. Discharge pressure of the steam from the turbine is 78.5 kPa.

(a) Determine the thermal efficiency of the cycle.
(b) If the pressure of the discharge steam is reduced to 15 kPa determine the thermal efficiency.

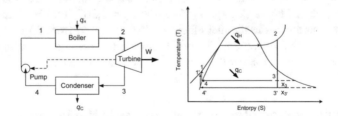

Solution:
Assume that the cycle is at steady-state flow and the changes in kinetic and potential energy are negligible. Efficiency of pump and turbine is 100%.

(a) $\dot{m}_s = 20.0$ kg/s. Using the data from the steam tables

Superheated steam: $P_2 = P_1 = 3500$ kPa, $H_2 = 3224.2$ kJ/kg,
$S_2 = 6.8443$ kJ/kg K, $T_2 = 400$ °C

Saturated steam: $P_3 = P_4 = 78.5$ kPa $(T_{sat} = 366.15$ K), $V_4 = 0.001038$ m^3/kg
$H_{3sat\ vap} = 2665.0$ kJ/kg, $H_4 = H_{3sat\ liq} = 389.6$ kJ/kg, $S_{3sat\ vap} = 7.4416$ kJ/kg K,
$S_{3sat\ liq} = 1.2271$ kJ/kg K
Basis: 1 kg/s steam. With a pump efficiency of $\eta_{pump} = 100\%$

$$W_{p,in} = V_1(P_1 - P_4) = (0.001038)(3500 - 78.5)\left(\frac{1\,kJ}{1\,kPa\,m^3}\right) = 3.55 \text{ kJ/kg}$$

$H_1 = H_4 + W_{p,in} = 393.1$ kJ/kg
Isentropic process $S_1 = S_4$ and $S_3 = S_2$.
The quality of the discharged wet steam $(S_2 < S_{3sat\ vap})$: $6.8443 < 7.4416$
$x_3 = (6.8463 - 1.2271)/(7.4416 - 1.2271) = 0.90$
$H_3 = 389.6(1 - 0.9) + 2665.0 \times 0.9 = 2437.5$ kJ/kg
Heat interactions: $q_{in} = H_2 - H_1 = 3224.2 - 393.1 = 2831.1$ kJ/kg
$q_{out} = -(H_4 - H_3) = -(389.6 - 2437.5) = 2048.0$ kJ/kg (heat discharged by the condenser)

The thermodynamic efficiency of the cycle: $\eta_{th} = 1 - \dfrac{|q_{out}|}{q_{in}} = \mathbf{0.276}$ **or 27.6%**
Therefore, the plant uses only 27.6% of the heat it received in the boiler.
Turbine work output: $W_{out} = H_3 - H_2 = 3224.2 - 2437.5 = -786.7$ kJ/kg

(b)   Steam properties: $P_3 = P_4 = 15$ kPa, $T_{sat} = 327.15$ K, $V_4 = 0.001014$ m^3/kg

$H_{3sat\ vap} = 2599.2$ kJ/kg, $H_4 = H_{3sat\ liq} = 226.0$ kJ/kg, $S_{3sat\ vap} = 8.0093$ kJ/kg K,
$S_{3sat\ liq} = 0.7550$ kJ/kg K
With a pump efficiency of $\eta_{pump} = 100\%$

$$W_{p,in} = V_1(P_1 - P_4) = (0.001014)(3500 - 15)\left(\frac{1\,kJ}{1\,kPa\,m^3}\right) = 3.53 \text{ kJ/kg}$$

$H_1 = H_4 + W_{p,in} = 226.0 + 3.53 = 229.5$ kJ/kg
Isentropic process $S_1 = S_4$ and $S_3 = S_2$.
The quality of the discharged wet steam $(S_3 < S_{3satvap})$: $6.8443 < 8.0093$
$x_{3'} = (6.8443 - 0.7550)/(8.0093 - 0.7550) = 0.84$
$H_{3'} = 226.0(1 - 0.84) + 2599.2 \times 0.84 = 2219.5$ kJ/kg
Heat interactions: $q_{in} = H_2 - H_1 = 3224.2 - 229.5 = 2994.7$ kJ/kg
$q_{out} = -(H_4 - H_{3'}) = -(226.0 - 2219.5.5) = 1993.5$ kJ/kg (heat discharged by the condenser)

The thermal efficiency of the cycle: $\eta_{th} = 1 - \dfrac{|q_{out}|}{q_{in}} = \mathbf{0.334}$ **or 33.4%**
Therefore, the plant uses only 33.4% of the heat it received in the boiler:
Turbine work output: $W_{out} = H_{3'} - H_2 = 2219.5 - 3224.2 = -1004.7$ kJ/kg (heat produced by the turbine)
Cycle work out: $W_{net} = (q_{out} - q_{in}) = (1993.5 - 2994.7) = -1001.2$ kJ/kg
For a 360 days of operation $= (360)(24) = 8640$ h/year
The annual increase in power output:
(20 kg/s)(1001.2 − 783.1) kJ/kg (8640) h/year = 37,687,680 kWh/year

The thermal efficiency increased from 0.276 to 0.334 by reducing the condenser pressure from 78.5 to 15.0 kPa. However, the quality of the discharged steam decreased from 0.9 to 0.84, which is not desirable for the blades of the turbine. Savings are considerable.

Example 9.3 Maximum possible efficiency calculation in Example 9.2
Estimate the maximum possible efficiency for parts (a) and (b) in Example 9.2 and compare them with those obtained in parts (a) and (b) in Example 9.2.

Solution:

(a) The thermal efficiency of a Carnot cycle operating between the same temperature limits:

$T_{min} = T_3 = 366.15$ K and $T_{max} = T_2 = 673.15$ K

$$\eta_{th,Carnot} = 1 - \frac{T_{min}}{T_{max}} = 1 - \frac{366.15 \text{ K}}{673.15 \text{ K}} = \textbf{0.456 or 45.6\%}$$

The difference between the two efficiencies occurs because of the large temperature differences. The ratio of the efficiencies, $\eta_{th}/\eta_{th,Carnot} = 0.276/0.456 = 0.60$ shows that only 60% of the possible efficiency is achieved in the cycle.

(b) The thermal efficiency of a Carnot cycle operating between the same temperature limits

$T_{min} = T_{3'} = 327.15$ K and $T_{max} = T_2 = 673.15$ K

$$\eta_{th,Carnot} = 1 - \frac{T_{min}}{T_{max}} = 1 - \frac{327.15}{673.15} = \textbf{0.514 of 51.4\%}$$

The ratio of the efficiencies, $\eta_{th}/\eta_{th,Carnot} = 0.334/0.51 = 0.65$ shows that only 65% of the possible efficiency is achieved in the cycle.

Example 9.4 Increasing the efficiency of a Rankine cycle by increasing the boiler pressure
A steam power plant is operating on the simple ideal Rankine cycle. The steam mass flow rate is 20 kg/s. The steam enters the turbine at 3500 kPa and 400 °C. Discharge pressure of the steam from the turbine is 78.5 kPa. If the pressure of the boiler is increased to 9800 kPa while maintaining the turbine inlet temperature at 400 °C, determine the change in the thermal efficiency.

Solution:
Assume that the cycle is at steady-state flow and the changes in kinetic and potential energy are negligible. Pump efficiency of $\eta_{pump} = 1$
Basis: 1 kg/s steam. Using the data from the steam tables
From Example 9.2 part a: $\eta_{th} = 0.276$ and $W_{net} = 783.1$ kJ/kg at 3500 kPa

(a) Superheated steam: $P_2 = P_1 = 9800$ kPa; $H_2 = 3104.2$ kJ/kg:
    $S_2 = 6.2325$ kJ/kg K, $T_2 = 400$ °C

Saturated steam: $P_3 = P_4 = 78.5$ kPa, $T_{sat} = 366.15$ K, $V_4 = 0.001038$ m³/kg
$H_{3sat\ vap} = 2665.0$ kJ/kg; $H_4 = H_{3sat\ liq} = 389.6$ kJ/kg; $S_{3sat\ vap} = 7.4416$ kJ/kg K;
$S_{3sat\ liq} = 1.2271$ kJ/kg K

$$W_{p,in} = V_1(P_1 - P_4) = (0.001038)(9800 - 78.5)\left(\frac{1\,kJ}{1\,kPa\,m^3}\right) = 10.1\ kJ/kg$$

$H_1 = H_4 + W_{p,in} = 389.6 + 10.1 = 399.7$ kJ/kg
Isentropic process $S_1 = S_4$ and $S_3 = S_2$.
The quality of the discharged wet steam x_3: ($S_3 < S_{3sat\ vap}$): 6.2325 < 7.4416
$x_3 = (6.2325 - 1.2271)/(7.4416 - 1.2271) = 0.80$
$H_3 = 389.6(1 - 0.8) + 2665.0 \times 0.8 = 2210.0$ kJ/kg
Heat interactions:
$q_{in} = H_2 - H_1 = 3104.2 - 399.7 = 2704.5$ kJ/kg
$q_{out} = -(H_4 - H_3) = -(389.6 - 2210.0) = 1820.4$ kJ/kg (heat discharged by the condenser)

The thermodynamic efficiency of the cycle: $\eta_{th} = 1 - \dfrac{q_{out}}{q_{in}} = \mathbf{0.326\ or\ 32.6\%}$

Therefore, the plant uses only 32.6% of the heat it received in the boiler.
The thermal efficiency increased from 0.276 to 0.326 by increasing the boiler pressure from 3500 to 9800 kPa. However, the quality of the discharged steam decreased from 0.9 to 0.80, which is not desirable for the blades of the turbine.
Turbine work output: $W_{out} = H_3 - H_2 = (2210.0 - 3104.2)$ kJ/kg $= -894.2$ kJ/kg
Cycle work out: $W_{net} = (q_{out} - q_{in}) = (1820.4 - 2704.5)$ kJ/kg $= -884.1$ kJ/kg

(b) $\dot{m}_s = 20.0$ kg/s, and for a 360 days of operation $= (360)(24) = 8640$ h/year

Annual increase in power output: (20 kg/s)(884.1 − 783.1) kJ/kg (8640) h/year = 17,452,800 kWh/year

---

Example 9.5 Increasing the efficiency of a Rankine cycle by increasing the boiler temperature
A steam power plant is operating on the simple ideal Rankine cycle. The steam mass flow rate is 20 kg/s. The steam enters the turbine at 3500 kPa and 400 °C. Discharge pressure of the steam from the turbine is 78.5 kPa. If the temperature of the boiler is increased to 525 °C while maintaining the pressure at 3500 kPa, determine the change in the thermal efficiency.

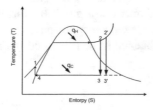

Solution:
Assume that the cycle is at steady-state flow and the changes in kinetic and potential energy are negligible.

(a) Basis: 1 kg/s steam. Using the data from the steam tables:

From Example 9.2 part a: $\eta_{th} = 0.276$ and $W_{net} = 783.1$ kJ/kg at 400 °C
Superheated steam: $P_2 = P_1 = 3500$ kPa; $H_2 = 3506.9$ kJ/kg:
$S_2 = 7.2297$ kJ/kg K, $T_2 = 525$ °C
Saturated steam: $P_3 = P_4 = 78.5$ kPa, $T_{sat} = 366.15$ K, $V_4 = 0.001038$ m^3/kg
$H_{3sat\ vap} = 2665.0$ kJ/kg, $H_{3sat\ liq} = 389.6$ kJ/kg, $S_{3sat\ vap} = 7.4416$ kJ/kg K,
$S_{3sat\ liq} = 1.2271$ kJ/kg K

$$W_{p,in} = V_1(P_1 - P_4) = (0.001038)(3500 - 78.5)\left(\frac{1\,kJ}{1\,kPa\,m^3}\right) = 3.55\ kJ/kg$$

$H_1 = H_4 + W_{p,in} = 389.6 + 3.55 = 393.15$ kJ/kg
Isentropic process $S_1 = S_4$ and $S_3 = S_2$.
The quality of the discharged wet steam ($S_2 < S_{3sat\ vap}$): 7.2297 < 7.4416
$x_{3'} = (7.2297 - 1.2271)/(7.4416 - 1.2271) = 0.96$
$H_{3'} = 389.6(1 - 0.96) + 2665.0 \times 0.96 = 2574.0$ kJ/kg
Heat interactions: $q_{in} = H_2 - H_1 = 3506.9 - 393.1 = 3113.8$ kJ/kg
$q_{out} = -(H_4 - H_{3'}) = -(389.6 - 2574.0) = 2184.4$ kJ/kg (heat received by the cooling medium)

The thermodynamic efficiency of the cycle is $\eta_{th} = 1 - \dfrac{q_{out}}{q_{in}} = \mathbf{0.298\ or\ 29.8\%}$

Therefore, the plant uses only 29.8% of the heat it received in the boiler.
The thermal efficiency increased from 0.276 to 0.298 by increasing the boiler temperature from 400 to 525 °C. The quality of the discharged steam increased from 0.9 to 0.96, which is desirable for the protection of the turbine blades.
Turbine work out $W_{out} = H_{3'} - H_2 = (2574.0 - 3506.9) = -932.9$ kJ/kg
Cycle work out: $W_{net} = (q_{out} - q_{in}) = (2184.4 - 3113.8) = -929.4$ kJ/kg

(b) $\dot{m}_s = 20.0$ kg/s and for a 360 days of operation = $(360)(24) = 8640$ h/year

Annual increase in power output: (20 kg/s)(929.4 − 783.1) kJ/kg (8640) h/year = 25.280,640 kWh/year

---

Example 9.6 Estimation of maximum possible efficiencies in Example 9.5
Estimate the maximum possible efficiency for parts (a) and (b) in Example 9.5 and compare them with those obtained in parts (a) and (b) in Example 9.5.

Solution:

(a) The thermal efficiency of a Carnot cycle operating between the temperature limits of

$T_{min} = 366.15$ K and $T_{max} = 673.15$ K:

$$\eta_{\text{th,Carnot}} = 1 - \frac{T_{\min}}{T_{\max}} = 1 - \frac{366.15 \text{ K}}{673.15 \text{ K}} = \mathbf{0.45 \text{ or } 45\%}$$

The ratio of the efficiencies, $\eta_{\text{th}}/\eta_{\text{th,Carnot}} = 0.276/0.45 = 0.61$ shows that only 61% of the possible efficiency is achieved in the cycle with the boiler temperature of 400 °C.

(b) The thermal efficiency of a Carnot cycle operating between the temperature limits of

$T_{\min} = 366.15$ K and $T_{\max} = 798.15$ K:

$$\eta_{\text{th,Carnot}} = 1 - \frac{T_{\min}}{T_{\max}} = 1 - \frac{366.15 \text{ K}}{798.15 \text{ K}} = \mathbf{0.54 \text{ or } 54\%}$$

The ratio of the efficiencies, $\eta_{\text{th}}/\eta_{\text{th,Carnot}} = 0.298/0.54 = 0.55$ shows that only 55% of the possible efficiency is achieved in the cycle with the boiler temperature of 525 °C.

9.2.3 Energy Conservation in Compression and Expansion Work

It is possible to save energy in the compression work by minimizing the friction, turbulence, heat transfer, and other losses. A practical way of energy conservation is to keep the specific volume of the gas small during the compression work. This is possible by maintaining the temperature of the gas low as the specific volume is proportional to temperature. Therefore, cooling the gas as it is compressed may reduce the cost of compression work in a multistage compression with intercooling as seen in Fig. 9.2. The gas is cooled to the initial temperature between the compression stages by passing the gas through a heat exchanger called the intercooler. Energy recovery by intercooling may be significant especially when a gas is to be compressed to very high pressure. Example 9.7 illustrates the energy conservation in a two-stage compression work by intercooling.

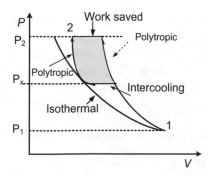

Fig. 9.2 Energy conservation in the compression work by intercooling; the work saved appears between two polytropic compressions starting at the second stage with the pressure P_x

As Fig. 9.2 shows the work saved varies with the value of intermediate pressure P_x, which needs to be optimized. The total work input for a two-stage compression process is the sum of the work inputs for each stage

$$
\begin{aligned}
W_{comp} &= W_{comp1} + W_{comp2} \\
&= \frac{\gamma R T_1}{MW(\gamma - 1)} \left[\left(\frac{P_x}{P_1} \right)^{(\gamma-1)/\gamma} \right] + \frac{\gamma R T_1}{MW(\gamma - 1)} \left[\left(\frac{P_2}{P_x} \right)^{(\gamma-1)/\gamma} \right]
\end{aligned} \tag{9.4}
$$

In Eq. (9.4), P_x is the only variable. The optimum value of P_x is obtained by differentiation of Eq. (9.4) with respect to P_x and setting the resulting expression equal to zero. Then, the optimum value of P_x becomes

$$
P_x = (P_1 P_2)^{1/2} \quad \text{or} \quad \frac{P_x}{P_1} = \frac{P_2}{P_x} \tag{9.5}
$$

Therefore, energy conservation will be maximum, when the pressure ratio across each stage of the compressor is the same and compression work at each stage becomes identical

$$
W_{comp1} = W_{comp2} \tag{9.6}
$$

$$
W_{comp} = \frac{2\gamma R T_1}{MW(\gamma - 1)} \left[\left(\frac{P_x}{P_1} \right)^{(\gamma-1)/\gamma} \right] \tag{9.7}
$$

Example 9.8 illustrates the estimation of minimum and actual power required by a compressor. Example 9.9 illustrates how to produce power out of a cryogenic expansion process. Reduction of pressure by using throttling valve wastes the energy. Replacing the throttling valve with a turbine produces power and hence conserves electricity [1].

Example 9.7 Energy conservation in a two-stage compression work by intercooling

Air with a flow rate of 2 kg/s is compressed in a steady state and reversible process from an inlet state of 100 kPa and 300 K to an exit pressure of 1000 kPa. Estimate the work for (a) polytropic compression with $\gamma = 1.3$, and (b) ideal two-stage polytropic compression with intercooling using the same polytropic exponent of $\gamma = 1.3$, (c) estimate conserved compression work by intercooling and electricity per year if the compressor is operated 360 days per year.

Solution:

Assumptions: steady-state operation; air is ideal gas; kinetic and potential energies are negligible.

$P_2 = 1000$ kPa, $P_1 = 100$ kPa, $T_1 = 300$ K, $\gamma = 1.3$, $MW_{air} = 29$ kg/kmol

Basis 1 kg/s air flow rate

(a) Work needed for polytropic compression with $\gamma = 1.3$

$$W_{comp} = \frac{\gamma R T_1}{MW(\gamma - 1)} \left[\left(\frac{P_2}{P_1}\right)^{(\gamma-1)/\gamma} \right] = \frac{1.3(8.314 \text{ kJ/kmol K})(300 \text{ K})}{29 \text{ kg/kmol}\,(1.3 - 1)} \left[\left(\frac{1000}{300}\right)^{(1.3-1)/1.3} - 1 \right]$$

$$= 261.3 \text{ kJ/kg}$$

(b) Ideal two-stage polytropic compression with intercooling ($\gamma = 1.3$)

$$P_x = (P_1 P_2)^{1/2} = (100 \times 1000)^{1/2} = 316.2 \text{ kPa}$$

$$W_{comp} = \frac{2\gamma R T_1}{MW(\gamma - 1)} \left[\left(\frac{P_x}{P_1}\right)^{(\gamma-1)/\gamma} \right] = \frac{2(1.3)(8.314 \text{ kJ/kmol K})(300 \text{ K})}{29 \text{ kg/kmol}\,(1.3 - 1)} \left[\left(\frac{316.2}{100}\right)^{(1.3-1)/1.3} - 1 \right]$$

$$= 226.8 \text{ kJ/kg}$$

Recovered energy = $(261.3 - 226.8)$ kJ/kg = 34.5 kJ/kg

Reduction in energy use: $\dfrac{261.3 - 226.8}{261.3} = 0.13$ or 13%

(c) Conservation of compression work = $(2 \text{ kg/s})\,(261.3 - 226.8)$ kJ/kg = 69 kW

Yearly conserved work: $(69 \text{ kW})\,(8640 \text{ h/year}) = 596{,}160$ kWh/year

The compression work has been reduced by 13% when two stages of polytropic compression are used instead of single polytropic compression and conserved 596,160 kWh/year.

Example 9.8 Compressor efficiency and power input

An adiabatic compressor is used to compress air from 100 kPa and 290 K to a pressure of 900 kPa at a steady-state operation. The isentropic efficiency of the compressor is 80%. The air flow rate is 0.4 kg/s. Determine the minimum and actual power needed by the compressor.

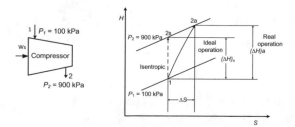

Solution:

Assume: Steady-state adiabatic operation. Air is ideal gas. The changes in kinetic and potential energies are negligible.

Enthalpies of ideal gas depends on temperature only.

The air mass flow rate = 0.4 kg/s, $\eta_C = 0.8$

Inlet conditions (Table D1): $P_1 = 100$ kPa, $T_1 = 290$ K, $H_1 = 290.16$ kJ/kg, $P_{r1} = 1.2311$

Exit conditions (Table D1): P_2 = 900 kPa

$$P_{r2} = P_{r1} \frac{P_2}{P_1} = 1.2311 \left(\frac{900\,\text{kPa}}{100\,\text{kPa}} \right) = 11.1$$

From Table D1: for P_{r2} = 11.1, we find T_2 = 540 K and H_{2s} = 544.35 kJ/kg

$$\eta_C = \frac{H_{2s} - H_1}{H_{2a} - H_1} = 0.8 \rightarrow H_{2a}$$

H_{2a} = 607.9 kJ/kg and T_2 = 600 K (Table D1) (Approximate)

As seen from the TS diagram above, H_{2a} is the actual enthalpy at the exit

Actual power required:

$$\dot{m}\Delta H_a = \dot{W}_{\text{net,in}} = \dot{m}(H_{2a} - H_1) = 0.4\,\text{kg/s}\,(607.9 - 290.16)\,\text{kJ/kg} = \textbf{127.1 kW}$$

(work received by the compressor).

Minimum power required:

$$\dot{m}\Delta H_s = \dot{W}_{\text{net,in}} = \dot{m}(H_{2s} - H_1) = 0.4\,\text{kg/s}\,(544.3 - 290.1)\,\text{kJ/kg} = \textbf{101.7 kW}$$

If the operation was ideal the rate of conserved energy would be 25.4 kW or 20% savings.

Example 9.9 Energy conservation in expansion by replacing a throttle valve with a turbine

A cryogenic manufacturing plant handles liquid methane at 115 K and 5000 kPa at a rate of 0.3 m³/s. In the plant a throttling valve reduces the pressure of liquid methane to 1000 kPa. A new process considered replaces the throttling valve with a turbine in order to produce power while reducing the pressure to 1000 kPa at 110 K. Using the data for the properties of liquid methane below estimate the power that can be produced by the turbine.

T (K)	P (kPa)	H (kJ/kg)	S (kJ/kg K)	C_p (kJ/kg K)	ρ (kg/m³)
110	1000	209.0	4.875	3.471	425.8
110	2000	210.5	4.867	3.460	426.6
110	5000	215.0	4.844	3.432	429.1
120	1000	244.1	5.180	3.543	411.0
120	2000	245.4	5.171	3.528	412.0
120	5000	249.6	5.145	3.486	415.2

Source [1]

Solution:

Assumptions: steady state and reversible operation; adiabatic turbine, methane is ideal gas; kinetic and potential energies are negligible.

(a) P_1 = 5000 kPa, T_1 = 115 K, Q_1 = 0.30 m³/s

H_1 = 232.3 kJ/kg = (215.0 + 249.6)/2, ρ_1 = 422.15 kg/m³ = (429.1 + 415.2)/2

P_2 = 1000 kPa, H_2 = 209.0 kJ/kg

Unit cost of electricity = \$0.09/kWh

Mass flow rate: $\dot{m} = \rho Q_1 = 422.15\,\text{kg/m}^3\,(0.3\,\text{m}^3/\text{s}) = 126.6\ \text{kg/s}$

Power produced:

$\dot{W}_{\text{out}} = \dot{m}(H_1 - H_2) = 126.6\,\text{kg/s}\,(232.5 - 209.0)\,\text{kJ/kg} = \textbf{2949.8 kW}$

Annual power production:

$\dot{W}_{\text{out}}\Delta t = (2949.8\,\text{kW})\,(360)(24)\,\text{h/year} = 25{,}486{,}099\ \text{kWh/year}$

Reduction of pressure by using throttling valve wastes the potential of power production. Replacing the valve with a turbine will produce power and hence conserve electricity.

9.2.4 Conservation of Energy by High-Efficiency Electric Motors

Compressors are powered by electric motors. Electric motors cannot convert the electrical energy they consume into mechanical energy completely. Electric motor efficiency is defined by

$$\eta_{\text{Motor}} = \frac{\text{Mechanical power}}{\text{Electrical power}} = \frac{\dot{W}_{\text{comp}}}{\dot{W}_{\text{elect}}} \tag{9.8}$$

Motor efficiency range: $0.7 < \eta_{\text{motor}} < 96$. The portion of electric energy that is not converted to mechanical power is converted to heat, which is mostly unusable. For example, assuming that no transmission losses occur [6, 7]:

- A motor with an efficiency of 80% will draw an electrical power of $1/0.8 = 1.25$ kW for each kW of shaft power it delivers.
- If the motor is 95% efficient, then it will draw $1/0.95 = 1.05$ kW only to deliver 1 kW of shaft work.
- Therefore, between these two motors, electric power conservation is

$1.25\ (\eta_{\text{motor}} = 95\%)\ \text{kW} - 1.05\ (\eta_{\text{motor}} = 95\%) = 0.20\ \text{kW}$.

High efficiency motors are more expensive but its operation may save energy. Saved energy is estimated by

$$\dot{W}_{\text{elect.saved}} = \dot{W}_{\text{elect.std}} - \dot{W}_{\text{elect.efficient}} = \dot{W}_{\text{comp}}\left(\frac{1}{\eta_{\text{std}}} - \frac{1}{\eta_{\text{efficient}}}\right) \tag{9.9}$$

$$\dot{W}_{\text{elect.saved}} = (\text{Rated power})(\text{Load factor})\left(\frac{1}{\eta_{\text{std}}} - \frac{1}{\eta_{\text{efficient}}}\right) \tag{9.10}$$

where the *rated power* is the nominal power delivered at full load of the motor and listed on its label. *Load factor* is the fraction of the rated power at which the motor normally operates. Annual saving is

$$\text{Annual energy saving} = \left(\dot{W}_{\text{elect.saved}}\right)(\text{Annual operation hours}) \tag{9.11}$$

- A compressor that operates at partial load causes the motor to operate less efficiently. The efficiency of motor will increase with the load.

- Using the cold outside air for compressor intake lowers the compressor work and conserves energy.

9.3   Energy Conservation in Home Heating and Cooling

Space heating and cooling as well as domestic water heating may account around 55% of the utility bill. To conserve energy, one should identify from where the home is losing energy, assign priorities, and form an efficiency plan that improves efficiency and reduces costs. For example, the potential energy savings from reducing drafts in a home may range from 5 to 30% per year. Heat loss through the ceiling and walls in your home could be very large if the insulation levels are less than the recommended minimum. Inspect heating and cooling equipment annually, or as recommended by the manufacturer. If the unit is more than 15 years old, the newer and energy-efficient units may reduce the cost [8, 9]. In colder climates, windows that are gas filled with low emissivity coatings on the glass reduces heat loss. In warmer climates, windows with selective coatings may reduce heat gain. Setting the thermostat low in the winter and high in the summer as comfortable as possible may reduce the cost of heating and cooling. Also, the followings can reduce the cost:

- Clean or replace filters on furnaces once a month or as needed.
- Clean baseboard heaters and radiators and make sure that they are not blocked.
- During summer, keep the window coverings closed during the day to lower solar gain.
- Select energy-efficient products when you buy new heating and cooling equipment.

Water heating typically accounts for about 12% of the utility bill. Insulating the electric, natural gas, or oil hot-water storage tanks as recommended can help conserve energy. Most water heaters can last around 15 years. However, replacing the units older than 7 years may reduce the cost of energy [6].

9.3.1   Home Heating by Fossil Fuels

Residential furnaces have a heat input rate of less than 225,000 Btu/h (66,000 W) and residential boilers have a heat input rate of less than 300,000 Btu/h (88,000 W). The residential furnace provides heated air with a blower to circulate air through the duct distribution system. The residential boiler is a cast-iron, steel, aluminum, or copper pressure vessel heat exchanger designed to burn fossil fuels to heat to a suitable medium such as water. Hot water can be distributed via baseboard radiators or radiant floor systems, while steam is distributed through pipes to steam radiators. Furnaces heat air and distribute the heated air through the house using ducts, while boilers heat water, providing either hot water or steam for heating. Most furnaces and boilers operate on natural gas, oil, or propane. Steam boilers operate at a higher temperature than hot water boilers. Oil-fired furnaces and boilers can use heating oil blended with biodiesel, which produce less pollution than pure heating oil [6, 8].

A condensing furnace or boiler condenses the water vapor produced in the combustion process and uses the heat from this condensation. Although condensing units cost more than non-condensing units, the condensing unit can reduce the consumption of fuel and the cost over the 15–20-year life of the unit. Old furnaces and boilers can be retrofitted to increase their efficiency. Some retrofitting options include installing programmable thermostats, upgrading ductwork in forced-air systems, and adding zone control for hot-water systems. Still the costs of retrofits should be compared with the cost of a new boiler or furnace.

9.3.2   Home Heating by Electric Resistance

An all-electric furnace or boiler has no energy loss carried out by the flue gas through the chimney. Electric resistance heating converts nearly 100% of the energy in the electricity to heat. However, most electricity is produced from oil, gas, or coal by converting only about 30% of the fuel's energy into electricity. Because of production and transmission losses, electric heat is often more expensive than heat produced using combustion appliances, such as natural gas, propane, and oil furnaces. Heat pumps are preferable in most climates, as they easily cut electricity use by 50% when compared with electric resistance heating. It is also possible to use heat storage systems to avoid heating during times of peak power demand [6, 8].

Blowers (large fans) in electric furnaces move air over a group of three to seven electric resistance coils, called elements, each of which are typically rated at five kilowatts. A built-in thermostat prevents overheating and may shut the furnace off if the blower fails or if a dirty filter blocks the airflow.

When operated in heating mode, a heat pump is more efficient than operating resistance heaters. Because an electric heater can convert only the input electrical energy directly to output heat energy with none of the efficiency or conversion advantages of a heat pump. Likewise, when a heat pump operates near its most inefficient outside temperature, typically 0 °F, the heat pump will perform close to the same as a resistance heater. Example 9.10 illustrates a simple analysis of heating a house by heat pump. Example 9.11 discusses the energy conservation in house heating by Carnot heat pump [9].

9.3.3 Home Heating by Solar Systems

Active or passive solar systems can be used for residential heating. There are two basic types of active solar heating systems using either liquid or air heated in the solar collectors. The solar units are environmentally friendly and can now be installed on your roof to blend with the architecture of a house. *Liquid-based systems* use heat water or an antifreeze solution, whereas *air-based systems* heat air in an air collector. A circulating pump transports the fluid through the collector so its temperature increases. Both of these systems absorb solar radiation and transfer it directly to the interior space or to a storage system. Liquid systems are more often used when storage is included, and are well suited for boilers with hot water radiators and heat pumps. These collectors easily last decades.

Active solar heating systems may reduce the cost more when they are used for most of the year. The economics of an active space heating system improve if it also heats domestic water. Heating homes with active solar energy systems can significantly reduce the fossil fuel consumptions, air pollution, and emission of greenhouse gases [5, 8].

Passive solar heaters do not have fans or blowers. In passive solar building design, windows, walls, and floors are made to collect, store, and distribute solar energy in the form of heat in the winter and reject solar heat in the summer. A passive solar building takes advantage of the local climate. Elements to be considered include window placement, thermal insulation, thermal mass, and shading.

Example 9.10 Heating a house by heat pump
A heat pump is used to heat a house and maintain it at 18 °C. On a day where the outside temperature is −2 °C, the house is losing heat at a rate of 75,000 kJ/h. The heat pump operates with a coefficient of performance (COP) of 2.8. Determine:

(a) Power needed by the heat pump,
(b) The rate of heat absorbed from to the surrounding cold air.

Solution:
Assume: Steady-state operation

(a) House is maintained at 18 °C.

Heat pump must supply the same amount of lost heat from the cold source:
$q_H = 75{,}000$ kJ/h $= 20.83$ kW

Power required by the heat pump:

$$\text{COP}_{\text{HP}} = \frac{\dot{q}_H}{\dot{W}_{\text{net,in}}} \rightarrow \dot{W}_{\text{net,in}} = \frac{\dot{q}_H}{\text{COP}_{\text{HP}}} = \frac{20.83 \text{ kW}}{2.8} = \textbf{7.44 kW}$$

(b) Energy balance: $\dot{q}_H - \dot{q}_C = \dot{W}_{\text{net,in}} = 7.44$ kW

$\dot{q}_C = \dot{q}_H - \dot{W}_{\text{net,in}} = (20.83 - 7.44)$ kW = **13.39 kW**

13.39 kW is extracted from the outside. The house is paying only for the energy of 7.44 kW that is supplied as electrical work to the heat pump.

If we have to use electricity in a resistance heater we have to supply 20.83 kW, so the energy conserved by using heat pump instead of electric heater is 13.39/20.83 = 0.64 (or 64%).

Example 9.11 Energy conservation in house heating by Carnot heat pump
A Carnot heat pump is used to heat a house during the winter. The house is maintained 20 °C. The house is estimated to be losing heat at a rate of 120,000 kJ/h when the outside temperature is −4 °C. Determine the minimum power needed by the heat pump and the rate of heat absorbed from to the surrounding cold air.
Solution:
Assume: Steady-state operation.
Temperatures of hot and cold sources: T_H = 273.15 + 20 °C = 293.15 K, T_C = 273.15 − 4 °C = 269.15 K
The amount of heat to be supplied to warm inside room:
q_H = 120,000 kJ/h = 33.33 kW
Minimum amount of power is possible only for a fully reversible heat pump. This heat pump is Carnot heat pump.
Coefficient of performance:

$$\text{COP}_{\text{HP}} = \frac{\dot{q}_H}{\dot{W}_{\text{net,in}}} = \frac{1}{1 - (T_C/T_H)} = \frac{1}{1 - (269.15 \text{ K}/293.15 \text{ K})} = 12.2$$

Since the house is losing heat at a rate of 120,000 kJ/h to maintain the house at 20 ° C, the heat pump must supply the same amount of heat from the cold source:

Energy balance: $\text{COP}_{\text{HP}} = \frac{\dot{q}_H}{\dot{W}_{\text{net,in}}} \rightarrow \dot{W}_{\text{net,in}} = \frac{\dot{q}_H}{\text{COP}_{\text{HP}}} = \textbf{2.73 kW}$

$\dot{q}_C = \dot{q}_H - \dot{W}_{\text{net,in}} = (33.3 - 2.73)$ kW = **30.57 kW**

The house pays for only 2.73 kW. If we use electricity for heating by an electric resistance heater, the rate of heat necessary is 33.3 kW. 30.6 kW of the necessary heat is supplied from the surrounding cold air. Therefore, the energy conserved is 30.6/33.3 = 0.92 (or 92%).

9.4 Energy Efficiency Standards

Thermal efficiency of residential furnaces and boilers is measured by annual fuel utilization efficiency (AFUE) . Annual fuel utilization efficiency is the ratio of heat output of the furnace or boiler compared to the total energy consumed by them over a typical year. Annual fuel utilization efficiency does not account the circulating air and combustion fan power consumptions and the heat losses of the distributing systems of duct or piping. An AFUE of 90% means that 90% of the energy in the fuel becomes heat for the home and the other 10% escapes up the chimney and elsewhere. Heat losses of the duct system or piping can be as much as 35% of the energy for output of the furnace. Table 9.1 shows some typical values of AFUE for furnace and boiler using various fossil fuels and electricity [8, 9]. Some of the minimum allowed AFUE ratings in the United States are:

- Noncondensing fossil-fueled, warm-air furnace is 78%.
- Fossil-fueled boiler is 80%.
- Gas-fueled steam boiler is 75%.

The annual savings from replacement of heating system with more efficient one may be estimated by using Table 9.2 assuming that both systems have the same heat output. For older units, actual savings in upgrading to a new system could be much higher than indicated in Table 9.2. AFUE is calculated using ASHRAE Standard 103 (ASHRAE 2007, https://webstore.ansi.org/standards/ashrae/ansiashrae1032007). A furnace with a thermal efficiency of 78% may yield an AFUE of only 64% or so, for example, under the standards' test conditions.

Table 9.1 Some typical values of annual fuel utilization efficiency (AFUE) for furnace and boilers [8, 9]

Fuel	Furnace/boiler	AFUE %
Heating oil	Retention head burner	70–78
	Mid-efficiency	83–89
Electric heating	Central or baseboard	100
Natural gas	Conventional	55–65
	Mid-efficiency	78–84
	Condensing	90–97
Propane	Conventional	55–65
	Mid-efficiency	79–85
	Condensing	88–95
Firewood	Conventional	45–55
	Advanced	55–65
	State-of-the-art	75–90

Table 9.2 Assuming the same heat output, estimated savings for every $100 of fuel costs by increasing an existing heating equipment efficiency [8]

Existing AFUE	New and upgraded system AFUE							
	60%	65%	70%	75%	80%	85%	90%	95%
55%	$8.3	$15.4	$21.4	$26.7	$31.2	$35.3	$38.9	$42.1
60%	–	$7.7	$14.3	$20.0	$25.0	$29.4	$33.3	$37.8
65%	–	–	$7.1	$13.3	$18.8	$23.5	$27.8	$31.6
70%	–	–	–	$6.7	$12.5	$17.6	$22.2	$26.3
75%	–	–	–	–	$6.5	$11.8	$16.7	$21.1
80%	–	–	–	–	–	$5.9	$11.1	$15.8
85%	–	–	–	–	–	–	$5.6	$10.5

9.4.1 Efficiency of Air Conditioner

The Energy Efficiency Ratio (EER) of a particular cooling device is the ratio of *output* cooling (in Btu/h) to *input* electrical power (in Watts) at a given operating point. The efficiency of air conditioners is often rated by the *Seasonal Energy Efficiency Ratio* (SEER). The SEER rating of a unit is the cooling output in Btu during a typical cooling-season divided by the total electric energy input in watt-hours during the same period. The coefficient of performance (COP) is an instantaneous measure (i.e. a measure of power divided by power), whereas both EER and SEER are averaged over a duration of time. The time duration considered is several hours of constant conditions for EER, and a full year of typical meteorological and indoor conditions for SEER. Typical EER for residential central cooling units = $0.875 \times$ SEER. A SEER of 13 is approximately equivalent to a COP of 3.43, which means that 3.43 units of heat energy are removed from indoors per unit of work energy used to run the heat pump. SEER rating more accurately reflects overall system efficiency on a seasonal basis and EER reflects the system's energy efficiency at peak day operations.

Air conditioner sizes are often given as "tons" of cooling where 1 ton of cooling is being equivalent to 12,000 Btu/h (3500 W). This is approximately the power required to melt one ton of ice in 24 h. Example 9.12 illustrates the estimation of electric cost of an air conditioner [3, 8, 9].

Example 9.12 Electricity cost of air conditioner
Estimate the cost of electricity for a 5000 Btu/h (1500 W) air-conditioning unit operating, with a SEER of 10 Btu/Wh. The unit is used for a total of 1500 h during an annual cooling season and the unit cost of electricity is $0.14/kWh.

Solution:
Air conditioner sizes are often given as 'tons' of cooling, where 1 ton of cooling is being equivalent to 12,000 Btu/h (3500 W).
The unit considered is a small unit and the annual total cooling output would be:

(5000 Btu/h) × (1500 h/year) = 7,500,000 Btu/year
With a seasonal energy efficiency ratio (SEER) of 10, the annual electrical energy usage is:
(7,500,000 Btu/year)/(10 Btu/Wh) = 750,000 Wh/year = 750 kWh/year
With a unit cost of electricity of $0.14/kWh, the annual cost is: (750 kWh/year) ($0.14/kWh) = **$105/year**
The average power usage may also be calculated more simply by:
Average power = (Btu/h)/(SEER, Btu/Wh) = 5000/10 = 500 W = 0.5 kW
With the electricity cost of $0.14/kWh, the cost per hour is: (0.5 kW) ($0.14)/kWh = $0.07/h
For 1500 h/year, the total cost is: ($0.07/h) (1500 h/year) = **$105/year**

9.4.2   High Efficiency for Cooling

The refrigeration process with the maximum possible efficiency is the Carnot cycle. The coefficient of performance (COP) of an air conditioner using the Carnot cycle is:

$$\text{COP}_{\text{Carnot}} = \frac{T_C}{T_H - T_C} \tag{9.12}$$

where T_C is the indoor temperature and T_H is the outdoor temperature in K or R. The EER is calculated by multiplying the COP by 3.413, which is the conversion factor from Btu/h to Watts:

$$\text{EER}_{\text{Carnot}} = 3.413(\text{COP}_{\text{Carnot}}) \tag{9.13}$$

For an outdoor temperature of 100 °F (311 K) and an indoor temperature of 95 °F (308 K), the above equation gives a COP of 103, or an EER of 350. This is about 10 times as efficient as a typical home air conditioner available today. The maximum EER decreases as the difference between the inside and outside air temperature increases. For example:
For T_H = 120 °F (49 °C) = 322.15 K, and T_C = 80 °F (27 °C) = 300.15 K, we have
$\text{COP}_{\text{Carnot}}$ = 300.15 K/(322.15 − 300.15) K = 13.6 or
EER = (3.413) (13.6) = 46.4
The maximum SEER can be calculated by averaging the maximum values of EER over the range of expected temperatures for the season.
Central air conditioners should have a SEER of at least 14. Substantial energy savings can be obtained from more efficient systems. For example:
By upgrading from SEER 9 to SEER 13
Reduction in power consumption = (1 − 9/13) = 0.30
This means that the power consumption is reduced by 30%. Residential air condition units may be available with SEER ratings up to 26. Example 9.13

illustrates the calculation of the annual cost of power used by an air conditioner. Example 9.14 discusses possible saving in cooling by using a unit operating at a higher SEER rating [3, 8, 9].

Example 9.13 Calculating the annual cost of power for an air conditioner

Estimate the annual cost of electric power consumed by a 6 ton air-conditioning unit operating for 2000 h/year with a SEER rating of 10 and a power cost of $0.16/kWh.

Solution:

Air conditioner sizes are often given as 'tons' of cooling:

1 ton of cooling = 12,000 Btu/h (3500 W).

This is approximately the power required to melt one ton of ice in 24 h.

The annual cost of electric power consumption: (6)(12,000 Btu/h) = 72,000 Btu/h

$$\text{Cost} = \frac{(\text{Size, Btu/h})(\text{time, h/year})(\text{Cost of energy, \$/kWh})}{(\text{SEER, Btu/Wh})(1000\,\text{W/kW})}$$

(72,000 Btu/h)(2000 h/year)($0.16/kWh)/[(10 Btu/Wh) (1000 W/kW)] = **$2304/year**

For the temperatures of hot and cold sources:

T_H = 273.15 + 20 °C = 293.15 K, T_C = 273.15 − 4 °C = 269.15 K, and $T_H - T_C$ = 24 K

$$\text{EER}_{\text{Carnot}} = 3.41 \left(\frac{T_C}{T_H - T_C} \right) = 3.41\,(269.13/24) = \mathbf{38.24}$$

The maximum EER decreases as the difference between the inside and outside air temperature increases.

Example 9.14 Saving the cost of cooling with a unit operating at a higher SEER rating

A 4 ton of current residential air conditioner operates with a seasonal energy efficiency ratio (SEER) rating of 10. This unit will be replaced with a newer unit operating with a SEER rating of 22. The unit operates 130 days with an average 10 h/day. Average inside and outside temperatures are 21 and 3 °C, respectively. The unit cost of energy is $0.16/kWh. Estimate the saving in the cost of electricity and the maximum energy efficiency ratio.

Solution:

Cooling load of 4 ton: (4)(12,000 Btu/h) = 48,000 Btu/h (14,000 W).

(130 days/year)(10 h/day) = 1300 h/year

The estimated cost of electrical power:

SEER = 10, and an energy cost of $0.16/kWh, using 130 days of 10 h/day operation:

Cost = (48,000 Btu/h)(1300 h/year)($0.16/kWh)/[(10 Btu/Wh) (1000 W/kW)] = $998.4/year

SEER = 22, and an energy cost of \$0.16/kWh, using 130 days of 10 h/day operation:

Cost = (48,000 Btu/h)(1300 h/year)(\$0.16/kWh)/[(22 Btu/Wh)

(1000 W/kW)] = \$453.8/year

Annual saving = \$998.4/year − \$453.8/year = **\$44.6/year**

The ratio of typical EER to maximum EER/EER_{max} = 19.25/52.31 = 0.368

Typical EER for the current cooling unit:

EER = 0.875 (SEER) = 0.875 (10) = 8.75

Typical EER for the new cooling unit: EER = 0.875 (SEER) = 0.875 (22) = 19.25

Maximum value of EER:

The temperatures of hot and cold sources: T_H = 273.15 + 21 °C = 294.15 K, T_C = 273.15 + 3 °C = 276.15 K

$T_H − T_C$ = 18 K

$$EER_{Carnot} = 3.41\left(\frac{T_C}{T_H - T_C}\right) = 3.41\,(276.13/18) = \mathbf{52.31}$$

The maximum EER decreases as the difference between the inside and outside air temperature increases. The annual saving is significant after using a more efficient unit. The value of EER for the new unit is only 36.6% of the maximum value of EER.

9.4.3 Fuel Efficiency

Fuel efficiency means the efficiency of a process that converts chemical energy contained in a fuel into kinetic energy or work. The increased fuel efficiency is especially beneficial for fossil fuel power plants or industries dealing with combustion of fuels. In the transportation field, fuel efficiency is expressed in miles per gallon (mpg) or kilometers per liter (km/l). Fuel efficiency is dependent on many parameters of a vehicle, including its engine parameters, aerodynamic drag, and weight. Hybrid vehicles offer higher fuel efficiency using two or more power sources for propulsion, such as a small combustion engine combined with electric motors [9, 10]. Example 9.15 compares the heating by electricity and by natural gas. Examples 9.16 and 9.17 discuss the amounts of coal necessary in a coal-fired steam power plant in two different efficiency values for the combustion and generator efficiency. The examples show that with the increased combustion efficiency and generator efficiency, the required amount of coal is reduced to 23.9 ton/h from 36.6 ton/h. This leads to around 35% savings of coal.

Example 9.15 Comparison of energy sources of electricity with natural gas for heating

The efficiency of electric heater is around 73%, while it is 38% for natural gas heaters. A house heating requires 4 kW. The unit cost of electricity is \$0.1/kWh, while the natural gas costs \$0.60/therm. Estimate the rate of energy consumptions for both the electric and gas heating systems.

Solution:

Energy supplied by the electric heater: $q_{used} = q_{in} \eta_{th} = (4\ kW)\ (0.73) = 2.92\ kW$

The unit cost of energy is inversely proportional to the efficiency:

Unit cost of electric energy =

(unit cost of energy)/(efficiency) = $0.1/0.73 = **$0.137/kWh**

Therm = 29.3 kWh (Table 1.9)

Energy input to the gas heater (at the same rate of used energy that is 2.92 kW):

$q_{in,\ gas} = q_{used}/\eta_{th} = 2.92/0.38 = 3.84\ kW\ (= 13,100\ Btu/h)$

Therefore, a gas burner should have a rating of at least 13,100 Btu/h to have the same performance as the electric unit.

Unit cost of energy from the gas = cost of energy/efficiency

= [($0.60/therm)/(29.3 kWh/therm)]/0.38 = **$0.054/kWh**

Ratio of unit cost of gas to electric energy; $0.054/$0.137 = 0.39

The cost of utilized natural gas is around 39% of the electricity cost; therefore, heating with an electric heater will cost more for heating.

Example 9.16 Overall efficiency and required amount of coal in a coal-fired steam power plant

An adiabatic turbine is used to produce electricity by expanding a superheated steam at 4100 kPa and 350 °C. The power output is 50 MW. The steam leaves the turbine at 40 kPa and 100 °C. If the combustion efficiency is 0.75 and the generator efficiency is 0.9, determine the overall plant efficiency and the amount of coal supplied per hour.

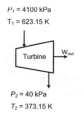
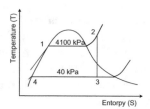

Solution:

Assume: Steady-state adiabatic operation. The changes in kinetic and potential energies are negligible.

Basis: stems flow rate = 1 kg/s

Combustion efficiency $\eta_{comb} = 0.75$, Generator efficiency $\eta_{gen} = 0.9$

Power output = 50 MW = 50,000 kW

From steam tables. Turbine inlet conditions:

$P_2 = 4100\ kPa$, $T_2 = 623.15\ K$, $H_2 = 3092.8\ kJ/kg$, $S_2 = 6.5727\ kJ/kg\ K$

Turbine outlet conditions: $P_4 = P_3 = 40\ kPa$, $T_3 = 373.15\ K$,

$V_4 = 1.027\ cm^3/g = 0.001027\ m^3/kg$

$S_{4\text{sat vap}} = 7.6709$ kJ/kg K, $S_{4\text{sat liq}} = 1.2026$ kJ/kg K at 40 kPa
$H_{4\text{sat vap}} = 2636.9$ kJ/kg, $H_{4\text{sat liq}} = 317.6$ kJ/kg at 40 kPa
$W_{p,\text{in}} = V_4(P_1 - P_4) = (0.001027$ m^3/kg$)(4100 - 40)$ kPa
$(1$ kJ/1 kPa m$^3) = 4.2$ kJ/kg
$H_1 = H_4 + W_{p,\text{in}} = 317.6 + 4.2 = 321.8$ kJ/kg
For the isentropic operation $S_3 = S_2 = 6.5727$ kJ/kg K
Since $S_2 < S_{3\text{sat vap}}$: $6.5727 < 7.6709$ the steam at the exit is saturated liquid–vapor mixture.

Quality of that mixture: $x_3 = \dfrac{S_2 - S_{2\text{sat liq}}}{S_{2\text{sat vap}} - S_{2\text{sat liq}}} = \dfrac{6.5727 - 1.2026}{7.6709 - 1.2026} = 0.83$

$H_3 = (1 - x_3)H_{3\text{sat liq}} + x_3 H_{3\text{sat vap}} = 2243.1$ kJ/kg
Heat interactions: $q_{\text{in}} = \dot{m}(H_2 - H_1) = (3092.8 - 321.8)$ kJ/kg $= 2771.0$ kJ/s
$q_{\text{out}} = -\dot{m}(H_4 - H_3) = -(317.6 - 2243.1)$ kJ/kg $= 1925.5$ kJ/s (heat received by the cooling medium that is the system in the condenser)

Thermal efficiency: $\eta_{\text{th}} = 1 - \dfrac{q_{\text{out}}}{q_{\text{in}}} = 0.305$ or 30.5%

Overall plant efficiency:
$\eta_{\text{overall}} = \eta_{\text{th}}\eta_{\text{comb}}\eta_{\text{gen}} = (0.305)(0.75)(0.9) = \textbf{0.205 or 20.5\%}$

The rate of coal energy required: $\dot{E}_{\text{coal}} = \dfrac{|\dot{W}_{\text{net}}|}{\eta_{\text{overall}}} = \dfrac{50,000\,\text{kW}}{0.205} = 243,900$ kJ/s

Energy of coal (bituminous) = 24,000 kJ/kg (Table 2.6)
$\dot{m}_{\text{coal}} = \dfrac{\dot{E}_{\text{coal}}}{\eta_{\text{overall}}} = \dfrac{243,900}{24,000}\left(\dfrac{1\,\text{ton}}{1000\,\text{kg}}\right) = 0.0101$ ton/s $= \textbf{36.6 ton/h}$

Example 9.17 Required amount of coal in a coal-fired steam power plant
An adiabatic turbine is used to produce electricity by expanding a superheated steam at 4100 kPa and 350 °C. The steam flow rate is 42 kg/s. The steam leaves the turbine at 40 kPa and 100 °C. If the combustion efficiency is 0.77 and the generator efficiency is 0.95, determine the overall plant efficiency and the amount of coal supplied per hour.

Solution:
Assume: Steady-state adiabatic operation. The changes in kinetic and potential energies are negligible.
Basis: stems flow rate = 1 kg/s
Combustion efficiency $\eta_{\text{comb}} = 0.77$
Generator efficiency $\eta_{\text{gen}} = 0.95$
Turbine inlet: $P_2 = 4100$ kPa, $T_2 = 623.15$ K, $H_2 = 3092.8$ kJ/kg,
$S_2 = 6.5727$ kJ/kg K (steam tables)
Turbine exit: $P_4 = P_3 = 40$ kPa, $T_3 = 373.15$ K,
$V_4 = 1.027$ cm^3/g $= 0.001027$ m^3/kg
$S_{4\text{sat vap}} = 7.6709$ kJ/kg K, $S_{4\text{sat liq}} = 1.2026$ kJ/kg K at 40 kPa
$H_{4\text{sat vap}} = 2636.9$ kJ/kg, $H_{4\text{sat liq}} = 317.6$ kJ/kg at 40 kPa
$W_{p,\text{in}} = V_4(P_1 - P_4) = (0.001027$ m^3/kg$)(4100 - 40)$ kPa
$(1$ kJ/1 kPa m$^3) = 4.2$ kJ/kg

$H_1 = H_4 + W_{p,in} = 317.6 + 4.2 = 321.8$ kJ/kg

For the isentropic operation $S_3 = S_2 = 6.5727$ kJ/kg K

Since $S_2 < S_{3sat\ vap}$: $6.5727 < 7.6709$ the steam at the exit is saturated liquid–vapor mixture.

Quality of the mixture: $x_3 = \dfrac{S_2 - S_{2satliq}}{S_{2satvap} - S_{2satliq}} = \dfrac{6.5727 - 1.2026}{7.6709 - 1.2026} = 0.83$

$H_3 = (1 - x_3)H_{3satliq} + x_3 H_{3satvap} = 2243.1$ kJ/kg

Heat interactions: $q_{in} = (H_2 - H_1) = (3092.8 - 321.8)$ kJ/kg $= 2771.0$ kJ/kg

$q_{out} = -(H_4 - H_3) = -(317.6 - 2243.1)$ kJ/kg $= 1925.5$ kJ/kg (heat received by the cooling medium)

Cycle work out: $W_{net} = \dot{m}(q_{out} - q_{in}) = (42$ kg/s$)(1925.5 - 2771.0)$ kJ/kg $= -35{,}511$ kW (work produced)

The thermal efficiency: $\eta_{th} = 1 - \dfrac{q_{out}}{q_{in}} = 0.305$ or 30.5%

The overall plant efficiency: $\eta_{overall} = \eta_{th}\eta_{comb}\eta_{gen} = (0.305)(0.77)(0.95) = 0.223$

The rate of coal energy supply: $\dot{E}_{coal} = \dfrac{|\dot{W}_{net}|}{\eta_{overall}} = \dfrac{35{,}511}{0.223} = 159{,}165$ kJ/s

Energy of coal (bituminous) $= 24{,}000$ kJ/kg (Table 2.6)

$\dot{m}_{coal} = \dfrac{\dot{E}_{coal}}{\eta_{overall}} = \dfrac{159{,}165}{24{,}000}\left(\dfrac{1\ ton}{1000\ kg}\right) = 0.0066$ ton/s $=$ **23.8 ton/h**

With the increased combustion efficiency and generator efficiency, the required amount of coal is reduced to 23.9 ton/h from 36.6 ton/h. This leads to around 35% savings of coal.

9.4.4 Fuel Efficiency of Vehicles

The fuel efficiency of vehicles can be expressed in miles per gallon (mpg) or liters (l) per km. Diesel engines generally achieve greater fuel efficiency than gasoline engines. Passenger car diesel engines have energy efficiency of up to 41% but more typically 30%, and petrol engines of up to 37.3%, but more typically 20%. The higher compression ratio is helpful in raising the energy efficiency, but diesel fuel also contains approximately 10% more energy per unit volume than gasoline which contributes to the reduced fuel consumption for a given power output [7, 10].

Fuel efficiency directly affects emissions causing pollution. Cars can run on a variety of fuel sources, such as gasoline, natural gas, liquefied petroleum gases, biofuel, or electricity. All these create various quantities of atmospheric pollution. A kilogram of carbon produces approximately 3.63 kg of CO_2 emissions. Typical average emissions from combustion of gasoline and diesel are [10]:

- Gasoline combustion emits 19.6 lb CO_2/US gal or (2.32 kg CO_2/l)
- Diesel combustion emits 22.5 lb CO_2/US gal or (2.66 kg CO_2/l)

These values are only the CO_2 emissions of the final forms of fuel products and do not include additional CO_2 emissions created during the drilling, pumping, transportation, and refining steps of the fuel production. Examples 9.18 and 9.19

illustrate the estimation of fuel consumption of a car and show that the conservation in fuel and reduction in emission of CO_2 are significant when the fuel efficiency of car increases.

Example 9.18 Fuel consumption of a car

The overall efficiencies are about 25–28% for gasoline car engines, 34–38% for diesel engines, and 40–60% for large power plants [2]. A car engine with a power output of 120 hp has a thermal efficiency of 24%. Determine the fuel consumption of the car if the fuel has a higher heating value of 20,400 Btu/lb.

Solution:

Assume: the car has a constant power output.

Gasoline car engine: $\eta_{th} = \dfrac{|\dot{W}_{net}|}{\dot{q}_{in}} \rightarrow \dot{q}_{in} = \dfrac{|\dot{W}_{net}|}{\eta_{th}} = \dfrac{120\,hp}{0.24}\left(\dfrac{2545\,Btu/h}{hp}\right)$

$\dot{q}_{in}$ = 1,272,500 Btu/h

Net heating value = higher heating value $(1 - 0.1)$ = 18,360 Btu/lb (approximately)

Fuel consumption = $\dot{q}_{in}$/net heating value =

1,272,500 Btu/h/18,360 Btu/lb = 69.3 lb/h

Assuming an average gasoline density of 0.75 kg/l:

ρ_{gas} = (0.75 kg/l)(2.2 lb/kg)(l/0.264 gal) = 6.25 lb/gal

Fuel consumption in terms of gallon: (69.3 lb/h)/(6.25 lb/gal) = **11.1 gal/h**

Diesel engine with an efficiency of 36%:

$\eta_{th} = \dfrac{|\dot{W}_{net}|}{\dot{q}_{in}} \rightarrow \dot{q}_{in} = \dfrac{|\dot{W}_{net}|}{\eta_{th}} = \dfrac{120\,hp}{0.36}\left(\dfrac{2545\,Btu/h}{hp}\right)$ = 848,333 Btu/h

$\dot{q}_{in}$ = **848,333 Btu/h**

9.4.5 Energy Conservation in Driving

Some possible energy conservation steps are:

- Speeding, rapid acceleration, and braking waste gas. It can lower your gas mileage by 33% at highway speeds and by 5% around town. Gas mileage usually decreases rapidly at speeds above 60 mph and observing the speed limit may lead to fuel saving of 7–23% [10].
- An extra 100 lb in your vehicle could reduce the fuel efficiency by up to 2%.
- Using cruise control helps save gas.
- Fixing a serious maintenance problem, such as a faulty oxygen sensor, can improve your mileage by as much as 40%.
- Gas mileage may be improved by up to 3.3% by keeping your tires inflated to the proper pressure.

The most efficient machines for converting energy to rotary motion are electric motors, as used in electric vehicles. However, electricity is not a primary energy source so the efficiency of the electricity production should also to be taken into account. Hydrogen cars powered either through chemical reactions in a fuel cell that creates electricity to drive electrical motors or by directly burning hydrogen in a combustion engine. These vehicles have near zero pollution from the exhaust pipe. Potentially the atmospheric pollution could be minimal, provided the hydrogen is made by electrolysis using electricity from non-polluting sources such as solar, wind, or hydroelectricity. In addition to the energy cost of the electricity or hydrogen production, transmission and/or storage losses to support large-scale use of such vehicles should also be accounted for [10].

Example 9.19 Fuel conservation with a more fuel efficient car
Assume two cars one with 11 l/100 km city and 9 l/100 km highway, and the other 6.5 l/100 km in city traffic and at 5 l/100 km motorway are used. Estimate the annual fuel saving and emission reduction achieved with the more fuel-efficient car traveling an average 15,000 km per year.
Solution:
Fuel saving for every 100 km city driving in liters = $(11 - 6.5) l = 4.5 l$
Fuel saving for every 100 km highway driving in liters = $(9 - 5) l = 4.0 l$
Average saving for 100 km is $(4.0 + 4.5)l/2 = 4.2 l$
A fuel-efficient car can conserve an average of about 4.2 l gasoline for every 100 km.
For a car driving 15,000 km per year, the amount of fuel conserved would be:
(15,000 km/year) (4.2 l/100 km) = **630 l/year (167 gal/year)**
Gasoline combustion emits 19.4 lb CO_2/US gal or (2.32 kg CO_2/l)
Annual emission reduction = (2.32 kg CO_2/l) (630 l/year) = **1462 kg/year (or 3294 lb/year)**
Both the conservation in fuel and reduction in emission of CO_2 are significant with increased fuel efficiency of cars globally.

9.4.6 Regenerative Braking

In conventional braking systems, the excess kinetic energy is converted to heat by friction in the brake pads and therefore wasted. Regenerative braking is the conversion of the vehicle's kinetic energy into chemical energy, which is stored in the battery and used in driving again in battery-powered and hybrid gas/electric vehicles. The recovery is around 60% [7, 9]. The advanced algorithms in the motor control the motor torque for both driving and regenerative braking. A torque command is derived from the position of the throttle pedal. The motor controller converts this torque command into the appropriate three-phase voltage and current waveforms to produce the commanded torque in the motor in the most efficient way. When the torque serves to slow the vehicle then energy is returned to the battery.

Negative torque applied to the rear wheels can cause a car to become unstable since regenerative braking is a source of negative torque. The traction control system limit regenerative break if the rear wheels start to slip. The control system provides the driver instant positive and negative torque command. Regenerative braking is limited when the batteries are fully charged, and motor controller will limit regenerative torque in this case. It is possible that with the increased use of battery-powered vehicles, a safe regenerative braking will be an efficient way of converting and recovering the kinetic energy [7].

9.4.7 Energy Conservation in Lighting

Energy for lighting accounts for about 10% of the household electric bill. Compact fluorescent bulbs use about 75% less energy than standard lighting, produces 75% less heat, and lasts up to 10 times longer. Compact fluorescent bulbs contain a very small amount of mercury sealed within the glass tubing. Although linear fluorescent, compact fluorescent bulbs, and light emitting diodes cost a bit more than incandescent bulbs initially, over their lifetime they are cheaper because of the less electricity they use [4, 8, 9]. Example 9.20 illustrates the energy conservation by using the compact fluorescent bulbs.

Example 9.20 Conservation of energy with compact fluorescent bulbs
Assume that an average residential rate of electricity is \$0.14/kWh and a household consumes about 10,000 kWh per year. If the lighting is provided by compact fluorescent bulbs only, estimate the conservation of energy and saving per year.

Solution:
Assume that 11% of the energy budget of household is for lighting, and compact fluorescent bulbs consume about 75% less energy than standard lighting.
Annual energy use and cost with standard lighting:
(10,000 kWh/year) (0.11) = 1100 kW/year
(1100 kW/year)(\$0.14/kWh) = \$154/year

Annual energy use and cost with compact fluorescent bulbs:
(1100 kW/year) (1 − 0.75) = 275 kWh/year
(275 kWh/year) (\$0.14/kWh) = \$38.5/year
Annual energy and cost savings:
(1100 − 275) kWh = **825 kWh**
(154 − 38.5)\$/year = **\$115 \$/year**
Conservation of energy and cost with compact fluorescent bulbs are considerable.

9.5 Energy Conservation in Electricity Distribution and Smart Grid

A *smart grid* is a form of electricity network using digital technology. Smart grid delivers electricity to consumers to control appliances at homes, optimize power flows, reduce waste, and maximize the use of lowest-cost power production resources. The smart grid is envisioned to overlay the ordinary electrical grid with an information and net metering system that includes smart meters. The increased data transmission capacity has made it possible to apply sensing, measurement and control devices with two-way communications to electricity production, transmission, distribution, and consumption parts of the power grid. These devices could communicate information about grid condition to system users, operators, and automated devices. Therefore, the average consumer can respond dynamically to changes in grid condition [8, 9].

A smart grid includes an intelligent monitoring system that keeps track of all electricity flowing in the system. It also has the capability of integrating renewable electricity such as solar and wind. When power is least expensive the user can allow the smart grid to turn on selected home appliances such as washing machines or some processes in a factory. At peak times, it could turn off selected appliances to reduce demand [2, 4]. Some smart grid functions are:

- Motivate consumers to actively participate in operations of the grid.
- Provide higher quality power that will save energy wasted from outages.
- Accommodate all generation and storage options.
- Enable electricity markets to flourish by running more efficiently.
- Enable higher penetration of intermittent power generation sources.

9.5.1 Standby Power

Standby power refers to the electricity consumed by many appliances when they are switched off or in standby mode. The typical power loss per appliance is low (from 1 to 25 W) but when multiplied by the billions of appliances in houses and in commercial buildings, standby losses represent a significant fraction of total world electricity use. Standby power may account for consumption between 7 and 13% of household power-consumption. Technical solutions to the problem of standby power exist in the form of a new generation of power transformers that use only 100 mW in standby mode and thus can reduce standby consumption by up to 90%. Another solution is the 'smart' electronic switch that cuts power when there is no load and restores it immediately when required [7, 9].

9.6   Energy Harvesting

Despite the high initial costs, energy harvesting is often more effective in cost and resource utilization when done on a group of houses, co-housing, local district, or village rather than on an individual basis. This leads to the reduction of electricity transmission and distribution losses. The net zero fossil energy consumption requires locations of geothermal, micro hydro, solar, and wind resources to sustain the concept. One of the key areas of debate in zero energy building design is over the balance between energy conservation and the distributed point-of-use harvesting of renewable energy such as solar and wind energy. Wide acceptance of zero or close to zero energy building technology may require more government incentives, building code regulations, or significant increases in the cost of conventional energy [2, 4, 5]. Some factors of energy harvesting are:

- Isolation for building owners from future energy price increases.
- Increased comfort due to more-uniform interior temperatures.
- Reduced total cost of ownership due to improved energy efficiency.

9.7   Energy Conservation and Exergy

Energy conservation through exergy concepts for some steady-state flow processes are [2]:

- Exergy is lost by irreversibilities associated with pressure drops, fluid friction, and stream-to-stream heat transfer due to temperature differences.
- In a steam power plant, exergy transfers are due to work and heat, and exergy is lost within the control volume.
- In a waste-heat recovery system, we might reduce the heat transfer irreversibility by designing a heat recovery steam generator with a smaller stream-to-stream temperature difference, and/or reduce friction by designing a turbine with a higher efficiency.
- A cost-effective design may result from a consideration of the trade-offs between possible reduction of exergy loss and potential increase in operating cost.

9.8   Energy Recovery on Utilities Using Pinch Analysis

A process may have available hot and cold streams. When the hot and cold streams exchange heat between them, heat is recovered and the hot and cold utility requirements may be minimized. Pinch analysis yields optimum energy integration of process heat and utilities by matching the hot and cold streams with a network of heat exchangers. In the pinch analysis, hot and cold streams can only exchange

Fig. 9.3 Optimum ΔT_{min} from energy cost and capital cost changes

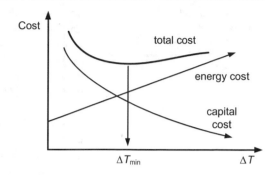

energy up to a minimum allowable temperature difference ΔT_{min}, which is called the pinch point leading to the minimum driving force for heat transfer [2, 3].

An increase in ΔT_{min} leads to higher energy and lower capital costs (a smaller heat exchanger area as seen in Fig. 9.3). For example, an increase of 5 °C from a value of $\Delta T_{min} = 10$ °C decreases the heat exchanger area by 11%, and increases the required minimum energy by about 9%. To find the optimum value of ΔT_{min}, the total annual cost is plotted against. An optimum ΔT_{min} appears at the minimum total annual cost of energy and capital. The optimum value for ΔT_{min} is generally in the range of 3–40 °C and needs to be established for a process.

9.8.1 Composite Curves

Temperature–enthalpy diagrams called *composite curves* represent the thermal characteristics of hot and cold streams and the amount of heat transferred between them (see Fig. 9.4). The enthalpy change rate for each stream is

$$q = \Delta H = \dot{m}C_p\Delta T = MC\Delta T \tag{9.14}$$

where ΔH is the enthalpy change rate, $\dot{m}$ is the mass flow rate, C_p is the heat capacity, ΔT is the temperature change in a stream, and MC the heat capacity rate $\dot{m}C_p$. The enthalpy change rates are added over each temperature interval that includes one or more of the streams. This leads to hot and cold composite curves of the streams shown in Fig. 9.4. If $\dot{m}C_p$ is constant, q versus T would be a straight line

$$\Delta T = \frac{1}{\dot{m}C_p}q \tag{9.15}$$

Enthalpy changes rather than absolute enthalpies are estimated, and the horizontal location of a composite line on the diagram is arbitrarily fixed. One of the two curves is moved horizontally until the distance of the closest vertical approach matches the selected value of ΔT_{min}. The pinch point is the location of ΔT_{min} on the

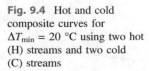

Fig. 9.4 Hot and cold composite curves for $\Delta T_{min} = 20\ °C$ using two hot (H) streams and two cold (C) streams

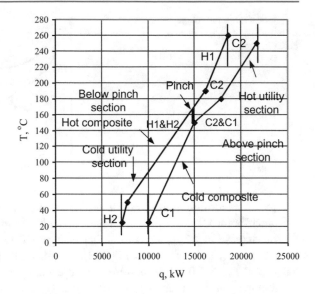

adjusted composite diagram where the hot and cold curves most closely approach to each other in temperature. The overshoot of the hot composite curve represents the minimum cold utility ($q_{c,min}$) required, and the overshoot of the cold composite curve represents the minimum hot utility ($q_{h,min}$) required (see Figs. 9.4 and 9.5). Example 9.21 illustrates the use of pinch analysis in process heat integration and minimizing the hot and cold utilities.

Above the pinch, only the hot utility is required, while only the cold utility is required below the pinch. These diagrams enable engineers to minimize the expensive utilities. Pinch analysis may also lead to optimum integration of

Fig. 9.5 Principle of pinch technology [3]

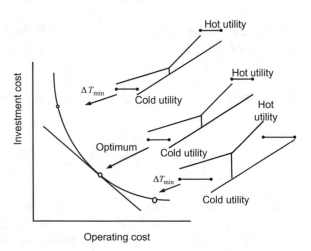

evaporators, condensers, furnaces, and heat pumps by reducing the utility requirements. Pinch analysis is utilized widely in industry leading to considerable energy savings. Figure 9.5 displays the importance of minimum temperature approach between the minimum hot and cold utilities. As the values of ΔT_{min} increase, the amount of hot and cold utilities increases together with the operating cost. On the other hand, as the values of ΔT_{min} decreases, the investment cost increases. So, these two opposing costs should be optimized, as seen in Fig. 9.3.

Example 9.21 Energy conservation by the pinch analysis

In a process available hot and cold process streams and their heat capacities are shown below.

Hot and Cold Process Stream Conditions

Streams		T_{in} (°C)	T_{out} (°C)	$C = \dot{m}C_p$ (kW/°C)	$q = \dot{m}C_p \Delta T$ (kW)
C1	Cold 1	25	180	40	6200
C2	Cold 2	150	250	55	5500
H1	Hot 1	260	50	35	−7350
H2	Hot 2	190	25	25	−4125

Construct the balanced composite curves for the process with $\Delta T_{min} = 20$ °C and $\Delta T_{min} = 10$ °C, and compare the amounts of hot and cold utilities needed.

Solution:

Assume that heat capacities of hot and cold streams are constant.

A starting enthalpy change rate is chosen as 10,000 kW at 25 °C for the streams to be heated, while for the streams to be cooled, the base value chosen is 15,000 kW at 260 °C. Initial temperature interval with overlaps is shown as follows:

Initial temperature interval			C (kW/°C)	q (kW)	Initial enthalpy selection	
Streams	Temperature interval (°C)				T (°C)	q (kW)
					25	10,000 (arbitrary)
C1	25	150	40	5000	150	15,000
C1 and C2	150	180	95	2850	180	17,850
C2	180	250	55	3850	250	21,700
				11,700		
					260	15,000 (arbitrary)
H1	260	190	35	−2450	190	12,550
H2 and H1	190	50	60	−8400	50	4150
H2	50	25	25	−625	25	3525
				−11,475		

This table may be converted to the hot and cold composite curves as below

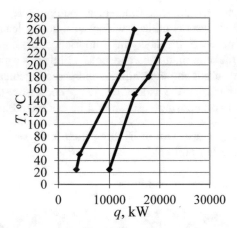

Here the composite curves move to each other so that at the pinch point minimum temperature difference would be ΔT_{min} = 20 °C, an the diagram is called a *balanced composite diagram*.

Revised temperature interval		C (kW/°C)	q (kW)	Revised enthalpy selection	
Stream	Required temperature interval (°C)			q (kW) 10,000	T (°C) 25
C1	25	150	40	5000	15,000	150
C1 and C2	150	180	95	2850	17,850	180
C2	180	250	55	3850	21,700	250
				11,700		
					18,650	260
H1	260	190	35	−2450	16,200	190
H2 and 1	190	50	60	−8400	7800	50
H2	50	25	25	−625	7175	25
				−11,475		

Composite diagram with ΔT_{min} = 20 °C and ΔT_{min} = 10 °C approach temperatures are shown as follows:

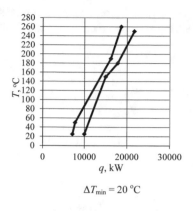

$\Delta T_{min} = 20\ ^\circ$C

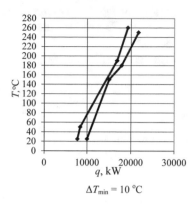

$\Delta T_{min} = 10\ ^\circ$C

For approach temperatures of $\Delta T_{min} = 10$ and 20 °C, the estimated minimum hot and cold utilities are:

Approach temperatures: ΔT_{min} (°C)	10	20
Hot utility (i.e., steam): $q_{hot,min}$ (kW)	2450	3050
Cold utility (i.e., cooling water): $q_{cold,min}$ (kW)	2225	2825

Energy savings = (3050 − 2450) kW = 600 kW
Energy savings = (2825 − 2225) kW = 600 kW
This simple analysis shows that the smaller approach temperature reduces the utilities needed considerably.

Summary

- *Energy conservation* mainly refers to reducing energy consumption and increasing efficiency in energy usage, leading to increased security, financial gain, and environmental protection. *Energy conservation in power production* underlines the importance of energy conservation of heat engines, such as Rankine cycle and Brayton cycle.
- Under the cold-air-standard temperature assumptions, *thermal efficiency of an ideal Brayton cycle* with regeneration depends on the ratio of minimum to maximum temperatures and the pressure ratio:

$$\eta_{th.regen.} = 1 - \left(\frac{T_1}{T_3}\right)(r_p)^{(\gamma-1)/\gamma}$$

where r_p is the compression ratio (P_2/P_1) and $\gamma = C_p/C_v$.

- Increasing the *efficiency of a Rankine cycle* is possible by reducing the condenser pressure, by increasing the boiler pressure, and by increasing the boiler temperature.
- Electric motor efficiency is

$$\eta_{Motor} = \frac{\text{Mechanical power}}{\text{Electrical power}} = \frac{\dot{W}_{comp}}{\dot{W}_{elect}}$$

- Motor efficiency range: $0.7 < \eta_{motor} < 96$. High efficiency motors save energy, which is estimated by

$$\dot{W}_{elect.saved} = (\text{Rated power})(\text{Load factor})\left(\frac{1}{\eta_{std}} - \frac{1}{\eta_{efficient}}\right)$$

where the *rated power* is the nominal power delivered at full load of the motor and listed on its label. *Load factor* is the fraction of the rated power at which the motor normally operates. Annual saving is estimated by

$$\text{Annual energy saving} = \left(\dot{W}_{elect.saved}\right)(\text{Annual operation hours})$$

- A compressor that operates at partial load causes the motor to operate less efficiently. The efficiency of motor will increase with the load. Using the cold outside air for compressor intake lowers the compressor work and conserves energy.
- *Space heating and cooling* as well as water heating at home accounts around 55% of the utility bill. To conserve energy, one should identify from where the home is losing energy, assign priorities, and form an efficiency plan that improves efficiency and reduces costs.
- *Water heating* typically accounts for about 12% of your utility bill. Insulate the electric, natural gas, or oil hot-water storage tank.
- *Residential furnaces* have a heat input rate of less than 225,000 Btu/h (66,000 W) and residential boilers have a heat input rate of less than 300,000 Btu/h (88,000 W). The residential furnace is an appliance that provides heated air with a blower to circulate air through the duct distribution system.
- *An all-electric furnace or boiler* has no flue loss through a chimney. Electric resistance heating converts nearly 100% of the energy in the electricity to heat. However, most electricity is produced from oil, gas, or coal by converting only about 30% of the fuel's energy into electricity. Because of production and transmission losses, electric heat is often more expensive than heat produced using combustion appliances, such as natural gas, propane, and oil furnaces.
- *Heat pumps* are preferable in most climates, as they easily cut electricity use by 50% when compared with electric resistance heating. When operated in heating mode, a heat pump is more efficient than operating resistance heaters. Because an electric heater can convert only the input electrical energy directly to output

heat energy with none of the efficiency or conversion advantages of a heat pump. Likewise, when a heat pump operates near its most inefficient outside temperature, typically 0 °F, the heat pump will perform close to the same as a resistance heater.

- *Active or passive solar systems* can be used for residential heating. There are two basic types of active solar heating systems using either liquid or air heated in the solar collectors. The solar units are environmentally friendly and can now be installed on your roof to blend with the architecture of a house. Active solar heating systems may reduce the cost more when they are used for most of the year. Passive solar heaters do not have fans or blowers. In passive solar building design, windows, walls, and floors are made to collect, store, and distribute solar energy in the form of heat in the winter and reject solar heat in the summer.
- *Thermal efficiency of residential furnaces* and boilers is measured by annual fuel utilization efficiency (AFUE) . Annual fuel utilization efficiency is the ratio of heat output of the furnace or boiler compared to the total energy consumed by them over a typical year. Annual fuel utilization efficiency does not account the circulating air and combustion fan power consumptions and the heat losses of the distributing systems of duct or piping. Some of the minimum allowed AFUE ratings in the United States are:

  - Noncondensing fossil-fueled, warm-air furnace is 78%.
  - Fossil-fueled boiler is 80%.
  - Gas-fueled steam boiler is 75%.

- *The Energy Efficiency Ratio (EER)* of a particular cooling device is the ratio of *output* cooling (in Btu/h) to *input* electrical power (in Watts) at a given operating point. The efficiency of air conditioners is often rated by the *Seasonal Energy Efficiency Ratio* (SEER). The SEER rating of a unit is the cooling output in Btu during a typical cooling-season divided by the total electric energy input in watt-hours during the same period. The coefficient of performance (COP) is an instantaneous measure (i.e. a measure of power divided by power), whereas both EER and SEER are averaged over a duration of time.
- Air conditioner sizes are often given as 'tons' of cooling where 1 ton of cooling is being equivalent to 12,000 Btu/h (3500 W).
- The refrigeration process with the maximum possible efficiency is the Carnot cycle. The coefficient of performance (COP) of an air conditioner using the Carnot cycle is:

$$\text{COP}_{\text{Carnot}} = \frac{T_C}{T_H - T_C}$$

where T_C is the indoor temperature and T_H is the outdoor temperature in K or R. The EER is calculated by multiplying the COP by 3.413, which is the conversion factor from Btu/h to Watts:

$$EER_{Carnot} = 3.413(COP_{Carnot})$$

- The maximum SEER can be calculated by averaging the maximum values of EER over the range of expected temperatures for the season.
- *Fuel efficiency* means the efficiency of a process that converts chemical energy contained in a fuel into kinetic energy or work.
- A *smart grid* is a form of electricity network using digital technology. Smart grid delivers electricity to consumers to control appliances at homes, optimize power flows, reduce waste, and maximize the use of lowest-cost production resources. The smart grid is envisioned to overlay the ordinary electrical grid with an information and net metering system that includes smart meters.
- In a steam power plant, exergy transfers are due to work and heat, and exergy is lost within the control volume.
- In a waste-heat recovery system, we might reduce the heat transfer irreversibility by designing a heat recovery steam generator with a smaller stream-to-stream temperature difference, and/or reduce friction by designing a turbine with a higher efficiency.
- A cost-effective design may result from a consideration of the trade-offs between possible reduction of exergy loss and potential increase in operating cost.
- *Pinch analysis* yields optimum energy integration of process heat and utilities by matching the hot and cold streams with a network of heat exchangers.

Problems

9.1. A power plant is operating on an ideal Brayton cycle with a pressure ratio of $r_p = 9$. The fresh air temperature at the compressor inlet is 295 K. The air temperature at the inlet of the turbine is 1300 K. The cycle operates with a compressor efficiency of 80% and a turbine efficiency of 80%. The flow rate gas is 3 kg/s. The cycle operates 360 days per year.

(a) Using the standard-air assumptions, determine the thermal efficiency of the cycle.

(b) If the power plant operates with a regenerator with an effectiveness of 0.80, determine the thermal efficiency of the cycle.

9.2. A power plant is operating on an ideal Brayton cycle with a pressure ratio of $r_p = 9$. The fresh air temperature at the compressor inlet is 300 K. The air temperature at the inlet of the turbine is 1300 K. The cycle operates with a compressor efficiency of 85% and a turbine efficiency of 85%. The flow rate gas is 5 kg/s. The cycle operates 360 days per year.

(a) Using the standard-air assumptions, determine the thermal efficiency of the cycle.

(b) If the power plant operates with a regenerator with an effectiveness of 0.80, determine the thermal efficiency of the cycle and annual conservation of fuel.

9.3. A power plant is operating on an ideal Brayton cycle with a pressure ratio of $r_p = 8$. The fresh air temperature at the compressor inlet is 300 K. The air temperature at the inlet of the turbine is 1350 K. The cycle operates with a compressor efficiency of 85% and a turbine efficiency of 80%. The flow rate gas is 5 kg/s. The cycle operates 360 days per year.

(a) Using the standard-air assumptions, determine the thermal efficiency of the cycle.

(b) If the power plant operates with a regenerator with an effectiveness of 0.7, determine the thermal efficiency of the cycle and annual conservation of fuel.

9.4. A power plant is operating on an ideal Brayton cycle with a pressure ratio of $r_p = 15$. The fresh air temperature at the compressor inlet is 290 K. The air temperature at the inlet of the turbine is 1400 K. The cycle operates with a compressor efficiency of 90% and a turbine efficiency of 80%. The flow rate gas is 4.5 kg/s. The cycle operates 360 days per year.

(a) Using the standard-air assumptions, determine the thermal efficiency of the cycle.

(b) If the power plant operates with a regenerator with an effectiveness of 0.75, determine the thermal efficiency of the cycle and annual conservation of fuel.

9.5. A power plant is operating on an ideal Brayton cycle with a pressure ratio of $r_p = 14$. The fresh air temperature at the compressor inlet is 290 K. The air temperature at the inlet of the turbine is 1400 K. The cycle operates with a compressor efficiency of 90% and a turbine efficiency of 80%. The flow rate gas is 4.5 kg/s. The cycle operates 360 days per year.

(a) Using the standard-air assumptions, determine the thermal efficiency of the cycle.

(b) If the power plant operates with a regenerator with an effectiveness of 0.75, determine the thermal efficiency of the cycle and annual conservation of fuel.

9.6. A power plant is operating on an ideal Brayton cycle with a pressure ratio of $r_p = 7.5$. The fresh air temperature at the compressor inlet is 290 K. The air temperature at the inlet of the turbine is 1250 K. The cycle operates with a compressor efficiency of 90% and a turbine efficiency of 85%. The flow rate gas is 6.5 kg/s. The cycle operates 360 days per year.

(a) Using the standard-air assumptions, determine the thermal efficiency of the cycle.
(b) If the power plant operates with a regenerator with an effectiveness of 0.75, determine the thermal efficiency of the cycle and annual conservation of fuel.

9.7. A power plant is operating on an ideal Brayton cycle with a pressure ratio of $r_p = 11.0$. The fresh air temperature at the compressor inlet is 290 K. The air temperature at the inlet of the turbine is 1350 K. The cycle operates with a compressor efficiency of 85% and a turbine efficiency of 80%. The flow rate gas is 6.0 kg/s. The cycle operates 360 days per year.

(a) Using the standard-air assumptions, determine the thermal efficiency of the cycle.
(b) If the power plant operates with a regenerator with an effectiveness of 0.76, determine the thermal efficiency of the cycle and annual conservation of fuel.

9.8. A power plant is operating on an ideal Brayton cycle with a pressure ratio of $r_p = 10.0$. The fresh air temperature at the compressor inlet is 290 K. The air temperature at the inlet of the turbine is 1350 K. The cycle operates with a compressor efficiency of 85% and a turbine efficiency of 80%. The flow rate gas is 6.0 kg/s. The cycle operates 360 days per year.

(a) Using the standard-air assumptions, determine the thermal efficiency of the cycle.
(b) If the power plant operates with a regenerator with an effectiveness of 0.76, determine the thermal efficiency of the cycle and annual conservation of fuel.

9.9. A steam power plant is operating on the simple ideal Rankine cycle. The steam mass flow rate is 20 kg/s. The steam enters the turbine at 3500 kPa and 400 °C. Discharge pressure of the steam from the turbine is 78.5 kPa.

(a) Determine the thermal efficiency of the cycle.
(b) If the pressure of the discharge steam is reduced to 15 kPa determine the thermal efficiency.

9.10. A steam power plant is operating on the simple ideal Rankine cycle. The steam mass flow rate is 35 kg/s. The steam enters the turbine at 4000 kPa and 350 °C. Discharge pressure of the steam from the turbine is 101.3 kPa.

(a) Determine the thermal efficiency of the cycle.

(b) If the pressure of the discharge steam is reduced to 15 kPa determine the thermal efficiency.

9.11. A steam power plant is operating on the simple ideal Rankine cycle. The steam mass flow rate is 30 kg/s. The steam enters the turbine at 3000 kPa and 400 °C. Discharge pressure of the steam from the turbine is 67.5 kPa.

(a) Determine the thermal efficiency of the cycle.

(b) If the pressure of the discharge steam is reduced to 15 kPa determine the thermal efficiency.

9.12. A steam power plant is operating on the simple ideal Rankine cycle. The steam mass flow rate is 30 kg/s. The steam enters the turbine at 2500 kPa and 300 °C. Discharge pressure of the steam from the turbine is 62.5 kPa.

(a) Determine the thermal efficiency of the cycle.

(b) If the pressure of the discharge steam is reduced to 15 kPa determine the thermal efficiency.

9.13. A steam power plant is operating on the simple ideal Rankine cycle. The steam mass flow rate is 35 kg/s. The steam enters the turbine at 3000 kPa and 350 °C. Discharge pressure of the steam from the turbine is 78.5 kPa.

(a) Determine the thermal efficiency of the cycle.

(b) If the pressure of the boiler is increased to 10,000 kPa while maintaining the turbine inlet temperature at 350 °C, determine the thermal efficiency.

9.14. A steam power plant is operating on the simple ideal Rankine cycle. The steam mass flow rate is 42 kg/s. The steam enters the turbine at 2500 kPa and 300 °C. Discharge pressure of the steam from the turbine is 15.0 kPa.

(a) Determine the thermal efficiency of the cycle.

(b) If the pressure of the boiler is increased to 9000 kPa while maintaining the turbine inlet temperature at 300 °C, determine the thermal efficiency.

9.15. A steam power plant is operating on the simple ideal Rankine cycle. The steam mass flow rate is 50 kg/s. The steam enters the turbine at 4000 kPa and 400 °C. Discharge pressure of the steam from the turbine is 15.0 kPa.

(a) Determine the thermal efficiency of the cycle.

(b) If the pressure of the boiler is increased to 11,000 kPa while maintaining the turbine inlet temperature at 400 °C, determine the thermal efficiency.

9.16. A steam power plant is operating on the simple ideal Rankine cycle. The steam mass flow rate is 35 kg/s. The steam enters the turbine at 3500 kPa and 300 °C. Discharge pressure of the steam from the turbine is 78.5 kPa.

(a) Determine the thermal efficiency of the cycle.

(b) If the temperature of the boiler is increased to 500 °C while maintaining the pressure at 3500 kPa, determine the thermal efficiency.

9.17. A steam power plant is operating on the simple ideal Rankine cycle. The steam mass flow rate is 40 kg/s. The steam enters the turbine at 3000 kPa and 300 °C. Discharge pressure of the steam from the turbine is 15 kPa.

(a) Determine the thermal efficiency of the cycle.

(b) If the temperature of the boiler is increased to 500 °C while maintaining the pressure at 3000 kPa, determine the thermal efficiency.

9.18. A steam power plant is operating on the simple ideal Rankine cycle. The steam mass flow rate is 40 kg/s. The steam enters the turbine at 6000 kPa and 350 °C. Discharge pressure of the steam from the turbine is 15 kPa.

(a) Determine the thermal efficiency of the cycle.

(b) If the temperature of the boiler is increased to 500 °C while maintaining the pressure at 6000 kPa, determine the thermal efficiency.

9.19. A steam power plant is operating on the simple ideal Rankine cycle. The steam mass flow rate is 40 kg/s. The steam enters the turbine at 9000 kPa and 400 °C. Discharge pressure of the steam from the turbine is 15 kPa.

(a) Determine the thermal efficiency of the cycle.

(b) If the temperature of the boiler is increased to 550 °C while maintaining the pressure at 9000 kPa, determine the thermal efficiency.

9.20. Estimate the maximum possible efficiency for parts (a) and (b) in Problem 9.18 and compare them with those obtained in parts (a) and (b) in Problem 9.18.

9.21. Estimate the maximum possible efficiency for parts (a) and (b) in Problem 9.19 and compare them with those obtained in parts (a) and (b) in Problem 9.19.

9.22. A steam power plant is operating on the simple ideal Rankine cycle. The steam mass flow rate is 20 kg/s. The steam enters the turbine at 3000 kPa and 400 °C. Discharge pressure of the steam from the turbine is 78.5 kPa. If the pressure of the boiler is increased to 9000 kPa while maintaining the turbine inlet temperature at 400 °C, determine the thermal efficiency.

9.23. A steam power plant is operating on the simple ideal Rankine cycle. The steam mass flow rate is 20 kg/s. The steam enters the turbine at 4000 kPa and 400 °C. Discharge pressure of the steam from the turbine is 15 kPa. If the pressure of the boiler is increased to 9500 kPa while maintaining the turbine inlet temperature at 400 °C, determine the thermal efficiency.

9.24. A steam power plant is operating on the simple ideal Rankine cycle. The steam mass flow rate is 20 kg/s. The steam enters the turbine at 5000 kPa and 350 °C. Discharge pressure of the steam from the turbine is 15 kPa. If the pressure of the boiler is increased to 10,000 kPa while maintaining the turbine inlet temperature at 350 °C, determine the thermal efficiency.

9.25. A steam power plant is operating on the simple ideal Rankine cycle. The steam mass flow rate is 20 kg/s. The steam enters the turbine at 9800 kPa and 350 °C. Discharge pressure of the steam from the turbine is 78.5 kPa. If the temperature of the boiler is increased to 600 °C while maintaining the pressure at 9800 kPa, determine the thermal efficiency.

9.26. A steam power plant is operating on the simple ideal Rankine cycle. The steam mass flow rate is 20 kg/s. The steam enters the turbine at 8000 kPa and 325 °C. Discharge pressure of the steam from the turbine is 78.5 kPa. If the temperature of the boiler is increased to 550 °C while maintaining the pressure at 8000 kPa, determine the thermal efficiency.

9.27. Estimate the maximum possible efficiency for parts (a) and (b) in Problem 9.25 and compare them with those obtained in parts (a) and (b) in Problem 9.25.

9.28. Estimate the maximum possible efficiency for parts (a) and (b) in Problem 9.26 and compare them with those obtained in parts (a) and (b) in Problem 9.26.

9.29. Air with a flow rate of 4 kg/s is compressed in a steady state and reversible process from an inlet state of 100 kPa and 300 K to an exit pressure of 1000 kPa. Estimate the work for (a) polytropic compression with $\gamma = 1.3$, and (b) ideal two-stage polytropic compression with intercooling using the same polytropic exponent of $\gamma = 1.3$, (c) estimate conserved compression work by intercooling if the compressor is operated 360 days per year.

9.30. Air is compressed in a steady state and reversible process from an inlet state of 100 kPa and 285 K to an exit pressure of 800 kPa. The mass flow rate of air is 8 kg/s. Estimate the work for (a) polytropic compression with $\gamma = 1.35$, and (b) ideal two-stage polytropic compression with intercooling using the same polytropic exponent of $\gamma = 1.35$, (c) estimate conserved compression work by intercooling and electricity per year if the compressor is operated 360 days per year.

9.31. Air is compressed in a steady state and reversible process from an inlet state of 110 kPa and 290 K to an exit pressure of 900 kPa. The mass flow rate of air is 10 kg/s. Estimate the work for (a) polytropic compression with

$\gamma = 1.3$, and (b) ideal two-stage polytropic compression with intercooling using the same polytropic exponent of $\gamma = 1.3$, (c) estimate conserved compression work by intercooling and electricity per year if the compressor is operated 360 days per year.

9.32. Air is compressed in a steady state and reversible process from an inlet state of 110 kPa and 290 K to an exit pressure of 900 kPa. The mass flow rate of air is 10 kg/s. Estimate the work for (a) polytropic compression with $\gamma = 1.2$, and (b) ideal two-stage polytropic compression with intercooling using the same polytropic exponent of $\gamma = 1.2$, (c) estimate conserved compression work by intercooling and electricity per year if the compressor is operated 360 days per year.

9.33. Air is compressed in a steady state and reversible process from an inlet state of 100 kPa and 290 K to an exit pressure of 900 kPa. The mass flow rate of air is 5 kg/s. Estimate the work for (a) polytropic compression with $\gamma = 1.25$, and (b) ideal two-stage polytropic compression with intercooling using the same polytropic exponent of $\gamma = 1.25$, (c) estimate conserved compression work by intercooling and electricity per year if the compressor is operated 360 days per year.

9.34. Air is compressed in a steady state and reversible process from an inlet state of 100 kPa and 290 K to an exit pressure of 900 kPa. The mass flow rate of air is 3.5 kg/s. Estimate the work for (a) polytropic compression with $\gamma = 1.3$, and (b) ideal two-stage polytropic compression with intercooling using the same polytropic exponent of $\gamma = 1.3$, (c) estimate conserved compression work by intercooling and electricity per year if the compressor is operated 360 days per year.

9.35. Natural gas contains mostly the methane gas. In a steady state and reversible process, natural gas is compressed from an inlet state of 100 kPa and 290 K to an exit pressure of 1000 kPa. The mass flow rate of natural gas is 8 kg/s. Estimate the work for (a) polytropic compression with $\gamma = 1.3$, and (b) ideal two-stage polytropic compression with intercooling using the same polytropic exponent of $\gamma = 1.3$, (c) estimate conserved compression work by intercooling and electricity per year if the compressor is operated 350 days per year.

9.36. Natural gas contains mostly the methane gas. In a steady state and reversible process, natural gas is compressed from an inlet state of 100 kPa and 290 K to an exit pressure of 900 kPa. The mass flow rate of natural gas is 5 kg/s. Estimate the work for (a) polytropic compression with $\gamma = 1.2$, and (b) ideal two-stage polytropic compression with intercooling using the same polytropic exponent of $\gamma = 1.2$, (c) estimate conserved compression work by intercooling and electricity per year if the compressor is operated 360 days per year.

9.37. In a steady state and reversible process, propane gas is compressed from an inlet state of 100 kPa and 300 K to an exit pressure of 900 kPa. The mass flow rate of propane is 3 kg/s. Estimate the work for (a) polytropic compression with $\gamma = 1.3$, and (b) ideal two-stage polytropic compression with intercooling using the same polytropic exponent of $\gamma = 1.3$, (c) estimate conserved work by intercooling and electricity per year if the compressor is operated 340 days per year.

9.38. In a steady state and reversible process, hydrogen gas is compressed from an inlet state of 100 kPa and 300 K to an exit pressure of 1100 kPa. The mass flow rate of hydrogen is 3 kg/s. Estimate the work for (a) polytropic compression with $\gamma = 1.3$, and (b) ideal two-stage polytropic compression with intercooling using the same polytropic exponent of $\gamma = 1.3$, (c) estimate conserved work by intercooling and electricity per year if the compressor is operated 350 days per year.

9.39. In a steady state and reversible process, carbon dioxide gas is compressed from an inlet state of 100 kPa and 290 K to an exit pressure of 1000 kPa. The mass flow rate of carbon dioxide is 4 kg/s. Estimate the work for (a) polytropic compression with $\gamma = 1.3$, and (b) ideal two-stage polytropic compression with intercooling using the same polytropic exponent of $\gamma = 1.3$, (c) estimate conserved compression work by intercooling and electricity per year if the compressor is operated 360 days per year.

9.40. An adiabatic compressor is used to compress air from 100 kPa and 290 K to 900 kPa at a steady-state operation. The isentropic efficiency of the compressor is 80%. The air flow rate is 0.55 kg/s. Determine the minimum and actual power needed by the compressor.

9.41. An adiabatic compressor is used to compress air from 100 kPa and 290 K to 1100 kPa at a steady-state operation. The isentropic efficiency of the compressor is 80%. The air flow rate is 0.35 kg/s. Determine the minimum and actual power needed by the compressor.

9.42. An adiabatic compressor is used to compress air from 100 kPa and 290 K to 1400 kPa at a steady-state operation. The isentropic efficiency of the compressor is 85%. The air flow rate is 0.5 kg/s. Determine the minimum and actual power needed by the compressor.

9.43. An adiabatic compressor is used to compress air from 100 kPa and 290 K to 1600 kPa at a steady-state operation. The isentropic efficiency of the compressor is 83%. The air flow rate is 0.4 kg/s. Determine the minimum and actual power needed by the compressor.

9.44. Estimate the power conservation when an electric motor with an efficiency of 78% is replaced with another motor operating at 88% efficiency. Both the motors drive compressor and must deliver a power of 24 kW for an average 2500 h/year.

9.45. Estimate the power conservation when an electric motor with an efficiency of 74% is replaced with another motor operating at 89% efficiency. Both the motors drive compressor and must deliver a power of 36 kW for an average 8000 h/year.

9.46. Estimate the power conservation when an electric motor with an efficiency of 74% is replaced with another motor operating at 89% efficiency. Both the motors drive compressor and must deliver a power of 18 kW for an average 80 h/day.

9.47. A cryogenic manufacturing plant handles liquid methane at 115 K and 5000 kPa at a rate of 0.15 m³/s. In the plant a throttling valve reduces the pressure of liquid methane to 2000 kPa. A new process considered replaces the throttling valve with a turbine in order to produce power while reducing the pressure. Using the data for the properties of liquid methane below estimate (a) the power that can be produced by the turbine, and (b) the savings in electricity usage per year if the turbine operates 360 days per year.

T (K)	P (kPa)	H (kJ/kg)	S (kJ/kg K)	C_p (kJ/kg K)	ρ (kg/m³)
110	1000	209.0	4.875	3.471	425.8
110	2000	210.5	4.867	3.460	426.6
110	5000	215.0	4.844	3.432	429.1
120	1000	244.1	5.180	3.543	411.0
120	2000	245.4	5.171	3.528	412.0
120	5000	249.6	5.145	3.486	415.2

Source [1]

9.48. A cryogenic manufacturing plant handles liquid methane at 115 K and 5000 kPa at a rate of 0.2 m³/s. In the plant a throttling valve reduces the pressure of liquid methane to 2000 kPa. A new process considered replaces the throttling valve with a turbine in order to produce power while reducing the pressure. Using the data for the properties of liquid methane given in Problem 9.47, estimate (a) the power that can be produced by the turbine, (b) the savings in electricity usage per year if the turbine operates 300 days per year.

9.49. A heat pump is used to heat a house and maintain it at 18 °C. On a day where the outside temperature is −2 °C, the house is losing heat at a rate of 79,200 kJ/h. The heat pump operates with a coefficient of performance (COP) of 3.5. Determine (a) power needed by the heat pump, (b) the rate of heat absorbed from to the surrounding cold air.

9.50. A heat pump is used to heat a house and maintain it at 20 °C. On a day where the outside temperature is 0 °C, the house is losing heat at a rate of 34,500 kJ/h. The heat pump operates with a coefficient of performance (COP) of 3.0. Determine (a) power needed by the heat pump, (b) the rate of heat absorbed from to the surrounding cold air.

9.51. A heat pump is used to heat a house and maintain it at 20 °C. On a day where the outside temperature is 4 °C, the house is losing heat at a rate of 65,500 kJ/h. The heat pump operates with a coefficient of performance (COP) of 3.9. Determine (a) power needed by the heat pump, (b) the rate of heat absorbed from to the surrounding cold air.

9.52. A Carnot heat pump is used to heat a house during the winter. The house is maintained 20 °C. The house is estimated to be losing heat at a rate of

108,000 kJ/h when the outside temperature is −4 °C. Determine the minimum power needed by the heat pump and the rate of heat absorbed from to the surrounding cold air.

9.53. A Carnot heat pump is used to heat a house during the winter. The house is maintained 20 °C. The house is estimated to be losing heat at a rate of 78,000 kJ/h when the outside temperature is 2 °C. Determine the minimum power needed by the heat pump and the rate of heat absorbed from to the surrounding cold air.

9.54. Estimate the cost of electricity for a 10,000 Btu/h (3000 W) air-conditioning unit operating with a SEER of 10 Btu/Wh. The unit is used for a total of 1500 h during an annual cooling season.

9.55. Estimate the cost of electricity for a 12,000 Btu/h air-conditioning unit operating with a SEER of 14 Btu/Wh. The unit is used for a total of 2500 h during an annual cooling season.

9.56. Estimate the cost of electricity for a 9000 Btu/h air-conditioning unit operating with a SEER of 12 Btu/Wh. The unit is used for a total of 2000 h during an annual cooling season.

9.57. (a) Estimate the annual cost of electric power consumed by a 6-ton air-conditioning unit operating for 1000 h/year with a SEER rating of 10.
(b) Estimate the value of EER for hot and cold temperatures of 20 °C and −4 °C, respectively.

9.58. (a) Estimate the annual cost of electric power consumed by a 4-ton air-conditioning unit operating for 2500 h/year with a SEER rating of 12.
(b) Estimate the value of EER for hot and cold temperatures of 21 °C and −10 °C, respectively.

9.59. (a) Estimate the annual cost of electric power consumed by a 9-ton air-conditioning unit operating for 3000 h/year with a SEER rating of 14.
(b) Estimate the value of EER for hot and cold temperatures of 21 °C and −5 °C, respectively.

9.60. A 4-ton current residential air conditioner operates with a SEER rating of 10. This unit will be replaced with a newer unit operating with a SEER rating of 22. The unit operates 130 days with an average 10 h/day. Average inside and outside temperatures are 20 °C and −4 °C, respectively. Estimate the saving in the electricity and the maximum energy efficiency ratio.

9.61. A 4-ton current residential air conditioner operates with a SEER rating of 10. This unit will be replaced with a newer unit operating with a SEER rating of 22. The unit operates 120 days with an average 9 h/day. Average inside and outside temperatures are 20 °C and −0 °C, respectively. Estimate the saving in the electricity and the maximum energy efficiency ratio.

9.62. A 4-ton current residential air conditioner operates with a SEER rating of 10. This unit will be replaced with a newer unit operating with a SEER rating of 20. The unit operates 120 days with an average 7 h/day. Average inside and outside temperatures are 22 °C and −10 °C, respectively. Estimate the saving in the electricity and the maximum energy efficiency ratio.

9.63. The efficiency of an open burner is around 70% for electric heater units and 40% for natural gas units. We operate a 4 kW electric burner. Estimate the rate of energy consumption by the burner and the utilized energy for both electric and gas burners.

9.64. The efficiency of an open burner is around 72% for electric heater units and 39% for natural gas units. We operate a 6 kW electric burner at a location. Estimate the rate of energy consumption by the burner and the utilized energy for both electric and gas burners.

9.65. The efficiency of an open burner is around 69% for electric heater units and 42% for natural gas units. We operate a 10 kW electric burner at a location. Estimate the rate of energy consumption by the burner and unit costs of the utilized energy for both electric and gas burners.

9.66. An adiabatic turbine is used to produce electricity by expanding a superheated steam at 4100 kPa and 350 °C. The power output is 60 MW. The steam leaves the turbine at 40 kPa and 100 °C. If the combustion efficiency is 0.70 and the generator efficiency is 0.9, determine the overall plant efficiency and the amount of coal supplied per hour.

9.67. An adiabatic turbine is used to produce electricity by expanding a superheated steam at 4100 kPa and 350 °C. The power output is 60 MW. The steam leaves the turbine at 40 kPa and 100 °C. If the combustion efficiency is 0.70 and the generator efficiency is 0.9, determine the overall plant efficiency and the amount of coal supplied per hour.

9.68. An adiabatic turbine is used to produce electricity by expanding a superheated steam at 5800 kPa and 400 °C. The power output is 55 MW. The steam leaves the turbine at 40 kPa and 100 °C. If the combustion efficiency is 0.72 and the generator efficiency is 0.9, determine the overall plant efficiency and the amount of coal supplied per hour.

9.69. An adiabatic turbine is used to produce electricity by expanding a superheated steam at 4100 kPa and 350 °C. The steam flow rate is 42 kg/s. The steam leaves the turbine at 40 kPa and 100 °C. If the combustion efficiency is 0.75 and the generator efficiency is 0.90, determine the overall plant efficiency and the amount of coal supplied per hour.

9.70. An adiabatic turbine is used to produce electricity by expanding a superheated steam at 4100 kPa and 350 °C. The steam flow rate is 42 kg/s. The steam leaves the turbine at 40 kPa and 100 °C. If the combustion efficiency is 0.75 and the generator efficiency is 0.90, determine the overall plant efficiency and the amount of coal supplied per hour.

9.71. The overall efficiencies are about 25–28% for gasoline car engines, 34–38% for diesel engines, and 40–60% for large power plants. Compare the energy necessary for gasoline and diesel engines. The efficiency for the diesel is 36%. A car engine with a power output of 240 hp has a thermal efficiency of 24%. Determine the fuel consumption of the car if the fuel has a higher heating value of 20,400 Btu/lb.

9.72. The overall efficiencies are about 25–28% for gasoline car engines, 34–38% for diesel engines, and 40–60% for large power plants [2]. Compare the energy necessary for gasoline and diesel engines. The efficiency for the diesel is 35%. A car engine with a power output of 180 hp has a thermal efficiency of 26%. Determine the fuel consumption of the car if the fuel has a higher heating value of 20,400 Btu/lb.

9.73. Fuel consumption of the two cars are one with 11 l/100 km city and 9 l/100 km highway, and the other 6.5 l/100 km in city traffic and at 5 l/100 km highway. Estimate the annual fuel saving and emission reduction achieved by the more fuel-efficient car traveling at an average 7500 km per year.

9.74. Fuel consumption of the two cars are one with 10 l/100 km city and 8 l/100 km highway, and the other 6.0 l/100 km in city traffic and at 5 l/100 km highway. Estimate the annual fuel saving and emission reduction achieved by the more fuel-efficient car traveling at an average 10,000 km per year.

9.75. Fuel consumption of the two cars are one with 12 l/100 km city and 9 l/100 km highway, and the other 7.0 l/100 km in city traffic and at 6 l/100 km highway. Estimate the annual fuel saving and emission reduction achieved by the more fuel-efficient car traveling at an average 12,000 km per year.

9.76. Assume that a household consume about 5000 kWh per year. If the lighting is provided by compact fluorescent bulbs only, estimate the conservation of energy and saving per year.

9.77. Assume that a household consume about 14,000 kWh per year. If the lighting is provided by compact fluorescent bulbs only, estimate the conservation of energy and saving per year.

9.78. In a process available hot and cold process streams and their heat capacities are shown below.

Hot and Cold Process Stream Conditions

Streams		T_{in} (°C)	T_{out} (°C)	$C = \dot{m}C_p$ (kW/°C)
C1	Cold 1	20	180	40
C2	Cold 2	160	250	55
H1	Hot 1	280	60	35
H2	Hot 2	190	20	25

Construct the balanced composite curves for the process with $\Delta T_{min} = 20$ °C and $\Delta T_{min} = 10$ °C, and compare the amounts of hot and cold utilities needed.

References

1. Çengel YA, Boles MA (2014) Thermodynamics: an engineering approach, 8th edn. McGraw-Hill, New York
2. Demirel Y (2018) Energy conservation. In: Dincer I (ed) Comprehensive energy systems, vol 5. Elsevier, Amsterdam, pp 45–90
3. Demirel Y, Gerbaud V (2019) Nonequilibrium thermodynamics: transport and rate processes in physical, chemical and biological systems, 4th edn. Elsevier, Amsterdam
4. Dincer I, Midilli A (2014) Energy conservation. In: Anwar S (ed) Encyclopedia of energy engineering and technology, 2nd edn. CRC Press, Boca Raton
5. Guan L, Chen G (2014) Solar energy: building operations use. In: Anwar S (ed) Encyclopedia of energy engineering and technology, 2nd edn. CRC Press, Boca Raton
6. Herold KE (2014) Waste heat recovery applications: absorption heat pumps. In: Anwar S (ed) Encyclopedia of energy engineering and technology, 2nd edn. CRC Press, Boca Raton
7. Jaffe RL, Taylor W (2018) The physics of energy. Cambridge University Press, Cambridge
8. Krigger J, Dorsi C (2008) The homeowner's handbook to energy efficiency: a guide to big and small improvements. Saturn Resource Management, Helena
9. McCardell SB (2014) Effective energy use: rewards and excitement. In: Anwar S (ed) Encyclopedia of energy engineering and technology, 2nd edn. CRC Press, Boca Raton
10. Uhrig RE (2014) Greenhouse gas emissions: gasoline, hybrid-electric, and hydrogen-fueled vehicles. In: Anwar S (ed) Encyclopedia of energy engineering and technology, 2nd edn. CRC Press, Boca Raton

Energy Coupling

10

Introduction and Learning Objectives: Various mechanisms and devices can facilitate coupling between energy-producing process and energy-requiring processes. For example, in living systems, synthesis of adenosine triphosphate requires energy and therefore must couple with an energy-providing process of electron transport chain. On the other hand, the hydrolysis of adenosine triphosphate supplies the energy needed in living systems. This chapter discusses bioenergetics briefly as a representative analysis of energy coupling in biochemical cycles, and energy expenditure.

The learning objectives of this chapter are to discuss:

- Bioenergetics,
- Energy coupling in biochemical cycles,
- Control of energy coupling,
- Energy metabolism,
- Energy expenditure.

10.1 Energy Coupling and Gibbs Free Energy

The Gibbs free energy of a system is related to the enthalpy and entropy

$$G = H - TS \qquad (10.1)$$

The change in the Gibbs free energy is

$$\Delta G = \Delta H - T \Delta S \qquad (10.2)$$

© Springer Nature Switzerland AG 2021
Y. Demirel, *Energy*, Green Energy and Technology,
https://doi.org/10.1007/978-3-030-56164-2_10

The Gibbs free energy is a state function, hence its change between two states does not depend on how this change takes place between the states. Any reaction for which ΔG is negative would be favorable, or spontaneous, while any reaction for which ΔG is positive would be unfavorable

$$\Delta G < 0. \quad \text{Favorable, or spontaneous reactions} \tag{10.3}$$

$$\Delta G > 0 \quad \text{Unfavorable, or non - spontaneous reactions} \tag{10.4}$$

Reactions are classified as either *exothermic* ($\Delta H < 0$) releasing energy or *endothermic* ($\Delta H > 0$) requiring energy. Reactions can also be classified as *exergonic* ($\Delta G < 0$) decreasing free energy or *endergonic* ($\Delta G > 0$) increasing free energy during the reaction. Consider the following two reactions:

1. $A \ \rightarrow B \qquad \Delta G_{AB} = (G_B - G_A) < < 0, \text{exergonic - releases energy} \tag{10.5}$

2. $C \ \rightarrow D \qquad \Delta G_{AB} = (G_B - G_A) > 0, \text{endergonic - requires energy} \tag{10.6}$

Then the first reaction can drive the second reaction if they are coupled and satisfy the following inequality

$$|\Delta G_{AB}| > |\Delta G_{CD}|. \tag{10.7}$$

Then the coupled reactions would have a total Gibbs energy with a negative value, hence become exergonic and favorable reaction

$$(\Delta G_{AB} + \Delta G_{CD}) < 0 \tag{10.8}$$

Energy coupling may require some special mechanisms, such as enzymes and configurations in living systems [1, 2], which are briefly discussed in the following section.

10.2 Energy Coupling in Living Systems

Coupling can occur when a flow proceeds without its primary driving force, or opposite to the direction imposed by its primary driving force. For example, the living cells can pump ions flow from low to high concentration levels. Since the favorable transport of materials takes place always from high to low concentrations, this opposite transportation of substance is possible only if it is coupled with an energy-providing process. In living systems, this energy is provided by the hydrolysis of adenosine triphosphate (ATP). Energy coupling in living systems means that the metabolic pathways intersect in such a way that energy released from the favorable reactions of catabolism can be used to drive the energy requiring reactions of the anabolic pathways. This transfer of energy from catabolism to anabolism would be possible through the energy coupling.

The chloroplasts in plants use the free energy of the sun to initiate electron transfer cycles and proton gradients to produce ATP in the photosynthesis. In animals, respiration cycle initiates electron transfer cycle and proton gradient to produce the ATP in the oxidative phosphorylation. ATP is an energy-rich biomolecule, which stores and provides energy necessary in living systems. Photosynthesis and oxidative phosphorylation represent some of the major energy coupling systems in living systems [3].

10.3 Bioenergetics

Bioenergetics is concerned with the energy production, conservation, and conversion processes in living systems [4, 6]. The outside energy is used in the synthesis of an energy-rich molecule called adenosine triphosphate (ATP) from adenosine diphosphate (ADP) and molecular phosphorous (Pi). The chemical formula of ATP is shown in Fig. 10.1. ATP has three phosphate groups attached to adenosine, in which the terminal phosphate has a weak linkage and can break spontaneously whenever ATP forms a complex with an enzyme and releases chemical energy. ATP therefore acts as energy source of the living cell.

ATP synthesis is an endothermic reaction as it needs energy from outside. So the ATP synthesis is coupled to electron transport chain. Therefore, the electron transport chain drives the ATP synthesis. Synthesizing of ATP is matched and synchronized to cellular ATP utilization through the hydrolysis (combining with water) of ATP

$$ADP + P_i + nH_{in}^+ = ATP + H_2O + nH_{out}^+, \qquad (10.9)$$

where 'in' and 'out' denote two phases separated by a membrane, and n is the ratio H^+/ATP, showing the level of transmembrane proton transport for each ATP to be synthesized. ATP utilization is coupled to synthesizing proteins, enzymes,

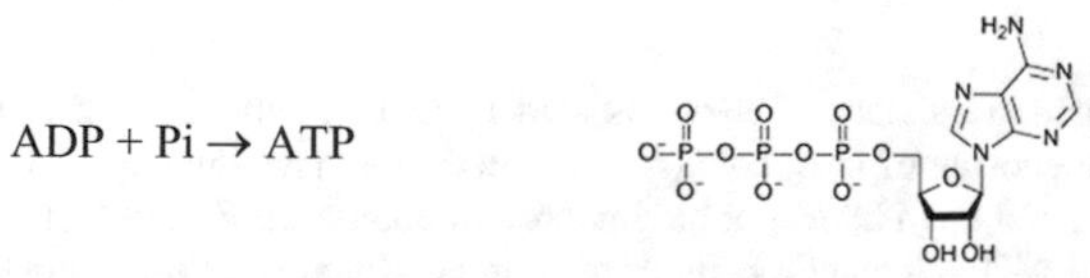

$$ADP + Pi \rightarrow ATP$$

Fig. 10.1 Chemical formula of adenosine triphosphate (ATP). ATP is an energy-rich compound having three phosphate groups attached to adenosine. Of the three phosphate groups, the terminal one has a weak linkage. This phosphate group can break spontaneously whenever ATP forms a complex with an enzyme. The breaking up of this bond releases chemical energy needed in all living systems

transporting ions, substrates, and producing mechanical work. ATP molecules, therefore, store and transfer energy in the living systems. For example, some of the internal mechanical work involves the pumping blood by the heart. Continuous biochemical cycles and transport processes can maintain a stationary state by the regulated production and transfer of energy of the ATP.

Most biochemical reactions in bioenergetics occur in pathways, in which other reactions continuously add substrates and remove products. The rate of these reactions depends on the enzymes that catalyze the reactions [4, 6].

10.3.1   Mitochondria

Mitochondria are organelles typically ranging in size from 0.5 to 1 μm in length, found in the cytoplasm of eukaryotic cells. Mitochondria contain inner and outer membranes constructed with phospholipids into which specific proteins are embedded. Mitochondrial membranes produce two compartments called the inter-membrane space and matrix space enclosed by the inner membrane (Fig. 10.2a). The inner membrane has numerous folds called *cristae* where the synthesis of ATP takes place [4].

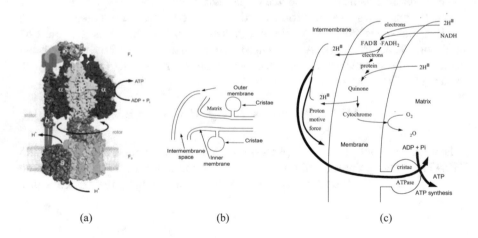

(a)   (b)   (c)

Fig. 10.2 a ATPases are a class of enzymes with F_o and F_1 subunits that catalyze the synthesis and hydrolysis of adenosine triphosphate (ATP). The cristae have the major coupling factors F_1 (a hydrophilic protein) and F_o (a hydrophobic lipoprotein complex). F_1 and F_o together comprise the ATPase (also called ATP synthase), **b** Inner membrane structure of the mitochondria, **c** Electron transport chain and oxidative phosphorylation of adenosine triphosphate [4, 5]

10.3.2   Electron Transport Chain and Adenosine Triphosphate (ATP) Synthesis

The inner membrane of the mitochondria houses the electron transport chain and ATP synthesis shown in Fig. 10.2c. The tricarboxylic acid cycle, also called the citric acid cycle, produces electron transport chain, which is the major energy-producing pathway. The electrons transport chain causes the proton flow across the inner membrane of the mitochondria and ATP is synthesized by the enzyme called the ATPases shown in Fig. 10.2a. ATPases have F_o and F_1 subunits that catalyze the synthesis and hydrolysis of ATP. The cristae have the major coupling factors F_1 (a hydrophilic protein) and F_o (a hydrophobic lipoprotein complex) [4].

In animals, the respiration chain generates energy by the oxidation of reducing equivalents of nutrients that are nicotinamide adenine nucleotides NADH and the flavin nucleotides $FADH_2$. When ADP levels are higher than ATP, the cell needs energy, and hence NADH is oxidized rapidly and the tricarboxylic acid cycle is accelerated. When the ATP level is higher than ADP, the electron transport chain slows down.

Photosynthesis, driven by the light energy, leads to the production of ATP through electron transfer and photosynthetic phosphorylation. Photosynthetic energy conservation takes place in the thylakoid membrane of plant chloroplasts. These membranes facilitate the interactions between the redox system and the synthesis of ATP, and are referred to as coupling membranes [4, 5].

10.3.3   Active Transport

Active transport is the transport of a substance from low to high concentration levels and hence against its concentration gradient. This process is also known as uphill transport and needs to couple with an energy-providing process. The hydrolysis of ATP provides energy for all cellular activity, including active transport of ions and molecules. If the transport process uses chemical energy, such as the energy released from the hydrolysis of adenosine triphosphate, it is termed *primary active transport*. *Secondary active transport* involves the use of an electrochemical gradient produced within the cell.

Hydrolysis of one mole of ATP is an exergonic reaction releasing 31 kJ/mol at pH = 7. This energy drives various energy-dependent metabolic reactions and the transport of various ions such as K^+ and Na^+ as seen in Fig. 10.3a. These concentration gradients of K^+ and Na^+ are established by the active transport of both ions. The same enzyme, called the Na^+/K^+ ATPase uses the energy released from the hydrolysis of ATP to transport 3 Na^+ ions out of the cell for each 2 K^+ ions pumped into the cell against their concentration gradients (see Fig. 10.3a). Almost one-third of all the energy generated by the mitochondria in animal cells is used in active transport. In *indirect active transport* other transporter molecules use the energy already stored in the gradient of a directly pumped ions [4, 5].

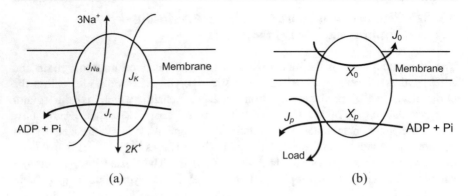

Fig. 10.3 **a** Active transport of Na⁺ and K⁺ ions coupled to hydrolysis of ATP [2]; **b** Energy coupling between respiration cycle and oxidative phosphorylation taking place in the inner membrane of mitochondria. Here J_o and J_p are the oxygen flow rate and the ATP production rate, respectively, while X_o and X_p show the redox potential for oxidizable substances and the phosphate potential, respectively. The flow J represents a load (i.e. osmotic work) coupled to ATP production

10.4 Simple Analysis of Energy Coupling

Figure 10.3b illustrates electron transportation cycle driving the ATP production through the energy coupling. The approximate representative linear phenomeno-logical relations of the oxidative phosphorylation are: [4]

$$J_o = L_oX_o + L_{op}X_p \tag{10.10}$$

$$J_p = L_{po}X_o + L_pX_p \tag{10.11}$$

Here J_o is the oxygen flow rate, J_p is the ATP production rate, X_o is the redox potential for oxidizable substances and X_p is the phosphate potential [2]. The J_o and J_p are called the flows, while X_o and X_p are called the forces. The L_{op} and L_{po} represent the cross-coefficients, while L_o and L_p are called the primary coefficients. The cross-coefficients, L_{op} and L_{po}, obey Onsager's reciprocal rules, which states that $L_{op} = L_{po}$.

The *degree of coupling q* is

$$q = \frac{L_{op}}{(L_pL_o)^{1/2}} \tag{10.12}$$

and indicates the extent of overall coupling of the ATP production driven by the respiration cycle in the inner membrane of mitochondria. By defining the *phe-nomenological stoichiometry Z* by

$$Z = \left(\frac{L_o}{L_p}\right)^{1/2} \tag{10.13}$$

and by dividing Eq. (10.10) by Eq. (10.11), we obtain the flow ratio j and the force ratio x in terms of Z

$$j = \frac{J_o}{J_p Z} \tag{10.14}$$

$$x = \frac{X_o Z}{X_p} \tag{10.15}$$

The flow ratio j and force ratio x can be related to each other by the degree of coupling q as follows

$$j = \frac{x + q}{qZ + 1} \tag{10.16}$$

The energy dissipation Ψ is expressed by [3, 4]

$$\Psi = J_o X_o + J_p X_p = \text{input   power} + \text{output   power} > 0 \tag{10.17}$$

The efficiency of the energy coupling of oxidative phosphorylation may be defined as the ratio of output power to input power and may be expressed in terms of the degree of coupling

$$\eta = -\frac{\text{output   power}}{\text{input   power}} = \frac{\text{driven   process}}{\text{driving   process}} = -\frac{J_p X_p}{J_o X_o} = -\frac{x + q}{q + (1/x)} \tag{10.18}$$

In a two flow and two force system, the value of optimum efficiency is a function of the degree of coupling only and is expressed by

$$\eta_{opt} = \frac{q^2}{\left(1 + \sqrt{1 - q^2}\right)^2} = \tan^2(\alpha/2) \tag{10.19}$$

where $\alpha = \arcsin(q)$. The value of x at η_{opt} becomes

$$x_{opt} = -\frac{q}{1 + \sqrt{1 + q^2}} \tag{10.20}$$

Therefore, the optimum value of ratio of forces x depends only on the value of degree of coupling [4.5]. Example 10.1 illustrates a simple representative analysis of energy coupling in photosynthesis.

Example 10.1 Efficiency of energy conversion of photosynthesis
Consider a model process with an energy exchange between a photon and a composite particle. In this over-simplified model, energy is exchanged through an excited state of the chloroplast by which energy-rich electron/proton pairs from the water react with the carbon dioxide. This produces carbohydrate and oxygen molecules, and heat is dissipated away. The flows are related by an approximation [4]

$$\left(\frac{J_D}{J_q}\right)^2 = \frac{k_B n D}{k} \tag{10.21}$$

where $n \sim 3.3 \times 10^{28}$ molecules/m^3, which is a typical value for water and condensed matter in general, J_D mass flow of carbon dioxide (rate of diffusion), J_q is the heat flow, D is the diffusion coefficient, k is the thermal conductivity, and k_B is the Boltzmann constant.

Only a small part of the free energy of photons is available for photosynthesis, and the rest is dissipated. The efficiency of energy conversion in photosynthesis is low and varies in the range 2.4–7.5%. The efficiency of energy conversion is

$$\eta = \frac{J_D}{J_q} \frac{\Delta G}{h v} \tag{10.22}$$

where ΔG is the Gibbs energy per molecule and hv is the energy per photon. Some approximate values for these driving forces and the coefficients are:

- $\Delta G \sim 7.95 \times 10^{-19}$ J/one unit carbohydrate.
- $(hv) \sim 2.92 \times 10^{-19}$ J/a solar photon (pertaining to red light with a wavelength of 680 nm, which is best absorbed by chlorophyll-α).
- Based on the thermal conductivity of water: $k = 0.607$ W/m K.
- Assuming that the intercellular diffusion of carbon dioxide could be the limiting process in photosynthesis, we have $D \sim 1.95 \times 10^{-9}$ m^2/s based on carbon dioxide in water or $D \sim 0.67 \times 10^{-9}$ m^2/s based on glucose in water.
- $k_B = 1.3806503 \times 10^{-23}$ m^2 kg/s^2 K.

Using these approximate values in the following equation

$$\eta = \frac{J_D}{J_q}\left(\frac{7.95 \times 10^{-19}}{2.92 \times 10^{-19}}\right) = 2.72\left(\frac{k_B n D}{k}\right)^{1/2} \tag{10.23}$$

the approximate predicted values of efficiency of energy conversion become [4]

η (predicted), %	η (measured), %	D (m^2/s)
2.4	4.9	0.10×10^{-9}
6.1	6.2	0.67×10^{-9}
7.5	~7	1.0×10^{-9}

Any plant growing under ideal conditions, the efficiency is expected to be close to 7%.

10.5 Variation of Energy Coupling

The value of degree of coupling q can be calculated from the measurements of oxygen flows at static head (sh) $(J_o)_{sh}$ in which the net rate of ATP vanishes, and at uncoupled state (unc) $(J_o)_{unc}$ where the proton gradient vanishes and the respiration cycles is uncoupled from the oxidative phosphorylation [4–6]

$$q = \sqrt{1 - (J_o)_{sh}/(J_o)_{unc}} \tag{10.24}$$

For example, the overall degree of energy coupling is higher for a rat liver mitochondrion (0.955 ± 0.021) than for brain (0.937 ± 0.026) or heart (0.917 ± 0.037).

Required degrees of coupling of oxidative phosphorylation vary when the ATP production is coupled to a load such as hydrolysis of the ATP (see Fig. 10.3b) [4]. The optimum production functions f for the ATP and output power production are given in terms of the degree of coupling

$$f = \tan^m(\alpha/2) \cos(\alpha) \tag{10.25}$$

The ATP production and power output occur at certain values of degrees of coupling:

- Maximum ATP production at optimal efficiency occurs at $q_f = 0.786$, $f = (J_p)_{opt}$.
- Maximum power output at optimal efficiency occurs at $q_p = 0.91$, $f = (J_p X_p)_{opt}$.
- Efficient ATP production at minimal energy cost occurs at $q_f^{ec} = 0.953$, $f = \eta(J_p)_{opt}$.
- Economic power output at minimal energy cost occurs at $q_p^{ec} = 0.972$, $f = \eta (J_p X_p)_{opt}$

With the consideration of conductance matching, [5] there are four production functions, which are given in Table 10.1. Figure 10.4 shows the effect of degree of coupling on the optimum efficiency. The sensitivity of oxidative phosphorylation to a fluctuating ATP utilization is minimal at a degree of coupling $q = 0.95$.

Table 10.1 Production functions with the consideration of conductance matching [4]

Production function	Loci of the optimal efficiency states	q	Energy cost
Optimum rate of ATP production: $J_p = (q+x)ZL_oX_o$	From the plot of J_p versus x: $(J_p)_{\mathrm{opt}} = \tan(\alpha/2)\cos\alpha ZL_oX_o$	$q_f = 0.786$ $\alpha = 51.83$	No $\eta = \mathrm{cons}$
Optimum output power of OP: $J_pX_p = x(q+x)L_oX_o^2$	From the plot of J_pX_o versus x: $(J_pX_p)_{\mathrm{opt}} = \tan^2(\alpha/2)\cos\alpha L_oX_o^2$	$q_p = 0.910$ $\alpha = 65.53°$	No $\eta = \mathrm{cons}$
Optimum rate of ATP production at minimal energy cost: $J_p\eta = -\dfrac{x(q+x)^2}{xq+1}ZL_oX_o$	From the plot of $J_1\eta$ versus x: $(J_p\eta)_{\mathrm{opt}} = \tan^3(\alpha/2)\cos\alpha ZL_oX_o$	$q_f^{ec} = 0.953$ $\alpha = 72.38°$	Yes
Optimum output power of OP at minimal energy cost: $J_pX_p\eta = -\dfrac{x^2(q+x)^2}{xq+1}L_oX_o^2$	From the plot of $J_1X_1\eta$ versus x: $(J_pX_p\eta)_{\mathrm{opt}} = \tan^4(\alpha/2)\cos\alpha L_oX_o^2$	$q_p^{ec} = 0.972$ $\alpha = 76.34°$	Yes

OP: Oxidative phosphorylation; q_f, q_p are the degree of couplings at optimized ATP production and power output, respectively, and q_f^{ec}, q_p^{ec} are the degree of couplings at optimized ATP production and power output with minimal energy cost, respectively

This means that the phosphate potential is highly buffered with respect to fluctuating energy demands at the degree of coupling, which is very close to the value of q_f^{ec}, at which net ATP production of oxidative phosphorylation occurs at the minimal energy cost [4].

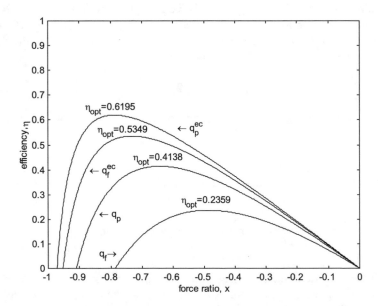

Fig. 10.4 Efficiency versus force ratio at various levels of energy couplings; the change of efficiencies η, given in Eq. (10.18), in terms of flow ratio x (X_oZ/X_p) and for the degrees of couplings q_f, q_p, q_f^{ec}, and q_p^{ec} [4]

The degree of coupling may be regulated based on the nature of the output required from the energy coupling system. In the heart and brain, the experimental value of q for the cellular respiration pathway is close to the value of $q_f^{ec} = 0.953$, which suggests that the pathway is optimized to economical ATP production. In the brain, the value of $q_p = 0.91$ suggests maximized cellular energy state. However, in the heart, the degree of coupling is 0.786, which is between q_p and q_f, and is consistent with the maximum ATP production necessary for preserving the cellular energy state [4, 5].

The optimum output power $(J_1X_1)_{opt}$ and the efficiency $(J_1X_1 \eta)_{opt}$ are calculated from the plots of J_1X_1 versus x and $J_1X_1\eta$ versus x, respectively. A transition from q_p to q_p^{ec} causes a 12% drop in output power (J_1X_1) and a 51% increase in efficiency.

10.5.1 Regulation of Energy Coupling

Regulation implies a physiological outcome as a result of manipulating a function. The physiological reasons for these regulations are mainly to synchronize to ATP supply efficiently to ATP demand in workload in bioenergetics, as well as to respond to the external stimuli [4, 5]. The electrochemical proton gradient across the membrane is one of the important mechanisms for regulating the rate of respiration and ATP synthesis. Various substrates regulate the metabolism of energy; fatty acids may regulate and tune the degree of coupling by inducing uncoupling and leading to optimum efficiency of oxidative phosphorylation. A fluctuating ATP/ADP ratio and deviations from the optimal efficiency of oxidative phosphorylation are largely overcome by some buffering enzymes. For example, the mitochondrial creatine kinase is a key enzyme of aerobic energy metabolism, and is involved in buffering, transporting, and reducing the transient nature of the system [5].

The sensitivity of the force (the phosphate potential) to the cellular ATP utilization is minimal at $q = 0.95$. This indicates that the phosphate potential is highly buffered with respect to the changing energy demand to maximize the kinetic stability and efficiency at the same degree of coupling. The standard Gibbs energy of ATP production is 31.3 kJ/mol at $T = 20$ °C, pH = 8.0, pMg = 2.5, and 0.08 M ionic strength. The standard enthalpy of the reaction is 28.1 kJ/mol [4].

10.5.2   Uncoupling

ATP synthesis is matched to cellular ATP utilization for osmotic work of (downhill and uphill) transport, or mechanical work such as muscle contraction and rotation of bacterial flagellum. The uncoupling of the mitochondrial electron transport chain from the phosphorylation of ADP is physiological and optimizes the efficiency and fine tunes the degree of coupling of oxidative phosphorylation. Fatty acids facilitate the net transfer of protons from intermembrane space into the mitochondrial matrix, hence lowering the proton electrochemical potential gradient and mediating weak uncoupling. Uncoupling proteins generally facilitate the dissipation of the trans-membrane electrochemical potentials of H^+ or Na^+ produced by the respiratory chain, and result in an increase in the H^+ and Na^+ permeability of the coupling membranes. Some uncoupling is favorable for the energy-conserving function of cellular respiration [5, 8].

10.5.3   Slippages and Leaks

Slippage results when one or two coupled reactions in a cyclic process proceed without its counterpart, which is also called intrinsic uncoupling. On the micro-scopic level, individual enzymes cause slippage by either passing a proton without contributing to ATP synthesis, or hydrolyzing ATP without contributing to proton pumping. On the macroscopic level, the measured degree of coupling may be different from the expected coupling.

In terms of the energy conversion, a slip may decrease efficiency; it may, however, allow dynamic control and regulation of the enzyme over the varying ranges of driving and driven forces. Mitochondrial energy metabolism may be regulated by the slippage of proton pumping. Slips and leaks may occur in parallel. The rate of leakage of the coupling ions depends on the magnitude of the particular thermodynamic force operating within the system in the absence of the other force. In oxidative phosphorylation, leaks cause a certain uncoupling of two consecutive pumps, such as electron transport and ATP synthase, and may be described as the membrane potential-driven backflow of protons across the bilayer [4].

10.6   Metabolism

Metabolism is the set of biochemical reaction networks that occur in living organisms to maintain life. Metabolism consists of two pathways: catabolism and anabolism. In catabolism organic matter is broken down, for example to harvest energy in cellular respiration. In anabolism, the energy is used to construct components of cells such as proteins and nucleic acids. In metabolic pathways, one chemical is transformed through a series of steps into another chemical by a sequence of enzymes. Enzymes allow some processes to drive desirable reactions that require energy by coupling them to spontaneous processes that release energy. One central coenzyme is ATP and used to transfer chemical energy between different chemical reactions. ATP in cells is continuously regenerated and acts as a bridge between catabolism and anabolism, with catabolic reactions generating ATP and anabolic reactions consuming it through energy coupling [6–8].

10.6.1 Catabolism

Catabolism is a set of metabolic cyclic processes that break down and oxidize large molecules, including food molecules to simpler molecules, such as carbon dioxide and water. The catabolic reactions provide the energy and components needed by anabolic reactions.

The most common set of catabolic reactions in animals can be separated into three main stages (see Fig. 10.5). In the first stage, large organic molecules such as proteins, polysaccharides, or lipids are digested into their smaller components outside cells. Next, these smaller molecules are taken up by cells and converted to yet smaller molecules, usually acetyl coenzyme A (acetyl-CoA), which releases some energy. Finally, the acetyl group on the CoA is oxidized to water and carbon dioxide in the citric acid cycle and electron transport chain, releasing the energy that is stored by reducing the coenzyme nicotinamide adenine dinucleotide (NAD^+) into NADH [7].

These digestive enzymes include proteases that digest proteins into amino acids, as well as glycoside hydrolases that digest polysaccharides into monosaccharides as shown in Fig. 10.5.

The major route of breakdown is glycolysis, where sugars such as glucose and fructose are converted into pyruvate and some ATP molecules are produced. This oxidation releases carbon dioxide as a waste product. Fatty acids release more energy upon oxidation than carbohydrates because carbohydrates contain more oxygen in their structures. Amino acids are either used to synthesize proteins and other biomolecules, or oxidized to urea and carbon dioxide as a source of energy [5, 6].

Fig. 10.5 A simplified outline of the catabolism of biofuels proteins, carbohydrates and fats leading to energy coupling systems [2, 3, 6]

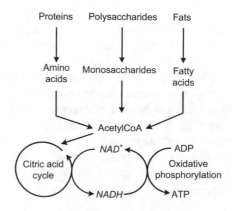

10.6.2   Anabolism

Anabolism is a set of constructive metabolic processes where the energy released by catabolism is used to synthesize complex molecules into three basic stages. First, the production of precursors such as amino acids, monosaccharides, and nucleotides; second, their activation into reactive forms using energy from ATP; and third, the assembly of these precursors into complex folded molecules such as proteins, polysaccharides, lipids, and nucleic acids [2, 6].

10.7   Bioenergy Sources

Carbohydrates, fats, and proteins are the major sources of chemical energy of the body and are obtained from the daily diet. Heat released by the oxidation of these fuels is used to maintain body's temperature:

- Carbohydrates release about 4 kcal/g (17 kJ/g).
- Fats release about 9 kcal/g (39 kJ/g).
- Proteins release about 4 kcal/g (17 kJ/g).

Excess of the body's immediate energy needs is stored mainly by:

- Fat-triacylglycerol (triglyceride): adipose triacylglycerol is the major energy store of the body.
- Glycogen: Energy stored by glycogen is relatively small but it is critical. For example, muscle glycogen is oxidized for muscle contraction.

Daily energy expenditure is the amount of energy required daily, including for the basal metabolic rate and physical activity. Basal metabolic rate for a person is approximately 24 kcal/kg body weight per day and the basic physical activity requires approximately 30% of basal metabolic rate [7].

Carbohydrates are straight-chain aldehydes or ketones with many hydroxyl groups that can exist as straight chains or rings. Carbohydrates are the most abundant biological molecules, and play numerous roles, such as the storage and transport of energy (starch, glycogen), and structural components such as cellulose in plants, chitin in animals.

The fuel value is equal to the heat of reaction of combustion (oxidation). Carbohydrates and fats can be completely oxidized while proteins can only be partially oxidized and hence has lower fuel values. The energy expenditure may be calculated from the energy balance. Assume that (i) carbohydrate (CH), fat (F), and protein (Pr) are the only compounds involved in the oxidation process, (ii) the other compounds are stationary, and (iii) the uptake and elimination of oxygen, carbon dioxide, and nitrogen is instantaneous. Then, from the energy balance, we have the energy expenditure $\dot{E}$ as

$$\dot{E} = \sum_i (\dot{m}\Delta H_r)_i = (\dot{m}\Delta H_r)_{\text{CH}} + (\dot{m}\Delta H_r)_{\text{F}} + (\dot{m}\Delta H_r)_{\text{Pr}} = \dot{q} + \dot{W} \qquad (10.26)$$

where the specific enthalpy values are: $\Delta H_{r,\text{CH}} = -17$ kJ/g, $\Delta H_{r,\text{Fat}} = -39$ kJ/g, $\Delta H_{r,\text{CH}} = -17$ kJ/g. Example 10.2 illustrates the oxidation of glucose, while Examples 10.3 to 10.5 discuss the energy expenditure estimations.

Example 10.2 Oxidation of glucose

(a) Estimate the reaction enthalpy for the isothermal and isobaric oxidation of glucose at 310 K and 1 atm.
(b) Estimate the energy expenditure for oxidation of 390 g/day glucose at steady conditions.

Solution:
Assume: The energy for transferring of gaseous components to aqueous solutions is small. The temperature correction for the heat of reaction is negligible.
$$C_6H_{12}O_6(aq) + 6O_2(g) \rightarrow 6CO_2(g) + 6H_2O(l)$$
where the symbol (aq) denotes aqueous, (g) is the gas, and (l) is the liquid. If the control volume is a piece of tissue, the reaction above may take place in an aqueous solution (aq).
From Table C1, we obtain the enthalpy of formations for the components of the reaction above at the standard state (298 K and 1 atm)
$\Delta H_r^o(298 \text{ K}) = 6(-393) + 6(-286) - (-1264) - 6(0) = -2810$ kJ/mol $\sim \Delta H_r(310K)$.
$MW(\text{glucose}) = 180$ g/gmol
The energy expenditure $\dot{E}$ at a glucose consumption of 390 g/day is
$$n_G = \frac{(390 \text{ g/day})/(180 \text{g/mol})}{(24 \text{ hr/day})(3600 \text{ s/hr})} = 0.025 \times 10^{-3} \text{ mol/s}$$
$\dot{E} = n_G(-\Delta H_r) = (0.025 \times 10^{-3} \text{ mol/s})(2810 \text{ kJ/mol}) = \mathbf{70.2 \times 10^{-3}}$ **W = 70.2 mW**

Example 10.3 Daily energy expenditure
A dietary history of an adult of 146 lb shows that he consumes approximately 150 g of carbohydrate, 50 g of protein, and 60 g of fat daily. The adult exercises regularly and consumes an average 480 kcal daily. Estimate the energy expenditure of this adult and comment if the energy intake is excessive.

Solution:
Daily energy expenditure is the amount of energy required daily and includes the energy required for the basal metabolic rate and for physical activity. Basal metabolic rate for a person is approximately 24 kcal/kg (101 kJ/kg) body weight per day [7].

Energy needed for physical exercise $= 480$ kcal/day

Energy expenditure $=$ Basal metabolic rate $+$ Physical exercise

Energy expenditure daily $= (24$ kcal/kg day$)(146$ lb$)($kg$/2.2$ lb$) + 480$ kcal/day

$$= 1594 \text{ kcal/day} + 480 \text{ kcal/day} = \textbf{2074 kcal/day}$$

Energy intake:

Carbohydrates: 4 kcal/g $\rightarrow$ (150 g/day)(4 kcal/g) = 600 kcal/day.

Fats: 9 kcal/g $\rightarrow$ (60 g/day)(9 kcal/g) = 540 kcal/day.

Proteins: 4 kcal/g $\rightarrow$ (50 g/day)(4 kcal/g) = 200 kcal/day.

Total energy intake = (600 + 540 + 200) kcal/day = **1,340 kcal/day.**

Total energy intake of 1,340 kcal/day is less than basal metabolic rate of 2,074 kcal.

Example 10.4 Energy expenditure in small organisms

Consider a small organism with a body weight of 10 μg consuming 4.0×10^{-9} mol oxygen every hour at steady state and eliminating 3.6×10^{-9} mol carbon dioxide, 0.4×10^{-9} mol N (as ammonia). The external work of the organism is 50×10^{-9} W. Estimate the energy expenditure and the heat loss using the heat of reactions given below.

Hydrocarbon(Glucose)(G) : $C_6H_{12}O_6 + 6O_2(g) \rightarrow 6CO_2 + 6H_2O$ -2870 kJ/mol

Fat(F) : $C_{55}H_{104}O_6 + 78O_2 \rightarrow 55CO_2 + 52H_2O$ -34300 kJ/mol

Protein(Pr) : $C_{32}H_{48}O_{10}N_8 + 33O_2 \rightarrow 32CO_2 + 8NH_3 + 12H_2O$ -14744 kJ/mol

Solution:

The energy expenditure may be calculated with the given reaction enthalpies:

$\dot{E} = 2870(\dot{n})_G + 34300(\dot{n})_F + 14744(\dot{n})_{Pr}$ (kJ/h)

The number of mole flow rates is obtained from the reaction stoichiometric coefficients:

For example, for oxygen:

- one mol/h of glucose consumes 6 mol/h of O_2
- one mol/h of fat consumes 78 mol/h of O_2
- one mol/h of protein consumes 33 mol/h of O_2

With the similar procedure for carbon dioxide and nitrogen, we have.

$\dot{n}_{O_2} = 6\dot{n}_G + 78\dot{n}_F + 33\dot{n}_{Pr} = 4.0 \times 10^{-9}$ mol/h.

$\dot{n}_{CO_2} = 6\dot{n}_G + 55\dot{n}_F + 32\dot{n}_{Pr} = 3.6 \times 10^{-9}$ mol/h.

$\dot{n}_N = 8\dot{n}_{Pr} = 0.4 \times 10^{-9}$ mol/h.

After solving the equations above, we have the energy expenditure:

$\dot{E} = [2870(0.194)_G + 34300(0.0152)_F + 14744(0.05)_{Pr}] \times 10^{-9}$

$= 1820 \times 10^{-9}$ kJ/h = **0.505 μW.**

With $W = 50 \times 10^{-9}$ W $= 0.05$ μW.

Heat loss: $\dot{q} = \dot{E} - \dot{W} = (0.505 - 0.050)$ μW = **0.46 μW.**

Example 10.5 Energy expenditure in an adult organism
An adult organism has an oxygen uptake of about 21.16 mol over 24 h, and the associated elimination of carbon dioxide and nitrogen is 16.95 mol and 5.76 g, respectively. If the adult has performed 1.2 MJ of external work over the same period and his energy expenditure at rest is $\dot{E}_o = 65$ W, estimate his energy expenditure, heat loss, and net efficiency for the external work using the biological fuel parameters given below [1, 7].

Fuel	Specific reaction enthalpy (kJ/g)	Specific turnover		
		O_2 (mmol/g)	CO_2 (mmol/g)	N (g/g)
CH	−17	33.3	33.3	–
Fat	−39	90.6	63.8	–
Pr	−17	43.3	34.43	0.16

CH: carbohydrate; Pr: Protein

Solution:
The energy expenditure may be calculated from the energy balance. Assume that (i) carbohydrate (CH), fat (F), and protein (Pr) are the only compounds involved in the oxidation process, (ii) the other compounds are stationary, and (iii) the uptake and elimination of oxygen, carbon dioxide, and nitrogen is instantaneous. From the first law of thermodynamics, we have

$$\dot{E} = \sum_i (\dot{m}\Delta H_r)_i = (\dot{m}\Delta H_r)_{CH} + (\dot{m}\Delta H_r)_F + (\dot{m}\Delta H_r)_{Pr} = \dot{q} + \dot{W} \text{ (kJ/day)}$$

Using the specific enthalpy values of the fuels:

$$17\dot{m}_{CH} + 39\dot{m}_F + 17\dot{m}_{Pr} = \dot{q} + \dot{W} = \dot{E} \tag{a}$$

The conservation of mass with the parameters from above table is:
$\dot{n}_{O_2} = 33.3\dot{m}_{CH} + 90.6\dot{m}_F = 21,160$ mmol/day
$\dot{n}_{CO_2} = 33.3\dot{m}_{CH} + 63.8\dot{m}_F = 16,950$ mmol/day
From these equations, we obtain.
$\dot{m}_{Pr} = 5.76/0.16 = 36$ g/day
$\dot{m}_{CH} = 208$ g/day
$\dot{m}_F = 157$ g/day
From Eq. (a), we have the total energy expenditure: basal metabolic rate + physical exercise
$17(208) + 39(157) + 17(36) = -\dot{q} + (-\dot{W}) = \dot{E} = \textbf{10,271 kJ/day}$
Since the work (part of physical exercise) is 1.2 MJ/day = 1200 kJ/day
Heat loss: $\dot{q} = (10271 - 1200)$ kJ/day = **9071 kJ/day**
Basal metabolic rate (energy expenditure at rest): $\dot{E}_o = 65$ W = 5616 kJ/day
A net available energy for physical exercise: $(10,271 - 5616)$ kJ/day
Then the bioenergetics efficiency is defined by

$$\eta_B = \frac{\dot{W}}{(\dot{E} - \dot{E}_o)} = \frac{1200}{(10271 - 5616)} = \textbf{0.257 or 25.7\%}$$

$\dot{E}_o$ is the energy expenditure during resting and $\dot{E}$ is the total energy.

Summary

- *The Gibbs free energy* of a system is defined as the enthalpy minus the product of temperature and entropy: $G = H - TS$
- The change in the Gibbs free energy of the system that occurs during a reaction is related to the enthalpy and entropy of the system. Any reaction for which ΔG is negative should be favorable, or spontaneous, while any reaction for which ΔG is positive is unfavorable.
- Reactions are classified as either *exothermic* ($\Delta H < 0$) or *endothermic* ($\Delta H > 0$) on the basis of whether they give off or absorb heat. Consider the following two reactions:
- 1. A $\rightarrow$ B $\Delta G_{AB} = (G_B - G_A) < \, < 0$, exergonic - releases energy
- 2. C $\rightarrow$ D $\Delta G_{AB} = (G_B - G_A) > 0$, endergonic - requires energy
- Then the first reaction can drive the second reaction if they are coupled and as long as their absolute values satisfy the following inequality: $|\Delta G_{AB}| > |\Delta G_{CD}|$.
- Energy coupling may require some special mechanisms, such as enzymes and configurations.
- A *coupled process* can occur without its primary driving force, or opposite to the direction imposed by its primary driving force; the living cells can pump ions from low to high concentration regions.
- *Bioenergetics* is concerned with the energy production, conservation, and conversion processes in living systems. The outside energy is used in the synthesis of an energy-rich molecule called adenosine triphosphate (ATP) from adenosine diphosphate (ADP) and molecular phosphorous (Pi).
- *ATP synthesis* is an endothermic reaction as it needs energy from outside. So, the ATP synthesis is coupled to electron transport chain.
- *Mitochondria* contain inner and outer membranes constructed with phospholipids into which specific proteins are embedded. The inner membrane has numerous folds called *cristae* where the synthesis of ATP takes place. The inner membrane houses the electron transport chain and ATP synthesis.
- *Photosynthesis*, driven by the light energy, leads to the production of ATP through electron transfer and photosynthetic phosphorylation.
- *Active transport* is the transport of a substance from low to high concentration and hence against its concentration gradient. This process is also known as uphill transport and needs energy.
- The *degree of coupling q* is defined by $q = L_{op}/(L_p L_o)^{1/2}$, and indicates the extent of overall coupling of the ATP production driven by the respiration cycle in the inner membrane of mitochondria.
- *The energy dissipation* Ψ is: $\Psi = J_o X_o + J_p X_p =$ input power + output power > 0
- *The efficiency of the linear energy coupling* of oxidative phosphorylation may be defined as the ratio of output power to input power and in terms of the degree of coupling, and given by

$$\eta = -\frac{\text{output power}}{\text{input power}} = \frac{\text{driven process}}{\text{driving process}} = -\frac{J_p X_p}{J_o X_o} = -\frac{x+q}{q+(1/x)}$$

- The value of optimum efficiency is a function of the degree of coupling only, and is expressed by
- The value of degree of coupling q can be calculated from the measurements of oxygen flows at static head (sh) and at uncoupled state (unc) $(J_o)_{\text{unc}}$:$q = \sqrt{1 - (J_o)_{\text{sh}}/(J_o)_{\text{unc}}}$
- *Regulation of energy coupling* implies a physiological outcome as a result of manipulating a mitochondrial function by *uncoupling* and *slippages and leaks.*
- *Metabolism* consists of two pathways: catabolism and anabolism. In catabolism organic matter is broken down, for example to harvest energy in cellular respiration. In anabolism, the energy is used to construct components of cells such as proteins and nucleic acids.
- *Bioenergy sources* such as carbohydrates, fats, and proteins are the major sources of chemical energy of the body and are obtained from the daily diet.
- From the energy balance, we have the *energy expenditure*

$$\dot{E} = \sum_i (\Delta H_r)_i = (\dot{m}\Delta H_r)_{\text{CH}} + (\dot{m}\Delta H_r)_{\text{F}} + (\dot{m}\Delta H_r)_{\text{Pr}} = \dot{q} + \dot{W}$$

- where the specific enthalpy values are: $\Delta H_{r,\text{CH}} = -17$ kJ/g, $\Delta H_{r,\text{Fat}} = -39$ kJ/g, $\Delta H_{r,\text{CH}} = -17$ kJ/g.

Problems

10.1. Estimate the approximate daily energy expenditure of an adult of 55 kg.

10.2. Estimate the approximate daily energy expenditure of a child of 20 kg.

10.3. Estimate the approximate daily energy expenditure of an adult of 51 kg.

10.4. A dietary history of an adult of 65 kg shows that he eats approximately 250 g of carbohydrate, 120 g of protein, and 160 g of fat daily. The adult exercises regularly and consumes an average 600 kcal daily. Estimate the energy expenditure of this adult and comment if the energy intake is excessive.

10.5. A dietary history of a child of 25 kg shows that he eats approximately 150 g of carbohydrate, 80 g of protein, and 100 g of fat daily. The child exercises regularly and consumes an average 450 kcal daily. Estimate the energy expenditure of this adult and comment if the energy intake is excessive.

10.6. A dietary history of an adult of 75 kg shows that he eats approximately 275 g of carbohydrate, 160 g of protein, and 190 g of fat daily. The child exercises regularly and consumes an average 700 kcal daily. Estimate the energy expenditure of this adult and comment if the energy intake is excessive.

10.7. Discuss the consequences of hydrolyzing ATP rapidly.

10.8. Discuss the consequences of intake of uncouple to the human body.

10.9. Discuss the consequence of a lack of oxygen flow to the human body.

10.10. An organism lose heat at a rate of 1.64 W and performs an external work of
1.4 Nm/min. (a) Estimate the energy expenditure of the organism, (b) if the
organism is able to pump 100 ml/min of fluid against a pressure drop of
25 mm Hg (3.34 kPa) with an efficiency of 10%, estimate the efficiency of
pumping.

10.11. An organism lose heat at a rate of 2.1 W and performs an external work of 5
Nm/min. (a) Estimate the energy expenditure of the organism, (b) if the
organism is able to pump 52 ml/min of fluid against a pressure drop of
2.5 mm Hg (0.334 kPa) with an efficiency of 10%, estimate the efficiency
of pumping.

10.12 a. Estimate the reaction enthalpy for the isothermal and isobaric oxidation
of glucose at 310 K and 1 atm.

b. Estimate the energy expenditure for oxidation of 390 g/day glucose at
steady conditions.

10.13. Consider a small organism with a body weight of 10 μg consuming
8.0×10^{-9} mol oxygen every hour at steady state and eliminating
7.1×10^{-9} mol carbon dioxide, 0.8×10^{-9} mol N (as ammonia). The
external work of the organism is 80×10^{-9} W. Estimate the energy
expenditure and heat loss using the heat of reactions given below.

Hydrocarbon(Glucose)(G) : $C_6H_{12}O_6 + 6O_2(g) \rightarrow 6CO_2 + 6H_2O$ -2870 kJ/mol
Fat(F) : $C_{55}H_{104}O_6 + 78O_2 \rightarrow 55CO_2 + 52H_2O$ -34300 kJ/mol
Protein(Pr):$C_{32}H_{48}O_{10}N_8 + 33O_2 \rightarrow 32CO_2 + 8NH_3 + 12H_2O$ -14744 kJ/mol

10.14. An adult organism has an oxygen uptake of about 21.16 mol over 24 h, and
the associated elimination of carbon dioxide and nitrogen is 16.95 mol and
5.76 g, respectively [7]. If the adult has performed 1.25 MJ of external
work over the same period and his energy expenditure at rest is $\dot{E}_o = 72$ W,
estimate his energy expenditure, heat loss, and net efficiency for the external
work using the biological fuel parameters given below.

Fuel	Specific reaction enthalpy (kJ/g)	Specific turnover		
		O_2 (mmol/g)	CO_2 (mmol/g)	N (g/g)
CH	−17	33.3	33.3	−
Fat	−39	90.6	63.8	−
Pr	−17	43.3	34.43	0.16

CH: carbohydrate; Pr: Protein
Source (Garby and Larsen (1995) Bioenergetics, Cambridge Univ Press, Cambridge)

10.15. An adult organism has an oxygen uptake of about 22.5 mol over 24 h, and the associated elimination of carbon dioxide and nitrogen is 17.4 mol and 6.6 g, respectively [7]. If the adult has performed 1.02 MJ of external work over the same period and his energy expenditure at rest is $\dot{E}_o = 60$ W, estimate his energy expenditure, heat loss, and net efficiency for the external work using the biological fuel parameters given below.

References

1. Alberts B, Johnson A, Lewis J, Morgan D, Raff M, Roberts K, Walter P (2014) Molecular biology of the cell, 6th edn. Garland, New York
2. Allen JW, Tevatia R, Demirel Y, DiRusso CC, Black PN (2018) Induction of oil accumulation by heat stress is metabolically distinct from N stress in the green microalgae *Coccomyxa subellipsoidea* C169. PLoS ONE 13(e0204505):1–20
3. Demirel Y (2012) Energy coupling. In: Terzis G, Arp R (eds) Information and Living Systems in Philosophical and Scientific Perspectives. MIT Press, Cambridge, pp 25–53
4. Demirel Y, Gerbaud V (2019) Nonequilibrium thermodynamics: Transport and rate processes in physical, chemical and biological systems, 4th edn. Elsevier, Amsterdam
5. Demirel Y (2010) Nonequilibrium thermodynamics modeling of coupled biochemical cycles in living cells. J Non-Newtonian Fluid Mechanics 165:953–972
6. Küçük K, Tevatia R, Sorgüven E, Demirel Y, Özilgen M (2015) Bioenergetics of growth and lipid production in Chlamydomonas reinhardtii. Energy 83:503–510
7. Marks DB (1999) Biochemistry. Kluwer, Ney York
8. Singh R, White D, Demirel Y, Kelly R, Noll K, Blum P (2018) Uncoupling fermentative synthesis of molecular hydrogen from biomass formation in *Thermotoga maritime*. Appl Environ Microbiol 84:e00998-e1018

Sustainability in Energy Technologies

11

Introduction and Learning Objectives: 'Sustainability' is maintaining or improving the material and social conditions for human health and the environment over time without exceeding the ecological capabilities that support them. The dimensions of sustainability are economic, environmental, and societal. Energy demand continues to grow worldwide while extraction of fossil fuels becomes more difficult and expensive. The ways we produce, convert, store, and use energy are changing earth's climate. Nearly two-thirds of global GHG emissions are related to production and consumption of energy. This makes the sustainability in energy sector critical to mitigating global warming and climate change. Increasingly resource-intensive consumption across the world requires novel and sustainable energy technologies and management. Therefore, this chapter focuses on the measurable sustainability and life cycle analysis in energy sector and addresses the implications of depletion of natural energy and material sources on the environmental change.

The learning objectives of this chapter are to understand:

- The sustainability in energy systems,
- Life cycle analysis of energy systems,
- Exergy and sustainability.

11.1 Sustainability

Sustainability is maintaining or improving the material and social conditions for human health and the environment over time without exceeding the ecological capabilities that support them. Energy is one of the main drivers of technology and development and hence the demand for energy continues to grow worldwide. The ways we produce, convert, store, and use energy are changing Earth's climate and environment; hence the ways of human's life as well as the next generation's future. Therefore, energy management is a global challenge. Nearly two-thirds of global

© Springer Nature Switzerland AG 2021
Y. Demirel, *Energy*, Green Energy and Technology,
https://doi.org/10.1007/978-3-030-56164-2_11

GHG emissions are associated with the production and consumption of energy. This makes the sustainability of energy technology critical to mitigating adverse effects of global warming and climate change [17, 18]. Some major actions needed in reducing the growth of atmospheric GHG emissions are:

- Increase energy efficiency.
- Increase the usage of renewable energy.
- Develop technologies for carbon capture and storage (CCS).
- Develop effective energy storage technologies.
- Recognize leadership role in transforming the global energy sector.

Zero energy building and green building efforts use resources more efficiently and reduce a building's negative impact on the environment. Zero energy buildings may have a much lower ecological impact compared with other 'green' buildings that require imported energy and/or fossil fuel. 'Green building certification' programs require reducing the use of nonrenewable energy considerably [18].

11.1.1   Natural Earth Cycles

The Carbon Cycle: Plants use carbon dioxide, water, and energy from the sun to form carbohydrates in photosynthesis

$$6CO_2 + 6H_2O \leftrightarrow C_6H_{12}O_6 + 6O_2 \qquad (11.1)$$

Carbohydrates, in turn, are oxidized to produce water and carbon dioxide. Atmospheric carbon dioxide levels have increased considerably since the industrial revolution.

Nitrogen Cycle: Atmospheric nitrogen is converted to ammonia or ammonium ion by nitrogen-fixing bacteria that live in legume root nodules and in soil. Atmospheric nitrogen is converted to nitrogen oxides during combustion at high temperatures

$$N_2 \rightarrow NH_3 \text{ or } NH_4^+ \qquad (11.2)$$

$$NO_2^- + H_2O \rightarrow H_2NO_3 \qquad (11.3)$$

Ammonia and ammonium are oxidized by soil bacteria first to nitrite ions and then to nitrate ions. After plants take up nitrogen from the soil in the form of nitrate ions, the nitrogen is passed along the food chain. The remaining nitrogen is released as ammonium ions or ammonia gas.

NO_x compounds: Molecules of nitrogen and atmospheric oxygen combine at very high temperatures to form nitric oxide. Lightening or the combustion chambers of engines are effective in causing this conversion of nitrogen to NO_x compounds. Once in the atmosphere, NO_x compounds may react with additional oxygen to form

nitrogen dioxide, which is a red-brown toxic gas that causes irritation to the eyes and respiratory systems [18, 20].

11.1.2 Ozone Formation and Destruction

High-energy ultraviolet (UV) photons react with oxygen, splitting them into highly reactive oxygen atoms. These are free radicals in the stratosphere and can combine with oxygen molecules to form ozone. Each ozone molecule can absorb a UV photon with a wavelength of less than 320 nm and can prevent potentially harmful UV rays from reaching Earth's surface. High-energy UV photons in the stratosphere split chlorine radicals from chlorofluorocarbons (CFCs) by breaking their carbon-chlorine (C–Cl) bonds. The freed chlorine radicals are very reactive and can destroy ozone by converting it to diatomic oxygen. Some *ozone depleting substances* are hydrochlorofluorocarbons, chlorofluorocarbons, and halons (https://www.epa.gov/ozone-layer-protection/ozone-depleting-substances) [16].

11.1.3 Greenhouse Gases

Greenhouse gases (GHGs) can trap heat in the atmosphere. Some important GHGs are carbon dioxide (CO_2), methane (CH_4), nitrous oxide (N_2O), and fluorinated gases. Carbon dioxide emitted to the atmosphere after burning fossil fuels, through solid waste, biological systems, and certain chemical reactions. Production of coal, natural gas, and oil emit methane. Livestock, agricultural processes, and decompose of organic waste also emit methane. Many agricultural activities, industrial processes, burning of fossil fuels and solid waste, as well as wastewater treatment emit nitrous oxides. Some industrial processes emit fluorinated gases of hydrofluorocarbons, perfluorocarbons, sulfur hexafluoride, and nitrogen trifluoride. These gases are powerful GHGs, although emitted in small quantities [16].

Carbon Tracking allows the calculation of CO_2 equivalent emissions after specifying 'CO_2 emission factor data source' and 'ultimate fuel source'. The CO_2 emission factor data source can be from European Commission decision of '2007/589/EC' or United States Environmental Protection Agency (EPA) Rule of E9-5711. Table 11.1 shows the emission rates from some of the fuel sources estimated by the emission factor data source [8].

Global CO_2 emissions due to fossil fuel use can be calculated as

$$e_{CO_2} = (C_f/E_f)(MW_{CO_2}/MW_C) \tag{11.4}$$

where e_{CO_2} is the CO_2 emission in $kgCO_2/kWh$, C_f is the carbon content in the fuel (kg_C/kg_{fuel}) and E_f is the energy content of the fuel (kWh/kg_{fuel}). One tonne of carbon is equivalent to: $MW_{CO_2}/MW_C = 44/12 = 3.7$ tonnes of carbon dioxide. Table 11.2 shows typical emission of carbon dioxide from the combustion of various fuels.

Table 11.1 Emission rates (lb/MMBtu*) for various CO_2 emission factor data sources and fuel sources [8]

Fuel Source	US-EPA-Rule-E9-5711	EU-2007/589/EC
Natural gas	130.00	130.49
Coal bituminous	229.02	219.81
Coal anthracite	253.88	228.41
Crude oil	182.66	170.49
Biogas	127.67	0

*MMBtu = one million Btu

Table 11.2 Emission of carbon dioxide from the combustion of various fuels [4, 16, 24]

Fuel	Specific carbon (kg_c/kg_{fuel})	Specific energy (kWh/kg_{fuel})	Specific CO_2 emission (kg_{CO2}/kg_{fuel})	Specific CO_2 emission (kg_{CO2}/kWh)
Coal (bituminous/anthracite)	0.75	7.50	2.3	0.37
Gasoline	0.90	12.5	3.3	0.27
Light Oil	0.70	11.7	2.6	0.26
Diesel	0.86	11.8	3.2	0.24
LPG—Liquid petroleum gas	0.82	12.3	3.0	0.24
Natural gas, Methane	0.75	12.0	2.8	0.23
Crude oil				0.26
Kerosene				0.26
Wood*				0.39
Peat *				0.38
Lignite				0.36

*Commonly viewed as a biofuel

11.1.4 Greenhouse Effects

Greenhouse effect occurs when the atmosphere traps radiating from the Earth to space, and hence warms climate. Carbon containing gases in the upper atmosphere and small particles in the form of smoke in the lower atmosphere can trap heat escaping from Earth's surface, while they allow sunlight to pass. GHGs, therefore, affect environmental and health adversely, including climate change, respiratory disease from smog, and air pollution. Climate change causes extreme weather behavior, flooding, and wildfires [16, 17].

Climate Change: The effect greenhouse gases (GHGs) on climate depends on the concentration (i.e. parts per million, ppm) and the time of remaining of carbon in the atmosphere (years). Global warming potential is estimated based on how long GHGs remain in the atmosphere and how effectively trap heat. Much of the world is covered with ocean water, which heats up by the trapped heat and evaporates more into the clouds. Accumulated moisture in the atmosphere may form into heavy rains and energy-intensive storms. In addition, warmer atmosphere melts glaciers and polar ice caps and cause the rising sea levels [17].

Global Warming Potential: Global warming potential (GWP) measures how much energy of the emissions of 1 ton of gas would trap over a given period, relative to the emissions of 1 ton of carbon dioxide. The period is usually 100 years. GWP compares emission levels of various GHGs. Carbon dioxide remains for a long time and has a GWP of 1 as a reference substance. Methane has a GWP of 28–36 over 100 years that shows methane can trap more energy than that of carbon dioxide. Nitrous oxides have a GWP of 265–298 for a 100 years. Chlorofluoro-carbons, hydrochlorofluorocarbons, hydrofluorocarbons, perfluorocarbons, and sulfur hexafluoride gases have very high values of GWP. Chlorofluorocarbon-12 has a global warming potential of 8500, while chlorofluorocarbon-11 has a GWP of 5000. Various hydro chlorofluorocarbons and hydrofluorocarbons have GWPs ranging from 93 to 12,100. These values are calculated over a 100-year period [16].

Global warming potential can be estimated based on data from the three popular standards: (1) the IPCC's 2nd (SAR), (2) 4th (AR4) Assessment Reports, and (3) the U.S. EPA's (CO2E-US) proposed rules from 2009 [8].

Since the early 1800s, it is known that various atmospheric gases, acting like the glass in a greenhouse, transmit incoming sunlight but absorb outgoing infrared radiation, thus raising the average air temperature at Earth's surface. Carbon dioxide is clearly the most influential greenhouse gas because of its high level of emissions. The most compelling evidence we have for climate change lies in the so-called paleoclimatic data obtained from ancient ice core samples in Greenland and Antarctica. By analyzing air bubbles that were trapped in the ice when it formed, scientists can determine the content of greenhouse gases and even the average temperature at each point in time. Figure 11.1 shows that over the past 420,000 years, the CO_2 content in the atmosphere has varied cyclically between about 180 and 290 ppm by volume with a period of about 100,000 years in conjunction with variations in Earth's orbit. Earth's temperature has closely followed the greenhouse gas concentration [20].

Around 1850, when the atmospheric CO_2 level was about 280 ppm, the level began to increase and now reached the value of 380 ppm, which indicates a 36% increase over the pre-industrial value. Increase in temperature can release CO_2 from the ground and seawater, so the two effects reinforce each other. The possible consequences of these increases include ice melts, sea level rises, and severe storms because of the additional energy in the atmosphere. As the ice melts, the resulting water and ground absorb more sunlight, thus exacerbating the global warming. The melt water may turn into a river, causing rapid heat transfer and erosion [20].

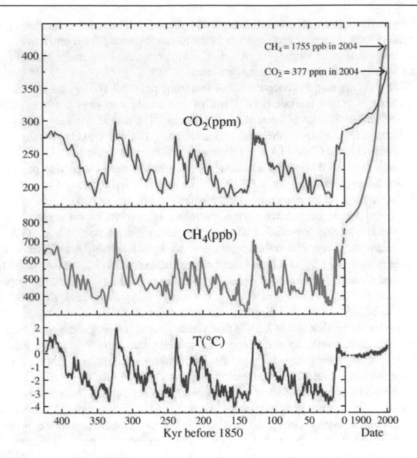

Fig. 11.1 Paleoclimatic data from ice cores shows recent increases in carbon dioxide and methane. The temperature, though increasing, has not yet reached record levels but will likely do so by midcentury [20]

11.1.5   Tackling Global Warming

Electricity production and transportation fuels accounts for a large part of total GHG emissions. Therefore, targeting electricity generation and transportation fuels will address about 70% of the carbon dioxide emissions. For example, the use of renewable energy for electricity generation does not cause additional carbon dioxide emissions and is sustainable. The major challenges with greatly expanded use of renewables are the cost, intermittency of supply, and distance between the resources and the end use. Some possible renewable energy sources are concentrating solar power, geothermal electric plants, wind power, distributed rooftop photovoltaic, and solar hot water heaters. Electric vehicles represent an important improvement as they can be plugged into the grid to be recharged and reduce the amount of gasoline use. In addition, using E85 (85–15% blend of ethanol and gasoline) may help reduce carbon dioxide emission [11, 16, 17].

11.1.6 Why the Sustainability Matters?

Use of renewable energy technologies (including solar, wind, biomass) and alternative ways of using traditional fossil and nuclear fuels are growing but constrained by various factors, including cost, infrastructure, public acceptance, and others. For example, biomass as a renewable resource can be used to produce biofuels and hence can reduce net GHG emissions by recycling most of the carbon [4, 5]. The following can have part of energy policies toward sustainability:

- Increase energy efficiency with energy integration.
- Increase the usage of renewable energy.
- Include technologies for carbon capture, storage, and utilization.
- Autothermal processes with heat recovery steam generation (HRSG).

Within the food–energy–water nexus, the world will need 80% more energy, 55% more water, and 60% more food in 2050 [11]. This projection emphasizes the role of energy in sustainability for global welfare and protection of the ecology with:

- Combination of the first and second-generation biomass will produce ethanol with less water and fertilizers and increase the yield,
- Use of waste lignin for producing valuable chemicals will help the second-generation-based biofuels to be more feasible,
- Municipal wastewater-based algal technology will use less water and nutrients to reduce the cost of algal-based biodiesel production.
- Consequently, biofuels will reduce the water intensity in power production sector and help decouple food production from fossil fuel usage for a more secure water and food supplies.

11.1.7 United Nation's Sustainable Development Goals

In 2015, UN General Assembly formally adopted the universal, integrated, and transformative 2030 Agenda for Sustainable Development Goals (SDGs) consisting of a set of 17 SDGs and 169 associated targets (https://sustainabledevelopment.un.org/sdgs). The SDGs are: (1) no poverty, (2) zero hunger, (3) good health and well-being, (4) quality education, (5) gender equality, (6) clean water and sanitation, (7) affordable and clean energy, (8) decent work and economic growth, (9) industry, innovation, and infrastructure, (10) reduced inequalities, (11) sustainable cities and communities, (12) responsible consumption and production,

(13) climate action, (14) life below water, (15) life on land, (16) peace justice and strong institutions, and (17) partnerships for the goals. The SDGs were adopted by 193 countries as the world's shared plan to protect the planet, end extreme poverty, and reduce inequality. Achieving the SDGs requires coherent and global efforts that need to be backed by practical commitments and actions [26].

11.1.8 Sustainability and Energy

The green energy transition can achieve environmental goals, improve economic development, job creation, and lead to social equity and welfare for society. Many renewable energy technologies are distributed and allow citizens to be more involved and empowered in the energy transition. The energy transition is at the heart of these objectives.

The IRENA's Transition Energy Scenario (TES) would result in a 70% reduction in carbon dioxide emissions by 2050 achieved by mobilizing resources and initiating policies. Additionally, deep decarbonization perspective emphasizes the important role of emerging technologies in reducing emissions stemming from energy production, transport, industrial processes, and wasteful behaviors. The energy transition is providing the interactions of the energy system with the wider economy, socioeconomics, and the natural environment [19].

Energy transition avoids the use of polluting fuels and creates a climate-resilient economy benefiting all including strong employment and welfare gains. However, the energy transition will generate diverse outcomes for regions and countries stemming from the depth, strength, and diversity of their national supply chains, as well as dependency on fossil fuels and other commodities, technologies, and trade patterns [17].

11.1.9 Sustainability and Ecology

Climate change refers to changes in temperatures, emissions, and climate. *Conservation ecology* addresses changes in populations related to the small population sizes of various species. *Conservation biology* applies science to the conservation of ecosystems, species, and populations. *Source depletion* refers to the exhaustion of raw materials. The collective environmental regulations and technical advances, such as pollution control and prevention, and waste minimization have greatly diminished adverse environmental impacts of energy production processes. From a sustainability viewpoint, the most important factors that determine the suitability of energy technology are [16, 17]:

* Energy use per unit of economic value-added product.
* Type of energy used (renewable or non-renewable).
* Materials use (or resource depletion).
* Fresh water use.

- Waste and pollutants production.
- Environmental impacts of product/process/service.
- Assessment of overall risk to human health and the environment.

Example 11.1 Carbon dioxide emission from natural gas combustion
When a hydrocarbon fuel is burned, almost all of the carbon in the fuel burns completely to form carbon dioxide CO_2, which causes the greenhouse effect. On average, 0.59 kg of CO_2 is produced for each kWh of electricity generated from a power plant that burns natural gas. A typical new household uses about 7000 kWh of electricity per year. Determine the amount of CO_2 production in a city with 200,000 households.

Solution:
A kWh is kW times hours. It is the amount of energy consumed in an hour by a device that uses 1 kW of power.
Data: m_{CO_2} = 0.59 kg CO_2/kWh, 200,000 houses, power = 7000 kWh/house per year
The total mass of CO_2 produced per year is the product of the rate of CO_2 production per kWh and the number of houses:

$$m_{totalCO_2} = \left(\frac{7000 \text{ kWh}}{\text{house year}}\right)\left(\frac{0.59 \text{ kg } CO_2}{\text{kWh}}\right)(200{,}000 \text{ Houses})$$

$m_{total \; CO2} = \mathbf{8.26 \times 10^8}$ **kg CO_2/year = 826.0 Mton CO_2/year**

11.2 Sustainability Metrics

The dimensions of sustainability are economic, environment, and societal as seen in Fig. 11.2. Thus, eco-efficiency metrics are indicative of changes in economic and environmental aspects, and are indicated by the intersection of the economic and environmental dimensions. The sustainability metrics, indicated by the intersection of all three dimensions, are truly representative of progress toward sustainability. One- and two-dimensional metrics, while useful, cannot alone certify progress toward sustainability. Economic and societal indicators may also be constrained with the environment, as seen in Fig. 11.2b. Suitable assessment tools are needed for the development and design of sustainable energy systems for the environment. A set of sustainability metrics that are applicable to a specific energy process are [8, 16]:

- Material intensity (nonrenewable resources of raw materials, solvents/unit mass, or power, or economic value produced)
- Energy intensity (nonrenewable energy/unit mass, or power, or economic value produced)
- Potential environmental impact (pollutants and emissions/unit mass, or power, or economic value produced)

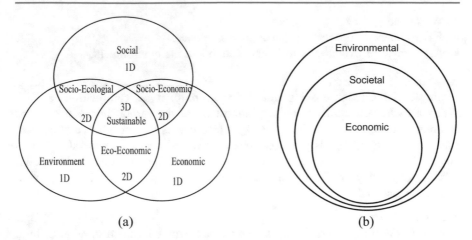

Fig. 11.2 a Three dimensions of sustainability; **b** economic and societal dimensions are constrained with the environment

- Potential chemical risk (toxic emissions/unit mass, or power, or economic value produced)

The first two metrics are associated with the process operation focusing on the inputs to the process. The remaining two metrics represent the outputs from the process as chemical risk to human health in the process environment, and the potential environmental impact of the process on the surrounding environment. The potential environmental impacts [8] can be measured by using the 'Carbon Tracking' and 'Global Warming Potential', which will be discussed briefly below.

The sustainability metrics reflect the three elements of sustainability: (i) environmental, (ii) economic, and (iii) social. Most of the sustainability metrics are calculated as ratios of a measure of impact independent of the scale of operation that is the ratio of impact per unit mass, or power, or economic value produced [15, 21].

Example 11.2 Consumption of Coal and Emission of Carbon Dioxide from Coal

A large public computer lab runs six days per week from Monday through Saturday. Each computer uses a power of around 240 W. If the computer lab contains 45 computers and each is on for 12 h a day, during the course of the year how much CO_2 will the local coal power plant have to release to the atmosphere in kg to keep these computers running?

Solution:
Data: Coal (bituminous/anthracite): 0.37 kg of CO_2/kWh. Data from Table 2.12. Electricity required:

$$\left(\frac{240 \text{ W}}{\text{computer}}\right)\left(\frac{\text{kW}}{1000 \text{ W}}\right)\left(\frac{52 \text{ weeks}}{\text{year}}\right)\left(\frac{6 \text{ days}}{\text{week}}\right)\left(\frac{12 \text{ h}}{\text{day}}\right)\left(\frac{45 \text{ computers}}{\text{lab}}\right)$$
$$= 40{,}435.2 \text{ kWh}$$

We know that 40,435.2 kWh is needed from the lab per year and since the power plant is using coal we can use its emission value of 0.37 kg of CO_2/kWh. (Table 11.2).
(40,435.2 kWh) (0.37 kg CO_2/kWh) = **14,961 kg CO_2 released**

Example 11.3 Reducing air pollution by geothermal heating
A district uses natural gas for heating. Assume that average NO_x and CO_2 emissions from a gas furnace are 0.0045 and 6.4 kg/therm, respectively. It is considered to replace the gas heating system with a geothermal heating system. The projected saving by the geothermal heating system would be 20×10^6 therms of natural gas per year. Determine the amount of NO_x and CO_2 emissions the geothermal heating system would save every year.

Solution:
therm = 29.3 kWh (Table 1.9).
Reduction in NO_x emission = (0.0045 kg/therm)(20×10^6 therm/year) = **9.0×10^4 kg/year**
Reduction in CO_2 emission = (6.4 kg/therm)(20×10^6 therm/year) = **$12.8.0 \times 10^7$ kg/year**
Atypical car produces about 8.5 kg NO_x and 6000 kg of CO_2 per year. Replacing the gas heating system by the geothermal heating system is equivalent to taking 10,600 cars off the road for NO_x emission and taking 21,000 cars of the road for CO_2 emission. Therefore, the proposed geothermal heating would have a positive impact on the air pollution.

11.3 Sustainability Index

The AIChE's Sustainability Index (https://www.aiche.org/ifs/resources/sustainability-index) has seven key elements that help determine how a company's sustainability efforts are perceived by the community, shareholders, customers, and peers.

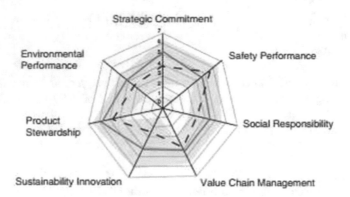

The indices benchmark well-defined performance metrics and indicators, including environment, health and safety performance, innovation, and societal measures to:

1. Benchmark a company's performance among peers
2. Assess a company's performance against well-defined metrics on an on-going basis
3. Measure progress toward best practices at regular intervals
4. Access unbiased, expert interpretation of publicly available technical data
5. Better understand public perception of your company's sustainability efforts

1. **Sustainability innovation** considers:

   - General research and development (R&D) commitment as evident in the amount of R&D expenditure per net sales
   - Sustainable products and processes with superior environmental, social, and economic performance
   - Sustainability approaches in R&D and innovation processes
   - R&D effectiveness as reflected in the number of patents issued and commercialization of new products that enhance environmental and social sustainability.

2. **Strategic commitment** considers:

   - Stated commitment to excellence in environmental and social performance throughout a company's value chain
   - Commitment to voluntary codes and standards, including responsible care, global compact, and others
   - Reporting of sustainability performance
   - Sustainability goals and programs that are specific and challenging
   - Third-party ratings with respected agencies' ratings on company-wide sustainability management and reporting.

3. **Environment performance** considers:

- Resource use with intensity of energy, material, and water consumption, and use of renewable sources of energy and materials for GHG emissions
- Other emissions with air emissions, wastewater, and hazardous waste releases
- Compliance management with environmental liability, fines and penalties, and environmental capital investment.

4. **Safety performance** considers:

- Employee safety with recordable and days-away-from work injury rates
- Process safety records incidents, normalized by number of employees, and occurrence of major safety incidents
- Plant security with plant security management system representing completion of a responsible care plant security audit.

5. **Product stewardship** considers:

- Assurance system incorporates a responsible care product safety process, and engagement of value-chain partners to assure product safety
- Risk communication incorporates a responsible care risk communication process
- Legal proceedings related to product safety, risk, and toxicity.

6. **Social responsibility** considers:

- Stakeholder partnerships and engagement at the project, facility, and corporate levels
- Social investment through employment, philanthropy, and community development projects
- Image in the community indicated by reputable awards and recognition programs, including 'most admired' and 'best employer' ratings

7. **Value-chain management** considers:

- Environmental management systems
- Supply chain management.

11.4   Sustainability Impact Indicators

There are sustainability impacts and impact indicators that are related to the depletion of resources and environmental change shown in Table 11.3. Defining and using of a sustainability framework and the associated potential impact indicators are essential for selecting and evaluating energy systems for the future [16, 17]. A set of indicators provides technical, sustainability, and societal

Table 11.3 Sustainability impacts and impact indicators

Impact	Potential Impact Indicators
Resource depletion	
Depletion of nonrenewable natural resources	Abiotic resource depletion potential
Depletion of nonrenewable primary energy resources	Nonrenewable primary energy consumption potential
Water usage	Water resource depletion potential
Environmental	
Greenhouse gas effect-CO_2e	Global warming potential
Air pollution	
Air acidification	Acidification potential
Photochemical oxidation	Photochemical ozone creation potential
Exhaustion of the ozone layer	Ozone layer depletion potential
Water pollution	
Eutrophication	Eutrophication potential (nitrification)
Soil pollution	
Soil contamination	Soil contamination potential
Toxic hazards	
For humans	Human toxicity potential
For aquatic ecosystems	Aquatic toxicity potential
For sedimentary ecosystems	Sedimentary toxicity potential
For terrestrial ecosystem	Terrestrial toxicity potential
Economics	
Feasibility	Net present value Payback period Rate of return
Earnings	Cash flow
Inflation	Interest rates
Cost of production	Consumer index
Development	Gross Domestic Product (GDP)
Societal	
Workplace	Employment
Society	Welfare
Health and quality of life	Health insurance, education, wellbeing

perspectives on energy system choices. Such indicators would allow comparison of diverse energy systems, would provide a common basis for evaluation, and would inform public and stakeholders on technology development and energy policy. A focus on a specific energy system such as biomass to energy may provide technical, sustainability, and societal perspectives. A simplified and specific representation of the biomass to energy may be biomass to ethanol energy system to address the aspects and impacts that need to be considered in developing metrics such as value chain of supply, production, use and fate, social and environmental dimensions of sustainability, context of resources, time and location. Sustainable energy includes:

- Renewable energy sources, such as biomass, solar, wind, wave, geothermal, hydro, and tidal.
- The strategies and technologies that improve energy efficiency and reduce the environmental impact.

11.4.1 Resource Depletion

These indicators display a balanced view of the environmental impact of resource usage, emissions, effluents, and waste produced [16, 17].

Energy: (i) total net primary energy usage rate = imports—exports (GJ/year), (ii) total net primary energy sourced from renewables (%), and (iii) total net primary energy usage per kg product (kJ/kg).

Material (excluding fuel and water): (i) total raw materials used, including packaging (ton/year), (ii) raw material recycled from other company operations (ton/year), (iii) raw material recycled from consumer (ton/year), (iv) raw material used which poses health, safety, or environmental hazard (ton/year), (v) total raw materials used per kg product (kg/kg), (vi) fraction of raw materials recycled within company (kg/kg), (vii) fraction of raw materials recycled from consumers (kg/kg), and (viii) hazardous raw material per kg product (kg/kg).

Water: (i) water used in cooling (ton/year), (ii) water used in process (ton/year), (iii) other water used, (ton/year), (iv) water recycled internally (ton/year), (v) net water consumed = total used—recycled (ton/year), and (vi) net water consumed per unit mass of product (kg/kg).

Land: (i) land occupied by operating unit (m^2) (include land needed for all activities), (ii) other land affected by unit's activities (m^2) (land used in mining raw material or in dumping waste product) total land (m^2), (iii) land restored to original condition (m^2/year), (iv) total land occupied + affected for value added (a) m^2/($/year), (v) rate of land restoration (restored per year/total) (b) (m^2/year)/m^2. The areas of land occupied and affected are those at the start of the reporting period, and the land restored is that area restored during the reporting period.

11.4.2 Environmental Burden

The Environmental Burden (EB) caused by the emission of a range of substances is calculated by adding up the weighted factor known as the 'potency factor'. A single substance contributes differently to different burdens; each substance will have several different potency factors. Environmental burden is estimated by [16]:

$$\text{Environmental Burden}(EB)_i = \sum m_n(\text{tons}) \times \text{Potency Factor}(pf)_{i,n} \quad (11.5)$$

where EB_i is the ith environmental burden, n is the substance index, and m is the mass amount of that substance.

Atmospheric Impact: (i) Global warming burden; (ii) Atmospheric acidification burden, (iii) Human health burden, (iv) Ozone depletion burden, (v) Photochemical ozone burden. The environmental burden of global warming is measured by ton/year carbon dioxide equivalent and atmospheric acidification is measured by ton/year sulfur dioxide equivalent (Table 11.4). The potential of certain gases released to form acid rain and acids is the potency factor for the atmospheric acidification. The potential factors of global warming are based on a 100-year integrated time period.

Table 11.4 Potency factors for global warming and atmospheric acidification [16]

Substance	Potency factor, pf
Global warming	
Carbon dioxide	1
Carbon monoxide	3
Carbon tetrachloride	1400
Chlorodifluoromethane, R22	1700
Chloroform	4
Chloropentafluoroethane, R115	9300
Dichlorodifluoromethane, R12	8500
Dichlorotetrafluoroethane, R114	9300
Difluoroethane	140
Hexafluoroethane	9200
Methane	21
Methylene chloride	9
Nitrous oxide	310
Nitrogen oxides (NO$_x$)	40
Tetrafluoroethane, R125	2800
Perfluoro methane	6500
Tetrafluoroethene	1300
Trichloroethane (1,1,1)	140
Trichlorofluoromethane, R11	4000

(continued)

Table 11.4 (continued)

Substance	Potency factor, pf
Trichloro trifluoroethane, R113	5000
Trifluoroethane, R143a	3800
Trifluoroethane, R23	11,700
Volatile organic compounds, VOC	11
Atmospheric acidification	
SO_2	1
Ammonia	1.88
HCL	0.88
HF	1.66
NO_2	0.7
$H2SO_4$ mist	0.65

Human Health (carcinogenic): There are no commonly accepted potency factors for human health. Carcinogenic effects are offered as a default set, but other sets can also be used if they are more appropriate. The potency factor for this category in Table 11.5 is from the reciprocal of the occupational exposure limits (OEL) set by the UK Health and Safety Executive. The OEL for benzene has been chosen as the normalizing factor for this category: Potency factor of substance = (OEL benzene/OEL substance). Chemicals with an OEL greater than 500 mg/m^3 have a minimal impact on the total weighted impact. The unit of EB is ton/year benzene equivalent [16].

Table 11.5 Human health potency factors [16]

Compound	CAS number	Potency factor, pf
Acrylamide	79-06-1	53.3
Acrylonitrile	107-13-1	3.6
Arsenic & compounds except arsine	7440-38-0	160
Azodicarbonate	123-77-3	16
Benzene	71-43-2	1
Bis (chloromethyl) ether	542-88-1	3,200
Buta-1,3-diene	106-99-0	0.73
Cadmium oxide fume	1306-19-0	640
Carbon disulfide -	136-23-6	0.5
1 Chloro-2,3-epoxypropane	106-89-8	8.4
1,2-dibromoethane	106-93-4	4.1
1,2-dichloroethane	107-06-2	0.76
Dichloromethane	75-09-2	0.05
2–2'-Dichloro-4,4'-methylene dianiline	101-14-4	3,200
Diethyl sulphate	64-67-5	50

(continued)

Table 11.5 (continued)

Compound	CAS number	Potency factor, pf
Dimethyl sulphate	77-78-1	3.8
2-Ethoxyethanol	110-80-5	0.43
2-Ethoxyethyl acetate	111-15-9	0.3
Ethylene oxide	75-21-8	1.7
Formaldehyde	50-00-0	6.4
Hydrazine	30-07-2	533.3
Iodomethane	74-88-4	1.3
Maleic anhydride	108-31-6	16
2-Methoxyethanol	109-86-4	1
2-Methoxyethyl acetate	110-49-6	0.64
4–4′-methylenedianiline	101-77-9	200
2-Nitropropane	79-46-9	0.8
Phthalic anhydride	85-44-9	4
Polychlorinated biphenyls (PCB)	1336-36-3	160
Propylene oxide	75-56-9	1.33
Styrene	100-42-5	0.04
o-Toluidine	95-53-4	18
Triglycidyl isocyanurate (TGIC)	2451-62-9	160
Trimellite anhydride 400	552-30-7	400
Vinylidene chloride	75-35-4	0.4

Stratospheric Ozone Depletion: The potency factor is based on the ozone depletion potential in the upper atmosphere relative to chlorofluorocarbon—CFC-11. The unit of Environmental Burden is ton/year CFC-11 equivalent (CFC-11 is trichlorofluoromethane) (see Table 11.6).

Table 11.6 Potency factors for ozone depletion [16]

Substance	Potency factor
CFC-11, CFC-12, CFC-13	1.0
CFC-113	0.8
CFC-114	1.0
CFC-115	0.6
CFC-111, CFC-112	1.0
CFC-112, CFC-213, CFC-214	1.0
CFC-215, CFC-216, CFC-217	1.0
halon-1211	3.0
halon-1301	10.0
halon-2402	6.0
Carbon tetrachloride	1.1
1,1,1-trichloroethane	0.1

(continued)

Table 11.6 (continued)

Substance	Potency factor
Methyl bromide	0.7
HCFC-21	0.04
HCFC-22	0.055
HCFC-31	0.02
HCFC-121	0.04
HCFC-122	0.08
HCFC-123	0.02
HCFC-124	0.022
HCFC-131	0.05
HCFC-132	0.05
HCFC-133	0.06
HCFC-141	0.07
HCFC-141b	0.11
HCFC-142	0.07
HCFC-142b	0.065
HCFC-151	0.005
HCFC-221	0.07
HCFC-222	0.09
HCFC-223	0.08
HCFC-224	0.09
HCFC-225	0.07
HCFC-226	0.1
HCFC-231	0.09
HCFC-232	0.1
HCFC-233	0.23
HCFC-234	0.28
HCFC-235	0.52
HCFC-241	0.09
HCFC-242	0.13
HCFC-243	0.12
HCFC-244	0.14
HCFC-252	0.04
HCFC-253	0.03

Photochemical Ozone (smog) Formation: Potency factors for photochemical ozone formation are obtained from the potential of substances to create ozone photochemically. The unit of Environmental Burden is ton/year ethylene equivalent (see Table 11.7).

Table 11.7 Potency factors for petrochemical ozone depletion [16]

Substance	Potency factors
Methane	0.034
Ethane	0.14
Propane	0.411
n-Butane	0.6
i-Butane	0.426
n-Pentane	0.624
i-Pentane	0.598
n-Hexane	0.648
2-Methylpentane	0.778
3-Methylpentane	0.661
2,2-Dimethylbutane	0.321
2,3-Dimethylbutane	0.943
n-heptane	0.77
2-Methylhexane	0.719
3-Methylhexane	0.73
n-Octane	0.682
2-Methylheptane	0.694
n-Nonane	0.693
2-Methyloctane	0.706
n-Decane	0.680
2-Methylnonane	0.657
n-Undecane	0.616
n-Dodecane	0.577
Cyclohexane	0.595
Methyl cyclohexane	0.732
Ethylene	1.0
Propylene	1.08
1-Butene	1.13
2-Butene	0.99
2-Pentene	0.95
1-Pentene	1.04
2-Methylbut-1-ene	0.83
3-Methylbut-1-ene	1.18
2-Methylbut-2-ene	0.77
Butylene	0.703
Isoprene	1.18
Styrene	0.077
Acetylene	0.28
Benzene	0.334
Toluene	0.771
o-Xylene	0.831

(continued)

Table 11.7 (continued)

Substance	Potency factors
m-Xylene	0.08
p-Xylene	0.948
Ethylbenzene	0.808
n-Propyl benzene	0.713
i-Propyl benzene	0.744
1,2,3-Trimethylbenzene	1.245
o-Ethyl toluene	0.846
m-Ethyl toluene	0.985
p-Ethyl toluene	0.935
3,5-Dimethylethylbenzene	1.242
3,5-Diethyltoluene	1.195
Formaldehyde	0.554
Acetaldehyde	0.65
Propionaldehyde	0.755
Butyraldehyde	0.77
i-Butyraldehyde	0.855
Valeraldehyde	0.887
Methylethylketone	0.511
Methyl-*i*-butyl ketone	0.843
Cyclohexanone	0.529
Methyl alcohol	0.205
Ethyl alcohol	0.446
i-Propanol	0.216
n-Butanol	0.628
i-Butanol	0.591
Diacetone alcohol	0.617
Cyclohexanol	0.622
Methyl acetate	0.046
Ethyl acetate	0.328
n-Propyl acetate	0.481
i-Propyl acetate	0.291
n-Butyl acetate	0.291
Formic acid	0.003
Acetic acid	0.156
Propionic acid	0.035
Butyl glycol	0.629
Propylene glycol methyl ether	0.518
Dimethyl ether	0.263
Methyl-*t*-butyl ether	0.268
Methyl chloride	0.035
Methylene chloride	0.031

(continued)

Table 11.7 (continued)

Substance	Potency factors
Methyl chloroform	0.002
Tetrachloroethylene	0.035
Trichloroethylene	0.075
Vinyl chloride	0.272
1,1-Dichloroethylene	0.232
cis 1,2-Dichloroethylene	0.172
trans 1,2-Dichloroethylene	0.101
Nitrogen dioxide	0.028
Sulphur dioxide	0.048
Carbon monoxide	0.027

Aquatic Impact

Aquatic acidification: Environmental burdens for emissions to water addresses aquatic acidification. The potency factor is the mass of hydrogen ion released by unit mass of acid that is the number of hydrogen ions released divided by the molecular weight. The unit of Environmental Burden EB is ton/year of H^+ ions released (see Table 11.8).

Aquatic Oxygen Demand The stoichiometric oxygen demand (StOD) may be the potency factor representing the potential of emissions to remove dissolved oxygen that would otherwise support aquatic life. An alternative potency factor is the chemical oxygen demand (COD). StOD is expressed as tons of oxygen required per ton of substance. The unit of environmental burden is ton/year oxygen (see Table 11.9).

Stoichiometric Oxygen Demand (StOD): From the knowledge of the chemical structure, we may calculate the empirical formula as follows: $C_cH_hN_nCl_{Cl}Na_{Na}O_o$ P_pS_s and the estimate the value of the StOD in ton equivalent (te) O_2 per ton of substance from the equation:

$$StOD = 16(2c + 0.5(h - Cl) + 2.5n + 3s + 2.5p + 0.5Na - o)/\text{Molecular Weight}$$

(11.6)

Table 11.8 Potency factor for aquatic acidification [16]

Substance	Potency factor
Sulfuric acid	0.02
Hydrochloric acid	0.027
Hydrogen fluoride	0.05
Acetic acid	0.02

Table 11.9 Potency factor for aquatic oxygen demand [16]

Substance	Potency factor
Acetic acid	1.07
Acetone	2.09
Ammonium nitrate in solution	0.8
Chlorotrifluoroethane	0.54
1,2-Dichloroethane (EDC)	0.81
Ethylene	1
Ethylene glycol	1.29
Ferrous ion	0.14
Methanol	1.5
Methyl methacrylate	1.5
Methylene Chloride	0.47
Phenol	0.238
Vinyl chloride	1.28

Here it is assumed that nitrogen is converted to nitrate ion (NO^{3-}) and carbon is converted to CO_2, hydrogen (H) to H_2O, phosphorus (P) to P_2O, sodium (Na) to Na_2O, sulfur (S) to SO_2, and halides (represented by Cl) to their respective acids. The compounds described after oxidation are those specified by international convention for calculating oxygen demand [16].

Example 11.4 Calculation of the stoichiometric oxygen demand
Estimate the values of StOD for acetic acid, phenol, and ammonium ion.

Solution:
Acetic acid CH_3COOH with a molecular weight of 60, we have
StOD = $16(2 \times 2 + 0.5 \times 4 - 2)/60 = 1.07$ ton equivalent O_2 per ton equivalent acetic acid
Another example, phenol C_6H_5OH with a molecular weight of 94
StOD = $16(2 \times 6 + 0.5 \times 6 - 1)/94 = 2.38$ te O_2 per ton equivalent of phenol
For ionic species, the charge of the ionic unit is considered. For the ammonium ion (NH^{4+}), for example, we remove an H^+ ion and calculate on the NH_3, so that the ionic balance is not disturbed.
StOD = $16(0.5 \times 3 + 2.5 \times 1)/17 = 3.76$ te O_2 per te of ammonia = 3.56 te O_2 per te of ammonium ion

---

Ecotoxicity to Aquatic Life (values for sea water conditions): The potency factor of metals is equal to the reciprocal of the Environmental Quality Standard (EQS)divided by the reciprocal of the EQS of copper. The unit of environmental burden is ton/year copper equivalent (see Table 11.10).

Potency factors of other substances (see Table 11.11) are equal to the reciprocal of the Environment Quality Standard (EQS) divided by the reciprocal of the EQS of formaldehyde. The unit of Environmental Burden is ton/year formaldehyde equivalent.

Table 11.10 Potency factor for ecotoxicity to aquatic life (values for sea water conditions) [16]

Substance	Potency factor
Arsenic	0.2
Cadmium	2.0
Chromium	0.33
Copper	1
Iron	0.005
Lead	0.2
Manganese	0.1
Mercury	16.67
Nickel	0.17
Vanadium	0.05

Table 11.11 Potency factors for ecotoxicity to aquatic life for other substances [16]

Substance	Potency factor
Ammonia	0.24
Benzene	0.17
Carbon tetrachloride	0.42
Chloride	0.5
Chlorobenzene	1
Chloroform	0.42
Cyanide	1.0
1,2-Dichloroethane	0.5
Formaldehyde	1
Hexachlorobenzene 1	166.67
Hexachlorobutadiene	50
Methylene chloride	0.5
Nitrobenzene	0.25
Nitrophenol	0.5
Toluene	0.125
Tetrachloroethylene	0.5
Trichloroethylene (TRI)	0.5
Xylenes	0.17

Eutrophication: Eutrophication is the potential for over fertilization of water and soil, which can result in increased growth of biomass. The species considered are to be responsible for eutrophication. The unit of environmental burden is ton/year phosphate equivalent (see Table 11.12).

Impact to Land: Two major impacts are the hazardous solid waste and non-hazardous solid waste.

Table 11.12 Potency factor for eutrophication [16]

Substance	Potency factors
NO_2	0.2
NO	0.13
NO_x	0.13
Ammonia	0.33
Nitrogen 0	42
PO4 (III-)	1
Phosphorus	3.06

11.4.3 Economics

Economic indicators determine if the process (plant) generates profit and is compatible with other existing plants and operations. Three profitability criteria with discounted cash flow approach with time value of the money are [6, 21]

(1) **Time**: Discounted Payback period (DPBP)

$$\text{Simple Payback Period} = \frac{\text{Total investment}}{\text{Average annual cash flows}} \quad (11.7)$$

DPBP $\leq n$ for profitable project; the shortest payback period is the most desirable. Here n is the number of years of competitive operation.

(2) **Cash**: Net present value (*NPV*); cumulative cash position at the end of the project

$$NPV = \sum_{j=1}^{n} \frac{CF_j}{(1+i)^j} \quad (11.8)$$

where CF_i is the discounted cash flow for year j, i is the interest rate for the bank loan, n is the total competitive operational time. *NPV* ≥ 0 for profitable project; the highest value of *NPV* is the most desirable.

(3) **Discount rate**: Discounted cash flow rate of return (*DCFROR*) is the interest rate for which the value of *NPV* is zero

$$0 = \sum_{j=1}^{n} \frac{CF_j}{(1+i*)^j} \quad (11.9)$$

where i^* is the discounted cash flow rate of return also known as the discount rate (percent). The discount rate is determined by the management and represents the minimum acceptable rate of return for a new project:

$$\text{Rate of return: } DCFROR = \frac{\text{Average annual net profit}}{FCI} \tag{11.10}$$

Discount rate: $DCFROR$ > internal interest rate for a profitable project. Here the FCI is the fixed capital investment and the net profit is the profit after tax.

11.4.4 Society

Indicators of social performance reflect the industry's attitude to treatment of public in the form of employees, suppliers, contractors, customers, and its impact on society at large [17].

Workplace: Employment situation: (i) number of employees who have resigned or been made redundant per year, (ii) number of direct employees promoted per year, (iii) working hours lost through absence per year (all unplanned causes—strikes, sickness, and absenteeism but not holiday or training), (iv) indicative wage and benefit package, (v) indicative wage and benefit package for lowest paid 10% of employees.

Society: Some indicators are: (i) coordination with external stakeholders concerning company operations annualy, (ii) indirect benefit to the community resulting from presence of operating unit ($/year), (iii) complications registered from members of the public concerning the process or products per year, (iv) number of successful legal actions taken against company or employees for work-related incidents or practices. External stakeholders include customers, residents and other community groups, local government, nongovernmental organizations.

Health and quality of life: A major social benefit arising from the presence of a successful process industry unit is the dissemination of skills and know-how which are used in the community to create wealth and enhance quality of life. It is difficult to quantify these benefits, but estimates may be made to include items such as (i) net value to community of freely published information and know-how, (ii) net value to community of training given to contractors and suppliers, (iii) net value to community of training given to employee should not include direct benefits. Value may be estimated by considering what it has cost the company to generate the benefit on the one hand, and what society might be willing to pay for it on the other [17].

11.5 Sustainability in Energy Systems

World energy requirements are expected to rise by roughly 50% by 2050. This level of energy requirement requires a sustainable energy production method. Renewable energy technologies like solar and wind power are the most environmentally friendly methods of energy production and can replace the fossil fuels-based power [18, 21].

Energy is a global challenge since the ways we produce, convert, store, and use energy are changing Earth's climate and environment, hence the ways of human's life as well as the next generation's future. Sustainable energy systems aim the largest triangle as seen in Fig. 11.3 for sustainable energy production with the inclusion of all the three dimensions of environment, social, and economics. Use of emerging renewable energy technologies (including solar, wind) and alternative ways of using traditional fossil and nuclear fuels are growing but constrained by various factors, including cost, infrastructure, public acceptance, and others.

Capturing and storing CO_2 from power plants and industrial processes add significant capital and operating costs without much economics return [9]. Operating costs of existing technologies for capturing the CO_2 from dilute flue gases are high; the flue gases of existing coal-fired power plants contain 7%–14% CO_2 and those from existing natural-gas-fired power plants contain 3–4% CO_2. One of the primary challenges is to make carbon capture and storage (CCS) viable for fossil fuel power plants; CO_2 must be captured and compressed to a supercritical state in order for CO_2 to be transported and stored. A few industrial processes, such as ethanol and ammonia productions, yield emissions that are nearly pure CO_2, mitigating the technical challenge and energy intensity of CO_2 capture. It is expected to further regulate emissions, solid waste, and cooling water intake that will affect the electric power sector, particularly the fleet of coal-fired power plants. In order to comply with those new regulations, existing coal-fired plants may need extensive emission and environmental control retrofits [17, 27].

Fig. 11.3 Targeting larger triangle for sustainable energy production with the inclusion of all the three dimensions

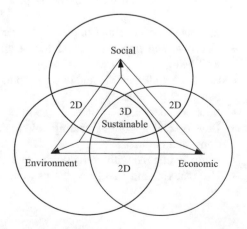

11.5.1 Sustainable Energy Systems Design

Sustainable design that takes into account environmental considerations can still be commercially successful, functional, and economical. Some important factors in the sustainability of energy system design are: (i) energy use per unit of economic value-added product, (ii) type of energy used (renewable or nonrenewable), (iii) resource depletion, (iv) fresh water use, (v) waste and pollutants production, and (vi) assessment of overall risk to human health and the environment. Some design concerns to address the environmental issues [1, 2, 3, 6] are:

- *The greenhouse effect*: We have an increase of the amount of CO_2 produced by the burning of fossil fuels and need to reduce it by developing alternative energy sources to fossil fuels and improving energy efficiency,
- Ozone layer: The ozone layer acts as a filter in the stratosphere, protecting the Earth from harmful ultraviolet radiation from the sun. Chlorofluorocarbons (CFCs) destroy stratospheric ozone and cause greenhouse effect.
- *Tropical deforestation*: Forests act as absorbers of carbon dioxide. Deforestation harms species, wide variety of animal and plant life, disrupt local climates, causes loss of habitat, and contributes to the emissions.
- *Waste*: Waste ends up in landfill sites. It is incinerated and may be simply dumped at sea. Most of this waste does not simply biodegrade into harmless substances. Incineration can generate energy but release toxic gases such as dioxins. Reuse, recycling, and re-manufacturing are essential.
- *Water pollution*: The growth of population and the increasing use of water for industrial purposes cause insufficient supply of clean water.

11.5.2 Sustainable Engineering Principles in Energy Systems

Sustainable engineering transforms existing engineering disciplines and practices to those that promote sustainability. To fully implement sustainable engineering solutions, engineers use the following principles in energy systems:

- Integrate environmental impact assessment tools.
- Conserve and improve natural ecosystems while protecting human health and well-being.
- Ensure that all material and energy inputs and outputs are as inherently safe and benign as possible.
- Minimize depletion of natural resources and waste.
- Develop and apply engineering solutions in line with local geography, aspirations, and cultures.
- Actively engage with communities and stakeholders.
- Use material and energy inputs that are renewable.

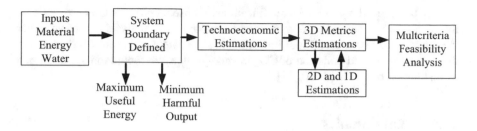

Fig. 11.4 Feasibility analysis with technoeconomic analysis and sustainability metrics [17, 22]

Figure 11.4 displays the major steps toward a comprehensive feasibility analysis incorporating the technoeconomic analysis and sustainability metrics. This may make it possible to count the damage caused because of harmful emissions from the fossil fuels used in energy systems [17, 22].

The plan–do–check–act management system provides a methodical structure for solving problems and implementing solutions. During the plan stage, the problem is identified and analyzed, and a plan is developed. Then, during the do stage, steps are taken to implement the planned actions. To ensure that the best possible solution has been developed, a series of verifications and analyses are conducted during the check stage, and potential improvements are identified. In the final step, the act stage, necessary changes are applied.

11.5.3 Thermodynamic Analysis in Design of Energy Systems

A typical distillation column resembles a heat engine delivering separation work by using heat at a high temperature in the reboiler and discharging most of it to the environment at a lower temperature in the condenser. The 'Column Targeting Tool' of Aspen Plus simulation package is based on the Practical Near-Minimum Thermodynamic Condition approximation representing a practical and close to reversible operation [7, 10]. The column targeting tool performs (i) thermal, (ii) exergy, and (iii) hydraulic analyses capabilities that can help identify the targets for appropriate column modifications in order to: (i) reduce utilities cost, (ii) improve energy efficiency, (iii) reduce emissions of GHG.

The column targeting tool exploits the capabilities for thermal and hydraulic analyses of distillation columns to identify the targets for possible column retrofits in: (1) feed stage location, (2) reflux ratio, (3) feed conditioning, and (4) side condensing and/or reboiling to reduce the cost and improve energy efficiency. The 'Carbon Tracking' (CT) and Global Warming Potential options can help quantify the reduction in CO_2 emission in a simulation environment. CTT produces 'Column Grand Composite Curves' and 'Exergy Loss Profiles,' which help the engineer suggest retrofits for column configurations and operating conditions. The exergy loss profiles represent inefficient use of available energy due to irreversibility and should be reduced by suitable modifications. Consequently, smaller exergy loss means less waste energy. A comparative thermodynamic analysis before and after the retrofits may be a useful tool for estimating the extent of possible reductions in the waste energy and emission of CO_2 in energy-intensive separation systems such as distillation columns [8, 10, 11].

11.5.4 Case Studies

Wind power-based electrolytic hydrogen production: The generation of renewable electricity suffers from intermittent and fluctuating character and necessitates the storage. Wind power-based electrolytic hydrogen may serve as a chemical storage for renewable electricity [14]. Figure 11.5 shows the schematic of wind power-based hydrogen production. Alkaline electrolysis technologies are the most mature commercial systems. These electrolyzers have the energy efficiencies of 57–75%. The typical current density is 100–300 mA/cm^2.

For producing one kg H_2, approximately 26.7 kg water is necessary. The total greenhouse gas emission is around 0.97 kg CO_2e/kg H2. The hydrogen production cost is highly dependent on the renewable electricity price [22].

Sustainable Methanol Production: Methanol synthesis needs carbon-rich feedstock, hydrogen, and a catalyst, mainly $Cu/ZnO/Al_2O_3$. Commercial process of methanol production from natural gas is the most efficient process with a typical energy efficiency of 75%. The coal-based syngas process has the highest emission of greenhouse gases (GHG), which is around 2.8–3.8 kg CO_2/kg methanol. Methanol from biomass or flue gas CO_2 is at least 2–3 times more expensive than the fossil-fuel-based methanol (Table 11.13) [23].

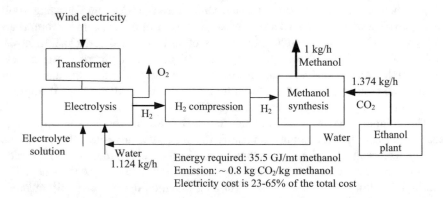

Fig. 11.5 Schematic for alkaline electrolysis of water for hydrogen production with compression, storage, and delivery [14, 23]

Fig. 11.6 Schematic of
methanol production using
renewable hydrogen and CO_2
[22]

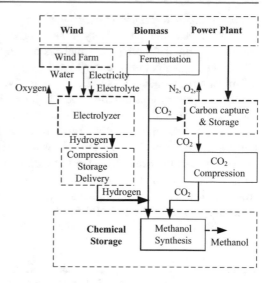

Converting CO_2 into chemicals is challenging, and inherently carries costs for
the energy and hydrogen supply. Figure 11.6 shows a schematic of wind
power-based hydrogenation of CO_2 to methanol. Methanol synthesis from water,
renewable electricity, and carbon may lead to chemical storage of renewable
energy, carbon recycle, fixation of carbon in chemical feedstock, as well as
extended market potential for electrolysis. Methanol synthesis can fix the CO_2 when
it is used as feedstock for producing various other chemicals Methanol can also be
used as transportation fuel and hence recycles CO_2 [19–21].

Fixation of carbon reduces the carbon dioxide equivalent emission by
around −0.85 kg CO_2e/kg methanol. Complete combustion increases the emission
by around +0.53 kg CO_2e/kg methanol. The electrolytic hydrogen production cost
is the largest contributor to the economics of the integral plant. Table 11.13 presents
the sustainability metrics, which show that the methanol facility requires 1.39 mt
CO_2/mt methanol. The environmental impact metrics show that the integral
methanol facility reduces emission by around −0.84 kg CO_2/kg methanol when
utilizing it as a chemical feedstock and recycles 0.53 kg CO_2/kg methanol after its
complete combustion [21–23].

Table 11.13 Sustainability
metrics for the integral
methanol plant [22]

Metrics	Integral methanol
Material intensity	1.39
CO_2 used/Unit product, mt/mt	
H2 used/Unit product, mt/mt	0.19
Energy intensity	
Net duty/unit product, MWh/mt	9.55
Net cost/Unit product, $/mt	828.67
Environmental impact	
Total CO_2e/Unit product, mt/mt	− 0.85

*For an assumed carbon fee of $2/mt CO_2

Sustainable Refinery Operation: Energy and environmental sustainability metrics for a crude oil refinery consisting of three distillation columns can be assessed by the thermodynamic analysis and energy analyzer. Distillation columns consumes large amount of energy used to operate the plants in petrochemical and chemical process industries [1, 2]. The objective is to explore the scope of reducing the thermal energy consumption and CO_2 emissions for a more sustainable refinery operation. Thermodynamic analysis is carried out by using the thermal analysis capability of 'Column Targeting Tool' to address the 'energy intensity metrics', and the 'Energy Analyzer' of the Aspen Plus design package to design and improve the performance of the heat exchanger network system for process heat integration and energy conservation [12]. Environmental pollution impact metrics are estimated from the 'Carbon Tracking' options of the Aspen Plus with a selected CO_2 emission data source of US-EPA-Rule-E9-5711 and using crude oil as a primary fuel source for the hot utilities [1].

11.6 Multicriteria Decision Matrix for Feasibility Analysis

Beside the economics analysis, sustainability metrics should also be an integral part of the feasibility of energy sector. For this purpose, a multicriteria Pugh decision matrix can be used to assess the feasibility of renewable and nonrenewable methanol production facilities (Table 11.14). With the weight factors adapted and the combined economic and sustainability indicators, the multicriteria decision matrix shows that overall weighted score is around +5.4 for the renewable integral methanol facility, which is higher than that of fossil fuel-based methanol with a score of +0.2. This may display the impact of multicriteria with sustainability indicators on evaluating the feasibility of processes requiring large investments and various types of energy resources [22, 25, 27].

11.7 Life Cycle Analysis

Life cycle analysis (LCA) is a standardized technique that tracks all material, energy, and pollutant flows of a system from raw material extraction, manufacturing, transport, and construction to operation and end-of-life disposal. LCA is used for product and process development, improvement, and comparison of technologies. LCA can lead to quantified assessment in order to [5, 23]:

- Minimize the impact of pollution by most effective technique
- Conserve nonrenewable energy and material resources
- Develop and utilize cleaner energy technologies
- Maximize recycling and optimize waste material production and treatment

Table 11.14 Multicriteria decision matrix for feasibility assessment of energy systems [22]

Economics and sustainability indicators	Weighting factor: 0–1	Nonrenewable methanol	Renewable methanol
Economic indicators			
Net present value (NPV)	1	+	−
Payback period (PBP)	0.8	+	−
Rate of return (ROR)	0.8	+	−
Economic constraint (EC)	0.9	+	−
Impact on employment	1	+	+
Impact on customers	1	+	+
Impact on economy	1	+	+
Impact on utility	0.7	−	+
Sustainability indicators			
Material intensity	0.7	−	+
Energy intensity	0.8	+	−
Environmental impact: GHG in production	0.8	−	+
Environmental impact: GHG in utilization	0.8	−	−
Toxic/waste material emissions-	1	−	+
Potential for technological improvements	0.8	−	+
Security/reliability	0.9	−	+
Political stability and legitimacy	0.8	−	+
Quality of life	0.8	−	+
Total positive score		**8**	**11**
Total minus score		**−9**	**−6**
Net score (positive-minus)		**−1**	**+5**
Weighted total score		**+0.2**	**+5.4**

LCA can help determine environmental impacts from 'cradle to grave' or 'cradle to gate' scenarios and can compare energy technologies (Fig. 11.7). Cradle-to-gate scenario estimates energy use (natural gas and electricity), GHG emissions, for feedstock production and conversion processes. LCA starts from the extraction of raw materials, production of product(s), and use of the product to reuse, recycle, and final disposal of wastes. Applying LCA provides a broader view of a product's environmental impact through the value chain, not just at the final manufacturing stage. LCA assesses the environmental aspects and potential impacts associated with a product, process, or service, by:

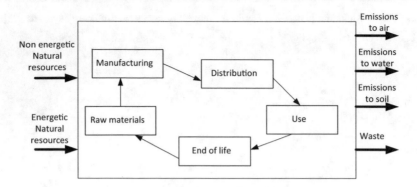

Fig. 11.7 Principles of life cycle assessment cradle-to-grave

- Compiling an inventory of relevant energy and material inputs and environmental releases,
- Evaluating the potential environmental impacts associated with identified inputs and releases,
- Interpreting the results to make a more informed decision

LCA examines the environmental impacts of a product by considering the major stages of a product's life, which are [18, 23]:

- Raw material harvesting and transportation to manufacturing site
- Raw materials processing and transportation to production sites
- Manufacturing product and assembly, packaging, and transportation to final distribution site
- Energy and emissions during normal product life, required maintenance, and product reuse (refurbishing, material reuse)
- Waste management/end of life, including recycling and landfills.

11.7.1   Life Cycle Assessment Principles

International standards in the ISO 14040 series define the principles of LCA. These standards are compiled within the ISO 14044 standard [18]. The LCA technique can be narrowed down to four main steps, which address one or more of the product's life stages at a time:

1. The definition and scope are determined along with information needs, data specificity, and collection methods and data presentation (ISO 14041).
2. The life cycle inventory (LCI) is completed through process diagrams, data collection, and evaluation of the data (ISO 14041). *Life cycle inventory* consists

of emissions data, the raw materials, and energy used during the life cycle of a product.

3. The life cycle impact assessment (LCIA) is determined with impact categories and their weights, as well as any subsequent results (ISO 14042). *Life cycle impact assessment* measures the impact of these emissions and nonrenewable raw materials depletion on environment.

4. The final report should include significant data, data evaluation and interpretation, final conclusions, and recommendations (ISO 14043). *Life cycle interpretation* consists of the results of the impact assessment in order to suggest improvements.

The first three steps of a life cycle analysis are related to one another. More importantly, however, data interpretation is an integral part of all three steps and should be done after each of the sub-analyses is completed.

1. **Raw material acquisition**: This stage includes the removal of raw materials and energy sources from the Earth, such as the extraction of crude oil and coal. Raw materials from the point of acquisition to the gate of processing are considered part of this stage.

2. **Manufacturing**: The manufacturing stage produces the product from the raw materials and delivers it to consumers:

   - Material manufacture stage converts raw materials into a form that can be used to fabricate a finished product. For example, the crude oil is refined to its various transportation liquid fuels.
   - Process stage includes the conversion of raw materials to a new product and make it ready to be stored.
   - Filling, packaging, and distribution includes all processes and transportation required to store, package, and distribute a finished product. Energy and environmental outputs caused by transporting the product to the consumer are accounted for a product's life cycle.

3. **Use**: The actual use, reuse, and maintenance of the product are considered with energy requirements and environmental wastes associated with product storage and consumption.

4. **Recycling and waste management**: Energy requirements and environmental wastes associated with product disposal are included in this stage, as well as post-consumer waste-management options, such as recycling, composting, and incineration.

The LCA model aims to cover all processes related to an energy system within a long timescale and with all effects anywhere in the world, and all relevant substances and environmental themes. An effective LCA allows analysts to:

- Identify and estimate the positive or negative environmental impact of a process or product.
- Find opportunities for process and product improvement.
- Compare and analyze several processes based on their environmental impacts.
- Quantitatively justify a change in a process or product.

Within the life cycle, inputs and outputs of material, energy and emissions of each process are quantified. These input and output flows are associated to various environmental impacts such as nonrenewable resource (material and/or energy) depletion, global warming, and toxicities. The assessments of input and output flows are called life cycle inventories (LCIs).

11.7.2 Benefits of International Organization for Standardization

ISO 50001 is an International Standard developed by the International Organization for Standardization (ISO) to provide organizations a framework to manage and improve their energy systems [18]. The improvements include operational efficiencies, decreasing energy intensity, and reducing environmental impacts. The standard addresses the following:

- Energy use and consumption
- Measurement, documentation, and reporting of energy use and consumption
- Design and procurement practices for energy-using equipment, systems, and processes
- Development of an energy management plan and other factors affecting energy performance that can be monitored and influenced by the organization.

ISO 50001 requires continuous energy performance improvement but it does not include prescriptive energy performance improvement goals. Rather, it provides a framework through which each organization can set and pursue its own goals for improving energy performance.

The International Organization for Standardization (ISO) established a protocol for performing an LCA study. While simple in concept, the conduct of an LCA can be complicated, mainly due to the large amount of data needed. However, the increasing availability of LCA databases and software programs makes it easier to conduct an LCA. In addition, The ISO 50001 Energy-Management System Standard is an international framework that helps industrial plants, commercial facilities, and organizations to manage energy, reduce costs, and improve environmental performance. Some of the specific benefits include:

- Assists organizations in optimizing their existing energy consumptions,
- Measures, documents, and reports energy intensity improvements and their projected impact on greenhouse gas emissions,

- Promotes energy-management best practices and reinforces better energy-management behaviors,
- Helps facilities evaluate and prioritize the implementation of new energy-efficient technologies,
- Provides a framework to promote energy efficiency throughout the supply chain.

11.7.3 Life Cycle Analysis Stages

Performing a life cycle analysis requires the following stages:

1. *The purpose of the LCA*: Eco-design,

   - Environmental impact (positioning products with respect to others in terms of their environmental balance),

   - Comparative LCA (positioning product by comparing with existing nonre-newable equivalents).

2. *The scope of the LCA*

   - *Cradle-to-gate.* This scope does not consider the end of product life scenarios, such as removal to landfill site, incineration, and recycling. If only energy consumption and GHG emission indicators are considered, the end of life impact would be less significant (see Fig. 11.8).

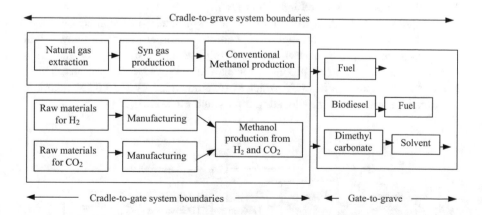

Fig. 11.8 Cradle-to-gate and cradle-to-grave system boundary definitions for methanol production using renewable and nonrenewable resources

- Cradle-to-grave. If the purpose is related to eco-design, then cradle-to-grave scope should be performed. When impact indicators such as eutrophication, air acidification, and human toxicity are assessed, it is vital that the end of life impact stage should be considered. The system boundaries describe which processes of the entire life cycle are included in the assessment [23]. Figure 11.8 shows the two common scopes for LCA.

3. The functional unit

A function unit, such as kg amount of product or other relevant unit such as energy MJ, or an economic value should be specified.

4. The indicators

Indicators may be energy consumption + climate change, and other local indicators.

Impact indicators consist of input and output flows generated by a product by using emission factors for each identified flow in order to represent the following impact indicators of this product:

- Change in GHG emissions.
- Change in toxicity.
- Change in nonrenewable energy resources.
- Change in nonrenewable material resources.
- Change in water usage.

One should assess all the possible impacts of the products by at least one of the indicators (see Table 11.3). Nonrenewable energy consumption and global warming potential are the best-assessed indicators [16].

Best practice in LCA is to include all the byproducts in the functional units within the system boundaries. Figure 11.9 shows a biofuel life cycle analysis with cradle-to-grave scope. The environmental impacts are estimated for the entire system and associated to all products, for example, kg CO_2eq/kg product. This is called system expansion in LCA, which is recommended by the ISO standards to avoid ambiguous choices in allocating inputs and emissions [23].

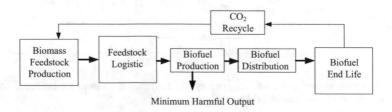

Fig. 11.9 Schematic of biofuel life cycle analysis with cradle-to-grave scope

11.7.4 Life Cycle Assessment

A life cycle sustainability assessment (LCSA) is: LCSA = LCA + LCC + SLC, where LCC is the life cycle costing and SLCA is social LCA. Here the quantification of SLCA is a major challenge. The comprehensive sustainability methods incorporate technoeconomic analysis and life cycle analysis. Both the methods are very broad, information and data intensive. A comparative assessment method for various processes at design stage would be an essential step toward sustainability. Such a method compares an innovative new process with a conventional process based on these selected parameters [5, 23]:

1. Economic constraint (EC)
2. Environmental impact of raw materials (EI)
3. Process cost and environmental impact (PCEI)
4. Environment, health, and safety index (EHSI)
5. Risk aspects (RA)

1. *Economic constraint (EC)* (**possible** weight factor: w = 0.3)

Economic feasibility is critical for economic sustainability. One of the definitions for EC is:

$$EC = \frac{\text{Average discounted annual cost of production}}{\text{Average discounted annual revenue}} \qquad (11.11)$$

For raw material case and for a single reaction, EC may be [22]

$$EC = \frac{\sum_i (c_i f_i)_{\text{RM}}}{\sum_i (c_i f_i)_{\text{Prod}}} \qquad (11.12)$$

where c and f are the cost and flow rates, respectively, RM and Prod stand for raw materials and products, respectively. For R number of multiple reactions, we have

$$EC = \sum_i^R EC_i \qquad (11.13)$$

With these definitions, one needs $EC \leq 1$ for economic sustainability. Fluctuations in prices and overall economic data affect the values of EC and hence an average over a time interval should be considered.

2. *Environmental impact of raw materials* (*EI*) (w = 0.2)

Environmental impact of raw materials is required for the production of a unit mass of product. Cumulative energy (renewable and nonrenewable) demand (CED) and GHG emission associated with all the raw materials from cradle to the relevant system boundary indicates the environmental impact (*EI*). Economic allocation factor A for the main product (mp) is

$$A_{mp} = \frac{c_{mp} f_{mp}}{\sum_i (c_i f_i)} \tag{11.14}$$

CED_{mp} (process energy + energy content of main product) is

$$CED_{mp} = \left(\frac{A_{mp}}{f_{mp}} \sum_i f_i (CED_i - E_i)\right) + E_{mp} \tag{11.15}$$

GHG_{mp} is

$$GHG_{mp} = \left(\frac{A_{mp}}{f_{mp}} \sum_i f_i GHG_i\right) + FC_{mp} \tag{11.16}$$

where E is the calorific value and FC is the embedded fossil carbon, and f is the flow rate. CED and GHG emissions are good indicators for environmental impact. Depending on the availability other data such as water and land use can be incorporated into the method. Limitation usually comes from toxicity inclusion.

3. *Process cost and environmental impact* (PCEI) (w = 0.2)

Energy loss in the reaction and separation sections of processing sequence can be used as an indicator for the expected cost and environmental impact based on energy loss index. Some possible indicators are:

- Presence of water at the reactor outlet and the difference in the boiling temperature of water and products.
- Product concentration (conversion rate, yield).
- Boiling point differences between the main product and other components.
- Total mass of all the components to mass of the product indicating the level of mass loss.
- Reaction duty and possibility of energy recovery.
- Number of coproducts for understanding the separation needs.
- Pretreatment of feedstock.

4. **Environment, health, and safety index** (EHSI)(w = 0.2–can be controlled)

This index considers the safety, health, and environment such as ecological toxicity aspects of a process. Each chemical present within the process is associated with persistency that is half-life in water or air. Heat is associated with irritation and chronic toxicity. Safety is associated with mobility (pressure, temperature), fire/explosion (flash point), reaction/decomposition, and acute toxicity. The property parameters and hazard classification of each chemical are taken into account to assign an index value.

$$EHSI = A_{mp}[Ew_E + Hw_H + Sw_S] \tag{11.17}$$

where w is the weights for each category, A is the allocation factor, and E,H,S denote environmental, health, and safety, respectively. For multiple reaction steps, the methodology may be applied separately for each step.

5. **Risk analysis** (RA) (w = 0.1 − high uncertainty)

RA is based on the external aspects such as economical and technical for commodity chemical and fuels with allocated weight factors:

- Feedstock supply risk: 0.25
- Regional feedstock availability: 0.15
- Market risk: 0.25
- Infrastructure risk: 0.2
- Technical aspects of application
- Chemicals (functional groups, retention of raw material functionality)
- Fuels (energy content, energy compatibility, storage, and transportation

Index ratio (*IR*) displays the potential benefit associated with the proposed novel process:

$IR =$ total weighted score of novel process/total weighted score of conventional process

$$\tag{11.18}$$

A lower index (<1) shows that the new process can provide certain benefits compared with the conventional process.

There may be a great deal uncertainty associated with the data input and weighting. The effect of these uncertainties can be analyzed using the Monte-Carlo analysis in terms of probability distributions based on deviations within the weight factor ranges shown in Table 11.15. A sensitivity analysis can be performed by estimating the impact of yield on the index ratio. This method combines the aspects of technoeconomic analysis, life cycle assessment, and green chemistry [21, 22].

Table 11.15 Distribution of deviations in weight factors [16]

	Weight	Minimum weight	Maximum weight
EC	0.3	0.25	0.6
PCEI	0.2	0.15	0.35
EI-CED	0.1	0.05	0.3
EI-GHG	0.1	0.05	0.3
EHSI	0.2	0.05	0.3
RA	0.1	0.05	0.25

11.7.5 Life Cycle Analysis Impact Categories

There are several environmental impacts and indicators [5] that are subject to LCA and shown previously in Table 11.3:

1. *Abiotic resource depletion potential (ARDP)*

ARDP includes depletion on nonrenewable resources such as fossil fuels, metals, and minerals. The total ARDP is estimated by

$$ARDP = \sum_i^n \frac{r_i}{twr_i} \tag{11.19}$$

where r_i is the quantity of resource i used per functional unit of product, n is the total number of resources, and twr is the total world reserves of the resource i. Burdens resource depletion (world reserves) are: coal reserves 8.72×10^{13} tons, oil reserves 1.24×10^{11} tons, and gas reserves 1.09×10^{14} m^3 [16]

2. *Global warming potential (GWP)*

GWP is the sum of emissions of the GHGs such as CO_2, N_2O, CH_4, and VOCs, multiplied by their respective GWP factors. GWP is estimated by

$$GWP = \sum_i^n e_i(pf_i) \tag{11.20}$$

where e_i is the emission rate of GHG at i and pf_i is the potency factor. The values of GWP depend on the time period for assessment of GWP; the short-term effect is for 20 and 50 years and long-term effect for 100 and 500 years. Some values for potency factors of GWP are: chlorinated hydrocarbons: 400, CFCs: 5000, and other VOCs: 11 [16] (see Table 11.4).

3. *Ozone depletion potential (ODP)*

ODP indicates the potential emissions of chlorofluorohydrocarbons (CFCs) and the chlorinated hydrocarbons (HCs) for depleting ozone layer and is estimated by

$$ODP = \sum_i^n e_i(pf_i) \qquad (11.21)$$

where e_i is the emission rate of ozone depleting gas i and pf_i is the ODP factor for gas i expressed relative to the ozone depleting potential of CFC-11. Some values of ODP are chlorinated hydrocarbons: 0.5, CFCs: 0.4, and other VOCs: 0.005 [16] (see Tables 11.6 and 11.7).

4. *Acidification potential (AP)*

AP indicates the contribution of SO_2, NO_x, HCl, NH_3, and HF to the potential acid formation and deposition such as their potential to form H^+ ions, and AP is estimated by

$$AP = \sum_i^n e_i(pf_i) \qquad (11.22)$$

The potency factor pf_i for gas i expressed relative to the AP of SO_2. Some values for AP are: NO_x: 0.7, SO_2: 1, HCl: 0.88, HF: 1.6, NH_3: 1.88 [16] (see Table 11.4).

5. *Eutrophication potential (EP)*

EP is the over-fertilization of water and soil resulting in increased growth of biomass. Eutrophication is characterized by excessive plant and algal growth due to the increased availability of one or more limiting growth factors needed, such as sunlight, carbon dioxide, and nutrient fertilizers. However, human activities have accelerated the rate and extent of eutrophication through both point-source discharges and nonpoint loadings of limiting nutrients, such as nitrogen and phosphorus, into aquatic ecosystems, with negative consequences for drinking water sources, fisheries, and recreational water bodies [16]. EP is estimated by

$$EP = \sum_i^n e_i(pf_i) \qquad (11.23)$$

The values of EP are relative to the PO_4^{-3} [16] (see Table 11.12).

6. *Photochemical oxidants creation potential (POCP)*

POCP or photochemical smog is usually expressed relative to the POCP factors of ethylene and calculated by

$$POCP = \sum_{i}^{n} e_i(pf_i) \tag{11.24}$$

Mainly, the values of VOCs are classified into the following categories: alkanes, halogenated hydrocarbons, alcohols, ketones, esters, ethers, olefins, acetylenes, aromatics, and aldehydes, and f_i is the respective POCP factors relative to ethylene [16].

7. *Human toxicity potential (HTP)*

HTP is estimated by adding the releases, which are toxic to humans, air, water, and soil

$$HTP = \sum_{i}^{n} (e_i pf_i)_A + \sum_{i}^{n} (e_i pf_i)_W + \sum_{i}^{n} (e_i pf_i)_S \tag{11.25}$$

where each term shows the emissions of different toxic substances and their toxicological classification factors for the effect of the toxic emissions to air, water, and soil, respectively. The factors are estimated using the acceptable or tolerable daily intake of the toxic substances. These factors are still at their developing stages so that HTP can only be taken as an indication and not as an absolute measure of the toxic potential. Some values of HTP are: CO: 0.012, NO_x: 0.78, SO_2: 1.2, HC (excluding CH4): 1.7, chlorinated hydrocarbons: 0.98, CFCs: 0.022, As: 4700, Hg: 120, F2: 0.48, HF: 0.48, NH_3: 0.02, Cr: 0.57, Cu: 0.02, Fe: 0.0036, Ni: 0.057, Pb: 0.79, Zn: 0.0029, fluorides: 0.041, nitrates: 0.00078, phosphates: 0.00004, pesticides: 0.14, phenols: 0.048 [16] (see Table 11.5).

8. *Aquatic toxicity potential*

Aquatic toxicity potential is estimated by

$$ATP = \sum_{i}^{n} e_i(pf_i) \tag{11.26}$$

The values of *ATP* are based on the maximum tolerable concentration of different toxic substance in water by aquatic organisms. Similar to HTP, classification factors for *ATP* are still developing so they can only be used as an indication of potential toxicity. Some values for *ATP* are: Cr: 9.07×10^8, Cu: 1.81×10^9, Ni: 2.99×10^8, Pb: 1.81×10^9, Zn: 3.45×10^8, pesticides: 1.18×10^9, phenols: 5.45×10^9 [16] (see Table 11.10).

11.8 Life Cycle Analysis of Energy Systems

LCA can provide a comprehensive framework to compare renewable energy technologies with fossil-based and nuclear energy technologies displayed in Table 11.16. Life cycle GHG emissions from renewable electricity generation technologies are generally less than those from fossil fuel-based technologies [22, 23]. Comparisons also show that the proportion of GHG emissions from each life cycle stage differs by technology:

- Fossil fuel combustion emits the vast majority of GHGs.
- For nuclear and renewable energy technologies, the majority of GHG emissions are upstream of operation.
- Most emissions for biopower are generated during feedstock production, where agricultural practices play an important role.
- For nuclear power, fuel processing stages are most important, and a significant share of GHG emissions is associated with construction and decommissioning.
- For other renewable technologies (solar, wind, hydropower, ocean and geothermal), most life cycle GHG emissions stem from component manufacturing and, to a lesser extent, facility construction.

Table 11.16 Life cycle analysis of various renewable and nonrenewable energy technologies [4, 5]

Generation technology	Photovoltaics (C-Si and thin film)	Concentrating solar power		Wind (onshore and offshore)	Nuclear (light water)	Coal (Sub- and supercritical, IGCC, fluidized bed)
Driving parameter	Solar irradiation (kWh/m^2/year)	Solar fraction (%)	Operating lifetime (years)	Capacity factor (%)	Operating lifetime (years)	CO_2 Emission Factor (kg CO_2/kWh)
Definition	Amount of solar energy incident per unit area of collector per year	Electricity produced only from solar energy	Assumed lifetime for the LCA or facility	Electricity generated per maximum potential for electricity generation	Assumed lifetime for the LCA or facility	CO_2 emitted per kWh of net electricity generated —a function of thermal efficiency, coal carbon content, and coal lower heating value

Comparison of published results for each specific electricity generation technology is challenging, due to variation of methods and assumptions and needs building on experience.

11.9 Economic Input–Output Life Cycle Assessment (EIO-LCA)

EIO-LCA method estimates the materials and energy resources required for processes and the environmental emissions resulting from energy production. Results from the EIO-LCA may provide guidance on the relative impacts of various products or industries with respect to resource use and emissions throughout the supply chain. Thus, the effect of producing a product would include not only the impacts at the final assembly facility, but also the impact from mining raw materials, transportation, storing, and others. LCA and EIO-LCA methods focus for the green design. The environmental impacts covered include global warming, acidification, energy use, nonrenewable ores consumption, eutrophication, conventional pollutant emissions and toxic releases to the environment. The EIO-LCA model is comprised of national economic input–output models and publicly available resource use and emissions data [4, 5].

11.10 Environment and Exergy

Since exergy is a measure of the departure of the state of the system from that of the environment, it relates the system to the environment. When a system is in thermal, mechanical, and chemical equilibrium with the environment, there are no processes taking place and the system is as the dead state. At dead state, the system has no motion and elevation relative to coordinates in the environment. Only after specifying the environment an exergy value can be estimated. Specifying environment usually refers to some portion of a system's surroundings. For example, in estimating exergy values, the temperature and pressure of the environment are usually the standard state values, such as 298.15 K and 101.31 kPa.

The environment is composed of large numbers of common species within Earth's atmosphere, ocean, and crust. The species exist naturally in their stable forms and do not take part in any chemical or physical work interactions between different parts of the environment. We mainly assume that the intensive properties of the environment are unchanging, while the extensive properties can change because of interactions with other systems. In the natural environment, however, there are components of states differing in their composition or thermal parameters from thermodynamic equilibrium state. These components can undergo natural thermal and chemical processes [7, 10].

11.10.1 Resource Depletion and Exergy

Resource depletion may cause environmental change. By reducing resource depletion, we can reduce ongoing environmental transformation. We may quantify the resource depletion by the *Depletion number Dp*, which is related to a nondimensional indicator Ex_{Dp} per unit consumption of exergy Ex_C

$$Dp = \frac{\dot{Ex}_{Dp}}{\dot{Ex}_C} \tag{11.27}$$

Depletion number provides a measure of system progress or maturity. It is a useful basis for studying the evolution of resource depletion patterns and the implementation of resource conservation strategies. The depletion number is a function of the following three indicators showing the level of implementation of resource conservation strategies [7]:

- The *exergy cycling fraction* ψ is a measure of recycling that accounts for both the throughput and quality change aspects of resource consumption and upgrading.
- The *exergy efficiency* η is a universal measure of process efficiency that accounts for the available energy.
- The *renewable exergy fraction* Ω is a measure of the extent to which resources supplied to an industrial system are derived from renewable sources. Industrial systems consume resources by supporting processes associated with supplying and removing resources.

The definition of depletion number in Eq. (11.27) is

$$Dp = \frac{\dot{Ex}_{Dp}}{\dot{Ex}_C} = 1 + \frac{\dot{Ex}_{Dsl}}{\dot{Ex}_C} + \psi \left[\frac{(1 - \Omega_{RU})}{\eta_{RU}} - 1 \right] + \frac{\dot{Ex}_{TV}}{\dot{Ex}_C} \left[\frac{(1 - \Omega_{VU})}{\eta_{VU}} - 1 \right] \tag{11.28}$$

where $\dot{Ex}_{Dsl}$ is the exergy dissipation rate, Ω_{RU} and η_{RU} are the renewable exergy fraction and transfer efficiency for the recovered resource upgrade process, respectively, $\dot{Ex}_{TV}$ is the exergy transfer rate to the nonrenewable source, and Ω_{VU} and η_{VU} are the renewed exergy fraction and transfer efficiency, respectively, for the nonrenewable resource upgrade process. Two structural constants $\alpha_{Ds\text{-}C}$ and $\alpha_{V\text{-}C}$ are defined by

$$\alpha_{Ds-C} = \frac{\dot{Ex}_{Dsl}}{\dot{Ex}_C(1 - \psi)}, \quad \alpha_{V-C} = \frac{\dot{Ex}_{TV}}{\dot{Ex}_C(1 - \psi)} \tag{11.29}$$

With these definitions, Eq. (11.28) becomes

$$Dp = 1 + \psi \left[\frac{(1 - \Omega_{RU})}{\eta_{RU}} - 1 \right] + (1 - \psi) \left\{ \alpha_{Ds-C} + \alpha_{V-C} \left[\frac{(1 - \Omega_{VU})}{\eta_{VU}} - 1 \right] \right\}$$

$$(11.30)$$

The depletion number is a function of a system's structural constant, the exergy efficiency, and renewed exergy fraction of the individual resource upgrade processes, and the extent of resource cycling.

The generalized depletion number may result from numerous consumption processes, such as incomplete cycling or partial upgrading, and the direct reuse of resources without upgrading. Recycling may reduce the need for resources and the exergy requirements of manufacturing processes. Generally, increasing resource cycling reduces depletion due to less exergy transfer from other sources. For example, producing one-ton aluminum from bauxite requires 27,400 MJ of exergy transfer, while converting the recycled aluminum to feedstock requires far less exergy transfer [4, 7].

11.10.2 Extended Exergy Analysis

The attribute 'extended' refers to the additional inclusion in the exergetic balance of previously neglected terms, such as labor, and to environmental remediation expenditures. Extended exergetics is defined by its raw state exergy, augmented by the sum of all the net exergetics inputs received, directly or indirectly, in various processes pertaining to its extraction, preparation, transportation, and pre-treatment, including the exergetic equivalents of labor, capital, and environmental costs. Therefore, extended exergy accounting may be capable of properly addressing environmental issues.

An extended representation of exergy flow diagrams constitutes a substantial generalization of Szargut's cumulative exergy consumption procedure, and may provide a coherent and consistent framework for including nonenergetic quantities like capital cost, labor cost, and environmental impact into an engineering optimization procedure. In this sense, the environmental approach represents already a significant extension of thermoeconomics, though its 'environmental penalty' functions suffer from their direct dependence on monetary cost. Exergetic approach for the calculation of environmental costs may establish exergy as one of the proper measures of environmental impact. Extended exergy accounting can be considered a further development of the pre-existing theories and methods of 'engineering cost analysis' [7, 10].

A natural extension of life cycle assessment may be the *exergetic life cycle analysis*, which performs a lifelong analysis of a plant or process using exergy as a quantifier. The 'exergetic life cycle analysis' includes labor, environmental damage, and recycle of energy and matter. Some of the efforts might be overstating the consequences of the laws of thermodynamics as it tries to combine the thermodynamic imperfections with the profitability of a product as well as the environmental and social issues [10].

11.11 Ecological Planning

Ecological planning for sustainable development should take into account the uncontrollable waste exergy into the environment. There are environmental impacts associated with the production and transmission of electricity. Emissions that result from the combustion of the fossil fuels include carbon dioxide (CO_2), carbon monoxide (CO), sulfur dioxide (SO_2), nitrogen oxides (NO_x), particulate matter, and heavy metals such as mercury. Nearly all combustion byproducts may have negative impacts on the environment and human health:

- CO_2 is a greenhouse gas and a source of global warming. Power plants that burn fossil fuels and materials made from fossil fuels and some geothermal power plants are the sources of carbon dioxide emissions.
- SO_2 causes acid rain, which is harmful to plants and to animals that live in water, and it worsens or causes respiratory illnesses and heart diseases, particularly in children and the elderly. SO_2 emissions are controlled by wet and dry scrubbers, which involve mixing lime in the fuel (coal) or by spraying a lime solution into the combustion gases. Fluidized bed combustion can also be used to control SO_2.
- NO_x contributes to ground level ozone, which irritates and damages the lungs. NO_x emissions can be controlled by several different techniques and technologies, such as low NO_x burners during the combustion phase or selective catalytic and noncatalytic converters during the post combustion phase.
- Particulate matter results in hazy conditions in cities and scenic areas, and, along with ozone, contributes to asthma and chronic bronchitis, especially in children and the elderly. Very small, or 'fine particulate matter' is also thought to cause emphysema and lung cancer. Heavy metals such as mercury can be hazardous to human and animal health. Particulate matter emissions are controlled with devices that clean the combustion gases that exit the power plant, such as 'Bag-houses' use large filters, electrostatic precipitators use charged plates, and wet scrubbers use a liquid solution [5].

11.12 Chemical-Looping Combustion for Sustainability

Chemical-looping combustion (CLC) is is a novel technology in which power production and CO_2 capture are intrinsically combined using an oxygen carrier that transfers oxygen from the air to the fuel preventing direct contact between them. Well-designed and operated systems of chemical-looping combustion of fossil fuel/biomass offer scalable, diverse, economical, and environmentally sustainable energy pathways with inherited carbon capture. The fuel may be coal, natural gas, and biomass. The oxygen carrier can be alternately oxidized and reduced. The product gas contains mainly CO_2 and water undiluted with nitrogen, and without the production of nitrogen oxides (NO_x) as the high temperatures associated with

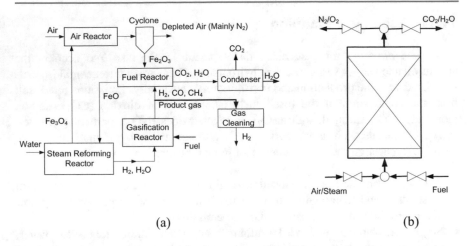

(a)   (b)

Fig. 11.10 Reactor configurations for chemical-looping technology: **a** Schematic of the chemical-looping combustion technology system, **b** Periodically operated chemical-looping technology in packed bed system [9]

the use of flame is avoided. The oxidation of the oxygen carrier is strongly exothermic and hence can be used to heat air flow to high temperatures (1000–1200 °C) and can drive a gas turbine to produce power [9].

In a new reactor concept, the oxygen carrier is not circulated between air and fuel reactors (see Fig. 11.10) but kept inside a packed bed reactor and is alternately exposed to oxidizing and reducing conditions by periodic switching of the feed streams of fuel and air.

Thermochemical conversions of fuel/biomass can be analyzed by using the systems of chemical-looping combustion, chemical-looping reforming for producing hydrogen, and chemical-looping steam gasification for producing liquid transportation fuels of gasoline and diesel by the Fischer-Tropsch synthesis. The outcomes of such analyses would be: (1) the lowest possible energy and economic costs for the fuel/biomass conversion systems without adverse environmental/societal consequences, (2) a reduction of carbon intensity from energy conversion and use, and (3) interactions of systems and patterns at the local/regional scale with systems/patterns at the global scale [9].

11.12.1   Decarbonization Technology

Decarbonization technology needs novel improvements in capturing, storing, and converting carbon into other chemicals and fuels [23–25]. This will lead to a wider use of fossil fuel deposits without carbon penalties. Hydrothermal reactions are as aqueous conversions under high temperature (200–350 °C) and high pressure

Fig. 11.11 Hydrothermal process of converting CO_2 to methanol and dimethyl ether [6]

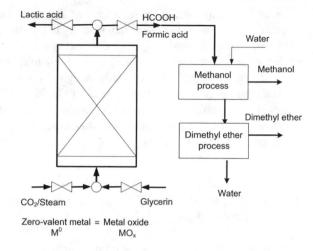

Zero-valent metal = Metal oxide
M^0 $\qquad$ MO_x

(around 15–20 MPa). Under hydrothermal conditions in periodically operated chemical-looping system with the following main reactions

$$M^0 + CO_2 + H_2O \rightarrow MO_x + HCOOH \tag{11.31}$$

Zero-valent metals have catalytic activity in CO_2 reduction and formic acid production. Because of in situ production of H_2, under hydrothermal conditions, no fresh supply, storage, or transportation of H2 would be required in such systems. Oxidized metal can be regenerated by a chemical such as crude glycerin, which is converted to lactic acid.

$$MO_x + C_aH_bO_c \rightarrow M^0 + C_aH_{b-2c}O_c + xH_2O \quad \text{Oxidation} \tag{11.32}$$

The overall reaction with glycerin is exothermic

$$CO_2 + C_3H_8O_3 \rightarrow HCOOH + C_3H_6O_3 \tag{11.33}$$

The formic acid can lead to methanol and dimethyl ether fuel productions as shown in Fig. 11.11 [9].

11.13 Nuclear Power Plants

Nuclear power plants are not a source of greenhouse gases or other emissions, but they do produce two kinds of radioactive waste [17]:

- *Low-level radioactive waste*: This includes items that have become contaminated with radioactive material, such as clothing, wiping rags, mops, filters, reactor water treatment residues, and equipment and tools. Low-level waste is stored at nuclear power plants until the radioactivity in the waste decays to a level where it can be sent to a low-level waste disposal site.

- *Spent (used) nuclear fuel*: The spent fuel assemblies are highly radioactive and must initially be stored in specially designed pools resembling large swimming pools (water cools the fuel and acts as a radiation shield) or in specially designed dry storage containers. An increasing number of reactor operators now store their older spent fuel in dry storage facilities using special outdoor concrete or steel containers with air cooling.

11.14 Projections on Energy and Environmental Protection

- It is more important than ever for policy-makers, industry, and other stake-holders to have a clear understanding of the state of the energy sector today, move toward a lower carbon and more efficient energy system, and to put the energy system on a more secure and sustainable footing. Energy use worldwide is set to grow by one-third by year 2040. Among the fossil fuels, natural gas— the least-carbon intensive—share grows. Policy preferences for lower carbon energy options are reinforced by trends in costs, as oil and gas gradually become more expensive to extract while the costs of renewables continue to fall. Cost reductions are the norm for more efficient equipment and appliances, and steady gains in technology.
- The newly agreed UN *Sustainable Development Goals* embrace a goal on energy including the target to achieve universal access to energy by 2030; the number of people without electricity falls to 800 million by 2030 and the number without access to clean cooking fuels declines only gradually to 2.3 billion in 2030 [17].
- Energy consumption in the *transportation sector* will decline slightly by developing fuel-efficient technologies for all vehicle fuel types [26].
- Delivered energy consumption in the *industrial sector* grows at a modest rate. Combined heat and power generation in the industrial sector grows considerably [27].
- Total primary energy consumption grows mostly because of the growth is in consumption of natural gas and renewable energy [2].
- *Hybrid renewable energy systems* usually consist of two or more renewable energy sources used together to provide increased system efficiency as well as greater balance in energy supply. For example, 60% from a biomass system, 20% from a wind energy system, and the remainder from fuel cells. Another example is the combination of a photovoltaic array coupled with a wind turbine. This would create more output from the wind turbine during the winter, whereas during the summer, the solar panels would produce their peak output [24].

- *Renewable resources* meet much of the growth in electricity demand. Wind and solar generation account for nearly two-thirds of the increase in total renewable generation [19].
- *Biomass* generation increases, led by co-firing at existing coal plants [2].
- Electricity prices are determined by a complex set of factors that include energy use and efficiency, the competitiveness of electricity supply, investment in new generation, transmission, and distribution capacity, and the operation and maintenance costs of plants in service [17].
- *Energy efficiency* plays a critical role in the global economy. Changing product design, reuse, and recycling ('material efficiency') also offers huge potential for energy saving in energy-intensive sectors, such as steel and cement [12].
- *Renewables* contribute considerably mainly due to mandatory energy efficiency regulation worldwide. Global energy transition is underway, but not yet at a pace that leads to a lasting reversal of the trend of rising CO_2 emissions [19].
- Injection of billions of tons of CO_2 to the atmosphere coupled with mass scale deforestation is responsible for the rise of global near surface air temperature to the present average of about 288 K. One proposed scheme to battle this problem is sequestration of CO_2 and its storage. Natural gas is a good fit for a gradually *decarbonizing energy system*. The power sector leads the way toward a decarbonized energy system with advanced control technologies to reduce air pollution and capture and store CO_2 safely and cost-effectively [17].

Summary

- *Sustainability* is maintaining or improving the material and social conditions for human health and the environment over time without exceeding the ecological capabilities that support them. Sustainability has the three dimensions of environment, social, and economics. Energy is one of the main drivers of technology and development and hence the demand for energy continues to grow worldwide even as extraction of fossil fuels becomes more difficult and expensive. Some major actions needed in reducing the growth of atmospheric GHG emissions are:
  - Increase energy efficiency
  - Increase the usage of renewable energy
  - Develop technologies for carbon capture and storage (CCS)
  - Develop effective energy storage technologies
  - Recognize leadership role in transforming the global energy sector
- There are two *sustainability impacts and impact indicators* that are related to the depletion of resources and environmental change
- The *sustainability metrics*, indicated by the intersection of all three dimensions, are truly representative of progress toward sustainability. A set of sustainability metrics that are applicable to a specific energy process:
  - Material intensity (nonrenewable resources of raw materials, solvents/unit mass of products)

- Energy intensity (nonrenewable energy/unit mass of products)
- Potential environmental impact (pollutants and emissions/unit mass of products)
- Potential chemical risk (toxic emissions/unit mass of products)

- The *global warming potential* (GWP) is a measure of how much a given mass of a chemical substance contributes to global warming over a given period.
- Sustainable energy systems aim the largest triangle for sustainable energy production with the inclusion of all the three dimensions of environment, social, and economics
- *Sustainable engineering* transforms existing engineering disciplines and practices to those that promote sustainability.
- Wind power-based electrolytic hydrogen may serve as a chemical storage for renewable electricity
- Methanol synthesis from water, renewable electricity, and carbon may lead to chemical storage of renewable energy, carbon recycle, fixation of carbon in chemical feedstock, as well as extended market potential for electrolysis. Methanol synthesis can fix the CO_2 when it is used as feedstock for producing various chemicals, and also be used as transportation fuel and hence recycles CO_2
- Besides the economics analysis, sustainability metrics should also be an integral part of the feasibility of energy sector. For this purpose, a multicriteria Pugh decision matrix can be used to assess the feasibility of renewable and nonrenewable methanol production facilities
- *Life cycle analysis* (LCA) is a standardized technique that tracks all material, energy, and pollutant flows of a system—from raw material extraction, manufacturing, transport, and construction to operation and end-of-life disposal. LCA is used for product and process development, improvement, and comparison of technologies. LCA can lead to quantified assessment to:

  - Minimize the impact of pollution by most effective technique
  - Conserve nonrenewable energy and material resources
  - Develop and utilize cleaner energy technologies
  - Maximize recycling and optimize waste material production and treatment
  - Raw material acquisition
  - Manufacturing
  - Use
  - Recycling and waste management

- ISO 50001 is an International Standard developed by the International Organization for Standardization (ISO) to provide organizations a framework to manage and improve their energy systems.
- A comparative assessment method for various processes at design stage would be an essential step toward sustainability. Such a method compares an innovative new process with a conventional process based on these selected parameters:

  - Economic constraint (EC)
  - Environmental impact of raw materials (EI)
  - Process cost and environmental impact (PCEI)

- Environment, health, and safety index (EHSI)
- Risk aspects (RA)

- LCA can provide a comprehensive framework to compare renewable energy technologies with fossil-based and nuclear energy technologies.
- Since *exergy* is a measure of the departure of the state of the system from that of the environment, it relates the system to the environment.
- *Resource depletion* may cause environmental change.
- *Extended exergetics* is defined by its raw state exergy, augmented by the sum of all the net exergetics inputs received, directly or indirectly, in various processes pertaining to its extraction, preparation, transportation, and pre-treatment, including the exergetic equivalents of labor, capital, and environmental costs.
- *Ecological cost* analysis may minimize the depletion of nonrenewable natural resources.

Problems

11.1. What are the advantages and disadvantages of using coal?

11.2. Does an electric car reduce the use of fossil fuels?

11.3. Is a fuel oil heater or an electric resistance heater the best for the environment?

11.4. Is a natural gas heater or a geothermal heating system the best for the house?

11.5. A car's daily traveling distance is about 80 miles/day. A car has a city-mileage of 20 miles/gal. If the car is replaced with a new car with a city-mileage of 30 miles/gal and the average cost of gasoline is $4.50/gal, estimate (a) the amount of fuel, energy, and money conserved with the new car per year, (b) reduction in CO_2 emission.

11.6. A car's daily traveling distance is about 80 miles/day. A car has a city-mileage of 10 miles/gal. If the car is replaced with a new car with a city-mileage of 32 miles/gal and the average cost of gasoline is $4.50/gal, estimate (a) the amount of fuel, energy, and money conserved with the new car per year, (b) reduction in CO_2 emission.

11.7. How can you control your carbon footprint?

11.8. A district uses natural gas for heating. Assume that average NO_x and CO_2 emissions from a gas furnace are 0.0045 kg/therm and 6.4 kg/therm, respectively. The district wants to replace the gas heating system with a geothermal heating system. The projected saving by the geothermal heating system would be 40×10^6 therms of natural gas per year. Determine the amount of NO_x and CO_2 emissions that the geothermal heating system would save every year.

11.9. How can using wind power help reduce pollution in the atmosphere and help conserve our oil supplies?

11.10. How can using thermal heat storage help reduce pollution in the atmosphere and help conserve our oil supplies?

References

1. Alhajji M, Demirel Y (2015) Energy and environmental sustainability assessment of crude oil refinery by thermodynamic analysis. Int J Energy Res 39:1925–1941
2. Alhajji M, Demirel Y (2016) Energy intensity and environmental impact metrics of the back-end separation of ethylene plant by thermodynamic analysis. Int J Energy Environ Eng 7:45–59
3. Allen JW, Unlu S, Demirel Y, Black P, Riekhof W (2018) Integration of biology ecology and engineering for sustainable algal based biofuel and bioproduct biorefinery. Bioresources Bioprocess 5(47):1–28
4. Cuellar-Franca RM, Azapagic A (2015) Carbon capture, storage and utilization technologies: a critical analysis and comparison of their life cycle environmental impacts. J CO_2 Utilization 9:82–102
5. Curran MA (2015) Life cycle assessment: a systems approach to environmental management and sustainability, CEP October
6. Demirel Y (2015) Sustainability and economic analysis of propylene carbonate and polypropylene carbonate production process using CO2 and propylene oxide. Chem Eng Process Technol 6:236
7. Demirel Y, Gerbaud V (2019) Nonequilibrium thermodynamics: transport and rate processes in physical, chemical and biological systems, 4th edn. Elsevier, Amsterdam
8. Demirel Y (2013a) Sustainable distillation column operations. Chem Eng Process Techniques 1005:1–15
9. Demirel Y (2013b) Thermodynamics analysis. Arab J Sci Eng 38:221–249
10. Demirel Y, Matzen M, Winters C, Gao X (2015) Capturing and using CO_2 as feedstock with chemical-looping and hydrothermal technologies and sustainability metrics. Int J Energy Res 39:1011–1047
11. Demirel Y (2017) Lignin for sustainable bioproducts and biofuels. J Biochem Eng Bioprocess Technol 1:1–3
12. Demirel Y (2018a) Biofuels. In: Dincer I (ed.), Comprehensive Energy Systems: Part B, vol 1. Elsevier, Amsterdam, pp 875–908
13. Demirel Y (2018b) Energy Conservation. In: Dincer I (ed.), Comprehensive energy systems, vol 5. Elsevier, Amsterdam, pp 45–90
14. Dincer I, Ratlamwala TAH (2013) Development of novel renewable energy-based hydrogen production systems: a comparative study. Int J Hydrogen Energy 72:77–87
15. Hall CAS, Lambert JG, Balogh SB, Gupta A, Arnold M (2014) EROI and quality of life. Energy policy 64153
16. IChemE-Institution of Chemical Engineers (2002) The sustainability metrics, institution of chemical engineers sustainable development progress metrics recommended for use in the process industries
17. IEA (2019) World energy outlook. IEA, Paris https://www.iea.org/reports/world-energy-outlook-2019
18. International Organization for Standardization (2006) Environmental management—life cycle assessment—principles and framework, 2006. Geneva, Switzerland, International Organization for Standardization
19. IRENA 2020 Global Energy Outlook: Energy Transformations 2050 (ed 2020) International Renewable Energy Agency, Abu Dabi
20. Kutscher CF (ed) (2007) Tackling climate change in the U.S., American solar energy society, in www.ases.org/climatechange. Accessed May 2014
21. Matzen M, Alhajji M, Demirel Y (2015a) Techno economics and sustainability of renewable methanol and ammonia productions using wind power—based hydrogen. Adv Chem Eng 5:128

22. Matzen M, Alhajji M, Demirel Y (2015b) Chemical storage of wind energy by renewable methanol production: feasibility analysis using a multi-criteria decision matrix. Energy 93:343–353
23. Matzen M, Demirel Y (2016) Methanol and dimethyl ether from renewable hydrogen and carbon dioxide: alternative fuels production and life-cycle assessment. J Cleaner Prod 139:1068–1077
24. Ptasinski KJ (2016) Efficiency of biomass energy. An exergy approach to biofuels, power, and biorefineries, Wiley, New York
25. Unlu S, Niu W, Demirel Y (2020) Bio-based adipic acid productions: feasibility analysis using a multi-criteria decision matrix. Biofuels, Bioproducts Biorefining, Biofpr,. https://doi.org/10.1002/bbb.2106
26. UNESCO and Earth Charter International Secretariat, (2007) https://earthcharter.org/library/good-practices-using-the-earth-charter-2/
27. Wang X, Demirel Y (2018) Feasibility of power and methanol production by an entrained-flow coal gasification system. Energy Fuels 32:7595–7610

Renewable Energy

12

Introduction and Learning Objectives: Energy demand continues to grow worldwide while the extraction of fossil fuels becomes more regulated and expensive. The ways we produce, convert, store, and use energy are changing the earth's climate and affecting the environment, hence the ways of human's life as well as the next generation's future. As the industrial revolution progressed, the use of electricity and the importance of liquid transportation fuels increased. Since the supply depended largely on fossil fuels plus some hydropower and nuclear energy, concerns kept increasing for greenhouse gas (GHG) emissions contributing to possible global warming. After advances, global efforts concentrated in replacing energy dense fossil fuels with the relatively dilute nature of wind and solar energy, which is very materials intensive. With the global consensus to utilize wind and solar energy, their costs have fallen and become comparable with the conventional sources because of the increased costs of fossil fuel technologies, especially with decarbonization charges. New focus may be on achieving a balance between supply and demand of renewables.

The learning objectives of this chapter are to understand:

- Solar and wind energy
- Bioenergy
- Renewable energy and ecology.

12.1 Renewable Energy

Renewable energy resources and are naturally replenished. Major renewable energy sources are hydroelectric, solar energy, biomass, wind, geothermal heat, and ocean. In its various forms, renewable energy comes directly from the sun, or from heat generated deep within the Earth. For each of these resources, the IRENA (2020)

© Springer Nature Switzerland AG 2021
Y. Demirel, *Energy*, Green Energy and Technology,
https://doi.org/10.1007/978-3-030-56164-2_12

reported the capacity factor from 2010 to 2019 as well as long-term projections [12, 13].

Solar and wind energy have properties different from conventional forms of power generation: (1) their maximum output fluctuates according to the real-time and location, (2) the fluctuations can be predicted accurately only for a short time, (3) they use converters in order to connect to the grid, (4) they are more modular and can be deployed in a distributed way, and (5) they frequently are located at a distance from load centers, requiring connection costs [10].

- There is global support for using solar and wind energy, which provides electricity without GHG emissions.
- Renewable electricity depends on the cost and efficiency of the wind and solar technology, which are improving and thus reducing costs at the source.
- Utilizing renewable electricity from solar and wind in a grid becomes challenging mainly due to the mismatch between supply and demand.
- Utility storage capacity is required due to the intermittent nature of solar and wind.
- In many countries, policy settings exist to support renewables with subsidies.

In an age where sustainability is an ethical norm, the focus on the environmental implications of sources and the cost imposed on carbon capture changed the economic scope of clean energy sources. This creates a unique challenge in generating technologies, which efficiently supply renewable power to the consumer at a competitive price.

12.1.1   Solar Energy

Photosynthesis as solar energy's main application in agriculture and forestry has been extended in many ways. Technologies for solar energy capture, conversion, and distribution may be either passive or active. *Active solar techniques* use solar thermal collectors to capture solar energy in the form of solar radiation into a heat transfer fluid, which may be water, air, or others for water and *space heating*. A typical water heating system includes a storage tank holding the hot liquid to manage the demand. *Passive solar systems* rely on gravity and the tendency for fluid to naturally circulate as it is heated.

Solar-based electricity relies on photovoltaics and heat engines that use parabolic trough collectors to collect the solar energy to generate steam to drive a conventional steam turbine [10]. The parabolic mirrors automatically track the sun throughout the day. The sunlight is directed to a central tube carrying a heat transfer

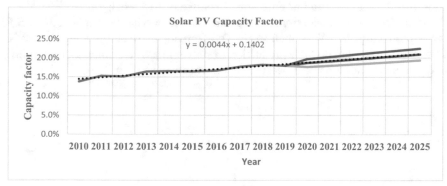

(a)

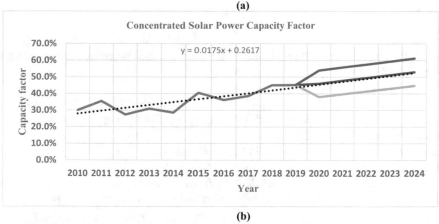

(b)

Fig. 12.1 Trend in capacity factor for **a** solar PV and **b** concentrated solar power using historical weighted average data (2010–2019) and forecasts with upper and lower confidence bounds; dotted line shows the linear trend of weighted average [12]

medium like synthetic oil, which heats around 400 °C. *Photovoltaic solar cell* is a solid-state electrical device that converts the energy of light directly into electricity using various semiconductors crystalline silicon, thin-film, and concentrator [14]. Photovoltaic systems produce direct current, which must be converted to alternating current to be used in grid. A major goal is to increase solar photovoltaic efficiency and decrease costs. Advances in technology have increased the capacity factors of solar power generation. Figure 12.1 shows that solar PV capacity factor will increase from 17% in 2019 to 22% in 2025 and concentrated solar power capacity factor will increase from 42% in 2019 to 52% in 2025 [11–13].

Example 12.1 Estimation of emissions because of lost work
A turbine discharges steam from 6 MPa and 400 °C to saturated vapor at 360.15 K
while producing 500 kW of shaft work. The temperature of surroundings is 290 K.
Determine maximum possible production of power in kW, the amount of work lost,
and emissions based on coal because of the waste energy.

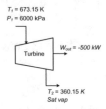

Solution:
Assume: The turbine operates at steady state. Kinetic and potential energy changes
are negligible.
Basis: 1 kg/s steam flow rate
Inlet: superheated steam: $H_1 = 3180.1$ kJ/kg, $S_1 = 6.5462$ kJ/kg K (steam tables)
Outlet: saturated steam: $H_2 = 2655.3$ kJ/kg, $S_2 = 7.5189$ kJ/kg K (steam tables)
$W_{out} = -500$ kW, $T_o = 290$ K
$\Delta H = H_2 - H_1 = (2655.3 - 3180.1)$ kJ/kg $= -524.8$ kJ/kg
The amount of heat transfer: $q = -W + (H_2 - H_1) = -24.8$ kJ/kg
We can determine the entropy production from an entropy balance on the turbine
operating at steady state that exchanges heat only with the surroundings:
$S_{prod} = S_2 - S_1 - \dfrac{q}{T_o} = (7.5189 - 6.5462)$ kJ/kg K $+ 24.8$ kJ/kg/(290) K $= 1.06$ kJ/kg K
Lost work: $W_{lost} = T_o S_{prod} = (290$ K$)(1.06$ kJ/kg K$)(1$ kg/s$) = 307.4$ kW
Maximum work output:
$W_{ideal} = W_{max} = W_{out} - T_o S_{prod} = -500 - 307.4 = -\mathbf{807.4\,kJ/kg}$
Annual emission for 1 kg/s steam flow rate:
$(0.37$ kg CO_2/kWh$)(307.4$ kW$)(24$ h/day$)(365$ days/year$) = 996,344.8$ kg CO_2/year

**Example 12.2 Avoided emissions in a compressor operated with a minimum
power**
A compressor receives air at 15 psia and 80 °F with a flow rate of 1.0 lb/s. The air
exits at 40 psia and 300 °F. Estimate the minimum power input to the compressor
and avoided emissions if it is operated with solar concentrated power instead of
power from natural gas-fired plant. The surroundings are at 520 R.

Solution:

Assume that potential energy effects are negligible, and steady process.

Basis: air flow rate = $\dot{m}$ = 1 lb/s. The surroundings are at 520 R

The properties of air from Table D1 after conversions from SI units:

State 1: P_1 = 15 psia, T_1 = 540 R, H_1 = 129.0 Btu/lb, S_1 = 0.6008 Btu/lb R

State 2: P_2 = 40 psia, T_2 = 760 R, H_2 = 182.0 Btu/lb, S_2 = 0.6831 Btu/lb R

Compressor work: $W_s = \dot{m}(H_2 - H_1) = (1 \text{ lb/s})(182.08 - 129.06)\text{Btu/lb} = 53.0$ Btu/s

The entropy production: $\dot{S}_{prod} = \dot{m}(S_2 - S_1) = 0.0823$ Btu/s R (Eq. 7.21)

Lost work: $W_{lost} = T_o S_{prod} = 42.8$ Btu/s

Minimum work required:

$W_{ideal} = W_{min} = \dot{m}\Delta H - T_o S_{prod} = (53.0 - 42.8)$ Btu/s = **10.2 Btu/s = 10.76 kW**

Annual emission for electricity from natural gas-fired power by replacing with solar power

Use Table 11.2: (0.23 kg CO_2/kWh)

(0.23 kg CO_2/kWh)(10.76 kW)(24 h/day)(365 days/year) = **21,679.2.8 kg CO_2/year**

12.1.2 Wind Energy

Despite to the low-capacity factor (30–40% utilization) and intermittency, wind power receives widespread installations and investments. According to Betz's law, no technological device can capture more than 59.5% of the kinetic energy in the wind or water. Utility-scale advanced wind turbines can achieve up to 80% of Betz's limit [2].

The differential heating drives a global atmospheric convection system reaching from Earth's surface to the stratosphere. Most of the energy stored in these wind movements can be found at high altitudes. Wind power is a totally renewable energy source with no greenhouse gas emissions, but due to its unpredictability has problems integrating with national grids. To mitigate the intermittency of wind power is to store it in the form of hydrogen by electrolysis and feed it into the gas grid and eventually electricity grid. The potential for wind power is considerable and integrating with existing transmission capacity helps large-scale deployment and reduces the cost [10, 11].

As the technology has improved, the capacity factors increased for wind power generation. Figure 12.2 shows that onshore wind capacity increases from 35% in 2019 to 40% in 2025 and offshore capacity factor increases from 42% in 2019 to 48% in 2025 [12, 13].

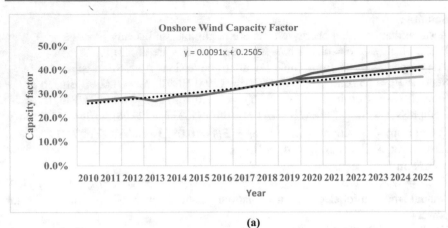

(a)

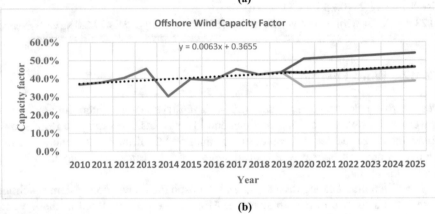

(b)

Fig. 12.2 Trend in capacity factor for **a** onshore wind and **b** offshore wind using historical weighted average data (2010–2019) and forecasts with upper and lower confidence bounds; dotted line shows the linear trend of weighted average [12]

Example 12.3 Avoided emissions by windmill power generation

A farm of windmills supplies a power output of 1 MW for a community. Each windmill has the blades with 10 m in diameter. At the location of the windmills, the average velocity of the wind is 11 m/s and the average temperature is 20 °C. Estimate the minimum number of windmills to be installed.

Solution:

Air is ideal gas and the pressure is atmospheric.

Inlet: $v = 11$ m/s, $R = 8.314$ kPa m^3/kmol   K, $T = 293$ K, $D = 10$ m, $MW_{air} = 29$ kg/kmol

Power output = 1 MW

Density of air $\rho = (MW)\dfrac{P}{RT} = \dfrac{(29 \text{ kg/kmol})101.3 \text{ kPa}}{(8.314 \text{ kPa m}^3/\text{kmol K})(293\text{K})} = 1.2 \text{ m}^3/\text{kg}$

Air mass flow rate: $\dot{m} = \rho A v = \rho \pi \dfrac{D^2}{4} v = 1036.2 \ \text{kg/s}$

Power from each windmill:

$$KE = -\dot{m}\frac{v^2}{2} = 1036.2 \ \text{kg/s} \frac{(11 \ \text{m/s})^2}{2} \left(\frac{\text{kJ/kg}}{1000 \ \text{m}^2/\text{s}^2} \right) = -62.7 \ \text{kW}$$

The minimum number of windmills to be installed:
$1000 \ \text{kW}/|62.7| \ \text{kW} = \textbf{16 windmills}$

12.1.3 Hydropower

Hydroelectric energy from the potential energy of rivers is the best-established electricity generation from renewable sources. The largest power plants in the world use dams on rivers. Different from wind and solar energy, hydro generating has significant mechanical inertia and is synchronous that supplies reliable energy to the grid. Hydro energy can be applied to peak-load demand easily as the individual turbines can be controlled from zero to full power in about 10 min. Therefore, hydro energy can complement the wind and solar power in a grid system.

Pollution is minimum from hydroelectricity as it is produced without fuel. Hydroelectric plants can operate long time and with less operating and maintenance cost. The sale of electricity from the station may cover the construction costs after 5–8 years of full operation.

In the transforming energy scenario (TES), hydropower capacity would need to increase 25% by 2030, and 60% by 2050, while pumped hydro storage capacity would need to double not only by new construction but by upgrading the existing turbines. Hydropower provides the cost effectiveness to counteract the variability of wind and solar generation, and seasonal complementarities in resource patterns besides its benefits in regulating river flows and reducing flooding. New hydropower investments need to consider local environmental impacts and engage in discussions with the stakeholders. Total hydropower power generation (GW): Historical Progress 1200 (2015) and 1300 (2019 estimated); in Transforming Energy Scenario (TES): 1670 (2030) and 2147 (2050) [11, 13]. Figure 12.3 shows that hydroelectric capacity increases from 50% in 2019 to 52% in 2025.

Example 12.4 Avoided emissions by hydroelectric power generation
A hydroelectric plant operates by water falling from a 200 ft height. The turbine in the plant converts potential energy into electrical energy, which is lost by about 5% through the power transmission, so the available power is 95%. If the mass flow rate of the water is 396 lb/s, estimate the power output of the hydro plant and the avoided emissions compared with coal-fired power plant.

Solution:

Equation: $\dot{W} = \dfrac{\dot{m}g\Delta z}{g_c}$

Data: $\Delta z = 200$ ft, water flow rate 396 lb/s; transmission loss = 5%

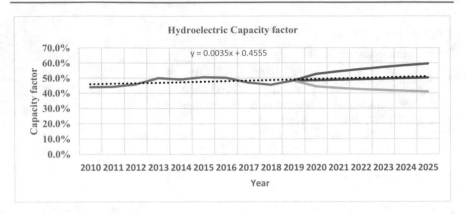

Fig. 12.3 Trend in capacity factor of hydroelectricity using historical weighted average data (2010–2019) and forecasts with upper and lower confidence bounds; dotted line shows the linear trend of weighted average [12]

$$\dot{W} = -\frac{(396 \text{ lb/s})(32.2 \text{ ft}^2/\text{s})(200 \text{ ft})}{32.2 \text{ ft lb /lb}_f\text{s}^2}\left(\frac{1.055 \text{ kW}}{778 \text{ lb}_f \text{ ft/s}}\right) = -107.4 \text{ kW (with the sign convention)}$$

With the transmission loss of 5% of the available power:

-107.4 kW $(1 - 0.05) = -\textbf{102.0 kW}$.

Annual avoided emission using Table 11.2:

$(0.37 \text{ kg } CO_2/\text{kWh})(102.0 \text{ kW})(24 \text{ h/day})(365 \text{ days/year}) = \textbf{330.602.4 kg } CO_2/\textbf{year}$

12.1.4   Geothermal Energy

Geothermal energy is the heat originating from radioactive decay of minerals, from volcanic activity, and from solar energy absorbed at Earth's surface. The hot water and steam from reservoirs can be used to drive generators and produce electricity. Geothermal power is cost-effective, reliable, sustainable, yet has declining capacity factor. Figure 12.4 shows that the geothermal capacity factor decreases from 82% in 2019 to 81% in 2022 [12]. The heat produced from geothermal is typically used in district heating in buildings.

Example 12.5 Avoided emission by using a geothermal energy source
A steam power plant is using a geothermal energy source. The geothermal source is available at 220 °C and 2320 kPa with a flow rate of 200 kg/s. The hot water goes through a valve and a flash drum. Steam from the flash drum enters the turbine at 550 kPa and 428.62 K. The discharged steam from the turbine has a quality of $x_4 = 0.96$. The condenser operates at 10 kPa. The water is a saturated liquid after passing through the condenser. Determine the work output of turbine and avoided emissions compared with coal-fired and natural gas-fired power productions.

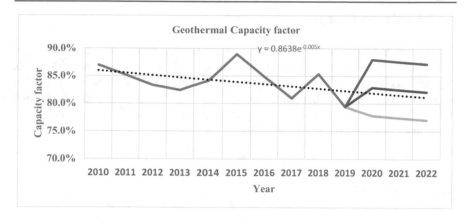

Fig. 12.4 Trend in capacity factor for geothermal energy using historical weighted average data (2010–2019) and forecasts with upper and lower confidence bounds; dotted line shows the trend of weighted average [12]

Solution:

Assume: The kinetic and potential energy changes are negligible, and this is a steady-state process.

$\dot{m}_1 = 200$ kg/s. From steam tables:

$T_1 = 493.15$ K, $P_1 = 2319.8$ kPa, $H_1 = H_2 = 943.7$ kJ/kg, $S_1 = 2.517$ kJ/kg K

$T_3 = 428.62$ K, $P_3 = 550$ kPa, $H_3 = 2751.7$ kJ/kg, $S_3 = 6.787$ kJ/kg K(saturated)

$H_{3sat\ vap} = 2551.7$ kJ/kg, $H_{3sat\ liq} = 655.80$ kJ/kg, $S_{3sat\ vap} = 6.787$ kJ/kg K,

$S_{3sat\ liq} = 1.897$ kJ/kg K, $P_4 = 10$ kPa, $H_{4sat\ vap} = 2584.8$ kJ/kg, $H_{4sat\ liq} = 191.8$ kJ/kg

In this geothermal power plant, the hot water is flashed, and steam is produced and used in the turbine.

The rate of vapor is estimated from the quality at state 2. The fraction of steam after flashing is:

$$x_2 = \frac{943.7 - 655.8}{2751.7 - 655.8} = 0.159$$

$S_2 = (1 - 0.159)1.897 + 0.159(6.787) = 2.6756$ kJ/kg K

The steam flow rate is: $\dot{m}_3 = x_2(\dot{m}_1) = 0.159(200) = 31.84$ kg/s

From the mass balance around the flash drum, we have

$\dot{m}_6 = \dot{m}_1 - \dot{m}_3 = 168.15$ kg/s

The discharged steam has the quality of: $x_4 = 0.96$

$H_4 = (1 - 0.96)H_{4sat\ liq} + (0.96)H_{4sat\ vap} = 2489.08$ kJ/kg

From the flash drum at state 6, we have: $\dot{W}_{net} = \dot{m}_3(H_4 - H_3) = -\mathbf{1993.82\,kW}$

Table 11.2: (0.37 kg CO_2/kWh)

Annual avoided mission:

(0.37 kg CO_2/kWh)(**1993.82**kW)(24 h/day)(365 days/year) = **6, 462, 369.4 kg CO_2/year**

12.1.5 Ocean Energy

Ocean energy falls into three categories that are tidal, wave, and temperature gradient. Ocean thermal energy conversion uses the temperature difference that exists between deep and shallow waters to run a heat engine system to harvest electrical power from ocean waves as a viable technology. The generator is powerful and the turbine has minimal environmental impact as the rotors pose little or no danger to wildlife as they turn quite slowly [11].

12.2 Bioenergy

Bioenergy is derived from biological sources, to be used for heat, electricity, or vehicle fuel. Biofuel derived from plant materials is among the most rapidly growing renewable energy. Bioenergy currently makes up a large share of renewable energy use and would remain a significant source for power and heat generation in industry and as a fuel in transportation sector. Sustainable bioenergy is the necessary pre-condition on existing farmlands, grasslands, and residues without encroaching upon rainforests and threatening food production. In the transforming energy scenario (TES), bioenergy plays an important role in shipping, aviation, and in industrial sectors. Bioenergy is increasing its share to 23% by 2050. Consequently, traditional uses of bioenergy must be phased out and replaced with cleaner options, such as modern bioenergy and other renewables. Share of total primary energy supply provided by bioenergy (%): Historical Progress 8.7 (2015) and 9.5 (2019 estimated); in (TES): 12 (2030) and 23 (2050). Figure 12.5 shows that bioenergy capacity factor increases gradually from 78% in 2019 to 82% in 2025 [11, 13].

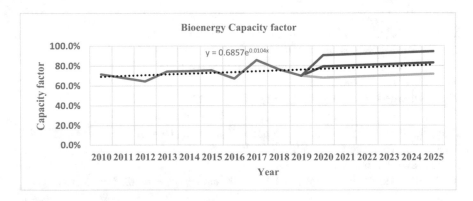

Fig. 12.5 Trend in capacity factor of bioenergy using historical weighted average data (2010–2019) and forecasts with upper and lower confidence bounds; dotted line shows the trend of weighted average [12]

Table 12.1 Biomass resources [21]

Generation	Type	Source	Examples
First	Food crops	Starch crops	Corn, wheat
		Sugar crops	Sugarcane, sugar beet, sweet sorghum
		Feed	Grass
Second	Lignocellulosic crops	Woody	Short-rotation crops, willow poplar
		Herbaceous	Miscanthus, switch grass. corn stover, bagasse
Third	Aquatic	Microalgae	*Chlamydomonas rheinhardii, chlorella, spirulina,*
		Macro algae	Seaweed
		Water plants	Salt marches, sea grass
Wastes	Natural	Agricultural	Animal manure, crop residues
		Forest	Logging residues, tree wastes
	Man-made	Municipal	Solid waste, sewage sludge, waste oil
		Industrial	Pulp and paper industry, sludge

12.2.1 Biomass Resources

Biomass feedstock originates from diverse resources including naturally growing terrestrial and aquatic plants, and man-made wastes. Biomass contains carbon, oxygen, and hydrogen in a variety of compounds including carbohydrate, lipids, and protein [21]. It has high moisture content and low energy density. Based on this diversity, biomass feedstock is classified as the first, second, and third generations, as well as wastes, as seen in Table 12.1. The first generation is food-based biomass such as corn, sugarcane, plant, vegetable oils, and fats. The second generation refers to nonfood-type biomass that includes lignocellulose, while the third-generation biomass mainly refers to algae and cyanobacteria. Cellulose, hemicellulose, and lignin are the structural components of lignocellulosic biomass. All these types of biomass can be converted to biofuels and bioproducts by using biochemical, thermochemical, and hydrothermal processes. Biofuel supply chain can promote sustainable energy and development [3–5].

Table 12.2 shows the proximate and ultimate analyses of various biomass that contains between 40 and 53% of carbon, 20 and 43% oxygen, 5 and 6% of hydrogen, and very little sulfur, while the percentage of ash varies between 0.25 and 12%. Lignocellulosic biomass contains cellulose (38–50%), hemicellulose (23–32%), and lignin (15–25%) [21].

12.2.2 Energy from Solid Waste

Garbage, often called municipal solid waste, contains biomass like paper, cardboard, food scraps, grass clippings, leaves, wood, and leather products, and other non-biomass combustible materials, mainly plastics and other synthetic materials made from petroleum. Recycling and composting programs may reduce the share of biomass in municipal solid waste that is landfilled or burned. Solid waste can be

Table 12.2 Proximate and ultimate analyses of the second generation of biomass in wt% and dry base [21]

Biomass type	Fixed Carbon	Volatiles	Ash	C	H	O	N	S	HHV[a] (MJ/kg)
Redwood	16.10	83.50	0.40	53.50	5.90	40.30	0.10	0.00	21.03
Wheat straw	19.80	71.30	8.90	43.20	5.00	39.40	0.61	0.11	17.51
Corn stover	19.25	75.17	5.58	43.65	5.56	43.31	0.61	0.01	17.65
Bagasse	14.95	73.78	11.27	44.80	5.35	39.55	0.38	0.01	17.33
Sawdust	14.33	76.53	0.25	40.00	5.98	44.75	0.01	0.01	19.95
Manure and sludge			38.8	29.9	4.3	22.40	2.90	0.62	13.1
Municipal waste			27.1	40.4	4.9	25.30	0.91	0.35	16.5

[a]HHV: higher heating values

burned in special waste-to-energy plants, which produce heat to generate steam or electricity. Such plants help reduce the amount of solid waste to be buried in landfills. There also are solid waste incinerators that simply burn the solid waste without electricity production [1].

12.3 Biofuels

Biofuels are biomass feedstock-based fuels and include hydrogen, ethanol, butanol, methanol, biooil, biogas, and biodiesel, while renewable electricity is called bio-power. Bioethanol and biodiesel have by far the largest share of the global biofuels markets. The biofuel supply chain involves the growing/production, harvesting, collecting, storing, and transporting of biomass to the biorefinery where it is converted to biofuel, bioproducts, heat, and power. Some agricultural products grown for biofuel production include corn, soybeans, willow switchgrass, rapeseed, wheat, sugar beet, palm oil, miscanthus, sorghum, cassava, and jatropha. Biomass and biodegradable outputs from industry, agriculture, forestry, and households can be used for biofuel production using: (i) fermentation to produce ethanol, (ii) transesterification for biodiesel production (iii) anaerobic digestion to produce biogas, (iv) gasification to produce bio syngas (H_2, CO, CO_2), or (vi) direct combustion to produce heat. The use of biomass, therefore, contributes to waste management as well as fuel [5, 8].

Biomass can be fermented directly to produce hydrogen, ethanol, and other high-value chemicals. Certain photosynthetic microbes produce hydrogen from water in their metabolic activities using light energy. Photo biological technology holds great promise, but because oxygen is produced along with the hydrogen, the technology must overcome the limitation of oxygen sensitivity of the hydrogen-evolving enzyme systems. A new system is also being developed that uses a metabolic switch (sulfur deprivation) to cycle algal cells between a photosynthetic growth phase and a hydrogen production phase. Global liquid biofuel production (billion liters): Historical Progress 129 (2015) and 136 (2017); in Transforming Energy Scenario (TES): 378 (2030) and 652 (2050) [12, 13, 15].

12.3.1 Biorefinery Systems

Biomass can be converted to biofuels, heat, and power in a biorefinery. Since the biomass comes from diverse resources and requires various conversion processes, biorefinery is beneficial in converting various biomass feedstocks into various valuable fuels. Conversion processes include biochemical, chemical, thermochemical, and hydrothermal. Biochemical processes refer to fermentation, anaerobic, and aerobic digestions. Chemical processes refer to transesterification of lipids, and Fischer-Tropsch (F-T) synthesis of biosyngas (mainly CO_2, CO, and H_2) into various liquid transportation fuels. Thermochemical processes refer gasification, pyrolysis, reforming, and hydrothermal processes involve hot water under high pressure for hydrogenation of various carbon sources. Biorefineries are built in locations where an adequate supply of biomass exists to maintain their operations [3–5].

12.3.2 Ethanol Production

Ethanol is a clear, colorless alcohol fuel made from the sugars found in grains, such as corn, sorghum, barley, sugarcane, and sugar beets. The most common processes to produce ethanol today use yeast to ferment the sugars (glucose). Fermentation is a natural microbiological process where sugars are converted to alcohol and carbon dioxide by yeast (*Saccharomyces cerevisiae*). The overall reaction within the fermentation may be represented by

$$C_6H_{12}O_6 \rightarrow 2(CH_3CH_2OH) + 2CO_2 \qquad (12.1)$$

Sugar $\rightarrow$ Alcohol + Carbon dioxide gas

Wet and dry milling routes are used to produce bioethanol from corn. Dry milling requires less investment and produces dried distiller's grain with solubles (DDGS) besides bioethanol, while the wet milling produces oil and animal feed besides the bioethanol. Figure 12.6 shows the basic steps of converting starch into bioethanol by biochemical process using 6-carbon sugars sources. Most corn is ground to a meal, and then the starch from the grain is hydrolyzed by enzymes to glucose (dry mill). The 6-carbon sugars are then fermented to ethanol by natural yeast and bacteria. The fermented mash is separated into ethanol and residue by distillation. Since water–ethanol mixture forms azeotropic mixture, molecular sieves produce fuel grade ethanol (0.4 vol.% water). The average yield of converting corn starch to ethanol is around 100 gallons bioethanol per dry ton corn. About one-third of every kilogram of corn grain is converted to ethanol, one-third to DDGS, and one-third to CO_2. Ethanol is produced at ASTM D4806 standards and shipped to the refiner or distributor for blending with conventional gasoline [5, 21].

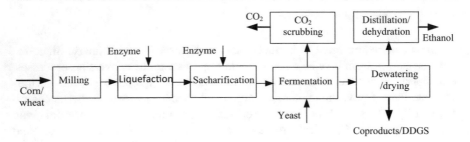

Fig. 12.6 Main steps for the first-generation biofuel production process producing bioethanol, DDGS, and CO_2

Lignocellulosic biomass and other nonfood biomass resources produce second-generation bioethanol using corn stover, corn cobs, sorghum stalks, wheat straw, cotton residue, alfalfa stems, wood, fast growing plants such as grass, bagasse of sugarcane, and sorghum stalks. As the cellulose is protected by lignin and hemicellulose, pretreatment is required to hydrolyze the biomass to 5-carbon and 6-carbon sugars (xylose and glucose) using enzymes. Unlike glucose, xylose is not readily fermented to ethanol. Genetically modified or metabolically engineered yeasts or bacteria are used to ferment both glucose and xylose to enhance the yield of ethanol from lignocellulose [5, 7, 8].

12.3.3 Biodiesel

Chemical process in biodiesel production is the transesterification of triglycerides of fatty acids into biodiesel (fatty acid methyl ester) using mainly methanol or ethanol and a catalyst (mainly acids and alkali-NaOH) and producing glycerin as a byproduct. General transesterification reaction using methanol is represented by

$$RCOOR' + CH_3OH = RCOOCH_3 + R'OH \qquad (12.2)$$

Triglyceride + Methanol = Fatty acid methyl esters (FAME) + Glycerin

Triglycerides are esters of glycerol and present in oilseed crops including soybean, rapeseed, palm, sunflower, and animal fat. Table 12.3 shows the typical oil content, and the number of carbon atoms. During the transesterification, methyl or ethyl esters of fatty acids are produced. Waste cooking oil contains free fatty acid and needs to be esterified with alkali or acid first before being converted to biodiesel by transesterification [5, 17].

The type of oil affects the quality of biodiesel. Soybean and palm oils are the two largest oil seed crops. On a dry-weight basis, soybean contains around 41% protein, 21% oil, and 29% carbohydrate, on average. Triacylglycerol (triglycerides $C_{55}H_{98}O_6$) (94%) is the primary component in the soybean oil, while phospholipids

Table 12.3 Fatty acid composition (wt%) of various oils and fats [21]

Oil or fat	Typical oil content %	Number of carbon atoms: number of double bonds						
		14:0	16:0	18:0	18:1	18:2	18:3	20:1
Rapeseed	39–43	–	4.3	1.3	59.9	21.1	13.2	
Soybean	16–18	–	6–10	2–5	20–30	50–60	5–11	
Sunflower	40–50	–	7.2	4.1	16.2	72.5	–	
Jatropha	28–38	–	11.3	17.0	12.8	47.3	–	1.8
Tallow	24–32	3–6	24–32	20–25	27–43	2–3	–	

content is around 3.7%. Other vegetable oils and animal fats such as canola, camelina, and jatropha constitute a small fraction of biodiesel feedstock [21].

12.3.4 Biodiesel from Algae

Oil-rich microalgae strains can produce biodiesel with (1) algae growth in open ponds or in photobioreactors, (2) algae harvesting and dewatering, (3) lipid extraction, and (4) transesterification of lipid to biodiesel. Algae require water, sunlight, carbon, and nutrients like nitrogen and phosphorus to grow. Many strains of algae can grow nearly in any water resource. The residue of lipid extraction can be fed into anaerobic digestion to produce biogas and biopower, or used as a feedstock for fermentation to alcohols, or used as fertilizer. Table 12.4 compares the properties of biofuels and diesel; fuel properties of biodiesel from soybean oil and the conventional diesel are comparable and biodiesel contains less sulfur and has higher Cetane numbers [1, 5].

Table 12.4 Properties of biodiesel from soybean and conventional diesel [5, 21]

Property	Green diesel	Biodiesel	Petroleum diesel
Density, g/cm^3	0.78	0.88	0.84
Cetane number	70–90	50–65	40
Heating value, MJ/kg	44	38	43
Sulfur, ppm	<1	<1	<10
Oxygen, %	0	11	0
Cold flow properties	Poor	Poor	Good
Oxidative stability	Good	Marginal	Good
Cloud point C	−20 to +20	−5 to +15	−5
NO$_x$	−10 to 0	+10	Baseline

12.3.5   Green Diesel

Triglycerides contain long, linear aliphatic hydrocarbon chains, which are partially unsaturated and have a carbon number range similar to the molecules found in petroleum diesel fuels. *Green diesel* production requires large volumes of H_2 and a catalyst to hydrogenate triglycerides into a high-cetane diesel fuel by removing all the oxygen from the triglyceride and saturating all the olefinic bonds in the fatty acids. The primary products from this hydrogenation are water, CO_2, propane, and a mixture of normal paraffin. Green diesel is fully compatible with petroleum-based diesel and is a drop-in-fuel and has a heating value equal to conventional diesel (see Table 12.4). Feedstocks rich in saturated fats, such as palm and tallow oil, require less hydrogen than feedstocks higher in olefin content, such as soybean or rapeseed oil. Diesel yield depends on both feedstock type and the level of hydro isomerization required to achieve product cloud point specification [5].

Biodiesel comes with several shortcomings that include NO_x and SO_x emissions, a cloud point with adverse effects, filter clogging, marginal stability, reduced energy content, and need to be blended with conventional petroleum diesel. Biodiesel has around 11% oxygen, whereas petroleum-based diesel and green diesel has no oxygen. Petroleum diesel has around 10 ppm sulfur and biodiesel and green diesel have less than 1 ppm sulfur [4].

12.3.6   Biogas and Renewable Natural Gas

Anaerobic digestion of wet biomass feedstock (<15% solid), such as animal manure, agricultural residue, and sewage sludge from municipal wastewater treatment produce methane-rich biogas using bacteria (methanogenic), archaea, and fungi at 37–55 °C. The chemical reactions in biochemical processes undergo at lower temperatures as well as at lower conversion rates in nonpolluting natural processes requiring low energy and other chemicals. Biogas contains around 60% methane and 40% carbon dioxide and hence a lower heating value of 20–25 MJ/m^3 at normal conditions. It can be used for heating, steam, and consequently electricity production. It can also be used as renewable natural gas after desulfurization and methane enrichment by removing 90% of carbon dioxide [5].

12.3.7   Bio Synthesis Gas

A biomass, represented by C_nH_{2m}, may be oxidized to CO_2 and water and releases heat of combustion, which can be used to produce steam and electricity

$$C_nH_{2m}(n + 0.5m)O_2 \rightarrow mH_2O + nCO_2 \tag{12.3}$$

In a conventional gasification process, biomass (or other carbon-containing feedstock) reacts with limited oxygen (or air), CO_2, at high temperatures (750–

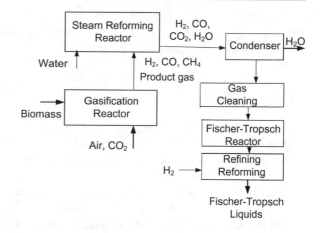

Fig. 12.7 Biomass steam gasification with production of liquid transportation fuels by Fischer-Tropsch synthesis [4, 5]

1100 °C) to produce synthesis (syngas) containing mainly H_2, CO, CO_2, methane, besides tars and a solid char [5, 17]. The following reaction represents the steam gasification (reforming) as seen in Fig. 12.7.

$$C_nH_m + H_2O = nCO + (m/2 + n)H_2 \qquad (12.4)$$

The water–gas shift reaction can increase the hydrogen content from 6 to 6.5% in the initial synthesis gas to 30–50 vol.% [21].

$$CO + H_2O = CO_2 + H_2 \qquad (12.5)$$

Syngas is cooled and sent to a gas cleaning unit to remove the CO_2 and H_2S. Purification of the syngas accounts for 60–70% of the total capital cost. Gasification process requires proper utilization of heat integration to reduce the operating costs. Various simulation and modeling studies of biomass (corn stover and distiller grain) gasifier can predict the flowrate and composition of product from given biomass composition and gasifier operating conditions.

Fischer-Tropsch (F-T) synthesis can produce various liquid biofuels from syngas. A representative Fischer-Tropsch (F-T) reaction is

$$(2n + 1)H_2 + nCO \rightarrow C_nH_{2n+2} + nH_2O$$
$$\rightarrow -170 \, kJ/mol \, (at \, 250 \, °C \, and \, 15 \, atm) \qquad (12.6)$$

In the production of diesel fuel 'n' can be in the range of 12–25 carbon; therefore, a H_2 to CO molar ratio of close to 2 is required. An iron-based catalyst and an operating temperature of 350 °C will produce mostly gasoline, while a cobalt base and an operating temperature of 200 °C will produce mostly diesel fuel. The crude biooil produced in the F-T synthesis may be distilled to naphtha, distillate, and wax [4].

Chemical-looping gasification and reforming produces syngas (CO and H_2) or a mixture of CO, H_2, CO_2, and CH_4 (gasification), or a relatively pure stream of H_2 with inherited carbon capture [17].

12.3.8 Biooil

Pyrolysis uses fast heating (700–800 °C) under anaerobic conditions to break down biomass into a volatile mixture of hydrocarbons and char. This mixture of hot gases is condensed into a bio-oil that is a rich mixture of hydrocarbons, some of which can be converted into biofuels.

Hydrothermal liquefaction of biomass also produces bio-oil (200–400 °C, 5–40 MPa). This liquefaction system can work with high-moisture biomass feedstocks and municipal waste streams such as sewage sludge and wet algae at much lower temperatures compared with the gasification and pyrolysis processes. Hydrogen required for hydrogenation of carbon comes from the water [4].

12.3.9 Butanol

Sugars can be converted into butyric, lactic, and acetic acids by fermentation process. Butyric acid is converted by fermentation into butanol. Corn starch also can be converted to biobutanol via the acetone–butanol–ethanol fermentation pathway. Co-products include alcohols with lower molecular weight than butanol and acetone. Butanol's toxicity to the microorganisms that ferment sugar creates obstacles. If corn grain is the source of the sugars for fermentation, a residue similar to dried distillers' grain is produced. This might require additional processing to remove any toxic biobutanol and acetone residue before it could be used as an animal feed. Gas stripping can be used to extract the butanol.

Isobutanol can be produced by anaerobic process using *E. coli* strains and continuous vacuum stripping for butanol fermentation. This process is still in the developing stage. The sugar to isobutanol conversion yield is around 85%. As the yield increases butanol purification improves considerably. Both butanol costs, water usage, and direct CO_2 emissions are higher than that of cellulosic bioethanol, while butanol is far superior to ethanol in energy efficiency [5].

12.3.10 Methanol

Renewable methanol is produced by using hydrogen from wind power in the electrolytic system and carbon dioxide from bioethanol plants (Fig. 12.8). The energy efficiency for the concentrated CO_2 and H_2-based methanol is around 46% [15].

Sewage sludge is a residue of a municipal wastewater treatment plant. Mechanically dewatered sludge contains 12–25% solid (LHV = 12.0 MJ/kg) with an organic fraction of 56% of dry solids rich in carbon (50%), hydrogen (7%), and

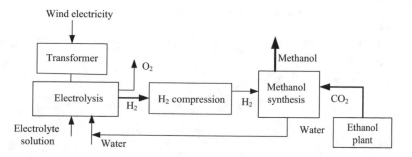

Fig. 12.8 The integral methanol production facility based on the feedstock of renewable hydrogen and CO_2 [16]

oxygen (31%). This may lead to energy recovery by producing biofuels as well as bioproducts besides the incineration commonly used. One way for energy recovery is the gasification of the sludge and conversion of syngas produced into methanol as well as other chemicals. This will help waste minimization as well as recover energy [6, 20]

12.3.11 Dimethyl Ether

Dimethyl ether (DME) is the simplest ether (CH_3OCH_3) and can be produced by catalytic methanol dehydration.

$$2CH_3OH \rightarrow CH_3OCH_3 + H_2O \tag{12.7}$$

DME is a colorless, nontoxic, highly flammable gas at ambient conditions, but can be handled as a liquid under slight pressure (0.5 MPa). The properties of DME are similar to those of liquefied petroleum gas with a lower heating value of 28.4 MJ kg^{-1} and a density of 0.67 kg/l. Tables 12.5 and 12.6 compare the properties and environmental impacts of DME with other fuels. It has the conventional diesel fuel equivalency of 0.59 as the lower heating value of diesel is 43.1 MJ/kg. DME is not a greenhouse gas and can be used as a substitute for diesel fuel, domestic gas. Methanol and DME from CO_2 hydrogenation may outperform conventional petroleum-based fuels, reducing greenhouse gas emissions 82–86%, minimizing other pollutants (SO_x, NO_x, etc.) and reducing fossil fuel depletion by 82–91% [16].

Table 12.5 Properties of bioethanol, biomethanol, biobutanol, bioisobutanol, and conventional gasoline [5, 21]

Properties	Methanol	Ethanol	Butanol	Isobutanol	Gasoline
C, %	37.5	52.1	64.8	64.8	85–88
H, %	12.6	13.1	13.5	13.5	12–15
O, %	49.9	34.7	21.6	21.6	–
Density (20 °C), kg/m	791	789	809	802	690–800
Normal boiling point, °C	65.0	78.5	117.7	107.9	27–225
Motor octane number	91	92	84	90	80–88
Higher Heating value (20 ° C), MJ/kg	22.3	29.8	37.3	37.2	47.2
Stoichiometric air/fuel ratio kg/kg	6.4	8.9	11.2	11.2	14.7
Flash point, °C	12	13	37	28	−43
Autoignition temperature, °C	470	363	340	415	250–300
Energy density, MJ/L	16	21.4	26.9	26.6	30–33
CO_2 production, MJ/kg fuel	15	13	15	15	14

Table 12.6 Environmental impacts for methanol or dimethyl ether [5, 21]

Indicator	Methanol	DME	Unit/mt product
Global warming potential	0.30	0.53	mt CO_2 eq
Acidification potential	0.67	0.97	kg SO_2 eq
Photochemical oxidant formation	0.69	1.17	kg NMVOC eq[a]
Particulate matter (PM) formation	0.29	0.44	kg PM_{10} eq
Human toxicity	0.10	8.18	kg 1,4-DB eq[b]

[a]NMVOC: Non-methane volatile organic compound
[b]1,4-DM: 1,4 dichlorobenzene

12.4   Hydrogen Production

Hydrogen is a zero-emission fuel with combustion products of water and a trace amount of NO_x. Steam methane reforming is a current and economic process of producing hydrogen in large scales with around 86% energy efficiency. Hydrogen can be produced mainly from the second generation of biomass feedstock by using thermochemical processes of gasification (steam reforming). Figure 12.9 shows a schematic of conventional hydrogen production by gasification of a biomass feedstock. Energy efficiency for biomass gasification for H_2 production is around 55–65% [9].

Figure 12.10 shows the schematic of wind energy-based renewable hydrogen production. The system includes the transformer, thyristor, electrolyzer unit, feed water demineralizer, hydrogen scrubber, gas holder, two compressor units,

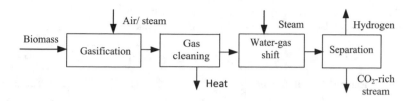

Fig. 12.9 Block flow diagram of biohydrogen production by gasification of biomass feedstock

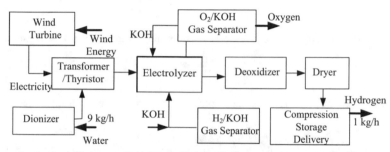

Production efficiency ~72%, Electrolyzer efficiency: ~62%; target: 76% (LHV)
Target cost: $0.3/kg H_2 = gasoline of $2.5/GJ; Cost: $3.74-5.86/kg H_2 0.97 kg
$CO_{2\text{-eq}}$/kg H_2

Fig. 12.10 Schematic for alkaline electrolysis of water for green hydrogen production with compression, storage, and delivery [16]

deoxidizer, and twin tower dryer. For producing one kg H_2, approximately 26.7 kg water is necessary. The total greenhouse gas emission is around 0.97 kg CO_2e/kg H_2. The hydrogen production cost largely depends on the electricity price. Therefore, electrolysis plants take advantage of low electricity prices at off-peak hours [9, 15].

Declining of coal-fired and nuclear electricity generation capacity may lead to gain in electricity generation by natural gas and renewables. Currently, 96% of H_2 is produced directly from steam reforming of natural gas, coal gasification, and partial oxidation of hydrocarbons such as biomass. The carbon is captured in the process than *blue hydrogen* is produced. Figure 12.11 shows some commercial processes for H_2 production from syngas carbon capture and storage [9].

Hydrogen can offer solutions supplied by electricity. Certain energy-intensive industries, such as iron making, and ammonia may relocate to areas with good renewable energy resources to produce cheap green hydrogen. Table 12.7 shows that by 2050, there would be 160 Mt (19 EJ) of green hydrogen produced annually in the transforming energy scenario (TES). Considerable scale-up of electrolyzers with increased efficiency would help produce between 50 and 60 GW per year of new capacity from now until 2050 [13].

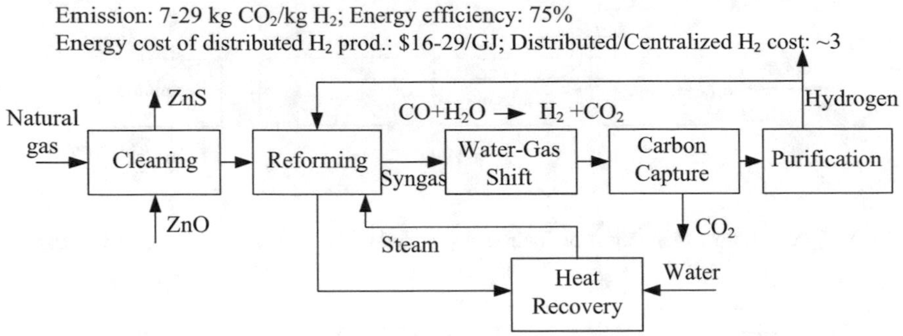

Emission: 7-29 kg CO_2/kg H_2; Energy efficiency: 75%
Energy cost of distributed H_2 prod.: \$16-29/GJ; Distributed/Centralized H_2 cost: ~3

Fig. 12.11 Blue hydrogen production by steam reforming of natural gas with carbon capture [5, 16]

Table 12.7 Renewable energy share in power generation transforming energy scenario (TES) [13]

	Historical progress	TES (2030)	TES (2050)
Blue hydrogen production (Mt)	0.6 (2015–2018)	30	80
Green hydrogen production (Mt)	1.2 (2015–2018)	25	160
Electrolyzers (GW)	0.04 (2016)	270	1700
Electricity demand to produce hydrogen from renewables (terawatt hours-TWh)	0.26 (2016)	1200	7500

12.5 Renewable Fuel Standards

The U.S. Congress created the Renewable Fuel Standard (RFS) program to reduce the GHG emissions and expand the nation's renewable fuels sector while reducing reliance on imported oil. The RFS program was authorized under the Energy Policy Act of 2005 and expanded under the Energy Independence and Security Act of 2007. This new renewable fuel standard is known as RFS2. The Clean Air Act requires EPA to set the RFS volume requirements annually.

The RFS program includes four categories of renewable fuel, each with specific fuel pathway requirements and RIN D-Codes [18]:

- Advanced biofuels (D5) are produced from any type of renewable biomass except corn starch ethanol. Required life cycle GHG emissions reduction is at least 50% compared to the petroleum baseline.

- Biomass-based diesel (D4) includes biodiesel and renewable diesel and required life cycle GHG emissions reduction of at least 50% compared to the diesel baseline.
- Cellulosic biofuel (D3 or D7) produced from cellulose or hemicellulose of biomass and required life cycle GHG emissions reduction is at least 60% compared to the petroleum baseline.
- Conventional renewable biofuel (D6) includes ethanol derived from corn starch, or any other qualifying renewable fuel with life cycle GHG emissions reduction of at least 20% compared to the average petroleum baseline.

12.6   Renewable Energy and Ecology

Energy is conserved in ecosystems. Carbon is linked to energy and water flow because of synthesis of C–H bonds in organic material through photosynthesis, which is the main source of energy in most ecosystems. Understanding how this link affects other elements and nutrients in native habitat will be important to protect ecosystems.

Extreme weather, flooding, rising sea levels, and other effects of climate change in ecosystems will have adverse effects on the quality of life, health, economy, and ecology. Therefore, governments, investors, and stakeholders should avoid new investments into the new fossil fuel facilities but improve the existing ones with promising carbon capture and utilization technologies. The Task Force on Climate Related Financial Disclosure encourages voluntary action on climate-related financial risks disclosures [13].

Eco-friendly energy systems need policy makers who would establish long-term targets and adapt policies and regulations accordingly. Energy scenarios including traditional fuels, power sector, and industrial sector can facilitate open dialogue among stakeholders and other agencies to reach a consensus of target and challenges for the eco-friendly energy transition.

Climate change is a global threat to ecology and the transforming energy sector may need different approaches for different countries depending on the priorities. Construction of some large solar projects, wind farms, and hydroelectric dams may harm habitat, including water quality, nutrient, and species flow representing critical natural balances in biodiversity in native ecosystems. Growing energy crops such as corn, soybean, and oil palms may affect tropical rainforests adversely through excessive land and water usage leaving less resources for growing food and for native species [6, 8].

Example 12.6 Avoided emissions in green energy production

A superheated steam (stream 1) expands in a turbine from 5000 kPa and 325°C to 150 kPa and 200 °C. The steam flow rate is 10.5 kg/s. If the turbine generates 1.1 MW of power, estimate the avoided emissions if the wind power instead of coal is used in the boiler to produce the steam.

$T_1 = 598.15$ K
$P_1 = 5000$ kPa

W_{out}

Turbine

$T_2 = 473.15$ K
$P_2 = 150$ kPa.

Solution:

Assume that the kinetic and potential energy effects are negligible; this is a steady process.

The properties of steam from the steam tables:

Stream 1: Superheated steam:

$P_1 = 5000$ kPa, $T_1 = 325\,°C(598.15$ K$), H_1 = 3001.8$ kJ/kg

Stream 2: Superheated steam:

$P_2 = 150$ kPa, $T_2 = 200\,°C(473.15$ K$), H_2 = 2872.9$ kJ/kg

$\dot{W}_{out} = -1100$ kW $= -1100$ kJ/s

The mass and energy balances for the turbine are:

Mass balance: $\dot{m}_{out} = \dot{m}_{in}$

Energy balance: $\dot{E}_{out} = \dot{E}_{in} \rightarrow \dot{m}_1(H_2 - H_1) = \dot{q}_{out} + W_{out}$

Heat loss from the energy balance

$\dot{q}_{out} = -\dot{W}_{out} + \dot{m}_1(H_2 - H_1) = 1100$ kJ/s $+ (10.5$ kg/s$)(2872.9 - 3001.8)$kJ/kg $= -253.45\,$kJ/s

The sign is negative as the heat is lost from the system.

Table 11.2: (0.37 kg CO_2/kWh)

Annual avoided emission from 1.1 MW power production:

$(0.37$ kg CO_2/kWh$)(1100$kW$)(24$ h/day$)(365$ days/year$) = 3,565,320\,$kg CO_2/year

Example 12.7 Avoided emissions in compression of air

Air enters a compressor at 100 kPa, 300 K, and a velocity of 2 m/s through a feed line with a cross-sectional area of 1.0 m². The effluent is at 500 kPa and 400 K. Heat is lost from the compressor at a rate of 5.2 kW. If the air behaves as an ideal gas, estimate the power requirement of the compressor in kW and the avoided emissions if the wind power instead of electricity from coal is used.

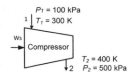

$P_1 = 100$ kPa
1 | $T_1 = 300$ K

Ws
Compressor

$T_2 = 400$ K
2 $P_2 = 500$ kPa

Solution:

Assume: The compressor is at steady state, air is ideal gas, and the change in the potential and kinetic energy of the fluid from the inlet to the outlet is negligible.

$R = 8.314$ Pa m^3/kmol K, $MW = 29$ kg/kmol; $v_1 = 2$ m/s, $A_1 = 1.0$ m^2
$P_1 = 100$ kPa, $T_1 = 300$ K, $P_2 = 500$ kPa, $T_2 = 400$ K, $q_{loss} = -5.2$ kW

For ideal gas: $V_1 = \dfrac{RT_1}{P_1} = 0.025$ m^3/mol, the specific volume

$(1/\rho)_{air} = \dfrac{V_1}{MW_{air}} = 0.86$ m^3/kg

The mass flow rate of air: $\dot{m} = \dfrac{v_1 A_1}{(1/\rho)} = 1.72$ kg/s

From Table D1 for ideal gas of air: $H_2 = 400.9$ kJ/kg, $H_1 = 300.2$ kJ/kg

Energy balance: $\dot{m}(H_2 - H_1) - \dot{q} = \dot{W}$

$(1.72$ kg/s$)((400.9 - 300.2)$kJ/kg $) + 5.2$ kW $= \dot{W} = \mathbf{178.4\,kW}$

This is the work needed by the compressor.

Table 11.2: (0.37 kg CO_2/kWh)

Annual avoided emission:

$(0.37$ kg CO_2/kWh$)(\mathbf{178.4}$kW$)(24$ h/day$)(365$ days/year$) = \mathbf{578{,}230.0\,kg\,CO_2/year}$

Example 12.8 Avoided emission in pumping

A pump increases the pressure in liquid water from 100 kPa at 25 °C to 4000 kPa. Estimate the minimum horsepower motor required and the avoided emissions if the wind power instead of electricity from coal is used in the pump for a flow rate of 0.5 m^3/s.

Solution:

Assume that the pump operates adiabatically and nearly isothermally, changes in potential and kinetic energy are negligible, and water is incompressible.

$P_1 = 100$ kPa, $T_1 = 298.15$ K, $P_2 = 4000$ kPa, $Q = 0.01$m^3/s

Energy balance: $\dot{q} + \dot{W}_{pump} = \dot{m}\Delta H$

For an adiabatic operation ($\dot{q} = 0$): $\dot{W}_{pump} = \dot{m}\Delta H$

Since the water is incompressible the internal energy is only the function of temperature.

$\Delta U = 0$ as the system is isothermal.

$\Delta H = \Delta U + \Delta(PV) = V\Delta P$

The specific volume of saturated liquid water at 25 °C is: $V_1 = 0.001003$ m^3/kg (steam tables)

The mass flow rate: $\dot{m} = \dfrac{Q}{V_1} = 9.97$ kg/s

$\dot{W}_{pump} = \dot{m}V_1(P_2 - P_1)$

$= (9.97$ kg/s$)(0.001003$ m^3/kg$)(4000 - 100)$ kPa $\left(\text{kJ/kPa m}^3\right)$

$= \mathbf{39kJ/s} = \mathbf{39kW} = \mathbf{52.3hP} = 39$ kW$(1$ hp $= 745.7$ W$)$.

Table 11.2: (0.37 kg CO_2/kWh)
Annual avoided emission:
(0.37 kg CO_2/kWh)(39kW)(24 h/day)(365 days/year) $= \mathbf{126,406.8\,kg\,CO_2/year}$

12.7   Impact of Biofuels on Sustainability

Greater demand for biomass from the biofuel industry may increase the dead zones that have overload of nitrogen and phosphorous which can harm the aquatic flora and fauna, and marine life. To clean the dead zones, one needs to purify and oxygenate the existent waterways by removing the excess fertilizer run-offs and fully restore the flora and fauna of the natural habitat.

It is a future challenge to increase the biofuel production from the second and third generations, and waste biomass feedstock resources. Improving the technologies of multiproduction would be essential for the economic feasibility of the biofuel sector. Within the biorefinery and multigeneration settings the unit cost of biofuels can be reduced because the current cost of biofuels is not competitive enough with those of fossil fuels. Distributed versus centralized biorefinery concepts should also be fully analyzed for an optimum process setting for sustainable biofuel production, regional economic development, and environmental protection [5–9].

The contribution of various biofuels to the transport sector keeps increasing. Some additional incentives in investments and electrification particularly in the transportation sector help increase the shares of renewables in power generation. This trend of increase is also projected in the 'Sustainable Development Scenario' (SDS) worldwide, complemented mainly by nuclear power and carbon capture technologies. By 2040, contributions from wind and solar PV will exceed hydropower if investments in renewables technologies would accelerate [10–13].

Higher energy efficiency guarantees secure, sustainable, and inclusive economic growth and reduces the environmental footprint of the energy systems. Half of these energy efficiency savings come from industry, with major contributions also from transport and buildings. The rate for primary energy demand is slower than in the past (1% versus 2.7%) as energy consumption and economic growth continue to decouple by 2040. The SDS depends upon fundamental changes to how energy is produced and consumed with rapid energy sector transformation. Because of its cost-effectiveness and attractive payback, thermal efficiency improvements are critical elements that brings the world closer to the SDS. Increases in the electrification of economies and the share of renewables can cause fluctuations in the carbon footprint of electricity use. Greater emphasis on circular economy with higher materials efficiency may reverse the trend of growing emissions from the industrial sector, such as steel and cement sectors [13].

IRENA's transforming energy scenario (TES) sees emissions fall at a compound rate of 3.8% per year to some 10 Gt, or 70% less than today's level, by 2050, keeping the expected temperature rise well below 2 °C. The TES shows how to achieve stable, climate-safe, sustainable long-term energy and economic development with the deeper decarbonization perspective (DDP) that would reduce emissions by 2050. In the TES, fossil-fuel use declines by 75% compared to today's level; use of coal declines by 41 and 87%, while petroleum oil by 31 and 70% in 2030 and 2050, respectively. In the baseline energy scenario (BES), energy-related emissions increase by a compound annual rate of 0.7% per year to 43 gigatons (Gt) by 2050 (up from 34 Gt in 2019), resulting in a likely temperature rise of 3 °C or more in the second half of this century [11, 13].

Example 12.9 Avoided emissions due to biofuel in a car

An average car consumes 50 gallons gasoline per month. Estimate the amount of energy and the avoided emissions if 85% ethanol blended gasoline is used instead of 100% gasoline consumed by the car per year.

Solution:

Assume that gasoline has an average density of 0.72 g/cm^3 and the heating value of 47.3 MJ/kg (Table 2.6). Cost of gasoline: \$3/gallon

Data: V = 50 gal/month = 189.25 l/month, 2271.0 l/year (3.785 l = 1 gal)

$\rho_{gas} = 0.72 \ g/cm^3 = 0.72 \ kg/l$

Mass of gasoline: $m_{gas} = \rho V = 1635.1 \ kg/year$

Energy consumed per year:

$E_{gas} = 1635.1 \ kg/year \ (47,300 \ kJ/kg) = \mathbf{77,340,230 \, kJ/year}$

$= \mathbf{77,340.2 \, MJ/year}$

Cost per year: C (gasoline): (\$3/gallon) (50 gallon) (12) = \$1800/year

---

Example 12.10 Energy consumed by low and a high-mileage cars

Assume that average daily traveling distance is about 40 miles/day. A car has a city-mileage of 20 miles/gal. If the car is replaced with a new car with a city-mileage of 30 miles/gal and the average cost of gasoline is \$3.50/gal, estimate the amount of fuels.

Solution:

Assume: The gasoline is incompressible with ρ_{av} = 0.75 kg/l.

Lower heating value (LHV) = 44,000 kJ/kg; 44,000 kJ of heat is released when 1 kg of gasoline is completely burned, and the produced water is in vapor state (Table 2.1).

Fuel needed for the old car: (40 miles/day)/(20 miles/gal) = 2 gal/day

Fuel needed for the new car: (40 miles/day)/(30 miles/gal) = 1.34 gal/day

Old car: 20 miles/gal

Mass of gasoline:

$m_{gas} = \rho_{av}$ (Volume) = (0.75 kg/l)(2.0gal/day)(3.785 l/gal) = **5.7 kg/day**

Energy of gasoline:

E_{gas}(LHV) = (5.7 kg/day)(44,000 kJ/kg) = 250800 kJ/day (365 days/year)

$\phantom{E_{gas}(LHV)}$ =   91,542,000 kJ/year = **91,542 MJ/year**

Cost: ($3.50/gal)(2 gal/day)(365 days/year) = $2555/year

New car: 30 miles/gal

Mass of gasoline:

$m_{gas} = \rho_{av}$ (Volume) = (0.75 kg/l)(1.34gal/day)(3.785 l/gal) = **3.8 kg/day**

Energy of gasoline:

E_{gas}(LHV) = (3.8kg/day)(44,000 kJ/kg) = 167,200 kJ/day (365 days/year)

$\phantom{E_{gas}(LHV)}$ =   61,028,000 kJ/year = **61,028 MJ/year**

Cost: ($3.50/gal)(1.34 gal/day)(365 days/year) = $1712/year

The new car reduces the fuel consumption by around 33%, which is significant.

Example 12.11 Avoided emissions by the biogas use

A city consumes natural gas at a rate of 500×10^6 ft^3/day. The volumetric flow is at standard conditions of 60 °F and 1 atm = 14.7 psia. If the natural gas is costing $6/GJ of higher heating value, estimate the yearly energy required and the emission avoided if biogas instead of natural gas is used in the city.

Solution:

Q = 500×10^6 ft^3/day at 60 °F and 1 atm = 14.7 psia.

The higher heating value is the heat of combustion of the natural gas when the water product is at liquid state. From Table 2.6, the value of HHV is: 50,000 kJ/kg ~ 1030 Btu/ft^3 (Table 2.7)

Heating value: (1030 Btu/ft^3)(500×10^6 ft^3/day) = 515.0×10^9 Btu/day

(515.0×10^9 Btu/day) (1055 J/Btu) = 543,325 GJ/day = 6.28 GW = 6288,483 kW

Yearly cost: (543,325 GJ/day) ($6/GJ) (365 days/year) = **11.9×10^8/year**

Table 11.2: (0.23 kg CO_2/kWh)

Annual avoided emission:

(0.23 kg CO_2/kWh) (6280,000 kW)(24 h/day)(365 days/year)

= **1.267 $\times$ 10^{10} kg CO_2/year**

Example 12.12 Avoided emission by a car

An average car consumes about 2 gallons (US gallon = 3.785 L) a day, and the capacity of the fuel tank is about 15 gallons. Therefore, a car needs to be refueled once every week. The density of gasoline ranges from 0.72 to 0.78 kg/l (Table 2.6). The lower heating value of gasoline is about 44,000 kJ/kg. Assume that the average density of gasoline is 0.75 kg/l. If the car was able to use 85% ethanol instead of gasoline, estimate the avoided emissions.

Solution:

Assume: The gasoline is incompressible with $\rho_{av} = 0.75$ kg/l. Cost of gasoline $3.5/gallon

Lower heating value (LHV) = 44,000 kJ/kg; 44,000 kJ of heat is released when 1 kg of gasoline is completely burned, and the produced water is in vapor state.

Mass of gasoline per day: $m_{gas} = \rho_{av}\ V = (0.75$ kg/l$)(2$ gal/day$)(3.785$ l/gal$)$ = 5.67 kg/day

Energy of gasoline per day:

$E_{gas} = m_{gas}$(LHV)=(5.67 kg/day)(44,000 kJ/kg) = 249,480 kJ/day = 2.88 kW

Energy of E85 per day:

$E_{gas} = m_{gas}$(LHV)=(5.67 kg/day)(31,816 kJ/kg) = 180,396 kJ/day 2.08 kW

Avoided energy: (2.88-2.08) kW = 0.8 kW

Avoided emission Table 11.2: (0.27 kg CO_2/kWh)

Annual avoided emission:

(0.27 kg CO_2/kWh) (0.8 kW)(24 h/day)(365 days/year) = **1892.1 kg CO_2/year**

12.8   Energy Management

For effective energy management, there are several basic steps that are management commitment, data analysis, analysis of energy conservation options, implementation energy conservation options, and continued feedback and analysis. Energy management program provides organizations with the information, tools, and assistance to reduce energy, water, nonrenewable energy use, and GHG emissions. The program provides the organizations with annual performance data, energy savings performance contracts, utility energy service contracts, and support with energy-efficient products. Within this framework organizations may prepare sustainability/energy score cards. The program offers expertise regarding all levels of project and policy implementation to reduce the energy intensity of operations [16]. For example, heat pumps efficiency gains ranging from two to four times higher than conventional heating systems may lead to increase 9-fold by 2050 [13].

Example 12.13 Thermal efficiency impact on emissions of an actual Brayton cycle

A power plant is operating on an ideal Brayton cycle with a pressure ratio of $r_p = 9$. The fresh air temperature is 295 K at the compressor inlet and 1300 K at the end of the compressor and at the inlet of the turbine. Assume the gas-turbine cycle operates with a compressor efficiency of 85% and a turbine efficiency of 85%. Determine the thermal efficiency of the cycle and impact of efficiency on emissions.

Solution:

Assume that the cycle is at steady-state flow and the changes in kinetic and potential energy are negligible. Heat capacity of air is temperature-dependent, and the air is an ideal gas. The standard-air assumptions are applicable

Basis: 1 kg air.

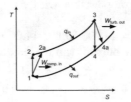

At $T_1 = 295$ K, $H_1 = 295.17$ kJ/kg; $P_{r1} = 1.3068$; (Appendix D Table D1)
P_r is the relative pressure (Eq. 7.36).
$$\frac{P_{r2}}{P_{r1}} = \frac{P_2}{P_1} = P_r \rightarrow P_{r2} = (9)(1.3068) = 11.76$$
Approximate values from Table D1 for compressor exit at $P_{r2} = 11.76$:
$T_2 = 550$ K, $H_2 = 555.74$ kJ/kg
Process 3-4 isentropic expansion in the turbine as seen on the TS diagram above
$T_3 = 1300$ K, $H_3 = 1395.97$ kJ/kg; $P_{r3} = 330.9$ (From Table D1)
$$\frac{P_{r4}}{P_{r3}} = \frac{P_4}{P_3} \rightarrow P_{r4} = \left(\frac{1}{9}\right)(330.9) = 36.76$$
Approximate values from Table D1 at the exit of turbine at 36.76: $T_4 = 745$ K and
$H_4 = 761.87$ kJ/kg
The work input to the compressor for a mass flow rate of 1 kg/s:
$W_{\text{comp.in}} = \dot{m}(H_2 - H_1) = (555.74 - 295.17)$ kW $= 260.57$ kW
Work output of the turbine: $W_{\text{turb.out}} = \dot{m}(H_4 - H_3) = (761.87 - 1395.97)$ kW
$= -634.10$ kW
From the efficiency definitions for compressor and turbine, we have:
$$\eta_C = \frac{W_{Cs}}{W_{Ca}} \rightarrow W_{Ca} = \frac{W_{Cs}}{\eta_C} = \frac{260.6 \text{ kJ/kg}}{0.85} = 306.6 \text{ kW (actual compression work)}$$
(where $W_{\text{comp.in}} = W_{Cs} = 260.6$ kW in Example 7.13)
Work output turbine with 85% efficiency: $\eta_T = \dfrac{W_{Ta}}{W_{Ts}} \rightarrow W_{Ta} = \eta_T(W_{Ts})$
$= 0.85(-634.1 \text{ kW}) = -539.0$ kW
(where $W_{\text{turb.out}} = W_s = -634.1$ kW in ideal operation in Example 7.13)
The net work out: $W_{\text{net}} = W_{\text{out}} - W_{\text{in}} = (-539.0 + 306.6) = -232.4$ kW
The back-work ratio r_{bw} becomes: $r_{bw} = \left|\dfrac{W_{\text{comp.in}}}{W_{\text{turb.out}}}\right| = \left|\dfrac{306.6}{539.0}\right| = 0.568$
This shows that the compressor now is consuming 56.8% of the turbine output. The
value of back-work ratio increased from 41 to 56.8% because of friction, heat
losses, and other nonideal conditions in the cycle.
Enthalpy at the exit of compressor:
$W_{Ca} = H_{2a} - H_1 \rightarrow H_{2a} = W_{Ca} + H_1 = (306.6 + 295.2)$kJ/kg $= 601.8$ kJ/kg
Heat added: $q_{\text{in}} = H_3 - H_{2a} = 1395.97 - 601.8 = 794.2$ kJ/kg
Thermal efficiency: $\eta_{th} = \dfrac{W_{\text{net}}}{q_{\text{in}}} = \dfrac{|232.4 \text{ kJ/kg}|}{794.2 \text{ kJ/kg}} = 0.292$ (or 29.2%)

Table 12.8 Electrification projections by IRENA 2020 in the Transforming Energy Scenario (TES) [13]

	Historical progress	TES (2030)	TES (2050)
Renewable share in electricity production (%)	23 (2015); 26 (2018)	57	86
Electrification share of final energy (%)	19 (2015); 20 (2017)	29	49
Renewable energy share in total final energy consumption (%)	9.5 (2015), 10.5 (2018)	28	66
Energy intensity improvement rate (%/yr)	1.8 (2015–2018)	3.6	3.2

The actual Brayton-gas cycle thermal efficiency drops to 0.292 from 0.444. Efficiencies of the compressor and turbine affects the performance of the cycle. Therefore, for a better cycle thermal efficiency, significant improvements are necessary for the compressor and turbine operations.

12.8.1 Electrification

The electrification will increase power demand, which needs to be met with increased efficiency and with renewables integrated into the grid. According to the TES, the number of electric vehicles (EVs) will increase from around 8 million in 2019 to over 1100 million in 2050. Despite falling renewable energy subsidies, renewable power generation technologies are increasing new capacity while lowering the cost. In the TES, electricity is projected as the main energy carrier, growing from a 20% to an almost 50% share by 2050 doubling the gross electricity consumption. Table 12.8 shows that the share of renewable power must increase from 23% currently to 86% by 2050 [13].

12.8.2 Power Flexibility

Flexibility in power systems is a key enabling the increased shares of renewable power in a decentralized and digitalized system at a share of over 30% or higher on an annual basis (see Table 12.9). In the TES 73% of the installed capacity and over 60% of all power generation would come from solar PV and wind. Power storage is essential for flexibility, and the amount of storage would need to expand from around 230 gigawatt-hours (GWh) today to over 23,000 GWh by 2050. Flexibility will also be achieved by grid expansion, smart charging of EVs, demand-side flexibility, and sector coupling. This would require considerable investment worldwide [12, 13].

Although variable source, offshore wind has the technical potential to meet today's electricity demand as it offers considerably higher capacity factors than solar PV and onshore wind with ever-larger turbines more reliable wind speeds.

Table 12.9 Renewable energy share in power generation by the transforming energy scenario (TES) [13]

	Historical progress	TES (2030)	TES (2050)
Variable renewable energy share in generation (%)	4.5 (2015) 10 (2018)	35	61
Solar power generation (GW)	222 (2015); 582 (2019e)	32,279	8828
Wind power generation (GW)	416 (2015); 624 (2019e)	2526	6044

Increasingly cost-competitive offshore wind projects can attract higher investment to 2040. In the sustainable development scenario, offshore wind rivals its onshore counterpart and helps achieve full decarbonization and the production of low-carbon hydrogen [13].

In the TES, half of the energy demand could be supplied by electricity by 2050. Of this, one-third is already supplied by end-use renewable sources, with the remaining two-thirds by fossil fuels. Direct use of bioenergy, solar thermal, geothermal heat, energy efficiency, and structural changes can reduce energy demand and lead to deeper electrification. However, more needed in shipping, aviation, and heavy industry to reduce emissions.

Industry, such as iron and steel, cement, and petrochemicals, is the dominant energy consumption sector in many countries, where the sector consumes around half of the final energy. Innovation is needed to find low-to-zero GHG emission solutions for the industrial process and to address transport modes of aviation and shipping sectors that are hard to electrify.

To achieve the TES, energy-related carbon emissions need to fall by 3.8% per year on average until 2050 that would lead to 70% below today's level. Outside the energy sector, emissions from non-energy use, land use, fugitive gases in the coal, oil and gas industries must be reduced. In addition. fossil-fuel investments need to be replaced with investments in renewables and energy efficiency [13].

12.8.3   Chemical Storage of Renewable Energy

Multichemical production from biorefineries can produce methane, methanol, ethanol, dimethyl ether, and others using renewable electricity and hence store it in various chemicals. For example, renewable power-based hydrogenation of carbon dioxide to methanol may lead to chemical storage of renewable energy. The produced methanol can be further converted to dimethyl ether, which is a valuable fuel and chemical feedstock. The required carbon may come from many sources including biomass fermentation and effluents of fossil-fuel-fired power plants. Hydrogen can come from biological processes or from the electrolysis of water using renewable energy. This may also help fix the carbon in chemical feedstock and extend the market potential for electrolysis. The variability of hydrogen sources allows for a continuous process to be achieved despite fluctuating conditions of hydrogen production because of peak electricity demands and fluctuating wind and solar power [16].

A biological process uses methanogenesis where microbes produce methane in a form of anaerobic fermentation. A chemical process is also possible with the well-known Sabatier reaction. This process uses a nickel catalyst and elevated temperatures between 300 and 400 °C to produce methane from hydrogen and carbon dioxide [9].

Example 12.14 Avoided emission in a cogeneration plant

A cogeneration plant is using steam at 8200 kPa and 773.15 K (see Fig. 6.15). One-fourth of the steam is extracted at 700 kPa from the turbine for cogeneration. After it is used for process heat, the extracted steam is condensed and mixed with the water output of the condenser. The rest of the steam expands from 8200 kPa to the condenser pressure of 10 kPa. The steam flow rate produced in the boiler is 60 kg/s. Determine the work output and process heat produced and estimate the avoided emissions if this cogeneration plant is replaced by a concentrated solar system producing power and heat.

Solution:

Assume that the kinetic and potential energy changes are negligible, and this is a steady state process.

Consider Fig. 6.15. From the steam tables:

$\dot{m} = 60$ kg/s, $z = 0.25$; basis: mass flow rate = 1 kg/s

$P_1 = P_8 = 10$ kPa, $H_{1 \text{sat vap}} = 2584.8$ kJ/kg, $H_{1 \text{sat liq}} = 191.83$ kJ/kg,

$V_1 = 0.00101$ m^3/kg, $S_{1 \text{sat vap}} = 8.1511$ kJ/kg K, $S_{1 \text{sat liq}} = 0.6493$ kJ/kg K,

$P_3 = P_7 = P_2 = P_4 = 700$ kPa, $H_3 = 697.06$ kJ/kg, $S_3 = 1.9918$ kJ/kg K, $z = 0.25$

$P_6 = 8200$ kPa, $T_6 = 773.15$ K, $H_6 = 3396.4$ kJ/kg, $S_6 = 6.7124$ kJ/kg K

In this cogeneration cycle, the steam extracted from the turbine is used as process heat. The liquid condensate from the process heat is combined with the output of the condenser.

$$W_{p1} = V_1(P_2 - P_1) = 0.00101(700 - 10)\left(\frac{1 \text{kJ}}{1 \text{ kPa m}^3}\right) = 0.697 \text{ kJ/kg}$$

$H_2 = H_1 + W_{p1} = 191.83 + 0.697 = 192.53$ kJ/kg

From the energy balance around the mixer, we have $\dot{m}_3/\dot{m}_6 = 0.25$

$\dot{m}_6 = \dot{m}_4 = 60$ kg/s, $\dot{m}_3 = \dot{m}_7 = 15$ kg/s, $\dot{m}_8 = \dot{m}_1 = 0.75(60) = 45.0$ kg/s

$\dot{m}_4 H_4 = \dot{m}_2 H_2 + \dot{m}_3 H_3 \rightarrow H_4 = [45(192.53) + 15(697.06)]/60 = 318.66$ kJ/kg

$T_4 = 349.15$ K, $V_4 = 0.001027$ kg/m^3

$$W_{p2} = V_4(P_5 - P_4) = 0.001027(8200 - 700)\left(\frac{1 \text{kJ}}{1 \text{ kPa m}^3}\right) = 7.70 \text{ kJ/kg}$$

$H_5 = H_4 + W_{p2} = 326.36$ kJ/kg

Isentropic processes: $S_6 = S_7 = S_8 = 6.7124$ and $P_7 = 700$ kPa, $H_7 = 2765.68$ kJ/kg.

We estimate the quality of the discharged wet steam at state 8:

$$x_8 = \frac{6.7124 - 0.6493}{8.1511 - 0.6493} = 0.808$$

$H_8 = 191.83(1 - 0.808) + 2584.80(0.808) = 2125.87 \text{ kJ/kg}$

The energy balance yields the fraction of steam extracted

$\dot{W}_{total} = \dot{m}_6(H_7 - H_6) + \dot{m}_8(H_8 - H_7) = -66634.44 \text{ kW}$

$\sum \dot{W}_{pi} = \dot{m}_1 W_{p1} + \dot{m}_4 W_{p2} = 493.51 \text{ kW}$

The net work output: $\dot{W}_{net} = \dot{W}_{total} - \sum \dot{W}_{pi} = -66{,}140.93 \text{ kW}$ (work output of the turbine

$\dot{q}_{process} = \dot{m}_7(H_3 - H_7) = \mathbf{-31{,}029.3 \text{ kW}}$ (heat discharged from the steam)

Table 11.2: (0.37 kg CO_2/kWh) for coal-based power

Annual avoided emission:

(0.37 kg CO_2/kWh) (66,140.93 kW)(24 h/day)(365 days/year) = **214,375,982 kg CO_2/year**

12.9   Global Socio-Economic Impact of Renewable Energy

Energy transition because of renewable energy technologies affects employment, GDP, and human welfare based on a model that integrates energy into sustainability. This is directly related to energy system planning and economic policy making. In the TES, jobs in the overall energy sector of renewable energy, fossil fuels, and nuclear power could reach 72% more than the total present employment by 2050. New jobs in solar PV, wind, and bioenergy would compensate job losses in fossil fuels and nuclear energy. This expanded renewable energy workforce will require growing investments, as well as focus on education, training, and social policies such as just and inclusive transition policies to limit the job losses. GDP gains can result by changes in consumer spending and other indirect and induced factors including increased investment and the prospects of economic growth. The primary goal of the transition would be to improve people's overall welfare through secure, clean energy supply, economic and social development, and mitigation of climate change [5, 11, 13, 20].

12.9.1   Welfare

Welfare, envisaged with the sustainability elements of economics, social, and environment would improve 13.5% by 2050 under the TES. The economic element is measured via household consumption and a composite of total investment and employment. Jobs gains would be linked to the transitions in renewable energy, energy efficiency, power grids, and energy flexibility under TES. The social element reflects spending on education and health. The environmental element entails the lower greenhouse gas emissions and the consumption of nonrenewable materials. Overall, health benefits and emission reductions influence welfare gains besides substantial socio-economic benefits globally.

However, owing to variations in regional and country-specific socio-economic structures, the welfare gains could vary at the country and regional levels and local solutions would be necessary. Regional socio-economic structures, resource endowment, industrial productive capacity and support policies, trade structures, and domestic supply chains would determine the opportunities offered by the energy transition. A successful transition requires that energy policies must be compatible with regional economic, industrial, labor, educational, and social policies [11, 13].

12.9.2 Transformative Decarbonization

Climate concerns besides economic inequality require a comprehensive global economic transformation with holistic solutions in which renewable energy plays a key role. The decarbonization of the global economy calls for large-scale policy interventions and substantial resource mobilization. Market mechanisms alone may not bring the desired emissions reductions.

Further, enabling policies include industrial policies, labor-market interventions, educational and skill development, and social protection measures. At the international level, the Climate Investment Platform announced in September 2019 by the International Renewable Energy Agency, the United Nations Development Program, the multi-partner Sustainable Energy for All initiative, and the Green Climate Fund aims to mobilize energy-transition investments on a scale commensurate with the climate goals. Investment decisions should be evaluated to the extent at which they accelerate the shift toward an inclusive low-carbon economy [10–13].

Summary

- *Renewable energy* comes from natural resources and are naturally replenished. Major renewable energy sources are hydroelectric, solar energy, biomass, wind, geothermal heat, ocean.
- Photosynthesis as solar energy's main human application is in agriculture and forestry. The application has been extended in many ways. Technologies for solar energy capture, conversion, and distribution may be either passive or active.
- The trend shows that concentrated solar power capacity ($\sim 40\%$) increases relatively higher than that of solar PV capacity ($\sim 18\%$).
- Despite to the low capacity factor (20–30% utilization) and intermittency, wind power receives widespread installations and investments (594 GWe onshore and 28 GWe offshore in 2019).
- As technology has increased in the past years, the capacity factors increased by increasing wind power generation. The following projections show the expected trends for onshore (30%) and offshore capacity factors (40%).
- Hydroelectric energy from the potential energy of rivers is the best-established electricity generation from renewable sources. The largest power plants in the world use dams on rivers. Capacity factor is around 50%.

- Pollution is minimal from hydroelectricity as it is produced without fuel. Hydroelectric plants can operate a long time and with less operating and maintenance costs.
- *Geothermal energy* is the heat originating from radioactive decay of minerals, from volcanic activity, and from solar energy absorbed at Earth's surface. The capacity factor is gradually decreasing from 85%.
- Ocean energy falls into three categories that are tidal, wave, and temperature gradient. Ocean thermal energy conversion uses the temperature difference that exists between deep and shallow waters to run a heat engine.
- *Bioenergy* is derived from biological sources, to be used for heat, electricity, or vehicle fuel. Biofuel derived from plant materials is among the most rapidly growing renewable energy technologies.
- Bioenergy currently makes up a large share of renewable energy use and would remain a significant source for power and heat generation in the industry and as a fuel in the transportation sector. The capacity factor is around 80% and increasing gradually.
- Biomass feedstock originates from diverse resources including naturally growing terrestrial and aquatic plants, and man-made wastes. Biomass contains carbon, oxygen, and hydrogen in a variety of compounds, such as carbohydrate, lipids, and protein.
- The first generation is food-based biomass such as corn, sugarcane, plant, vegetable oils and fats. The second generation refers to nonfood-type biomass that includes cellulose and hemicellulose, while third-generation biomass mainly refers to algae and cyanobacteria.
- Recycling and composting programs may reduce the share of biomass in municipal solid waste that is landfilled or burned.
- Some agricultural products grown for biofuel production include corn, soybeans, willow switchgrass, rapeseed, wheat, sugar beet, palm oil, miscanthus, sorghum, cassava, and jatropha.
- Biodegradable outputs from industry, agriculture, forestry, and households can be used for biofuel production, using for example: (i) anaerobic digestion to produce biogas, (ii) gasification to produce bio syngas (H_2, CO, CO_2), or (iii) by direct combustion.
- Biomass can be fermented directly to produce hydrogen, ethanol, and other high-value chemicals. Certain photosynthetic microbes produce hydrogen from water in their metabolic activities using light energy.
- Biofuels are biomass feedstock-based fuels and include biohydrogen, bioethanol, biobutanol, biomethanol, biooil, biogas, and biodiesel, while renewable electricity is called biopower. Bioethanol and biodiesel have by far the largest share of the global biofuels markets.
- Biorefinery is beneficial in converting various biomass feedstocks into various valuable fuels. Conversion processes include biochemical, chemical, thermochemical, and hydrothermal.

- Ethanol is a clear, colorless alcohol fuel made from the sugars found in grains, such as corn, sorghum, and barley, sugarcane, and sugar beets. Therefore, ethanol is a renewable fuel.
- Lignocellulosic biomass and other nonfood biomass resources produce second-generation bioethanol. Cellulose, hemicellulose, and lignin are the structural components of lignocellulosic biomass and includes corn stover, corn cobs, sorghum stalks, wheat straw, cotton residue, alfalfa stems, wood, and fast growing plants such as grass, and bagasse of sugarcane and sorghum stalks.
- Chemical process in biodiesel production is the transesterification of triglycerides of fatty acids into biodiesel (fatty acid methyl ester) using mainly methanol or ethanol and a catalyst (mainly acids and alkali-NaOH) and producing glycerin as a byproduct [5, 7].
- Oil-rich microalgae strains can produce biodiesel with (1) algae growth in open ponds or in photobioreactors, (2) algae harvesting and dewatering, (3) lipid extraction, and (4) transesterification of lipid to biodiesel.
- *Green diesel* production requires large volumes of H_2 and a catalyst to hydrogenate triglycerides into a high-cetane diesel fuel by removing all the oxygen from the triglyceride and saturating all the olefinic bonds in the fatty acids.
- Biodiesel (fatty acid methyl ester) comes with several shortcomings that include NO_x and SO_x emissions, a cloud point with adverse effects, filter clogging, marginal stability, reduced energy content, and the need to be blended with conventional petroleum diesel.
- Anaerobic digestion of wet biomass feedstock (<15% solid), such as animal manure, agricultural residue, and sewage sludge from municipal wastewater treatment to produce methane-rich biogas using bacteria (methanogenic), archaea, and fungi at 37–55 °C.
- *Fischer-Tropsch* (F-T) synthesis can produce various liquid biofuels from biosyngas.
- Chemical-looping gasification and reforming processes target syngas (CO and H_2) or a mixture of CO, H_2, CO_2, and CH_4 (gasification), or a relatively pure stream of H_2 as products with inherited carbon capture.
- *Pyrolysis* uses fast heating to high temperatures under anaerobic conditions to break down biomass into a volatile mixture of hydrocarbons and char. This mixture of hot gases is condensed into a bio-oil with a rich mixture of hydrocarbons, some of which can be converted into biofuels. Bio-oil must be upgraded to fuels.
- *Hydrothermal liquefaction* of biomass also produces biooil under at elevated temperatures (200–400 °C) and pressure (5–40 MPa).
- By fermentation process, sugars can be converted into butyric, lactic, and acetic acids. Butyric acid is converted by fermentation into biobutanol. Corn starch also can be converted to biobutanol via the acetone–butanol–ethanol fermentation pathway.
- Methanol is produced almost exclusively by the ICI, the Lurgi, and the Mitsubishi processes. These processes differ mainly in their reactor designs and the way in which the produced heat is removed from the reactor.

- Biodimethyl ether (DME) can be produced from methanol using wind power-based electrolytic hydrogen and CO_2 captured from an ethanol fermentation process. Methanol and DME from CO_2 hydrogenation may outperform conventional petroleum-based fuels, reducing greenhouse gas emissions 82–86%, minimizing other criteria pollutants (SO_x, NO_x, etc.) and reducing fossil fuel depletion by 82–91%.
- Hydrogen is a zero-emission fuel with combustion products of water and a trace amount of NO_x. Hydrogen can be produced mainly from the second generation of biomass feedstock by using thermochemical processes of gasification (steam reforming) and fast pyrolysis.
- Hydrogen can also be produced by microorganisms in biological process as well as electrolysis of water by using a renewable power source of wind, solar, or hydro.
- Hydrogen can offer a solution for types of energy demand not supplied by electricity and has a steep increase trend. Green hydrogen is produced by renewable electricity through electrolysis, and costs are falling and will become cost competitive with 'blue' hydrogen, which is produced from fossil fuels combined with carbon capture and storage.
- Extreme weather, flooding, rising sea levels and other effects of climate change in ecosystems will cause adverse effects on quality of life, health, economy, and ecology. Despite worldwide effort and research, decarbonization costs are still high and expanding, and the supply of fossil-fuel infrastructure may only keep increasing these adverse impacts.
- Eco-friendly energy systems need policy makers who would establish long-term targets and adapt policies and regulations.
- Climate change is a global threat to ecology and the transforming energy sector may need different approaches for different countries depending on the priorities.
- The contribution of various biofuels to the transport sector is continuous with an increasing trend.
- As the increase in the electrification of economies and the share of variable renewables, the carbon footprint of electricity use increasingly fluctuates.
- For *effective energy management,* there are several basic steps that are management commitment, data analysis, analysis of energy conservation options, implementation energy conservation options, and continued feedback and analysis.
- The electrification will increase power demand, which needs to be met with increased efficiency and with renewables for sustainable transformation of the number of electric vehicles (EVs) from around 8 million in 2019 to over 1100 million in the TES by 2050.
- *Flexibility in power systems* is a key enabling the increased shares of renewable power in a decentralized and digitalized system at a share of over 30% or higher on an annual basis.

- Wind power-based hydrogenation of CO_2 to methanol may lead to chemical storage of renewable energy, carbon recycle. This will lead to the fixation of carbon in chemical feedstock and extended market potential for electrolysis.
- Climate concerns besides economic inequality require a comprehensive global economic transformation with holistic solutions in which renewable energy plays a key role. The decarbonization of the global economy calls for large-scale policy interventions and massive resource mobilization.

Problems

12.1. The primary energy source for cells is the aerobic oxidation of sugar called glucose ($C_6H_{12}O_6$). Estimate the energy released from the oxidation 500 g of glucose:

$$C_6H_{12}O_6(s) + 6O_2(g) \rightarrow 6CO_2(g) + 6H_2O(g)$$

12.2. Synthesis gas is a mixture of H_2 and CO. One of the uses of synthesis gas is to produce methanol from the following reaction: $CO(g) + 2H_2(g) \rightarrow CH_3OH(g)$. Estimate the energy released after burning 100 gallons of methanol.

12.3. Estimate the energy released by the combustion of 500 ft^3 propane:

$$C_3H_8(g) + 5O_2(g) \rightarrow 3CO_2(g) + 4H_2O(g)$$

12.4. Estimate the energy released by the combustion of 1000 ft^3 ethane:

$$C_2H_6 + \frac{7}{2}O_2 \rightarrow 2CO_2(g) + 3H_2O(g)$$

12.5. Estimate the energy released from the combustion of 2000 ft^3 of methane:

$$CH_4 + 2O_2 \rightarrow CO_2(g) + 2H_2O(g)$$

12.6. What is the energy released by the combustion at 25 °C of 500 gallons of CH_3OH when the reaction products are $CO_2(g)$ and $H_2O(g)$? The reaction is: $6CH_3OH(g) + 9O_2 \rightarrow 6CO_2(g) + 12H_2O(g)$.

12.7. Estimate the energy released by the combustion of n-pentane gas at 25 °C when the combustion products are $CO_2(g)$ and $H_2O(l)$: $C_5H_{12}(g) + 8O_2(g) \rightarrow 5CO_2(g) + 6H_2O(l)$.

12.8. A heating oil with an average chemical composition of $C_{10}H_{18}$ is burned with oxygen completely in a bomb calorimeter. The heat released is measured as 43,900 J/g at 25 °C. The combustion on the calorimeter takes place at constant volume and produces liquid water. Estimate the energy released by the oil at 25 °C when the products are (g) H_2O and CO_2 (g).

12.9. A superheated steam (stream 1) expands in a turbine from 5500 kPa and 325 °C to 100 kPa and 100 °C. The steam flow rate is 12.0 kg/s. If the turbine generates 1.1 MW of power, estimate the total amount of avoided emissions if the wind power instead of coal is used in the boiler to produce the steam.

12.10. An air compressor compresses air from 100 to 600 kPa. The air flow rate is 900 kg/h, and specific volumes of air at the inlet and outlet are 0.95 and 0.19 m^3/kg, respectively. The jacket cooling water removes 100 kW of heat generated due to compression. Estimate the power input to the compressor and the total amount of avoided emissions if the wind power instead of electricity from coal is used.

12.11. An air compressor compresses air from 100 to 1000 kPa. The air flow rate is 400 kg/h, and specific volumes of air at the inlet and outlet are 0.95 and 0.19 m^3/kg, respectively. The jacket cooling water removes 100 kW of heat generated due to compression. Estimate the power input to the compressor and the avoided emissions if the wind power instead of electricity from coal is used.

12.12. A centrifugal pump delivers 1.8 m^3/h water to an overhead tank 20 m above the eye of the pump impeller. Estimate the work required by the pump and the avoided emissions if the wind power instead of electricity from natural gas is used.

12.13. A pump increases the pressure in liquid water from 100 kPa at 25 °C to 2000 kPa. Estimate the minimum horsepower motor required and the total amount of avoided emissions if the wind power instead of electricity from coal is used in the pump for a flow rate of 0.4 m^3/s.

12.14. A centrifugal pump delivers 1.8 m^3/h water to an overhead tank 20 m above the eye of the pump impeller. Estimate the work required by the pump and the total amount of avoided emissions if the wind power instead of electricity from natural gas is used.

12.15. An average car consumes about 2 gallons (US gallon = 3.785 L) a day, and the capacity of the fuel tank is about 15 gallons. Therefore, a car needs to be refueled once every week. The density of gasoline ranges from 0.72 to 0.78 kg/l (Table 2.6). The lower heating value of gasoline is about 44,000 kJ/kg. Assume that the average density of gasoline is 0.75 kg/l. If the car was able to use 85% ethanol instead of gasoline estimate the total amount of avoided emissions.

12.16. A city consumes natural gas at a rate of 500×10^6 ft^3/day. The volumetric flow is at standard conditions of 60 °F and 1 atm = 14.7 psia. If the natural gas is costing $6/GJ of higher heating value estimate the yearly energy required and the total amount of emission avoided if biogas instead of natural gas used in the city.

12.17. An average car consumes 50 gallons gasoline per month. Estimate the amount of energy and the total amount of avoided emissions if 85% ethanol blended gasoline is used instead of 100% gasoline consumed by the car per year.

12.18.   For a car consuming 2000 gallons gasoline per year, estimate the amount of energy and the avoided emissions if 85% ethanol blended gasoline is used instead of 100% gasoline consumed by the car.

12.19.   The overall efficiencies are about 25–28% for gasoline car engines, 34–38% for diesel engines, and 40–60% for large power plants [7]. A car engine with a power output of 120 hp has a thermal efficiency of 24%. Determine the fuel consumption of the car and avoided emissions if the gasoline is mixed with 85% ethanol and if the gasoline has a higher heating value of 20,400 Btu/lb.

12.20.   A steam power plant is using a geothermal energy source. The geothermal source is available at 220 °C and 2320 kPa with a flow rate of 150 kg/s. The hot water goes through a valve and a flash drum. Steam from the flash drum enters the turbine at 550 kPa and 428.62 K. The discharged steam from the turbine has a quality of $x_4 = 0.90$. The condenser operates at 10 kPa. The water is a saturated liquid after passing through the condenser. Determine the work output of turbine and the total amount of avoided emissions compared with coal-fired power productions.

12.21.   A steam power plant is using a geothermal energy source. The geothermal source is available at 180 °C and 2000 kPa with a flow rate of 100 kg/s. The hot water goes through a valve and a flash drum. Steam from the flash drum enters the turbine at 550 kPa and 428.62 K. The discharged steam from the turbine has a quality of $x_4 = 0.96$. The condenser operates at 10 kPa. The water is a saturated liquid after passing through the condenser. Determine the work output of turbine and the total amount of avoided emissions compared with coal-fired and natural gas-fired power productions.

12.22.   A cogeneration plant uses steam at 900 psia and 1000 °F to produce power and process heat. The steam flow rate from the boiler is 16 lb/s. The process requires steam at 70 psia at a rate of 3.2 lb/s supplied by the expanding steam in the turbine with a value of $z = 0.2$. The extracted steam is condensed and mixed with the water output of the condenser. The remaining steam expands from 70 psia to the condenser pressure of 3.2 psia. Determine the process heat and power generated and estimate the total amount of avoided emissions if this plant is replaced with the concentrated solar power system.

12.23.   A cogeneration plant uses steam at 900 psia and 1000 °F to produce power and process heat. The steam flow rate from the boiler is 25 lb/s. The process requires steam at 70 psia at a rate of 10.5 lb/s supplied by the expanding steam in the turbine with a value of $z = 0.2$. The extracted steam is condensed and mixed with the water output of the condenser. The remaining steam expands from 70 psia to the condenser pressure of 9.5 psia. Determine the process heat and power generated and estimate the total amount of avoided emissions if this plant is replaced with the concentrated solar power system.

12.24. A hydroelectric plant operates by water falling from a 150 ft height. The turbine in the plant converts potential energy into electrical energy, which is lost by about 8% through the power transmission, so the available power is 95%. If the mass flow rate of the water is 600 lb/s, estimate the power output of the hydro plant and the total amount of avoided emissions compared with coal-fired power plant.

12.25. A hydroelectric plant operates by water falling from a 90 ft height. The turbine in the plant converts potential energy into electrical energy, which is lost by about 8% through the power transmission, so the available power is 90%. If the mass flow rate of the water is 350 lb/s, estimate the power output of the hydro plant and the total amount of avoided emissions compared with coal-fired power plant.

12.26. In a hydropower plant, a hydro turbine operates with a head of 46 m of water. Inlet and outlet conduits are 1.80 m in diameter. If the outlet velocity of the water is 5.5 m/s, estimate the power produced by the turbine and the total amount of avoided emissions compared with natural gas-fired power plant.

12.27. A farm of windmills supplies a power output of 0.8 MW for a community. Each windmill has the blades with 10 m in diameter. At the location of the windmills, the average velocity of the wind is 14 m/s and the average temperature is 20 °C. Estimate the minimum number of windmills to be installed and the total amount of avoided emissions compared with natural gas-fired power plant.

12.28. A wind turbine with 10 m rotor diameter is to be installed on a hilltop where the wind blows steadily at an average speed of 15 m/s. Determine the maximum power that can be generated by the turbine operated with a 35% efficiency and the total amount of avoided emissions compared with natural gas-fired power plant.

12.29. A farm of windmills supplies a power output of 3 MW for a community. Each windmill has the blades with 11 m in diameter. At the location of the windmills, the average velocity of the wind is 14 m/s and the average temperature is 20 °C. Estimate the minimum number of windmills to be installed and the total amount of avoided emissions compared with natural gas-fired power plant.

12.30. A farm of windmills supplies a power output of 4.2 MW for a community. Each windmill has the blades with 10 m in diameter. At the location of the windmills, the average velocity of the wind is 15 m/s and the average temperature is 20 °C. Estimate the minimum number of windmills to be installed and the total amount of avoided emissions compared with natural coal-fired power plant.

12.31. A refrigerator using tetrafluoroethane (R-134a) as refrigerant operates with a capacity of 250 Btu/s. Cooling water enters the condenser at 70 °F. The evaporator is at 10 °F, and the condenser is at 80 °F. The refrigerated space is at 20 °F. Determine the ideal and actual power necessary and compare the avoided emissions if the wind power is used instead of power from coal-fired plant. Assume that the compressor efficiency is 80% and the kinetic and potential energy changes are negligible.

12.32. A refrigerator using tetrafluoroethane (R-134a) as refrigerant operates with a capacity of 2500 kW. Cooling water enters the condenser at 280 K. Evaporator is at 271.92 K, and the condenser is at 299.87 K. The refrigerated space is at 280 K. Determine the ideal and actual power necessary and compare the avoided emissions if the wind power is used instead of power from coal-fired plant. Assume that the compressor efficiency is 80% and the kinetic and potential energy changes are negligible.

References

1. Allen JW, Unlu S, Demirel Y, Black P, Riekhof W (2018) Integration of biology, ecology and engineering for sustainable algal based biofuel and bioproduct biorefinery. Biores Bioprocess 5(47):1–28
2. Betz A (1966) Introduction to the theory of flow machines (D. G. Randall, Trans.). Pergamon Press, Oxford
3. Cherubini F (2010) The biorefinery concept: using biomass instead of oil for producing energy and chemicals. Energy Convers Manage 51:1412–1421
4. Demirel Y, Matzen M, Winters C, Gao X (2015) Capturing and using CO_2 as feedstock with chemical-looping and hydrothermal technologies. Int J Energy Res 39:1011–1047
5. Demirel Y (2018) Biofuels. In: Dincer I (ed), Comprehensive energy systems, vol 1, Part B, pp 875–908. Elsevier, Amsterdam
6. Demirel Y (2018) Energy conservation. In: Dincer I (ed), Comprehensive energy systems, vol 5, pp 45–90. Elsevier, Amsterdam
7. Demirel Y (2018) Sugar versus lipid for sustainable biofuels. Int J Energy Res 42:881–884. https://doi.org/10.1002/er.3914
8. Demirel Y (2017) Lignin for sustainable bioproducts and biofuels. J Biochem Eng Bioprocess Technol 1:1–3
9. Dincer I, Ratlamwala T (2013) Development of novel renewable energy-based hydrogen production systems: a comparative study. Energy Convers Manag 72(2013):77–87
10. IEA (2019) World energy outlook 2019, IEA, Paris. https://www.iea.org/reports/world-energy-outlook-2019
11. IEO (2019) International energy outlook 2019 with projections to 2050. September 2019, U. S. Energy Information Administration, U.S. Department of Energy, Washington, DC 20585. https://www.eia.gov/ieo
12. IRENA (2020) Renewable power generation costs in 2019. International Renewable Energy Agency, Abu Dhabi
13. IRENA (2020) Global renewables outlook: energy transformation 2050 (edition: 2020). International Renewable Energy Agency, Abu Dhabi
14. Jaffe RL, Taylor W (2018) The physics of energy. Cambridge University Press, Cambridge
15. Küçük K, Tevatia R, Sorgüven E, Demirel Y, Özilgen M (2015) Bioenergetics of growth and lipid production in *Chlamydomonas reinhardtii*. Energy 83:503–510
16. Matzen M, Demirel Y (2016) Methanol and dimethyl ether from renewable hydrogen and carbon dioxide: alternative fuels production and life-cycle assessment. J Cleaner Product 139:1068–1077

17. Moghtaderi B (2012) Review of the recent chemical looping process developments for novel energy and fuel applications. Energy Fuels 26(2012):15–40
18. National Research Council (2011) Renewable fuel standard: potential economic and environmental effects of U.S. biofuel policy. The National Academies Press, Washington, DC
19. Nguyen N, Demirel Y (2011) A novel biodiesel and glycerol carbonate production plant. Int J Chem Reactor Eng 9:1–25
20. Olah GA, Goeppert A, Surya GK (2009) Chemical recycling of carbon dioxide to methanol and dimethyl ether: from greenhouse gas to renewable, environmentally carbon neutral fuels and synthetic hydrocarbons. J Org Chem 74:487–498
21. Ptasinski KJ (2016) Efficiency of biomass energy. An exergy approach to biofuels, power, and biorefineries. Wiley, New York

Energy Management and Economics

13

Introduction and Learning Objectives: Accurate estimation of the cost is possible by using techno economic analysis in planning and operation of energy production and usage toward proper energy management and economics. Inaccurate assumptions, scale-down/scale-up problems, noncompetitive utility costs, and waste treatment may lead to failures in employing the actual trend in energy management cost estimations with projections for upscale energy management operations. Risk analysis and sensitivity analysis help operate large energy systems investments successfully. In a comprehensive feasibility analysis, techno economic analysis is being complemented with sustainability metrics to account current socioeconomics, socio-ecology, and eco-economics changes. Besides, 'Circular Economy' is incorporated into sustainability through the element of economics as it replaces conventional linear economy by increasing efficiency in energy management manufacturing, recycling by recycling and reducing emissions.

Learning objectives of this chapter are:

- Energy production and consumption
- Economics of energy production and conversion
- Bioeconomy
- Circular economy

13.1 Energy Management and Economics

Cost estimation of energy production and usage is possible by using techno economic analysis. Inaccurate assumptions, scale-down/scale-up problems, noncompetitive utility costs, and waste treatment may lead to failures in employing the actual trend in energy management and economics for upscale energy system operations. Risk analysis and sensitivity analysis help for large investments to operate successfully. In a comprehensive feasibility analysis, techno economic

© Springer Nature Switzerland AG 2021
Y. Demirel, *Energy*, Green Energy and Technology,
https://doi.org/10.1007/978-3-030-56164-2_13

analysis is being complemented with sustainability analysis to account current socioeconomics, socio-ecology, and eco-economics changes in energy production and consumption. In addition, 'Circular Economy' should be incorporated into the economic sustainability in monitoring the trends in energy production and usage with improved energy efficiency, recycling, and reduced emissions [39].

13.2   Energy Production and Consumption

Advanced, efficient, and low-emission energy technologies utilizing renewable and nonrenewable resources are vital for a sustainable energy technology, which plays a major role in global climate change as well as in international politics and trade. Currently, fossil fuel-based power plants worldwide account for considerable portion of total energy production and of total carbon dioxide (CO_2) emissions. World energy requirements require a sustainable energy production method, such as energy production with carbon capturing and storing. However, this adds significant capital and operating costs without much economics return. One of the primary challenges for the fossil fuel-based power plants is to make the carbon capture and store (CCS) viable since the effluent gas from them contains low concentration of carbon dioxide [4, 39].

13.2.1   Nonrenewable Power Production

The sustainability of existing fossil-fuel-fired power plants, especially with the decarbonizing constrains, is decreasing [12, 18]. There are several methods of capturing the carbon, either post-conversion, pre-conversion, oxy-fuel combustion, or the use of chemical looping technologies [6]. Each method or process has benefits and drawbacks; some of the drawbacks are more severe than others and some technologies are not currently ready for industrial applications. Fossil-fuel-fired power plants may benefit multi-production including electricity, heat, and fuels such as methanol [39]. A further assessment would be needed to identify the proportion of the global coal-power risk by analyzing existing coal plants and new renewable power costs at a country level. Worldwide investments for electric vehicles, as well as heat pumps, would accelerate low-cost renewable electricity generation toward an energy sector transformation. These decreasing costs and the advances in the grid operating capability with high shares of variable renewables are helping the electrification. With falling battery costs and electrification of vehicles, most of total energy use in the transport sector would come from electricity. Lower total installed costs and higher capacity factors and higher storage capacity are driving the decline in the cost of electricity from concentrated solar power [18, 23].

Organic Rankine cycle operates between 50 and 150 °C using on organic, such as isopentane as the working fluid. These types of heat engines operate at lower temperature and pressure requiring low-temperature resources such as geothermal and cost much less compared with those steam Rankine cycles. They present sustainable energy solutions in rural areas. However, their efficiencies are low [20].

13.2.2 Gas Utilities in Energy Sector

In the business as usual, gas utilities are the energy source that strives to meet policy goals, but there is pushback from environmentalists to decarbonize the energy industry. The advancement of clean energy technologies is becoming global goal that is in line with how investments made in the energy sector to grow based on policy and reliability. Increased use of natural gas and the decrease in coal-based electricity generation helped reduce greenhouse gas (GHG) emissions. As decarbonization efforts are increased and more heavily enforced, energy utility companies are challenged to come up with near-term solutions and long-term goals, such as limits to the amount of allowable greenhouse gas emissions [4, 5, 6].

13.2.3 Gas Resource Planning

Gas utilities use a wide range of factors when planning resource allocation, which makes up their Integrated Resource Planning (IRP). This includes regional market prices, weather data, historical usage data, new regulations, and improving technology. The IRP process is broken down into four main stages starting with understanding the demands of the market in a large scope and using previous data to characterize specific areas. The next is to break the data into smaller sections and look for trends based on previous forecasts and current demand. The third is defining and predicting variance due to theoretical risk and event occurrences. The final step is then to develop a portfolio that can handle the daily energy demands and has contingencies in place for emergencies [27, 34, 36].

13.2.4 Grid Resilience

Resilience in utilities focuses on how quickly the system can become operational after failure. The degree of resilience is directly tied to investment costs and who manages it, because resiliency of a system directly involves first responders, government agencies, and others who restore the grid system immediately in the event of failure. Increasing resiliency is a goal that many utilities are undertaking and adopting affordable measures that provide incentives based on increased economic benefit and reliability in different capacities. Changing to 100% renewable energy generation offers the limited number of ways to be successful because of the reliability of renewable energy sources and the way these energy sources are connected to the existing power grid [20].

13.2.5 Coal-Fired Power Plants

Coal-fired power plants are required to meet standards that limit the amounts of some of the pollutants released into the air. These plants must use coal that is low in sulfur content and ash. Coal can also be pre-treated and processed to reduce the types and amounts of undesirable compounds in combustion gases. Smaller and lighter particulates called fly ash are collected in air emission control devices and are usually mixed with the bottom ash. Placing a fee on CO_2 capture may encourage the deployment of Carbon Capture & Storage (CCS) technology. Due to lower capital costs and relatively low prices, natural gas combined-cycle plants with carbon sequestration may be cheaper to build than advanced coal plants with carbon sequestration [20, 38].

Chemical-looping combustion with inherent CO_2 capture, if properly developed, can produce water and CO_2 stream, and helps reduce carbon capture costs. In this technology, a metal oxide is used as an oxygen carrier to transfer the oxygen from air to the fuel in the fuel reactor so that the direct contact between the fuel and air is avoided [6]. Pure CO_2 can be produced by condensing the water and sequestrated or used for other purposes. Once the fuel oxidation is complete, the reduced metal oxide is transported to the air reactor where it is reoxidized and recycled.

The same energy can be produced using different technologies and resources. Electricity, for example, can be generated by means of solar panels, wind, coal, hydroelectricity, geothermal, and nuclear power plants. They may offer certain unique features and newly emerging energy technologies may have some inherent shortcomings. Figure 13.1 shows the global power generation capacity and projections with the rise of renewables and natural gas, while the change in coal and nuclear energy sources will be almost flat. In 2040, half of the total electricity generation will be from renewables [18, 19, 20, 21] and wind power would be as affordable as hydropower. Besides, the adaptation of characteristics of the local resources in energy and efficiency, as well as dependence on nonrenewable sources should be analyzed. Timely availability of a workforce with required skills for installation, operation and maintenance, and Research and Development (R&D) facility would ensure the success of the new energy technologies.

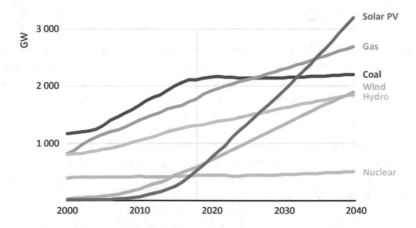

Fig. 13.1 Global power capacity of renewables and natural gas with projections. The projections display sharp increase in power generations by solar PV and wind, as well as by natural gas, while stagnation in power generation by coal and nuclear continues. Solar PV capacity catches wind power by 2023 and hydropower before 2030 and increases sharply to dominate as a source of power by 2040. However, coal and natural gas will keep their considerable contributions by 2040 [18, 19, 20, 21]

13.2.6 Renewable Energy Production

Renewable power generation is increasing and becoming affordable. The global weighted-average levelized cost of electricity (LCOE) from renewable resources have all been within the range of fossil-fuel-fired power generation costs [19, 20]. The conversion economics of biomass into energy may be more challenging compared with that of energy from fossil fuels. Lignocellulosic biofuel model contains all unit operations to transform biomass into ethanol using dilute acid pretreatment and co-fermentation of 5-carbon (C5) and 6-carbon (C6) sugars. Sections of the biorefinery model include: (i) feedstock handling (including transportation), (ii) pretreatment and saccharification, (iii) fermentation, (iv) ethanol recovery, (v) water recovery and recycling, energy recovery and electricity production, and (vi) wastewater treatment [3, 7, 24].

Advanced biofuels: More than half of the advanced biofuels projects based on lignocellulosic biomass are being developed and complemented with new innovations and coproductions. The main internal barriers include product development, byproduct and coproduct distribution and marketing, continuous project growth, management, strategy, and technology for conversion rate and yield. The main external barriers include competition, funding, supplies, pathway processes, tax credits, Renewable Fuel Standards (RFS) , waiver credits, renewable volume obligations, renewable identification numbers, energy costs, and third-party relationships [30].

Joint Bio Energy Institute (JBEI) of the Department of Energy of U.S. focuses on mitigating climate change, ensure future energy security, improve human and ecological health, and create local economic activity. JBEI's model (http://econ. jbei.org) estimates the economic, environmental and energetic performance of a lignocellulosic biorefinery process and allows users to model promising scenarios. The model combines the life cycle and techno economic analyses to determine the cost impacts of engineered feedstocks with reduced lignin content, as well as of various pre-treatment configurations. This may lead to a complete assessment of the economic and environmental tradeoffs to produce biofuels and bioproducts from lignocellulosic biomass.

Need for biofuel and abundant supply of fossil fuels may lead to these objectives: (1) sequester carbon to reduce GHG emissions, (2) achieve greater energy efficiency, (3) integrate rural programs into increasing energy security, (4) stimulate economic growth and development, and (5) obtain feasible conversion technologies [7].

Solar Energy: As a direct conversion device of solar radiation into electrical power, photovoltaic (PV) solar cells play an important role in worldwide power generation. Solar PV technologies have three categories: first-generation PVs contain conventional crystalline silicon cells, second-generation PVs contain thin-film PV cells with semiconductors having a direct band gap, and third-generation PVs include multi-function PVs, dye-sensitized cells, organic PVs, quantum dots and graphene. Most of the research efforts address the efficiency and cost of PVs. Photovoltaic energy is carbon free with no moving parts, and low maintenance requirements. PVs can be installed anywhere including remote locations [19].

Solar insolation in units of energy per unit area is around $I = 1366$ W/m^2, which is also called as solar constant. Solar radiation is mainly concentrated around the wavelength of visible light. Besides the absorption of solar radiation on Earth's surface, partial of that radiation is reflected. Greenhouse gases, such as CO_2 absorbs the reflected infrared radiation from the Earth and causes the greenhouse heating effect that can be created by a material that transmits short-wavelength light but absorbs long-wavelength light [23]. Since the distance from the sun varies throughout the year, solar insolation varies over time. The use of solar energy depends on the conversion device of energy from electromagnetic waves to thermal or electrical energy. Thermal energy from solar energy is widely used for heating spaces and water by low-temperature collectors operating below 80 °C. Intermediate-temperature solar energy applications would be in the range between 80 and 250 °C that would increase the industrial use of renewable energy for process heating considerably. For example, parabolic concentrator designs employ a system of mirrors that focus the incoming direct light onto a single focal point. The solar energy at high concentration can heat a fluid such as water, molten salt, oil, or other mixtures to be used in solar thermal electricity production [23, 27].

Commercial solar PV cells have an efficiency range of up to 26% conversion and can be connected to electricity grids. Efficiency is increased using concentrating solar systems. To mitigate grid integration problems with solar power, battery systems, and other means are being developed [18].

Wind Power: *Offshore wind operating and maintenance* (O&M) costs are higher than those for onshore wind due to the higher costs of access to the site and perform maintenance on towers and transmission cabling. These are estimated to be between US$ 0.02/kWh (for established markets for sites closer to shore) and US$ 0.05/kWh (for less-established offshore wind markets and harsher site conditions) [19].

The use of a high level of wind and solar energy will require costly enhancement of system integration measures including flexible power sources such as hydro and open cycle gas turbines, demand-side measures, electricity storage, smart transmission, and distribution grids. The costs may be well above the generation costs alone, depending on location and technology used with these components:

- Adequacy costs ensure that the power system has sufficient capacity to meet peak loads.
- Balancing costs ensure that the power system can respond to demand changes at any given time.
- Interconnection cost links sources of supply to sources of demand.

These costs vary depending on the location and technologies in place [20].

Hydropower: The average O&M cost of hydropower is around 2% of total installed costs per year and below 2% for larger plants, and around 3% for smaller capacities [19].

Geothermal Energy: Geothermal O&M costs are high because of the need to work over production wells on a periodic basis to maintain fluid flow and hence production [20].

Bioenergy: Any type of organic materials including trees, crops, grass, woody material, animal waste, and agricultural waste from biological systems are called biomass, which can be used to produce biofuels. Burning the waste organic material and biomass provides thermal energy and reduces waste biomass. More complex carbohydrates form polymers including cellulose, hemicellulose, and lignin that together comprise lignocellulosic biomass. Cellulose $(C_6H_{10}O_5)_n$ is a polymer of glucose monomers compared and can be broken into glucose units. Hemicellulose is composed of 5C sugars such as xylose that cannot be fermented by the yeast used for fermenting glucose. Lignin is a heterogeneous polymer [7, 10, 25, 26].

Bioethanol: Starch is the main energy-containing molecules and cannot be directly fermented. First, starch must be broken down into smaller units using the enzyme alpha-amylase through hydrolysis to break the bonds between adjacent glucose molecules. Saccharification follows to increase the amount of fermentable sugars [7].

Besides the value of EROI, one needs to consider the contributions of biofuel usage toward decarbonization. When they replace fossil fuels, biofuels help reduce the GHG emissions since the carbon emitted was originally captured by plants through the photosynthesis. However, one has to consider the excessive use of fossil fuels in intensive agricultural needs in producing fertilizers, planting, harvesting, and transportation of the crop, as well as energy input in the biochemical process to produce and purify ethanol in the overall energy balance [16, 17, 35].

The chlorophyll molecule absorbs light of the visible spectrum and excites electrons that power photosynthesis to produce organic compounds mainly of carbohydrates such as sugars and starches. The photosynthesis proceeds with light reactions involving the capture of photon energy and store in biological intermediates, and dark reactions to store energy in stable molecules by fixing CO_2 into sugars of glucose, sucrose, starch, and cellulose for the growth of the organisms [23, 24]. Photosynthesis in green plants produces energy-transporting molecule adenosine triphosphate (ATP) and reducing agent nicotinamide adenine dinucleotide phosphate while releasing oxygen from the water splitting reaction

$$2H_2O \rightarrow O_2 + 4H^+ + 4e^- \qquad (13.1)$$

Therefore, the overall process of photosynthesis converts water, carbon dioxide, and sun light into oxygen and stable carbohydrates that store solar energy

$$nCO_2 + nH_2O \rightarrow C_nH_{2n}O_n + nO_2 \qquad (13.2)$$

The conversion efficiency of photosynthesis is very low and varies in the range of 1–7%. Some plants known as 4-carbon C_4 plants, such as maize, sugarcane, and sorghum, perform photosynthesis at relatively higher efficiency. These plants are also called energy crops [24, 35].

Biogas: Biomass and waste biodegraded by bacteria in landfill produce methane, which can be used for heat and power production. Biogas consisting of mainly methane and CO_2 is produced from organic waste through anaerobic digestion that facilitates the breakdown of the waste without the presence of oxygen.

$$C_nH_{2n}O_n \rightarrow n/CH_4 \rightarrow n/2CO_2 \tag{13.3}$$

The typical composition of biogas contains 50–55% methane. Many worldwide domestic digesters produce a considerable amount of power from biogas.

Thermal gasification by heating biomass in the presence of oxygen, air, steam produces synthesis gas consisting of CO_2, CO, water, and hydrogen. Pyrolysis, on the other hand, takes place by fast heating without oxygen and releases gas, bio-oil, and char. Fischer-Tropsch process can convert synthesis gas into liquid transportation fuels [7].

13.2.7 Optimum Energy Production

Underutilization of capacity reduces profits, which may be a survival issue for new technology. Projections of the demand–supply gap in the market should be made based on technology, equipment, public policy, and finances. When a process technology is mature, large capacity produces more energy per unit time and more revenue can be earned. The availability of transportation and other infrastructure must be studied. A large part of the land area should be reserved for service roads, storm water mains, railways, overhead gas, steam, air pipelines, and water reservoirs [27].

13.2.8 Distributed Operation and Centralized Operation

Centralized plants need a reliable supply of a large amount of feedstock throughout the year. This may be the case for primary energy production in large-scale operation, which helps lower the unit cost of energy generated and create considerable economic activity. For renewable fuels from biomass, however, large-scale operation may adversely affect the economics because of the nature of biomass with low density and widely scattered around the plant. Water;dissolved organic and inorganic compounds, and solid particulates of various sizes can be present in biomass processes. Distributed process technology may offer economic solutions to areas where the biomass feedstock is not readily transportable in large quantities [2, 27].

13.3 Source Conservation

Energy conservation focuses on reducing energy consumption and increasing efficiency in energy usage for the useful energy output; for example, using less fuel or reducing the demand on a limited supply such as fossil fuels and replacing it with renewable energy may lead to source conservation [8]. Energy recovery also may be a part of energy conservation through captured and reduced waste energy. Energy conservation may lead to increased security, financial gain, and environmental protection. For example, electric motors consume a considerable amount of electrical energy and operate at efficiencies between 70 and 90% [23]. Therefore, using an electric motor operating with higher efficiency will conserve energy throughout its useful life.

Energy conservation and consequently higher energy efficiencies across most sectors of economy are becoming part of sustainable development [8]. The building sector uses almost 35–40% of all energy consumed and is responsible for 30–40% of the GHG emissions [18]. Residential air-conditioning is a main factor to peak demand on the power grids. The power at these peak demands is very expensive. This is one of the reasons why the residential energy efficiency and conservation programs focus on space heating and cooling as well as water usage. The industrial sector is a good candidate for energy conservation, improved energy efficiency, and reduced greenhouse gas (GHG) emissions.

Energy recovery may lead to the use of available energy from a subsystem in another part of the same system. There is a large potential for energy recovery in industries and utilities leading to reduced use of fossil fuels and hence less GHG emission. Some examples of energy recovery practices are [8]:

- Energy and water recycling
- Heat recovery steam generator (HRSG)
- Heat regenerative cyclone engine
- Thermal diode
- Thermoelectric modules
- Regenerative brake is used in electric cars and trains, where the part of kinetic energy is recovered and stored as chemical energy in a battery
- Active pressure reduction systems where the differential pressure in a pressurized fluid flow is recovered rather than converted to heat in a pressure reduction valve.

Reducing energy consumption through efficiency and conservation requires a combination of increased research and development on energy efficiency and policies. It will also require structural and behavior changes in residential and transportation systems such as replacing fossil fuels with renewable energy [20]. The *Technical Potential Scenarios* estimate the level of energy consumption that would occur if all industrial processes are upgraded with Energy Conservation Management (ECM) leading to technically feasible solutions. The *economic potential* scenario 1 estimates the level of energy consumption that would occur if all industrial processes are upgraded with ECM to become economically feasible with a 2-year simple payback period. The *economic potential* scenario 2 predicts a 5-year simple payback period at a predefined uptake rate and trend. The projections for economic and technical potentials are up to 2050 on a 5-year increment [8, 18, 19, 21].

13.3.1 Some Barriers to Energy Conserving

The barriers to energy efficiency and ECM often focus on a situation where internal and external perspectives diverge [8]. Internal perspective may consist of many-under evaluated behavioral elements. By overlaying the energy-saving potentials with the associated economic, organizational, and technical barriers, the following potential measures encourage ECMs uptake:

1. Improving energy efficiency through *energy management*, instead of replacing equipment
2. Mandatory implementation of energy management systems for large energy-intensive sectors
3. Mandatory sub-metering requirement for significant energy-consuming equipment
4. Facilitating the development of insurance products for energy-savings guarantee
5. Promoting and facilitating the further potential for resource sharing among industrial sectors.

13.4 Energy Usage

Energy is the main driver of technological development to deliver the prosperity of modern society by being able to produce all goods and services. The International Energy Outlook 2019 report [18, 19] projects a considerable increase in global energy use and despite the increased use of renewable energy, fossil fuels will continue to supply global energy through 2050. Therefore, GHG emissions will continue to rise.

The *green certificates* are increasingly becoming common and get awarded related to the units of green energy that they produce. The value of a green certificate is not fixed and can vary depending on market conditions. The *emission trading system* (ETS) requires big companies to acquire *emission certificates* for the GHG emissions produced. If a company can switch partly from fossil fuels to bioenergy, the avoided cost for buying certificates could be an additional benefit of the bioenergy plant.

Table 13.1 shows the forecast of global energy consumption by end-use sector for 2050 in comparison to the results of 2018. The growth of energy use in the industrial sector is driven by the increase in product demand.

Figure 13.2 clearly shows that industrial and transportation sectors will be the largest energy consumers in increasing trend. Energy use in residential and commercial sectors will grow gradually [18, 19].

Table 13.1 Global Energy consumption in quads (10^{15} Btu) by end-use sector [19]

End-use sector	2018		2050	
	Quads	Share (%)	Quads	Share (%)
Residential	58.9	13	90.6	15
Commercial	32.6	7	48.7	8
Industrial	238.8	53	316.7	51
Transportation	120.9	27	167.2	27
Total delivered energy	**451.2**	100	**623.2**	100

Fig. 13.2 World end-use energy consumption by sector: quadrillion (10^{15} Btu) [19]

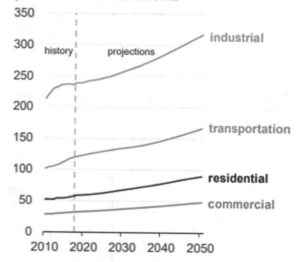

Figure 13.3 displays that liquid transportation fuels, natural gas, and electricity will increasingly dominate the world's end use of energy, while coal will remain stagnant and renewables will have the lowest share [18, 19].

Figure 13.4 shows that industrial, residential, and commercial sectors are the main users of electricity, while the transportation sector has the lowest share as electricity users. The projections show that the residential sector merges with the industrial sector by 2050 [18, 19].

Example 13.1 Energy cost of a car
An average car consumes 50 gallons gasoline per month. Estimate the amount and cost of energy consumed by the car per year.

Solution:
Assume that gasoline has an average density of 0.72 g/cm^3 and the heating value of 47.3 MJ/kg (Table 2.6). Cost of gasoline: $3/gallon

Fig. 13.3 World end use
energy consumption by
various fuels: quadrillion
(10^{15} Btu) [19]

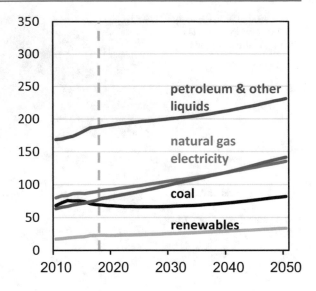

Fig. 13.4 World Electricity
use by sector: quadrillion
(10^{15} Btu) [19]

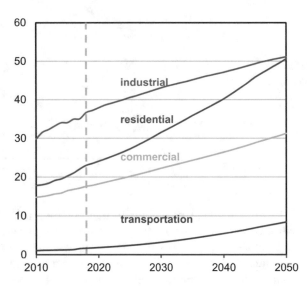

Data: $V = 50$ gal/month $= 189.25$ l/month, 2271.0 l/year $(3.785 \, \mathrm{l} = 1 \, \mathrm{gal})$; $\rho_{gas} = 0.72$ g/cm^3
$= 0.72$ kg/l
Mass of gasoline: $m_{gas} = \rho V = 1635.1$ kg/year
Energy consumed per year:
$E_{gas} = 1635.1$ kg/year $(47,300 \, \mathrm{kJ/kg}) = \mathbf{77,340,230 \, kJ/year}$
Cost per year: C (gasoline): ($3/gallon) (50 gallon)(12) = 1800/year

Example 13.2 Fuel consumption and emission reduction by a low and a high-mileage car

Assume that the average daily traveling distance is about 40 miles/day. A car has a city-mileage of 20 miles/gal. If the car is replaced with a new car with a city-mileage of 30 miles/gal and the average cost of gasoline is $3.50/gal, estimate the amount of fuel, energy, and money conserved with the new car per year.

Assume: The gasoline is incompressible with ρ_{av} = 0.75 kg/l.

Lower heating value (LHV) = 44,000 kJ/kg; 44,000 kJ of heat is released when 1 kg of gasoline is completely burned, and the produced water is in vapor state (Table 2.1).

Fuel needed for the old car: (40 miles/day)/(20 miles/gal) = 2 gal/day

Fuel needed for the new car: (40 miles/day)/(30 miles/gal) = 1.34 gal/day

Old car: 20 miles/gal

Mass of gasoline: m_{gas} = ρ_{av} (Volume) = (0.75 kg/l)(2.0gal/day)(3.785 l/gal) = **5.7 kg/day**

Energy of gasoline: E_{gas}(LHV) = (5.7 kg/day)(44,000 kJ/kg) = 250,800 kJ/day(365 day/year) = 91,542,000 kJ/year = **91,542 MJ/year**

Cost: ($3.50/gal)(2 gal/day)(365 day/year) = $2555/year

New car: 30 miles/gal

Mass of gasoline: m_{gas} = ρ_{av} (Volume) = (0.75 kg/l)(1.34gal/day)(3.785 l/gal) = **3.8 kg/day**

Energy of gasoline: E_{gas}(LHV) = (3.8 kg/day)(44,000 kJ/kg) = 167,200 kJ/day (365 day/year) = 61,028,000 kJ/year = **61,028 MJ/year**

Cost: ($3.50/gal)(1.34 gal/day)(365 day/year) = $1712/year

Annual fuel savings: **30,514 MJ/year** = (91,542–61,028) MJ/year

Annual cost savings: **$843/year** = ($2555 − $1712) MJ/year

Approximate annual emission reduction from Table 11.2: 3.3 kg CO_2/kg gasoline: (5.7 − 3.8) kg/day (365 day/year) (3.3 kg CO_2/kg gasoline) = **2288.5 kg CO_2/year**

The new car reduces the fuel consumption by around 33%, which is significant.

Example 13.3 Yearly consumption of natural gas by a city

A city consumes natural gas at a rate of 500×10^6 ft^3/day. The volumetric flow is at standard conditions of 60 °F and 1 atm = 14.7 psia. If the natural gas is costing $6/GJ of higher heating value, what is the yearly cost of the gas for the city.

Solution:

Q = 500×10^6 ft^3/day at 60 °F and 1 atm = 14.7 psia.

The higher heating value is the heat of combustion of the natural gas when the water product is at liquid state. From Table 2.6, the value of HHV is: 50,000 kJ/kg ~ 1030 Btu/ft^3 (Table 2.7)

Heating value: (1030 Btu/ft^3)(500 × 10^6 ft^3/day) = 515.0 × 10^9 Btu/day (515.0 × 10^9 Btu/day) (1055 J/Btu) = 543,325 GJ/day

Yearly cost: (543,325 GJ/day) ($6/GJ) (365 days/year) = **$11.9 × 10^8/year**

Example 13.4 Avoided fossil fuel by a car

An average car consumes about 2 gallons (US gallon = 3.785 L) a day, and the capacity of the fuel tank is about 15 gallons. Therefore, a car needs to be refueled once every week. The density of gasoline ranges from 0.72 to 0.78 kg/l (Table 2.6). The lower heating value of gasoline is about 44,000 kJ/kg. Assume that the average density of gasoline is 0.75 kg/l. If the car was able to use 0.2 kg of nuclear fuel of uranium-235, estimate the time in years for refueling and avoided cost of fossil fuel of gasoline.

Solution:

Assume: The gasoline is incompressible with ρ_{av} = 0.75 kg/l. Cost of gasoline $3.5/gallon

Lower heating value (LHV) = 44,000 kJ/kg; 44,000 kJ of heat is released when 1 kg of gasoline is completely burned, and the produced water is in vapor state.

Complete fission energy of U-235 = 6.73×10^{10} kJ/kg

Mass of gasoline per day:

$m_{gas} = \rho_{av} V = (0.75 \text{ kg/l})(2\text{gal/day})(3.785 \text{ l/gal}) = 5.67$ kg/day

Energy of gasoline per day:

$E_{gas} = m_{gas} \text{(LHV)} = (5.67 \text{ kg/day})(44,000 \text{ kJ/kg}) = 249,480$ kJ/day

Energy released by the complete fission of 0.2 kg U-235:

$E_{U\text{-}235} = (6.73 \times 10^{10} \text{ kJ/kg})(0.2 \text{ kg}) = 1.346 \times 10^{10}$ kJ

Time for refueling: $(1.346 \times 10^{10} \text{ kJ})/(249,480 \text{ kJ/day}) = \textbf{53,952 days} = \textbf{148 years}$

Therefore, the car will not need refueling for about 148 years.

Avoided cost of gasoline: ($3.5/gallon) (2 gallon/day) (365 days/year) (148 year) = **$378,140**

13.5 Implications of Energy Production and Usage

Environment Impact Assessment identifies the short- and long-term impact of energy generation on the environment within the liability to local and federal government agencies [36]. Possible impacts include on natural physical resources, natural biological/resources, and quality-of-life values. Without careful assessment of the mass and energy balances and ecological impacts, there is a danger that energy projects may end up as unsustainable. Dependence on non-renewables poses risks and challenges including:

- Depletion of fossil fuel resources
- Decrease of useful energy output
- Uneven distribution of fossil fuels around the globe
- Price volatility and increases on international energy markets
- Security of energy supply
- Damaging effects of fossil fuel use to the environment and public health.

13.5.1   Food–Energy and Water Nexus

By 2050 the demand for energy will nearly double with water and food increase by over 50% globally. The interlinkage (Fig. 13.5) between the food, energy, and water supply systems is a major consideration in sustainable development strategies. The complex interconnections show that: (i) water supply is influenced by demands from energy and food sectors, (ii) food production both depends on water and energy, and (iii) energy production and operations require water. Understanding the interconnections fully can change day-to-day practices of resource managers and policy-makers [11].

13.5.2   Energy Assessment

Country and industry-specific energy assessment can improve energy–efficiency considerably in a cost-effective way in the industrial sector. Several government programs have been established around the world to encourage, facilitate, or mandate industrial facilities to undertake energy assessment. The U.S. energy assessment program called the Industrial Assessment Centers (IACs) targets facilities with (1) gross annual sales below $100 million, (2) less than 500 employees, and (3) annual energy bills more than $100,000 and less than $2 million. The IAC is supported by the Industrial Technologies Program at the US Department of Energy (https://www.energy.gov/eere). The IAC team prepares 'Assessment Report' to help medium-sized manufacturers identify ways to save energy, reduce waste, and improve productivity. The IAC assessment protocol consists of pre-assessment information collection, contacting key plant personnel, pre-assessment analysis, assessment, and post-assessment activities [36, 38].

In pre-assessment information gathering stage, the client is sent a pre-assessment form that includes the size of a plant and plant layout, industry type and process

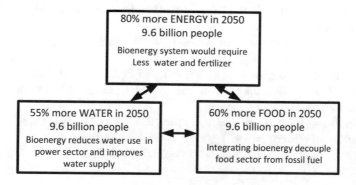

Fig. 13.5 Food, energy, water interdependency; sustainable bioenergy can positively affect water, energy, and food security, and slows the climate change [11]

description, production levels, units and costs, operating hours, history of utility bills, list of major energy consuming equipment, and key plant managers of plant, energy, environment, and maintenance.

The pre-assessment analysis involves an analysis of the manufacturing process, analysis of chart and graph of utility bills, establishing the unit cost of energy, plant profile, identifying key energy systems, review design and other technical documentations, identifying possible energy-saving potential recommendations using IAC database, and develop assessment day strategy. *Day of the assessment* involves the description of manufacturing, utility bills, and waste materials, developing a list of potential energy-saving opportunities with potential recommendations, discussing findings with management, preliminary estimate of potential savings, and prioritizing recommendations of analysis. *Post-assessment activities* conduct engineering and financial analyses, including implementation cost and deliver the report to the client. A follow-up to the report involves calling the client in 6–9 months for implementation data [36, 38].

Example 13.5 Energy densities and emissions of ethanol and gasoline
Assuming complete combustion of ethanol and gasoline (assuming 100% octane) estimate the HHVs by using the oxidations and carbon dioxide intensity in $kgCO_2$/MJ.

Solution:
$$C_2H_5OH + 3O_2 \rightarrow 2CO_2(g) + 3H_2O(l) \quad \Delta H_c = -1368\,kJ/mol \quad MW = 46\,g/mol$$
$$E_c = 29.7\,MJ/kg$$
$$C_8H_{18} + 25/2O_2 \rightarrow 8CO_2(g) + 9H_2O(l) \quad \Delta H_c = -5476\,kJ/mol \quad MW = 114\,g/mol$$
$$E_c = 47.5\,MJ/kg$$
We estimate the specific energy densities (E_c) by dividing the heats of combustions by their molecular weights (MW). Comparisons of the specific energy densities show that ethanol produces 63% (29/7/47.5) of energy per unit mass of octane.

Combustion of 46 g of ethanol produces two moles of carbon dioxide that is 88 g (2×44 g), while combustion of 114 g of octane produces eight moles that is 352 g (8×44 g). If we use the CO_2 intensity defined as amount of CO_2 released per unit of energy, we have

Amount of CO_2 produced/kg of ethanol = 44 g/mol x (number of moles of CO2 produced by 1000 g of ethanol)
= (44 g/mol) 2x (1000 g/46 g = 43.47) = 1.913 kg CO_2/kg ethanol
Amount of CO_2 produced/kg of octane = 44 g/mol x (number of moles of CO2 produced by 1000 g of octane)
= (44 g/mol) 8x (1000 g/114 g = 70.17) = 3087.7 kg CO_2/kg octane
CO_2 intensity of ethanol = 1913 g/29.7 = 64.4 g CO_2/MJ
CO_2 intensity of octane = 3087 g/47.5 = 65.0 g CO_2/MJ

The CO_2 intensities based on burning are almost the same for these two fuels, except the processes required for producing them would be different. This shows the estimation of environmental impact of fuels involved their feedstock, conversion, as well as their use.

13.5.3   Environmental Impact

Fossil fuels are well known for their detrimental environmental effects such as air pollution, greenhouse gas emissions, and natural habitat disruption associated with drilling practices. While biofuels also have negative effects, these effects are considered less costly compared to fossil fuels. To better compare the environmental effects of fossil fuels with biofuels, the life cycle assessments of the fuels may be helpful. For example, there are numerous studies comparing corn ethanol with gasoline and the results vary. The variance in results can be attributed to the differences in the studies, the process by which the study was carried out, and the scope of the study [18, 27].

The cultivation of different feedstocks for biofuel production utilizes different amounts of resources depending upon the chosen type. The use of short-rotation woody crops can also have different greenhouse gas emissions in comparison to annual crops such as corn or soybeans. The way in which the crops used as feedstock are grown and cultivated also may have adverse effects on the environment. Feedstock crops, like all agricultural crops, must be planted. The farm management practices used during this stage can also have varying effects on the environment. There are numerous factors producers must consider when growing crops, including to maximize yields and profits. These decisions play a part in the total impact that biofuels have on the environment during their lifetime, from plant to product. Crops require water and nutrients to grow, machinery to harvest and transport, industrial processes to refine. By minimizing the impact at each stage, biofuels provide a low carbon alternative to fossil fuels [3, 7, 9, 10].

The methods to assess the impact of biofuels over the duration of their lifecycle focus on different aspects of the biofuel lifecycle or utilize simulations that are very sensitive to variable inputs. This may cause a large margin of error that can be difficult to pinpoint at the end of the assessment. For example, the potential changes in the use of land to grow crops, driven by an increase in biofuel feedstock demand, can influence the carbon content contained in the soil. Based on the previous vegetation, large amounts of carbon can be released into the atmosphere or stored in the soil.

Once the crops are grown, the next steps are transportation, which releases emissions when moving biomass to the biorefineries, conversion processes with additional emissions released from the production and use of chemicals in the process, and from the generated electricity to run the plant. All these sources should be accounted for when attributing greenhouse gas emissions to biofuels or fossil fuels.

Life cycle analysis of biofuels may be challenging because of the sheer number of variables that need to be accounted for with reasonable assumptions and minimum errors possible. Many of these unknowns are from the agricultural practices adapted by producers. Soil nutrients vary drastically across regions and so do water and moisture levels. One of the keys for future policy decisions on bioenergy may be the comparison of greenhouse gas emissions from biofuel production and use with the emissions from fossil fuels [28].

Example 13.6 Cost estimations of electricity

A superheated steam (stream 1) expands in a turbine from 5000 kPa and 325 °C to 150 kPa and 200 °C. The steam flow rate is 10.5 kg/s. If the turbine generates 1.1 MW of power, estimate the emissions caused to produce the steam from natural gas and revenue of electricity.

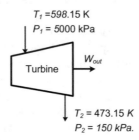

$$T_1 = 598.15 \text{ K}$$
$$P_1 = 5000 \text{ kPa}$$

W_{out}

Turbine

$$T_2 = 473.15 \text{ K}$$
$$P_2 = 150 \text{ kPa}.$$

Solution:

Assume that the kinetic and potential energy effects are negligible; this is a steady process.

The properties of steam from the steam tables:

Stream 1: Superheated steam: $P_1 = 5000\,\text{kPa}$, $T_1 = 325\,°\text{C}$ (598.15 K),
$H_1 = 3001.8\,\text{kJ/kg}$

Stream 2: Superheated steam: $P_2 = 150\,\text{kPa}$, $T_2 = 200\,°\text{C}(473.15\,\text{K})$,
$H_2 = 2872.9\,\text{kJ/kg}$

$\dot{W}_{out} = -1100\,\text{kW} = -1100\,\text{kJ/s}$

The mass and energy balances for the turbine are: Mass balance: $\dot{m}_{out} = \dot{m}_{in}$

Energy balance: $\dot{E}_{out} = \dot{E}_{in} \rightarrow \dot{m}_1(H_2 - H_1) = \dot{q}_{out} + W_{out}$

Heat loss from the energy balance

$\dot{q}_{out} = -\dot{W}_{out} + \dot{m}_1(H_2 - H_1) = 1100 \text{ kJ/s} + (10.5 \text{ kg/s})(2872.9 - 3001.8)\text{kJ/kg}$
$= \mathbf{-253.45\,kJ/s}$

The sign is negative as the heat is lost from the system.

Enthalpy If water entering the boiler is at 98 °C and 94 kPa + 410.6 kJ/kg (saturated steam table)

Energy to produce the steam:

(10.5 kg/s) (3001.8 − 410.6) kJ/kg = 27,207.6 kW = 27,207 kJ/s

From Table 2.8: 53.6 MJ/kg natural gas = 53,600 kJ/kg NG

Amount of natural gas needed:

(27,207 kJ/s)/(53,600 kJ/kg) = 0.507 kg/s = **1827.3 kg/h**

Every 1 kg natural gas produces 2.8 kg CO_2 (Table 11.2)

Emissions for every hour: (1827.3 kg/h) (2.8 kg CO_2/kg NG) = **5116.4 kg CO_2/h**

1.1 MW power plant = 1100 kWh each hour operating at full power and 26,400 kWh/day and 9636,000 kWh per year

For $0.14/kWh the total revenue from electricity would be:

($0.14/kWh) (9,636,000 kWh/year) = **$1,349,040/year**

Example 13.7 Cost of compression work

Air enters a compressor at 100 kPa, 300 K, and a velocity of 2 m/s through a feed line with a cross-sectional area of 1.0 m². The effluent is at 500 kPa and 400 K. Heat is lost from the compressor at a rate of 5.2 kW. If the air behaves as an ideal gas, estimate the power requirement of the compressor in kW and cost of power.

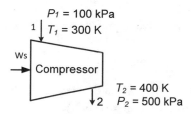

P_1 = 100 kPa
1 | T_1 = 300 K

Ws
→ Compressor

T_2 = 400 K
↓2 P_2 = 500 kPa

Solution:

Assume: The compressor is at steady state, air is ideal gas, and the change in the potential and kinetic energy of the fluid from the inlet to the outlet is negligible.

$R = 8.314 \, \mathrm{Pa\,m^3/kmol\,K}$, $MW = 29 \, \mathrm{kg/kmol}$; $v_1 = 2 \, \mathrm{m/s}$, $A_1 = 1.0 \, \mathrm{m^2}$

$P_1 = 100 \, \mathrm{kPa}, T_1 = 300 \, \mathrm{K}$, $P_2 = 500 \, \mathrm{kPa}$, $T_2 = 400 \, \mathrm{K}$, $q_{loss} = -5.2 \, \mathrm{kW}$

For ideal gas:

$V_1 = \dfrac{RT_1}{P_1} = 0.025 \, \mathrm{m^3/mol}$, the specific volume $(1/\rho)_{air} = \dfrac{V_1}{MW_{air}} = 0.86 \, \mathrm{m^3/kg}$

The mass flow rate of air: $\dot{m} = \dfrac{v_1 A_1}{(1/\rho)} = 1.72 \, \mathrm{kg/s}$

From Table D1 for ideal gas of air: H_2 = 400.9 kJ/kg, H_1 = 300.2 kJ/kg

Energy Balance: $\dot{m}(H_2 - H_1) - \dot{q} = \dot{W}$

$(1.72 \text{ kg/s})((400.9 - 300.2)\text{kJ/kg}) + 5.2 \text{ kW} = \dot{W} = \mathbf{178.4\,kW}$

Daily electricity: $(178 \text{ kW}) (24 \text{ h/day}) = 4281.6 \text{ kWh/day}$

For $0.14/kWh electricity

Annual cost $= \mathbf{\$218.939.6/year} = (\$0.14/\text{kWh})(4281.6 \text{ kWh/day})(365 \text{ day/year})$

Example 13.8 Cost of pumping

A pump increases the pressure in liquid water from 100 kPa at 25 °C to 4000 kPa. Estimate the minimum horsepower motor required and its cost to drive the pump for a flow rate of 0.01 m³/s.

Solution:

Assume that the pump operates adiabatically and nearly isothermally, changes in potential and kinetic energy are negligible, and water is incompressible.

$P_1 = 100\,\text{kPa}, \quad T_1 = 298.15\,\text{K}, \quad P_2 = 4000\,\text{kPa}, \quad Q = 0.01\,\text{m}^3/\text{s}$

Energy balance: $\dot{q} + \dot{W}_{\text{pump}} = \dot{m}\Delta H$

For an adiabatic operation ($\dot{q} = 0$): $\dot{W}_{\text{pump}} = \dot{m}\Delta H$

Since the water is incompressible the internal energy is only the function of temperature.

$\Delta U = 0$ as the system is isothermal.

$\Delta H = \Delta U + \Delta(PV) = V\Delta P$

The specific volume of saturated liquid water at 25 °C: $V_1 = 0.001003 \text{ m}^3/\text{kg}$ (steam tables)

The mass flow rate: $\dot{m} = \dfrac{Q}{V_1} = 9.97 \text{ kg/s}$

$\dot{W}_{\text{pump}} = \dot{m}V_1(P_2 - P_1) = (9.97 \text{ kg/s})(0.001003 \text{ m}^3/\text{kg})(4000 - 100)\,\text{kPa}\left(\text{kJ/kPa m}^3\right)$

$= \mathbf{39\,kJ/s} = \mathbf{39\,kW} = \mathbf{52.3\,hP}(1 \text{ hp} = 745.7 \text{ W})$

For $0.14/kWh electricity

Annual cost $= \mathbf{\$48.829.6/year} = (\$0.14/\text{kWh})(39 \text{ kW}) (24 \text{ h/day})(365 \text{ day/year})$

Example 13.9 Cost and emissions of pump work

A centrifugal pump delivers 1.8 m³/h water to an overhead tank 20 m above the eye of the pump impeller. Estimate the work required by the pump cost of power used and emissions from that power production using natural gas.

Solution:

Assume that the pump operates adiabatically and nearly isothermally, changes in potential and kinetic energy are negligible, and water is incompressible and steady frictionless flow.

Data: $Q = 1.8$ m^3/h and density of water: $\rho = 1000$ m^3/h

Energy balance: $\Delta\left(H + \frac{1}{2}v^2 + zg\right)\dot{m} = \dot{q} + \dot{W}_s$

For an adiabatic operation ($\dot{q} = 0$) with no kinetic and potential energy effects:

$\dot{W}_{pump} = \dot{m}g\Delta z = \left(\frac{1.8\,\text{m}^3/\text{h}}{3600\,\text{s/h}}\right)(1000)\,\text{kg/m}^3$

$(9.81\,\text{m/s})(30\,\text{m}) = 147.15\,\text{Nm/s} = \mathbf{147.15\,W}$

From Table 2.8: 53.6 MJ/kg natural gas = 53,600 kJ/kg NG

Amount of natural gas needed:

$(147.15\,\text{kJ/s})/(53,600\,\text{kJ/kg}) = 0.00274\,\text{kg/s} = \mathbf{9.88\,kg/h}$

Every 1 kg natural gas produces 2.8 kg CO_2 (Table 11.2)

Emissions for every hour: $(9.88\,\text{kg/h})(2.8\,\text{kg}\,CO_2/\text{kg NG}) = \mathbf{27.66\,kg\,CO_2/h}$

Example 13.10 Saving of fuel cost by regeneration in a Brayton cycle

A power plant is operating on an ideal Brayton cycle with a pressure ratio of $r_p = 9$. The fresh air temperature at the compressor inlet is 295 K. The air temperature at the inlet of the turbine is 1300 K. The cycle operates with a compressor efficiency of 80% and a turbine efficiency of 80%. The unit cost of fuel is \$0.14/kWh. The cycle operates 360 days per year.

(a) Using the standard-air assumptions, determine the thermal efficiency of the cycle.

(b) If the power plant operates with a regenerator with an effectiveness of 0.78, determine the thermal efficiency of the cycle and annual saving of fuel cost.

Solution:

Assume that the cycle is at steady-state flow and the changes in kinetic and potential energy are negligible. Heat capacity of the air is temperature-dependent, and the air is an ideal gas.

(a) Basis: 1 kg/s air. Data given: $\eta_{turb} = 0.8$, $\eta_{comp} = 0.8$, $r_p = P_2/P_1 = 9$

Process 1–2: isentropic compression

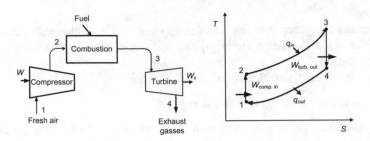

Data from (Appendix D Table D1): $T_1 = 295\,K$, $H_1 = 295.17\,kJ/kg$, $P_{r1} = 1.3068$

P_r shows the relative pressure defined in Eq. (7.36).

$$\frac{P_{r2}}{P_{r1}} = \frac{P_2}{P_1} = r_p \rightarrow P_{r2} = (9)(1.3068) = 11.76$$

Approximate values from Table D1 for the compressor exit: $P_{r2} = 11.76$, $T_2 = 550\,K$, and $H_2 = 555.74\,kJ/kg$

Process 3–4: isentropic expansion in the turbine as seen on the TS diagram above $T_3 = 1300\,K$, $H_3 = 1395.97\,kJ/kg$, $P_4/P_3 = 1/r_p = 1/9$, $P_{r3} = 330.9$

$$\frac{P_{r4}}{P_{r3}} = \frac{P_4}{P_3} \rightarrow P_{r4} = \left(\frac{1}{9}\right)(330.9) = 36.76$$

Approximate values from Table D1 at the exit of turbine: at $P_{r4} = 36.76$: $T_4 = 745\,K$, $H_4 = 761.87\,kJ/kg$

The work input to the compressor:

$$W_{comp.in} = \frac{H_2 - H_1}{\eta_{comp}} = \frac{(335.74 - 295.17)\ kJ/kg}{0.8} = 325.7\,kJ/kg$$

The work output of the turbine:

$$W_{turb.out} = \eta_{turb}(H_4 - H_3) = 0.8(761.87 - 1395.97) = -507.3\,kJ/kg$$

The back-work ratio r_{bw}: $r_{bw} = \dfrac{W_{comp.in}}{|W_{turb.out}|} = \dfrac{325.7}{507.3} = 0.64$

This shows that 64% of the turbine output has been used in the compressor.

Work output: $W_{net} = W_{out} - W_{in} = -181.6\,kJ/kg$

Actual values of air enthalpy at the compressor outlet:

$H_{2a} = H_1 + W_{comp.in} = (295.17 + 325.70)kJ/kg = 620.87\,kJ/kg$

The amount of heat added: $q_{in} = H_3 - H_{2a} = 1395.97 - 620.87 = 775.1\,kJ/kg$

The thermal efficiency: $\eta_{th} = \dfrac{|W_{net}|}{q_{in}} = \dfrac{181.6}{775.1} = \mathbf{0.234\ or\ 23.4\%}$

The temperature of exhaust air, T_4, and the actual enthalpy are estimated from the energy balance: $W_{turbout} = H_{4a} - H_3$; $H_{4a} = 1395.97 - 507.3 = 888.67\,kJ/kg \rightarrow T_4 = 860\,K$ (from Table D1).

Advances in the compressor and turbine designs with minimal losses increase the efficiency of these components. In turn, a significant increase in the thermal efficiency of the cycle is possible.

(b) $T_4 = 860$ K and $T_2 = 550$ K, since $T_4 > T_2$ regeneration is possible. Regeneration with effectiveness of $\varepsilon = 0.78$:

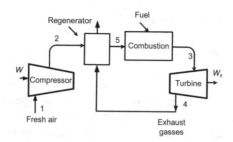

$$\varepsilon = \left(\frac{H_5 - H_2}{H_4 - H_2}\right) = \frac{H_5 - 620.9}{888.67 - 620.9} = 0.78 \rightarrow H_5 = 829.80 \text{ kJ/kg}$$

$q_{in} = H_3 - H_5 = 1395.97 - 829.8 = 566.2 \text{ kJ/kg}$

This represents a conservation of $840.2 - 566.2 = 274.0$ kJ/kg from the fuel required.

The thermal efficiency: $\eta_{th} = \dfrac{|W_{net}|}{q_{in}} = \dfrac{181.6}{566.2} = \textbf{0.32 or 32\%}$

Unit cost of fuel = $0.14/kWh

Days of operation 360, hours of operation per year = $360(24) = 8640$ h/year

Conserved fuel: $(274 \text{ kJ/kg})(1 \text{ kg/s}) (8640 \text{ h/year}) = 2{,}367{,}360$ kWh/year

Saved money: $(2{,}367{,}360 \text{ kWh/year}) (\$0.14/\text{kWh}) = \textbf{\$331,430/year}$

After the regeneration, the thermal efficiency has increased from 23 to 32% in the actual Brayton cycle operation. The addition of a regenerator (operating without thermal losses) does not affect the net work output of the cycle. Savings of fuel and costs are considerable.

13.6 Energy Policy

The relation between economic growth and energy consumption is the main factor for energy policy implication since the Gross Domestic Product (GDP) (one of the main indicators of economic growth) increases with the increasing usage of energy. Therefore, governments and energy policy makers design energy policies based on the data relating GDP to energy usage toward conservation hypothesis and growth hypothesis. According to the conservation hypothesis, in some developed nations economic growth has less reliance on energy usage, while the growth hypothesis predicts that in some developing nations economy is more dependent on energy and economic growth and energy usage are directly related to each other. Based on the hypotheses, policymakers may decide to reduce energy usage and hence the emissions [18, 19, 21, 30].

13.6.1   Global Energy Scenario

Energy policy recognizes the importance of energy efficiency and the necessary transition to renewables to tackle the risks and challenges of fossil fuel use. One effort called the 'Green Growth Strategy' helps policymakers and stakeholders to increase the renewables in energy development. In developed countries, growth in energy consumption is relatively due to slower population and economic growth, improvements in energy efficiency, and less growth in energy-intensive industries with the help of energy conservation policies [18]. Major progress in energy conservation is needed to achieve a sustainable world with the collaboration of organizations and policy makers around the globe.

The IEA's Sustainable Development Scenario (SDS) proposes the transformation of the global energy system toward reducing GHG emissions from 33 billion tons in 2018 to less than 10 billion tons by 2050. Energy production and use is the largest source of GHG emissions, hence the energy sector is crucial for achieving sustainable development [18, 20]. The deployment of new leading renewable resources is solar energy and wind energy for renewable power. Because of international consensus toward secure and sustainable energy policies, the contributions from solar photovoltaic, wind, and hydro energy are increasing as the cost of renewable energy is becoming more affordable [18, 19].

The contribution of various biofuels to the transport sector is increasing. Some additional incentives in investments and electrification particularly in the transportation sector help increase the shares of renewables in power generation. This trend of increase is also projected in the SDS worldwide, complemented mainly by nuclear power and carbon capture technologies.' By 2040, contributions from wind and solar PV will exceed hydropower if investment in renewables technologies would accelerate (Table 13.2). The SDS assumes that clean energy technologies will play a critical role in reaching sustainable energy goals [18, 19].

Higher energy efficiency guarantees secure, sustainable, and inclusive economic growth and reduces the environmental footprint of the energy system. Half of these energy efficiency savings come from industry, and major contributions also from transport and building sectors. The rate for primary energy demand is slower than in the past (1 vs. 2.7%) as energy consumption and economic growth continue to decouple by 2040 [18, 19].

The SDS depends upon fundamental changes to the way energy is produced and consumed with rapid energy sector transformation. Because of its cost-effectiveness and attractive payback, an increase in thermal efficiency improvements is a critical

Table 13.2 Global annual average investment in renewables by scenario (billion US$ 2018) (18,19)

	Stated policies			Sustainable development	
	2018	2019–30	2031–40	2019–30	2031–40
Renewables-based power generation	304	329	378	528	636
Wind	89	111	122	180	223
Solar PV	135	116	125	179	191
End-use sectors	25	117	139	124	145

element that brings the world toward the SDS. As the increase in electrification and the share of variable renewables continue, the carbon footprint of electricity may fluctuate. Greater emphasis on circular economy with higher materials efficiency may reverse the trend of growing emissions for steel and cement, leading to a stronger decline in industrial CO_2 emissions.

IRENA's *Transforming Energy Scenario* (TES) sees emissions fall at a compound rate of 3.8% per year to some 10 Gt, or 70% less than today's level, by 2050, hence keeping the expected temperature rise well below 2 °C. The TES shows how to achieve stable, climate-safe, sustainable long-term energy and economic development with the Deeper Decarbonization Perspective (DDP) that would reduce emissions to zero by 2050 or latest by 2060, consistent with holding the line at 1.5 °C. In the TES fossil-fuel use declines by 75% compared to today's level; the use of coal declines by 41% in 2030 and 87% in 2050, while petroleum oil by 31% in 2030 and 70% in 2050 [19–21] (Table 13.3).

In the *Baseline Energy Scenario* (BES), energy-related emissions increase by a compound annual rate of 0.7% per year to 43 gigatons (Gt) by 2050 (up from 34 Gt in 2019), resulting in a likely temperature rise of 3 °C or more in the second half of this century [19–21].

The *Planned Energy Scenario* (PES), or main reference case, sees emissions increase slightly by 2030 and then decline to 33 Gt, roughly today's level, by 2050. This would result in a likely global temperature rise of 2.5 °C in the second half of this century [19–21].

Table 13.3 Primary energy use and energy related emissions with historical progress and Transforming Energy Scenario (TES) (Gt: Gigatonnes: 10^9 metric tons; EJ: Exajoules 10^{18} J, e: estimated) [19–21]

	Historical progress	TES (2030)	TES (2050)
Total primary energy supply (EJ)	571 (2015); 599 (2018)	556	538
Fossil-fuel use (EJ)	468 (2015); 485 (2018)	313	130
Energy-related CO_2 emissions (Gt)	32 (2015); 34 (2019e)	25	9.5

13.7   Energy Economics

Expenditures of solar, wind and storage are very close to each other, while the hydro expenditure is relatively low. Annual operating and maintenance costs are high and similar for wind and storage, while lower for solar and much lower for hydro energy. Energy economies should be handled with extreme care due to its impact in the ecosystem [20].

Hydroelectric costs and average costs of wind, solar photovoltaic, biomass or geothermal energy are getting closer to each other [19, 20]. Wind and solar-based renewable energy has the optimum cost of replacing fossil fuels worldwide. Solar power may be used during the day and wind power during the night. In addition, reliable grid structure, utility-scale energy storage, and smaller wind turbines help increase the contribution of renewable energy without nuclear power. Battery storage cost would be reduced considerably if the energy is supplied from clean natural gas generation with carbon capture. Cheap and abundant natural gas can be an important back-up fuel for intermittent renewable energy besides the mixture of hydroelectricity and stored energy [19, 20, 25, 31, 34].

13.7.1   Renewable Energy Cost

Renewable power is continuously becoming more competitive each year in the order of affordability from highest to lowest: concentrated solar power, bioenergy, solar photovoltaic and onshore wind power, hydropower, geothermal and offshore wind power. This trend also follows low-cost decarbonization, which is encouraging in sustainability concerns. Especially onshore wind and solar power are becoming less expensive than any fossil-fuel options and proving that renewable power is cost-effective and more evenly distributed worldwide compared with fossil fuels [19, 20].

13.7.2   Techno Economic Analysis

Techno economic analysis (TEA) undergoes pre-investment, investment, and operation phases. The pre-investment phase involves the identification of investment opportunities, preliminary selection of the project ideas, project design, and final evaluation based on the expectations and requirements of various stakeholders. The investment phase involves several inter-disciplinary tasks, including negotiation and contracting for project plant, construction, detailed project design and cost estimates, feasibility, and startup of trial operation. The operation phase involves day-to-day operation and marketing the products [29, 40].

TEA produces the discounted cash flow diagrams (DCFD) prepared for 20–25 years of operation using the current economic data. Based on the equipment list from the process flow diagrams, fixed capital investments (FCI), operating and maintenance costs, and cost of manufacturing are estimated. Chemical Engineering Plant Cost Index (CEPCI) can be used to estimate and update the costs and capacity to the present date by [40].

$$\text{Cost}_{\text{New}} = \text{Cost}_{\text{New}} \frac{\text{CEPCI}_{\text{New}}}{\text{CEPCI}_{\text{Old}}} \left(\frac{\text{Capacity}_{\text{New}}}{\text{Capacity}_{\text{Old}}} \right)^x \tag{13.4}$$

where x is the factor, which is usually assumed to be 0.6. A possible depreciation method is the maximum accelerated cost recovery system. After estimating the revenue DCFDs are prepared to generate the three economic feasibility criteria that are Net Present Value, Payback Period, and Rate of Return. At least two out of three criteria should be favorable for the operation to be feasible. Plant capacity affects feasibility.

The sudden growth of renewables markets can lead to high energy prices in the short term if the supply does not match the demand. It is also possible that excess supply causes the energy prices falling below the production costs. This can make analyzing the cost of renewable power generation technologies unreliable in given locations and times [19].

Stochastic analysis: The estimated profitability criteria using discounted cash flow diagram may be affected by the changes in the economic data and market conditions, including fixed costs, raw material prices, revenue, interest rates, and tax rate. Therefore, a stochastic model such as Monte-Carlo method helps account the viability of the economic data due to the uncertainties. The stochastic model produces probability density distributions of important profitability criteria [28, 31, 40].

13.7.3   Levelized Cost of Electricity

Levelized cost of electricity (LCOE) is the price at which electricity must be produced from a specific source to break even. It is an economic assessment tool and includes initial investment, cost of operations and maintenance, cost of fuel, and cost of capital [18].

$$\text{LCOE} = \frac{\sum_i^n \frac{I_i + M_i + F_i}{(1+r)^i}}{\sum_i^n \frac{E_i}{(1+r)^i}} = \frac{\text{Total costs}}{\text{Total electricity produced}} \tag{13.5}$$

where LCOE is the average lifetime levelized electricity generation cost, I is the Investment expenditures, M is the operations and maintenance expenditures, F is the fuel expenditures, E is the electricity production, i is the interest rate, and n is the useful life of operation. Typically, levelized energy costs are calculated over 20- to 40-year operational lifetimes and are given in the units of $/kWh or $/MWh. The simple levelized cost of energy is calculated by:

$$\text{LCOE} = \{(\text{capital cost} * \text{capital recovery factor} + \text{fixed O\&M cost})/$$
$$(8760 * \text{capacity factor})\} + (\text{fuel cost} * \text{heat rate}) + \text{variable O\&M cost,} \quad (13.6)$$

In the denominator, 8760 is the number of hours of operation in a year and capacity factor is a fraction between 0 and 1 representing the portion of a year that the power plant is generating power. *Fixed O&M costs* include labor, scheduled maintenance, routine component/equipment replacement (for boilers, gasifiers, feedstock handling equipment), insurance, and others. The fixed O&M costs of larger plants are lower per kW due to economies of scale. *Variable O&M costs* include ash disposal, unplanned maintenance, equipment replacement, and incremental serving costs. A capital recovery factor is the ratio of a constant annuity to the present value of receiving that annuity for a given length of time. Using a discount rate i and time period of operation n the capital recovery factor *CRF* is estimated by

$$CRF = \frac{i(1+i)^n}{(1+i)^n - 1} \quad (13.7)$$

Fuel cost is expressed in ($/MMBtu) and heat rate is measured in Btu/kWh. Fuel cost is optional since some generating technologies like solar and wind do not have fuel costs. The LCOE indicates the average cost per unit of electricity generated at the actual plant and allows the recovery of all costs over the lifetime of the plant [18, 19]. Table 13.4 shows the levelized cost of new generation resources [19].

Table 13.4 Estimated levelized cost of electricity for new generation plants entering service in 2025 US$/MWh [18, 19]

Plant type	Capacity factor, %	U.S. Levelized Costs of Electricity[a] (LCOE) (2019) ($/MWh)			
		Capital cost	Op. & Maint.	Transmission investment	Total system cost
Combined cycle	87	7.5	28	1.2	36.6
Combustion turbine	30	45.8	75.1	3.5	68.7
Wind, onshore	40	23.5	7.5	3.1	34.1
Wind, offshore	45	84	27.9	3.2	115.1
Solar PV	30	24.1	5.8	2.9	32.8
Geothermal	90	20.4	15.6	1.5	37.5
Hydro	73	28.9	9.1	1.6	39.5

[a]Costs are expressed in terms of net AC power available to the grid for the installed capacity

The LCOE of renewable energy technologies varies by technology, location, and project, based on the renewable energy resource, capital and operating costs, and the efficiency/performance of the technology. Most renewable power generation technologies are capital-intensive that has a critical impact on the LCOE. The cost of debt and the required return on equity, as well as the ratio of debt-to-equity varies between individual projects and countries and hence impacts on the values of LCOE [18, 19].

13.7.4 Economic Assessment of Biofuels

The biochemical process of fermentation converts the first-generation biomass including corn, sugarcane, and wheat into bioethanol, while biochemical and thermochemical processes convert mostly the lignocellulosic and algal biomass into bioethanol, biodiesel, green diesel, and other fuels and chemicals. Conversion of lignocellulosic feedstock requires first the complex process of conversion of cellulose and hemicellulose into sugars. The cost of the conversion processes of biomass increases in this direction: triglycerides → starch → lignocellulosic, while the cost of biomass increases in this order: lignocellulosic → starch → triglycerides. Bioethanol producers have adopted various technologies such as high tolerance yeasts, continuous ethanol fermentation, cogeneration of steam and electricity, and molecular sieve driers to reduce ethanol production costs [7, 25].

Second-generation biomass resources are geographically more evenly distributed than fossil fuel resources, which may lead to energy security. Biofuels might create local economic activity and employment. Most agricultural biomass production, except of forest products, is seasonal and results in a large volume of feedstock that needs to be stored and transported to a biorefinery. Regional pre-processing infrastructure can be set up to clean, sort, chop, or grind, control moisture, densify, and package the feedstocks before transporting them to biorefineries. In contrast, forest products are available year-round so that long-term storage might not be necessary [35].

Table 13.5 shows the large amounts of water requirements of bioethanol and biodiesel productions from various biomass feedstocks. The effects of biofuel production on the world wide trade of grains, livestock, biomass, and crude oil are a part of economic assessment. The biofuel industry has also some economic effects related to the national budget spending such as tax credits, subsidies, incentives, and other policy matters. The diversion of land to corn and soybean production and a greater demand for biofuels may coincide with an increase in the price of the wheat, corn, and soybean.

13.7.5 Cost Trends in Renewable Energy

The International Renewable Energy Agency (IRENA) reported data showing how renewable energy generation has produced low-cost electricity and continued to

Table 13.5 Comparison of water requirements for ethanol and biodiesel productions [25, 35]

Biofuel Crop	Water use (m^3 water/kg crop)	Biofuel conversion (Liter fuel/kg crop)	Crop water use (m^3 water/kg fuel)	Crop water use Per unit energy (m^3 water/GJ)
Ethanol Corn (grain)	833	409	2580	97
Sugarcane	154	334	580	22
Corn Stover	634	326	2465	92
Switchgrass	525	336	1980	74
Grain sorghum	2672	358	9460	354
Sweet sorghum	175	238	931	35
Biodiesel Soybean	1818	211	9791	259
Canola	1798	415	4923	130

decrease over the past 10 years. Using this data, predictions can be made as to what the future costs may be based upon the trends. The categories of renewable energy sources are onshore and offshore wind, solar photovoltaic and concentrated solar power, hydroelectric, geothermal, and bioenergy. For each of these categories, the IRENA (2020) reported the global weighted average total installed costs and the levelized cost of electricity from 2010 to 2019. Using this data, the forecasted costs could be predicted [20].

As technology has improved in the past years, the cost of installing new onshore wind facilities and turbines has decreased. Electricity costs passed to the costumer are lowered. Offshore wind is a more volatile class on a year-to-year basis due to a more dispersed market (see Fig. 13.6). The projects associated with offshore wind use larger turbines that are more costly but in turn generate more power [19, 20, 21].

Figure 13.6 shows that wind total installed costs for both the onshore and offshore systems will keep falling from $1500/kW in 2019 to $1000/kW in 2025 for onshore wind and from $4000/kW in 2019 to $3200 in 2025 for offshore. This will also be the case for levelized cost of electricity from both the systems as Fig. 13.7 shows. The values of LCOE would fell from $0.06/kWh in 2019 to $0.04/kWh in 2025 for onshore wind and from $0.14/kW in 2019 to $0.09 in 2025 for offshore wind.

Advances in solar power technology have increased the power generation of photovoltaic cells at a lower cost and industrial level installation projects are cheaper per unit than smaller scale installations. The levelized cost of electricity is expected to continue to decrease as can be seen in the chart [19, 20].

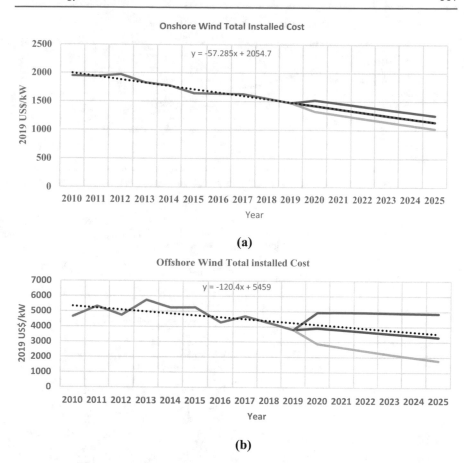

Fig. 13.6 Total installed costs for **a** onshore wind and **b** offshore wind using historical weighted average data (2010–2019) and forecasts with upper and lower confidence bounds; dotted line shows the linear trend of weighted average [20]

Figure 13.8 shows that total installed costs for both the solar PV and concentrated solar powers will keep decreasing from $1000 kW in 2019 to $400/kW in 2025 for solar PV and from $6000/kW in 2019 to $4300 in 2024 for concentrated solar power. The cost is relatively higher for concentrated solar power systems. Figure 13.9 shows very similar trends for the levelized cost of electricity falling from $0.20/kWh in 2019 to $0.15/kWh in 2024 for concentrated solar systems and from 0.10/kWh in 2019 to 0.05/kWh in 2024 for PV solar systems[19, 20].

Figure 13.10 shows that the installed cost and LCOE from bioenergy will remain high and flat and around $2500/kW. The values of LCOE from bioenergy will be comparable with those of wind and solar-based LCOEs around 2022 [19, 20] at $0.07/kWh.

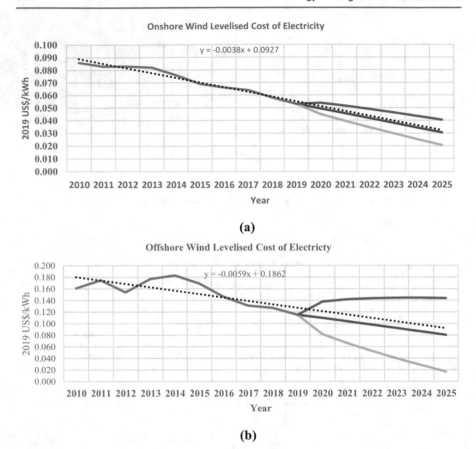

Fig. 13.7 Levelized cost of electricity for **a** onshore wind and **b** offshore wind using historical weighted average data (2010–2019) and forecasts with upper and lower confidence bounds; dotted line shows the linear trend of weighted average [20]

Bioenergy stems from a wide range of conversion technologies and feedstocks. Some of the technologies operate at high temperatures and pressures and therefore, energy intensive processes. In addition, these technologies mostly require costly feedstock pre-treatment process. This creates a large distribution of installation costs based on the type of plant, the necessary materials, and the location in which the plant is being built. This makes the projections challenging [7, 9, 20, 25].

Geothermal power production is a mature technology and continues to grow at a lesser pace. The total installation costs would remain around $4000/kW. The volatility can be associated with the additions and changes made to the geothermal grid. Figure 13.11 shows that the values of LCOE will be around $0.08/kWh and remain competitive in the long-time projections [19, 20].

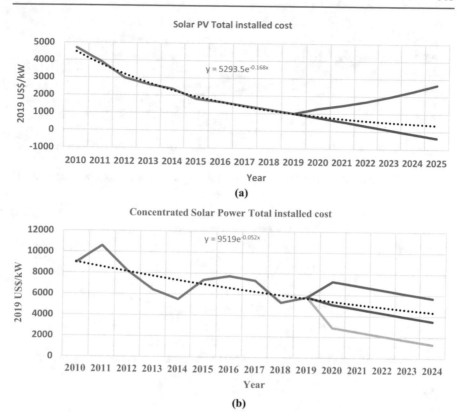

Fig. 13.8 Total installed cost for **a** solar PV and **b** concentrated solar power using historical weighted average data (2010–2019) and forecasts with upper and lower confidence bounds; dotted line shows the trend of weighted average [20]

Figure 13.12 shows that the total installed cost and the values of LCOE for hydropower. Hydroelectric power generation is a well-established and reliable source of electricity. The total installation costs have been driven up in recent years due to an increase in developing areas with a more geographical and engineering challenges. Due to the rising installation costs and relative consistency of the capacity factor, the levelized cost is not expected to decrease and would be around $0.05/kWh [19–21].

Renewable power generation increased by more than the increase in electricity demand, while fossil-fuel electricity generation decreased. The progress in the electrification of transport is causing the rapid cost reductions of solar PV and wind along with the key enabling technologies such as batteries and electric vehicles, as well as renewable hydrogen [19–21].

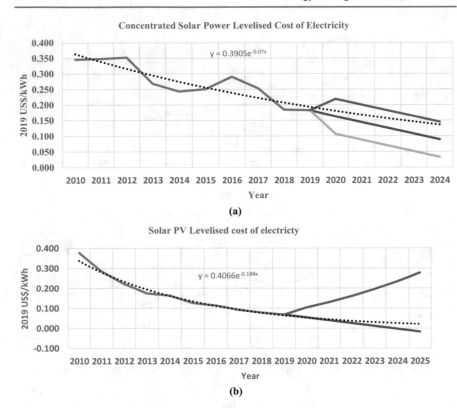

Fig. 13.9 Levelized cost of electricity for **a** concentrated solar power and **b** solar PV using historical weighted average data (2010–2019) and forecasts with upper and lower confidence bounds; dotted line shows the linear trend of weighted average [20]

13.7.6   Bio Break Model

The purchase price for feedstocks should be obtained by surveying biorefineries, and the costs of producing and delivering biomass feedstocks to a biorefinery should be based on observed production practices. The bio break model represents the regional feedstock supply system and biofuel biorefinery economics. The bio break model estimates willingness to pay (*WTP*), willingness to accept (*WTA*), and the price gap (*PG*) between them [30]:

(1) *WTP* is the maximum price that a biorefinery to pay for a dry ton of biomass delivered at the gate. *WTP* is a function of the price of bioethanol, the conversion yield, and the cost of processing biomass.

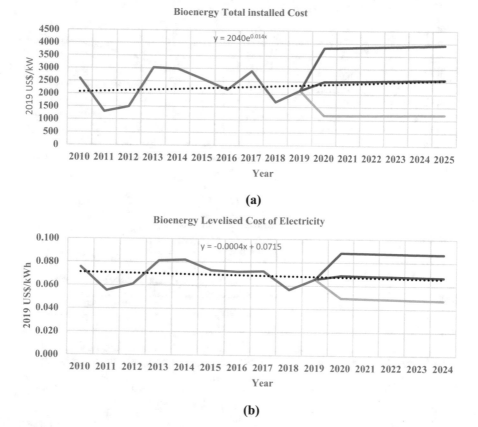

Fig. 13.10 Bioenergy **a** total installed cost and **b** levelized cost of electricity using historical weighted average data (2010–2019) and forecasts with upper and lower confidence bounds; dotted line shows the trend of weighted average [20]

(2) *WTA* is the minimum price that a biomass producer would accept for a dry ton biomass delivered at the gate of a biorefinery. *WTA* depends on the biomass opportunity cost, production, and delivery costs.

Assumptions used in the bio break model are:

- Producer minimizes costs on the long-run average cost curve
- A yield distribution for biomass crops is based on the expected mean yield
- A transportation cost is based on the average hauling distance for a defined circular capture region
- Biorefinery has annual capacity of 89 million liter or more to be competitive in the market

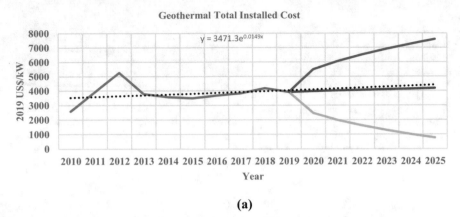

(a)

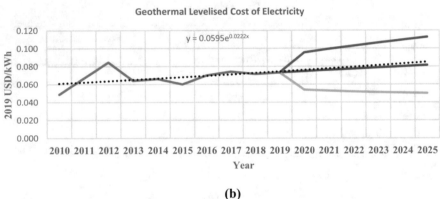

(b)

Fig. 13.11 Geothermal **a** total installed cost and **b** levelized cost of electricity using historical weighted average data (2010–2019) and forecasts with upper and lower confidence bounds; dotted line shows the linear trend of weighted average [20]

- Each biorefinery uses a single feedstock with no market disruptions
- Energy price uncertainty on biofuel investment is neglected.

Willingness to pay (WTP) for lignocellulosic bioethanol production: Eq. (13.8) shows the estimation of WTP for 1 dry ton of cellulosic material delivered to a biorefinery

$$WTP = \left(P_{gas}E_V + T + V_{CP} + V_O - C_I - C_O \right) Y_E \qquad (13.8)$$

The market price of bioethanol is estimated as the energy equivalent price of gasoline, where the P_{gas} denotes per gallon price of gasoline and E_V is the energy equivalent factor of gasoline to ethanol. Based on historical data gasoline and crude oil are related by $P_{gas} = 0.13087 + 0.023917 P_{oil}$. Beyond direct ethanol sales, the

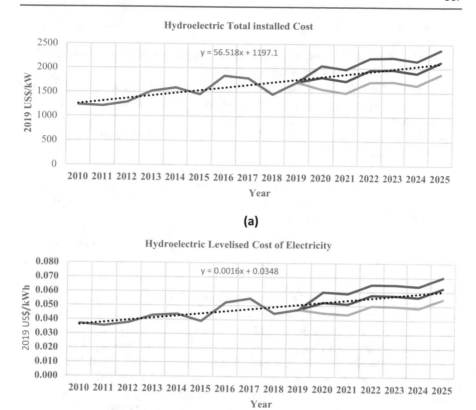

(a)

(b)

Fig. 13.12 Trends in hydrothermal energy for **a** total installed cost and **b** levelized cost of electricity using historical weighted average data (2010–2019) and forecasts with upper and lower confidence bounds; dotted line shows the linear trend of weighted average [20]

ethanol processor also receives revenues from the tax credits T, the coproduct production V_{CP}, and the octane benefits V_O per gallon of processed bioethanol. Biorefinery costs are the investment costs C_I and the operating C_O costs per gallon. A conversion ratio is used for gallons of ethanol produced per dry ton of biomass Y_E. Therefore, Eq. (13.8) provides the maximum amount the processor can pay per dry ton of biomass delivered to the biorefinery and still break even. The values of the variables in Eq. (13.8) are based on the following assumptions [30]:

- Energy equivalent factor (E_V) for ethanol to gasoline and the value of octane benefits (V_O) would be fixed based on the current economic data per gallon.
- Coproduct value (V_{CP}) is the excess energy as the only coproduct

- Conversion ratio (Y_E) is assumed as a mean value of 70 gallons per dry ton feedstock.
- Investment costs (C_I) are estimated for an optimized *nth* biorefinery for bio-chemical conversion of corn stover to ethanol.
- Operating costs (C_O) are separated into two components: enzyme costs and nonenzyme operating costs including salaries, maintenance, overhead, insurance, taxes, and other conversion costs.
- Biofuel production incentives and tax credits (T) for cellulosic ethanol producers would be determined locally.

Willingness to accept (WTA) *for lignocellulosic bioethanol production*: Eq. (13.9) shows the estimation of WTA for 1 dry ton of cellulosic material delivered to the biorefinery [30]

$$WTA = \{(C_{ES} + C_{Opp})/Y_B + C_{HM} + SF + C_{NR} + C_S + DFC + DVC * D\} - G$$
$$(13.9)$$

The value of WTA is equal to the total production costs less than the government incentives G (tax credits and production subsidies). Costs include the establishment and seeding C_{ES} per acre land and biomass opportunity costs C_{Opp} per acre, the harvest and maintenance C_{HM}, the stumpage fees SF, the nutrient replacement C_{NR}, the biomass storage C_S, the transportation fixed costs DFC, and the variable transportation costs calculated per mile DVC multiplied by the average hauling distance to the biorefinery. Therefore, the biomass yield per acre Y_B is used to convert the per acre costs into per dry ton costs. Equation (13.9) is used with the following assumptions:

- Nutrient replacement cost (C_{NR}) is based on the added value by the uncollected cellulosic material to the soil through enrichment and protection against rain, wind, and radiation, thereby limiting the loss of vital soil nutrients such as nitrogen, phosphorus, and potassium.
- Harvest and maintenance costs (C_{HM}) and stumpage fees (SF) are adjusted by the current economic data for short-rotation woody crops.
- Transportation costs $(DVC, DFC, \text{and } D)$: One-way transportation distance D has been evaluated up to around 140 miles for woody biomass and between 5 and 75 miles for all other feedstocks. The average hauling distance is between 13 and 53 miles.
- Biomass storage costs (C_S) depend on the feedstock, harvest technique, and storage area.
- Establishment and seeding costs (C_{ES}) are assumed to not incur for corn stover, wheat straw, and forest residue suppliers, whereas all other feedstock suppliers would have to be compensated for their establishment and seeding costs.

- Opportunity costs (C_{Opp}) of using biomass for ethanol production are assumed a mean opportunity cost per acre of switchgrass and *Miscanthus*.
- Biomass yield (Y_B) depends on the feedstock and the process selected.
- Biomass supplier government incentives (G) are the dollar/dollar matching payments for collecting, harvesting, storing, and transporting (CHST), which is a temporary (2-year) payment.

Bio break model estimates the price gap *(PG)* : *PG = WTA − WTP;* if the PG is negative or zero, a biomass market is economically feasible, otherwise the biomass market is not [35]. Policy incentives for carbon emissions could also affect the PG for a possible interaction with biofuel policy [30]. Biomass supplier government incentives are crucial to maintaining a moderate WTA cost. They can subsidize the delivery costs significantly, encouraging buyers to purchase more biomass for biorefinery purposes. They can be adjusted accordingly to manipulate the supply and demand of biomass in the market to ensure that both the suppliers' and buyers' benefit—maintains a reasonable market value for biomass. WTA does not take into consideration external changes in cost independent of energy costs, for example, drought, pest infestations, agricultural disease—this prevents it from being used as an accurate overall measure of the viability of biorefinery. WTA can be used as an approximation/indicator of profitability in the industry instead—high WTA suggests less chance of the process being profitable in the long term. Harvest and maintenance cost (C_{HM}) and the opportunity cost (C_{Opp}) are the main contributors to WTA and need to be minimized to reduce the WTA substantially (Table 13.6).

13.7.7 Thermoeconomics

Thermoeconomics assigns costs to exergy-related variables by using the *exergy cost theory* and *exergy cost balances* [12]. *Extended exergy* accounts for the environmental impact in a more systematic way by estimating the resource-based value of a commodity. Some concerns in thermoeconomics evaluations are: (i) costs of fuel and equipment change with time and location, (ii) optimization of an individual process does not guarantee an overall optimum for the system, and (iii) for the whole system, often several design variables should be considered and optimized simultaneously.

For any process or subsystem i, the specific cost of exergy c in \$/kW-unit time for a stream is

$$c = \dot{C}/\dot{Ex} \tag{13.10}$$

where $\dot{C}$ and $\dot{Ex}$ are the cost rate and the rate of exergy transfer for a stream, respectively. However, the cost of a product and other exiting streams would

Table 13.6 Summary of economics of biofuel conversion per dry ton of feedstock [7, 35]

	Ethanol by fermentation		Gasoline or diesel by gasification and F-T		Gasoline or diesel by pyrolysis with purchased H_2
	90 gal/ton	70 gal/ton	High temp.	Low temp.	High yield
Single plant capital, million US$	380	380	606	498	200
Million gallons of fuel per year	69.5	52.4	41.7	32.3	58.2
Million gallons of gasoline equivalent per year	46.3	34.9	41.7	32.3	58.2
Cost to produce					
nth plant million US$	375	500	430	480	210
Pioneer plant, million US$	650	850	800	750	350
Number of plants to meet 16 billion gallons of ethanol equivalent biofuels in 2022	230	305	256	331	183
Capital costs for RPS2, billion US$	88	116	155	165	37
Price gap, billion US$ per year					
At US$52 per barrel	25	39	31	37	8
At US$111 per barrel	10	24	16	21	−7
At US$191 per barrel	−10	3	−4	1	−28
Biomass feed requirements					
Million dry tons per year	178	236	175	226	133
Million acres at 5 tons per acre	36	47	35	45	27

include the fixed capital investment $\dot{C}_{FCI}$ and the annual operating cost of process $\dot{C}_{OP}$. This will be called the total cost of process

$$\dot{C}_P = \dot{C}_{OP} + \dot{C}_{FCI} \tag{13.11}$$

Then the cost rate balance for a single process is

$$\left(\sum_i c_i \dot{Ex}_i\right)_{out} = \left(\sum_i c_i \dot{Ex}_i\right)_{in} + \dot{C}_P \tag{13.12}$$

As seen in Example 13.11, the cost rate balance for a boiler (control volume 1) relates the total cost of producing high-pressure steam to the total cost of the entering streams plus the cost of the boiler $\dot{C}_B$, and from Eq. (13.12) we have

$$c_{HP}\dot{E}x_{HP} + c_{EG}\dot{E}x_{EG} = c_F\dot{E}x_F + c_A\dot{E}x_A + c_W\dot{E}x_W + \dot{C}_B \qquad (13.13)$$

where the symbol HP denotes the high-pressure steam, EG the exhaust gas, while F, A, and W are the fuel, air, and water, respectively. All the cost estimations are based on exergy as a measure of the true values of work, heat, and other interactions between a system and its surroundings. By neglecting the costs of air and water, and assuming that the combustion products are discharged directly into the surroundings with negligible cost, Eqs. (13.10) and (13.13) yield the specific cost of high-pressure steam

$$c_{HP} = c_F\left(\frac{\dot{E}x_F}{\dot{E}x_{HP}}\right) + \frac{\dot{C}_B}{\dot{E}x_{HP}} \qquad (13.14)$$

The ratio $(\dot{E}x_F/\dot{E}x_{HP}) > 1$ due to inevitable exergy loss is in the boiler, and hence $c_{HP} > c_F$.

Similarly, the cost rate balance for the turbine (control volume 2) is

$$c_E\dot{W}_E + c_{LP}\dot{E}x_{LP} = c_{HP}\dot{E}x_{HP} + \dot{C}_T \qquad (13.15)$$

where c_E, c_{LP}, and $\dot{C}_T$ are the specific costs of electricity, low-pressure steam, and the total cost of the turbine, respectively; $\dot{W}_E$ and $\dot{E}x_{LP}$ are the work produced by the turbine and the exergy transfer rate of low-pressure steam, respectively. Assuming that the specific costs of low and high-pressure steams are the same $c_{LP} = c_{HP}$, we have

$$c_E = c_{HP}\left(\frac{\dot{E}x_{HP} - \dot{E}x_{LP}}{\dot{W}_E}\right) + \frac{\dot{C}_T}{\dot{W}_E} \qquad (13.16)$$

Using the exergetic efficiency of turbine $\eta_t = \dot{W}_E/(\dot{E}x_{HP} - \dot{E}x_{LP})$, Eq. (13.16) becomes

$$c_E = \frac{c_{HP}}{\eta_t} + \frac{\dot{C}_T}{\dot{W}_E} \qquad (13.17)$$

As $\eta_t < 1$, the specific cost of electricity (product) will be higher than that of high-pressure steam. Example 3.11 illustrates the cost calculations in power production.

Example 13.11 Approximate thermoeconomics of power generation
Consider the exergy costing on a boiler and turbine system shown below:

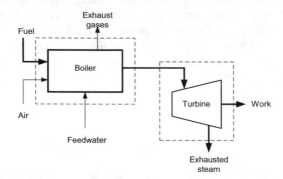

A turbine produces 30 MW of electricity per year. The average cost of the steam is
$0.017/(kWh) of exergy (fuel). The total cost of the unit (fixed capital investment
and operating costs) is $1.1 × 10^5. If the turbine exergetic efficiency increases from
84% to 89%, after an increase of 2% in the total cost of the unit, evaluate the change
of the unit cost of electricity [12].

Solution:
Assume that the heat transfer effects between the turbine and surroundings are
negligible. Also, kinetic, and potential energy effects are disregarded. From
Eq. (13.16), we have

$$c_{\mathrm{E}}(84\%) = \frac{c_{\mathrm{HP}}}{\eta_{\mathrm{t}}} + \frac{\dot{C}_{\mathrm{T}}}{\dot{W}_{\mathrm{E}}} = \frac{0.017}{0.84} + \frac{1.1 \times 10^5}{30 \times 10^6} = \$0.0239/(\mathrm{kW\,h})$$

$$c_{\mathrm{E}}(89\%) = \frac{0.017}{0.89} + \frac{(1.02)1.1 \times 10^5}{30 \times 10^6} = \$0.0228/(\mathrm{kW\,h})$$

The reduction in the unit cost of electricity after the increase in efficiency is about
4.4%. This simple example shows the positive effect of exergetic efficiency on the
unit cost of electricity.

13.7.8 Wind Power-Based Hydrogen Production

Wind power-based electrolysis production cost accounts for geographic factors, the
cost of electricity as well as the capital cost of the electrolyzer systems and their
operating efficiency. Higher efficiencies are possible with polymer electrolyte
membrane (PEM) and solid oxide electrolytic cell (SOEC) electrolyzers. The primary research challenge is to reduce the capital and operating costs of electrolysis
systems. Electrolytic H_2 may be more attractive for regions without access to
natural gas or if H_2 is used as an energy storage medium.

Integration with low-cost renewables and the flexibility to produce H_2 from the grid electricity during off-peak periods may help lower the production cost of H_2. Small systems are often built with PEM electrolyzer cell technology. The gas output streams from the electrolyzer are assumed to be 100% pure (typical real outputs are 99.9 to 99.9998% for H_2 and 99.2 to 99.9993% for O_2). Electricity cost is typically 70–80% of the total cost of H_2 production [7, 15, 29].

13.7.9 Energy Storage

One of the main concerns is the storage of renewable energy for utility-scale power generation, integrating into the grid, and for transport applications. Some important parameters are energy density (kJ/kg or kJ/m^3), storage efficiency, cost, safety, and cycle life. Batteries and fuel cells, composed of electrochemical cells with electrolyte, anode, and cathode, store electrical energy in chemical form and facilitate electrochemical reactions to convert energy to and from electrical form, such as lead-acid battery widely used in transportation sector. A battery directly converts stored chemical energy into electrical energy. If ΔH and ΔG are the enthalpy and Gibbs free energy of the net electrochemical reaction, then the conversion efficiency would be $\eta = \Delta G/\Delta H$ that is the fraction of the chemical energy that can be converted into work in the cell. Energy storage density of batteries is generally below 1 MJ/kg, while lithium-based batteries achieve relatively higher energy density [23, 29].

With the increasing share of renewable power, the estimating total system costs are becoming important for future planning and projections. A range of low-cost flexibility needs systemic approach of enabling technologies, business models, market design and system operation for higher shares of renewable power. However, global investment is necessary in grids and energy storage technology for this. The increasing share of renewable power leads to lower cost of electricity in markets, higher capacity factors for solar PV and wind, and the reduced cost of financing. Cost reductions in solar PV power were supported by crystalline silicon module price declines [19].

13.7.10 Algae-Based Energy Economy

The third-generation biofuels and bioproducts derived from algae continue to be a promising new low carbon economy for several reasons. Algae can be produced on non-arable areas such as lakes, oceans, or deserts, thus reducing competition with food supply chain. High photosynthetic carbon sequestration efficiencies and carbon capture percentages of 90% paired with the ability to harvest and use the totality of the biomass makes algae very well suited as a source for biofuels and bioproducts [3, 24].

Integration of engineering, ecology, and biology is essential since neither of them alone is likely to yield the desired solution. Biology will mostly involve the understanding of the capabilities of strain and/or communities of strain, cultivation with optimized growth and stability, understanding the lipid pathway, maintenance, and carbon capture, as well as the protection of algae culture from contamination. The engineering part will involve in supplying resources of CO_2, light, and energy, extraction and conversion of lipid into fuels and bioproducts, waste and co-product processing with possible recycling, mass and energy balances, cost analysis, and sustainability assessment. Applications of the science of aquatic ecology can play an important role: (i) in optimum nutrients supplied to algal cultivation systems, (ii) to design and construct biotic communities that will help to maximize algal biomass yields and minimize grazing losses, and biomass losses to infectious disease, (iii) to help guide the magnitude and frequency of algal crop harvests, and (iv) to create biologically adaptive algal biomass production systems that are both resistant and resilient to future climate change [3].

In general, the use of photosynthetic organisms as a feedstock will aid in mitigating ever-increasing anthropogenic CO_2 emissions. Microalgae can fix CO_2 10–50 times more efficient than other energy plants. Algae based biofuels fuels would have a positive impact on energy and environment and can improve energy security and reduce GHG emissions from the transportation sector provided: (i) research, development, and demonstration of algal strain selection, (ii) useful energy output that is comparable to other transportation fuels, and (iii) the use of wastewater in recycling nutrients for cultivating algae.

However, despite their excellent potential and the well-studied cultivation options, microalgae are not yet commercially viable feedstock for biofuels and bioproducts. The major technological constraints include the cost-effective harvesting of algae and the extraction of lipids. A national assessment of land requirements for algae cultivation that accounts climatic conditions, fresh water, inland and coastal saline water, wastewater resources, and CO_2 sources would help estimate the potential of algal biofuels that could be produced economically [7, 9].

13.7.11 Optimum Cost of Algae Biomass

Sustainable development of algal biofuels requires the favorable energy efficiency and emissions, water use, supply of nitrogen, phosphorus, CO_2, and appropriate land resources. The quantity of algal biomass (M_{AB}, tons) representing the energy equivalent of a barrel of crude petroleum is [3].

$$M_{AB} = \frac{E_{\text{petroleum}}}{Q(1-w)E_{\text{biogas}} + YwE_{\text{biodiesel}}} \qquad (13.18)$$

where $E_{\text{petroleum}}$ (~ 6100 MJ/brl) is the energy contained in a barrel of petroleum, Q (m^3/ton) is the biogas volume produced by anaerobic digestion (400 m^3/ton), E_{biogas} (MJ/m^3) is the energy content of biogas (~ 2.4 MJ/m^3), Y is the yield of

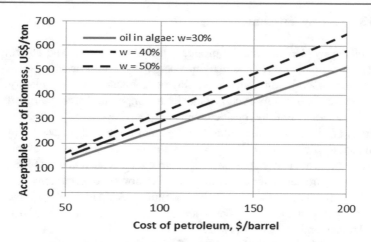

Fig. 13.13 Acceptable cost of biomass with oil percentages of $w = 30\%$, $w = 40\%$, and $w = 50\%$ [3]

biodiesel from algal oil (80% by weight), $E_{\text{biodiesel}}$ is the average energy content of biodiesel (37,800 MJ/ton), and w is the oil content of algae biomass. Assuming that a barrel of crude oil has the same energy of M tons of algae, acceptable cost of algae C_{Algae} becomes

$$C_{\text{Algae}} = \frac{C_{\text{petroleum}}}{M_{AB}} = \frac{C_{\text{petroleum}}}{E_{\text{petroleum}}}\left(Q(1 - w)E_{\text{biogas}} + YwE_{\text{biodiesel}}\right) \quad (13.19)$$

Figure 13.13 shows the acceptable cost of algae biomass with respect to crude oil prices. When the cost of petroleum is \$100/brl biodiesel produced from algae oil costing \$2.61/gal is likely to be competitive with petroleum diesel.

13.7.12 Risk Assessment for Renewable Energy

Risk analysis consists of financial risk, environmental risk, technical risk, and social risk. For a standalone plant, such as heat production in sawmill for drying wood, availability and failure risk will be of minor importance. However, for emerging technologies, some risk of unexpected failure or production breakdown should be taken into consideration. If the bioenergy plant is integrated into an industrial production, for example, then energy supply of a pulp and paper mill and availability of the feedstock is high ($\sim 99.5\%$) and while the risk of unexpected failure is low. Most developers will carry out some form of risk assessment as part of their project activities to identify key constrains of finance, environment, technical, and social [7, 16, 25].

13.7.13 Energy Efficiency

Energy efficiency standard (EES) produces techno economic, and environmental analyses for equipment price and markup, energy use, consumer life cycle cost and payback period, emission impact, employment impact, and regulatory. These analyses create standards that achieve maximum improvement in energy efficiency that are feasible and lead to considerable energy savings.

Thermal efficiencies of residential furnaces and boilers are measured by annual fuel utilization efficiency (AFUE) . AUFE is the ratio of heat output of the furnace or boiler to the total energy consumed by them over a typical year. AFUE does not account for the circulating air and combustion fan power consumptions and the heat losses of the distributing systems of duct or piping. An AFUE of 90% means that 90% of the energy in the fuel becomes heat for the home and the other 10% escapes up the chimney and elsewhere [18, 27].

The Energy Efficiency Ratio (EER) of a cooling device is the ratio of output cooling to input electrical power at a given operating point. The efficiency of air conditioners is often rated by the Seasonal Energy Efficiency Ratio (SEER) . The SEER rating of a unit is the cooling output in Btu during a typical cooling-season divided by the total electric energy input in Watt-hour during the same period. The coefficient of performance (COP) is an instantaneous measure (i.e. a measure of power divided by power), whereas both EER and SEER are averaged over a duration of time [18, 20].

13.7.14 Comparison of Energy-Efficiency Standards

The EES for residential appliances, equipment and lighting have been adopted globally and is contributing considerably in achieving energy conservation. Many countries have mandatory minimum energy efficiency standards (MEES) and labeling programs beside voluntary standards [30]. EES for equipment and appliances are the MEES in most countries.

Energy star programs are used in many countries to help industries, businesses, and consumers to adopt energy-saving products and practices, reduce energy waste and GHG emissions. Energy star program: (1) establishes specifications, testing procedures and verifications requirements for various consumer appliances and products, (2) combines research into residential energy use to promote energy-efficient homes, and (3) develops commercial building energy asset rating program to assess energy usage accurately [19, 27].

13.7.15 Rebound Effect and Energy Efficiency

Increased energy efficiency reduces the cost of energy, so consumer may use it more. An example is the energy-efficient light bulbs that reduce electricity bills, savings are used for other goods and services, while energy use increases slightly.

The direct rebound is the amount of energy savings from increased efficiency that are offset by increased use of energy. This is acknowledged in energy economics and interpreted as allocating the savings to other beneficial use. The indirect rebound results from the way the consumers use their savings; for example, savings can be used in clean energy resources. Overall, the rebound effect may redirect the benefits, but it does not cause the gains of efficiency to be lost [18, 23].

13.7.16 Energy Return on Investment (EROI)

Most of society is aware of the contribution and limitations of type of energy on the technological development and economic activity. The ratio of useful energy to input energy required is called the energy return on investment (EROI) that helps assess the energy cost on society and its impact on development. Most renewable energy resources have the lower values of EROI compared with conventional fossil fuels. This means that more output energy must be available to society with respect to energy needed to produce and hence funds can be redirected toward societal needs, such as health and educational expenses [16, 17].

Some examples of EROI are: EROI (sugarcane ethanol) ~ 8 and EROI (corn ethanol) ~ 1.2 to 1.6 (excluding the energy content of dried distillers grains with solubles) that shows a better overall economy for bioethanol production from sugarcane [35].

$$EROI_{Society} = \frac{Energy\ returned\ to\ society}{Energy\ invested\ to\ get\ this\ energy} = \frac{ER}{EI} \qquad (13.20)$$

In the case of $EROI_{Society}$, the numerator (ER) is composed of a nation's Gross Domestic Product multiplied by the MJ per unit of energy used in the generation of that GDP. The denominator (EI), the energy invested in order to produce the energy output, is composed of the total energy consumed by that nation in a given year (in MJ) multiplied by dollars per unit spent in the acquisition of that fuel.

The estimated minimum EROI of 3 is required for an energy source to be beneficial to society [16, 17]. There is a strong correlation between EROI and societal well-being [16, 17]. Indicators of quality of life may be percent children under weight, health expenditures, gender inequality index, literacy rate, and access to improved water. As the value of EROI increases, so does the quality of life. According to this trend, a minimum EROI of 14 might be necessary to maintain the qualities of an advanced society. The values of EROI are around 0.8–1.7 for bioethanol, and 1.3 for biodiesel [16].

The energy from the traditionally high EROI deposits may need to be supplemented or rapidly replaced by alternative energy sources to avoid future energy constraints and the potential effects of climate change. These 'new' energy sources must be sufficiently abundant and have a large enough EROI value to power society [16]. If the EROI values of traditional fossil fuel energy sources (e.g. oil) continue to decline and renewable energy resources fail to provide enough high EROI alternatives, then we may be moving toward the 'net energy cliff.' The human development index (HDI) is a commonly used composite index of well-being and is calculated using four measures of societal well-being: life expectancy at birth, adult literacy, combined educational enrollment, and per capita GDP [18].

Table 13.7 shows the energy ratio that is the biofuel energy output to fossil energy inputs in the production process with different feedstock. The energy ratios for all the biodiesels from various feedstocks are higher than 1, suggesting that biofuels are renewable energy with positive net energy outputs and reductions in GHG emissions (Table 13.7).

13.7.17 Circular Economy

The linear economy is characterized by 'take, make, and dispose,' while the circular economy closes the loop by collecting and recycling waste, and using it to manufacture new products and protect natural resources. Circular economy describes strategies for waste prevention and resource efficiency toward regenerative manufacturing model. This is achieved through design, maintenance, repair, and recycling (Fig. 13.14). In this bottom-up approach, the shareholders and investors are

Table 13.7 Energy ratio for bioethanol and biodiesel production processes from various feedstocks and change in life cycle GHG emissions per kilometer traveled by replacing diesel with 100% biodiesel fuel; the cost ratio (price of biodiesel/price of fossil fuel) for biodiesel is around 1.2 [16–18, 35]

Feedstock	Energy ratio[a]	EROI	GHG emissions change[b] (%)
Sugarcane	≈8	0.8–10	−87 to − 96
Sugar beets	≈2		−35 to − 56
Sweet sorghum	≈1		
Corn	≈1.5	0.84–1.65	−21 to − 38
Wheat	≈2		−19 to − 47
Lignocellulosic	≈2–36	0.69–6.61	−37 to − 82
Gasoline	≈0.8		
Rapeseed	≈2.5	1.0–1.5	−21 to −51
Soybeans	≈3	0.7–2.0	−63 to −78
Sunflower	≈3	0.4–1.2	
Castor	≈2.5		
Palm oil	≈2.5–9		
Jatropha	≈1.4		
Waste vegetable oil	≈5–6		−92
Diesel (crude oil)	0.8–0.9		

[a]Energy from biofuel/fossil energy used in production of biofuel
[b]Approximate avoided GHG emissions because of the biomass feedstock used in bioethanol production

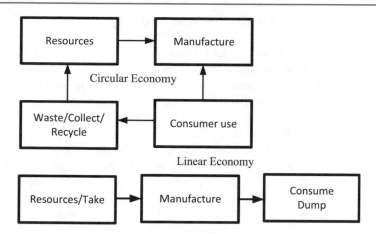

Fig. 13.14 Circular economy versus linear economy

raising concerns on about natural resource depletion, climate change, sustainability, energy consumption, water consumption, and regulatory compliance. Companies see that tackling these concerns is the future success of their business. Pressure to reform business practices is from many sectors including energy sector such as oil and gas and others [13, 40]. Some focused sustainability and circular initiatives with new models enable companies to:

- Reduce nonrenewable material/energy and water use such as replacing fossil fuels with renewable energy and wastewater treatment
- Reduce emissions such as CO_2 and NO_x and emphasize decarbonization with feasibility assessment
- Improve process thermal efficiency with processes intensification and product value chain integration
- Develop innovative product and process solutions including recycling/reusing
- Safe and reliable operation with predictive maintenance solutions.

Sustainability-related objectives are expanding and becoming more specific process metrics for emissions, resource use, waste reduction, and exploring new energy sources with lower carbon footprints. The circular economy forces companies to rethink how to design and manufacture products, as well as to interact with customers. Here, the customer's role is no longer just consume, but the use of a function. This demands the business community to build long-term relationships with the consumers by retaining the value and integrity of the product.

13.7.18   Circular Economy and Sustainability

The circular economy is viewed as a condition for sustainability. Both notions emphasize regeneration commitments, shared responsibilities motivated by environmental hazards and problems on a global scale. Both the concepts integrate noneconomic aspects into development through system design and innovations, and cooperation between stakeholders. Multiple and coexisting pathways of development are essential for both the concepts.

However, there are different goals associated with the circular economy and sustainability. Sustainability focuses on benefiting the environment, the economy, and society at large, while the circular economy emphasizes the efficient use of resources with less waste and emission benefiting the economic actors that implement the system. Society benefits also from environmental improvements and more economic activity. In the sustainability debate, responsibilities are not clearly defined but loosely shared with the focus on interest alignment between stakeholders, while private businesses, regulators, and policymakers are responsible for the transition to a circular economy [9, 10, 13].

13.8   Bioeconomy

Bioeconomy is based on all economic activity originated from invention, development, production, and use of biological products and processes in the production of food, energy, and materials [25, 41]. Bioeconomy respond to the diminishing fossil-based resources, climate change, growing world population, resource-efficient fuels, and processes for the well-being of societies.

Production of biofuels and bioproducts depends on the availability of biomass feedstock and biotechnology. The costs of different types of biomass for producers as well as the converting processes are key for the bioeconomy. The intersecting of newly emerging bioenergy market with established markets in agriculture, forestry, water, and energy is causing substantial impacts on the prices of agricultural commodities, food, feedstuffs, forest products, fossil fuel energy, and land values [7, 41].

Low-cost biomass feedstocks available as byproducts from agricultural or forestry can provide competitive electricity [19, 20]. Large-scale bioenergy generation plants (>50 MW) are unattractive compared to fossil fuel plants because of the logistical costs of transporting low-density feedstock from far production sites. In addition, large-scale productions of biofuels may have the adverse effect of land available for food crops to be allocated for energy crops. Biofuels are promoted worldwide by tax credits and subsidies to reduce petroleum imports and GHG emissions from vehicles [22, 34, 41].

Energy efficiency of a conversion process for producing biofuels from a biomass affects the bioeconomy. Energy efficiency for biomass to biofuel can be estimated by

$$\begin{pmatrix} \text{Energy efficiency} \\ \text{of biomass to biofuel} \\ \text{conversion} \end{pmatrix} = \frac{\left\{ \begin{pmatrix} \text{LHV of biofuel} \\ \text{produced per 1 kg} \\ \text{of biomass} \end{pmatrix} - \begin{pmatrix} \text{External energy} \\ \text{used in 1 kg biomass} \\ \text{to biofuel conversion} \end{pmatrix} \right\}}{\begin{pmatrix} \text{LHV of the 1 kg biomass} \\ \text{used in the conversion} \end{pmatrix}}$$

$$(13.21)$$

Here the LHV is the lower heating value. As the biofuel technology becomes more energy efficient and advanced with technological improvements, the production cost will be reduced by around 50% by 2030 [7, 34, 38]. In the transforming energy scenario, the share of renewables would increase 28% by 2030 and 66% by 2050 due to progress in biotechnology [21].

13.8.1 Bioeconomy and Circular Economy

Sustainability-related objectives are expanding and becoming more specific process metrics for emissions, resource use, waste, and discharge reduction from production units as well as exploring new energy sources with lower carbon footprints. Therefore, the benefits of the circular economy overlap with the elements of sustainability and align with the sustainable development policy. The inclusion of bioenergy into the circular economy needs the combination of interests and concerns of policymakers, researchers, technology developers, project developers, and society. In addition, new and innovative technologies are necessary to produce and consume without depleting natural resources, which will not be possible without society rethinking ecological balances in production–consumption chain [32, 33, 37]. This shows that bioenergy and circular economy are closely linked with multiple elements overlapping, in which mature circular bioeconomy can merge urban and rural communities, as well as cross-cutting policies, such as environmental protection, waste management, eradication of poverty, and manufacturing strategies [13, 35, 41].

13.8.2 Agricultural Economic Models

Three of the models FAPRI, FASOM, and POLYSYS are partial-equilibrium (PE) models, which means that not all sectors of the economy are included in the model [30]. These PE models mainly focus on the agricultural sector. General-equilibrium (GE) models, such as GTAP, cover all sectors of the economy and all regions of the world. Thus, GE models capture the interactions among sectors and between product and factor markets. However, GE models, especially global ones like GTAP, cannot model the interactions among detailed sectors of agriculture or regions as PE models can. PE are appropriate for responding to policy questions regarding the sector(s) of interest regarding cropping practices, land quality and use, and regional variations, and they can permit more in-depth analysis of sector-specific policies. However, the models implicitly assume that the agricultural sector can be analyzed without worrying explicitly about what happens in the rest of the economy. GTAP is more heavily focused on the trade dimension.

FASOM is a forward-looking model and solves all years simultaneously, whereas FAPRI and POLYSYS are recursive dynamic models. GTAP can be run either as a comparative static or dynamic model. Comparative static models compute the market equilibrium under one set of conditions; when conditions change, the models compute the new equilibrium without worrying about the path from one equilibrium to the other. The three PE models originally had a heavy focus on agricultural policy, although FASOM was designed to examine competition between forestry and agricultural sectors for land from its early stages. GTAP originally was developed to evaluate the effects of alternative trade policies in international trade negotiations and regional and bilateral trade agreements [30].

13.9   Hydrogen Economy

Hydrogen economy implies producing and using hydrogen as a clean fuel. Because of high energy density (kJ/kg), hydrogen is attractive for a large amount of energy storage. However, there are safety issues in hydrogen storage, which is a major challenge for commercial applications. Hydrogen burns with an invisible flame and can attract difficulty in mitigating hydrogen fires. It becomes extremely difficult to control hydrogen fires because of the fuel's higher flame temperature and speed.

The contribution of hydrogen toward clean energy depends on: (1) replacing conventional hydrogen production with low/carbon-free production with carbon capture for blue hydrogen production, (2) improved/new hydrogen storage technology, and (3) cost-efficient technologies for electrolysis and fuel cells [15, 18]. The cost of electrolytic hydrogen depends on the cost of electricity as well as the capital cost of the electrolyzer systems and their operating efficiency. The capital cost of the electrolyzer increases considerably as the wind farm availability and electrolyzer capacity decrease. The unit cost estimates of wind power-based electrolytic H_2 are limited geographically. Other factors such as large-scale storage, compression, and transport need separate analyses.

Hydrogen production is mainly based on natural gas, coal steam reforming, and water electrolysis. Large-scale processes, using natural gas and coal, are the most economical processes while biomass gasification still needs technological improvements before becoming competitive.

The average levelized cost of green hydrogen (LCOH) would be around US $6.8/kg from solar PV, US$4.2/kg from wind in 2020, US$3.2/kg from solar PV, US$2.8/kg from wind in 2030, and US$2/kg from solar PV and US$1.2/kg from wind in 2050 [20, 21]. This shows that the cost of green hydrogen from wind would remain lower than the cost from solar PV. The average levelized cost of blue

Table 13.8 Hydrogen and electrolyzer cost projections with transforming energy scenario (TES) [20, 21]

	Historical progress	TES (2030)	TES (2050)
Green hydrogen production cost (US$/kg):	4.0–8.0 (2015–2018)	1.8–3.2	0.9–2.0
Electrolyzer costs (US$/kW)	770	540	370

hydrogen from fossil fuels with carbon capture and storage would remain in the range of $2 to $3 between 2020 and 2050 [20]. For producing one kg H_2, approximately 26.7 kg water is necessary. The total greenhouse gas emission is around 0.97 kg CO_2e/kg H_2. The hydrogen production cost is highly dependent on the electricity price, which may be around 75% of the final cost. Table 13.8 shows the historical progress and transforming energy scenario (TES) on green hydrogen production and electrolyzer costs.

Figure 13.15 shows the schematic of wind energy-based green hydrogen production. Alkaline electrolysis technologies are the most mature commercial systems. The system includes the transformer, thyristor, electrolyzer unit, feed water demineralizer, hydrogen scrubber, gas holder, two compressor units to 30 bar, deoxidizer, and twin tower dryer. These electrolyzers have the energy efficiencies of 57–75% [15, 29].

13.9.1 Hydrogen Storage

At 1 atm and 300 K hydrogen gas has a density of 8.2×10^{-5} kg/L, an approximate volumetric energy density of 12 kJ/m^3, which is very low, and gravimetric energy density of 140 MJ/kg. This makes the storage very challenging as one needs either very large volume or compressions under very high pressure (up to 700 atm

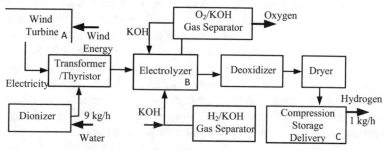

Production 72%, CSD 28% of total cost
Electrolyzer efficiency: ~62%; target: 76% (LHV)
Target cost: $0.3/kg H_2 = gasoline of $2.5/GJ; Cost: $3.74-5.86/kg H_2
0.97 kg CO_{2-eq}/kg H_2: A: 78%; B: 4.4%; C; 17.6%

Fig. 13.15 Schematic for alkaline electrolysis of water for hydrogen production with compression, storage, and delivery [29]

and 5.6 MJ/L) to store hydrogen. If hydrogen is liquefied at a temperature below 20 K, it reaches to the volumetric energy density of approximately 10 MJ/L. Besides, a significant amount of energy is necessary to liquefy hydrogen. Metal hydrides are considered to store hydrogen effectively at ambient conditions [23].

Fuel cells, different from batteries are fed continuously with flow of consumables. Hydrogen and oxygen react to produce water in hydrogen fuel cells: $2H_2 + O_2 = 2H_2O$. The standard enthalpy of combustion of hydrogen is $-\Delta H° = 285.8$ kJ/mol and $-\Delta G = 237.1$ kJ/mol at 1 atmosphere and 298 K. Limit of efficiency is around $\eta = \Delta G/\Delta H = 237.1/285.8 = 0.83$. One considerable challenge to the wide use of hydrogen fuel cells is the storage problems of hydrogen. Most current fuel cells use hydrogen and oxygen electrodes made of platinum, also acts as a catalyst for a high rate of reaction. This increases the cost of cells significantly [15, 23].

13.10   Methanol Economy

An affordable and viable alternative to hydrogen can be methanol, which has more hydrogen by mass in the same volume. Therefore, methanol may be a safe carrier fuel for hydrogen. The reported octane number of methanol is 108.7 as compared to that of gasoline, which is 91. This suggests that the methanol–air mixture can be compressed more before it is ignited. This boosts up the efficiency of the engine and ensures cleaner emissions. Unlike gasoline and diesel, methanol does not emit any soot or smoke on combustion. Also, methanol can acquire the same infrastructure as used for gas and diesel and thus can relieve infrastructure and dispensing costs [14, 15].

Methanol economy may solve many issues of climate change due to the followings:

(i) methanol is produced from synthesis gas (syngas) obtained from coal and natural gas, (ii) methanol can also be produced by reductive hydrogenation by utilizing CO_2, (iii) methanol has a higher 'flame speed' which enables faster and more complete fuel combustion in the cylinders, (iv) methanol has a higher-octane number than gasoline, which increases the efficiency of the internal combustion engines.

Methanol can be used in the direct methanol fuel cell. Renewable hydrogen-based methanol would recycle carbon dioxide as a possible alternative fuel to diminishing oil and gas resources [29, 40]. It is also used as a chemical feedstock to ultimately fix the carbon. There are already vehicles, which can run with M85, a fuel mixture of 85% methanol and 15% gasoline. Methanol can be used with the existing distribution infrastructure of conventional liquid transportation fuels.

13.10.1   Methanol and Environment

Methanol is a polar liquid and can rapidly degrade both in aerobic and anaerobic conditions, unlike crude oil. Hydrogen burns with an invisible flame and can attract

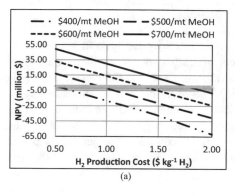
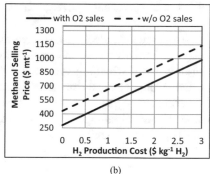

Fig. 13.16 The influence of H_2 production cost on: (**a**) net present value at constant methanol (MeOH) price, (**b**) Selling price of methanol for NPV = 0 with and without selling O_2 byproduct at $100/mt [29]

difficulty in mitigating hydrogen fires. It becomes extremely difficult to control hydrogen fires because of the fuel's higher flame temperature and speed as compared to methanol.

Methanol has some added disadvantages: (1) Methanol is severely harmful if consumed. Higher doses can lead to death. (2) In methanol production, infrastructural modifications are required to meet the diverse energy needs and it will also require time and research before methanol economy is put into practice but these efforts are worthy and viable in order to shift to an alternate fuel.

Methanol can be used as a fuel for blending or producing other fuels. Dimethyl ether with methanol can produce various liquid fuels, such as gasoline, kerosene, and gasoil. It may be produced both from fossil fuels (natural gas and coal) and renewables [28, 29, 34, 40]. Fig. 13.16 shows the influence of hydrogen cost on methanol economics. Net present value of a methanol plant decreases, while the selling price of methanol increases as the hydrogen cost increases [29].

13.10.2 Methanol Economy Versus Hydrogen Economy

Hydrogen energy is clean energy as it produces water only when oxidized. However, its production, storage, and transportations create unique and difficult problems including safety issues. High energy costs are needed for producing green and blue hydrogens, besides the considerable energy needed for their storage and transportation. Conventional liquid fuels have outweighed the available hydrogen/hydrogen storage methods mostly due to low volumetric energy density. Methanol can be produced in a cost-effective mature technology using abundant resource of natural gas or coal [29, 40] that will curb the emissions. Methanol can serve as a safer-energy carrier, a relatively less polluting fuel, and feedstock for other chemical productions [14].

13.11   Electricity Storage in Chemicals

The economics of electricity storage are influenced by the type of storage technology, electricity price, the requirements of each application, the frequency of charging and discharging cycles, and the system in which the storage facility is located. Main chemical storage of electricity involves the production of hydrogen, synthetic natural gas, and chemicals including methanol and ammonia. A combination of several storage applications together may help electricity storage to be more feasible and flexible. The initial investment requires a cost per unit of power and a cost per unit of energy capacity, which are technology-dependent [28, 29].

13.12   Ecological Cost

Ecological cost analysis may help minimize the excessive depletion of natural resources. Determining the extraction of raw materials and fossil fuels from natural resources is not sufficient in fully understanding the ecological impact of production processes. The production, conversion, and utilization of energy may lead to environmental problems, such as air and water pollution, impact on the use of land and rivers, thermal pollution due to mismanagement of waste heat, and global climate change. The influence of waste products discharged into the environment should also be considered. The waste products may be harmful to agriculture, plant life, human health, and industrial activity.

Several options exist for reducing CO_2 emissions, such as nonrenewable energy conservation, energy efficiency improvement, increasing reliance on nuclear and renewable energy, and carbon capture and storage (CCS) systems [4]. CCS is a process in which CO_2 is separated from effluent streams and injected into geologic formations, avoiding its release into the atmosphere. Some significant challenges to CSS are the cost of building and operating capture-ready industrial facilities, the feasibility of permanently storing CO_2 underground, and the difficulty of constructing infrastructure to transport CO_2 to injection sites [5].

13.12.1   Resource Depletion

We may measure the resource depletion by the *Depletion number Dp*, which is a nondimensional indicator Ex_{Dp} per unit consumption Ex_C.

$$Dp = \frac{\dot{Ex}_{Dp}}{\dot{Ex}_C} = 1 + \frac{\dot{Ex}_{Dsl}}{\dot{Ex}_C} + \psi \left[\frac{(1 - \Omega_{RU})}{\eta_{RU}} - 1 \right] + \frac{\dot{Ex}_{TV}}{\dot{Ex}_C} \left[\frac{(1 - \Omega_{VU})}{\eta_{VU}} - 1 \right] \quad (13.22)$$

where $\dot{Ex}_{Dsl}$ is the exergy dissipation rate, Ω_{RU} and η_{RU} are the renewable exergy fraction and transfer efficiency for the recovered resource upgrade process, respectively, $\dot{Ex}_{TV}$ is the exergy transfer rate to the nonrenewable source, and Ω_{VU} and η_{VU} are the renewed exergy fraction and transfer efficiency, respectively, for the nonrenewable resource upgrade process [12].

Depletion number provides a measure of system progress or maturity and is a function of the following three indicators:

- The *exergy cycling fraction* ψ is a measure of recycling that accounts for both the throughput and quality change aspects of resource consumption and upgrading.
- The *exergy efficiency*
- The *renewable exergy fraction* Ω. Boundary conditions determine which resources and processes constitute an industrial system.
- Spatial boundary conditions are mainly geographical and resource-specific, while temporal boundary conditions define the scope of time for the exergy transfer and loss in processes.

13.12.2 Cost of Pollution Control

After understanding the allowable pollution limits, the next step is to create a list of suitable technologies that abate pollution and comply with national regulations. To do that, the most common equipment, and systems for mitigating polluted air, waste, and wastewater by industry and pollutant must be identified [2, 18]:

- Annual waste disposal and wastewater treatment costs are significantly larger than new air pollution and greenhouse gas abatement costs.
- Air emissions regulations vary widely by country and region.
- The petroleum refining industry together with the food industry accounts for major greenhouse gas abatement costs and are responsible for waste disposal and wastewater treatment costs.

Pollution abatement cost analysis includes capital costs, direct and indirect installation costs, and direct and indirect operating and maintenance costs. For greenhouse gas emissions, each country's legislative requirements defined emissions reduction goals [4, 5]. For most of the countries, reduction targets are broken down into product- or energy-specific benchmarks. Actual average emissions per ton of product produced compared with the required benchmark represented the required reduction needed per ton of product. Annual waste disposal and wastewater treatment costs are significantly larger than new air pollution and greenhouse gas abatement costs because waste is properly disposed of and treated each year while the latter involves the marginal addition to existing equipment.

The production, conversion, and utilization of energy may lead to *ecological cost* including air and water pollution, impact on the use of land and rivers, thermal pollution due to mismanagement of waste heat, and global climate change. Some of the major disturbances are:

- Chaos due to the destruction of order is a form of environmental damage.
- Resource degradation leads to exergy loss.
- Uncontrollable waste exergy emission can cause a change in the environment.

13.12.3   Index of Ecological Cost

The exergy destruction number N_{Ex} is the ratio of the nondimensional exergy destruction number of the improved system to that of the conventional one

$$N_{\mathrm{Ex}} = \frac{\mathrm{Ex}_i^*}{\mathrm{Ex}_c^*} \tag{13.23}$$

where subscripts i and c denote the improved and conventional cases, respectively and Ex* is the nondimensional exergy destruction number, which is defined by

$$\mathrm{Ex}^* = \frac{\mathrm{ex}_{\mathrm{fd}}}{\dot{m} T_o C_p} \tag{13.24}$$

Here $\mathrm{ex}_{\mathrm{fd}}$ is the flow-exergy destruction and T_o is the reference temperature. The system will be more sustainable only if the N_{Ex} is less than unity.

The exhaustion of nonrenewable natural resources is called the *index of ecological cost*. To determine the domestic ecological cost c_{eco}, the impact of imported materials and fuels is considered. The degree of the negative impact of the process on natural resources can be characterized by means of the *ecological efficiency* η_e.

$$\eta_e = \frac{\mathrm{Ex}_c}{c_{\mathrm{eco}}} \tag{13.25}$$

Ex_c is the gross consumption of the domestic nonrenewable natural resource. Usually, $\eta_e < 1$, but sometimes values of $\eta_e > 1$ can appear if the restorable natural resources are used for the process [12].

13.13   Sustainability in Energy Systems

Energy management can impact Earth's climate and environment, hence the ways of human's life as well as the next generation's future. The use of emerging renewable energy technologies (including solar, wind) and alternative ways of using traditional fossil and nuclear fuels are growing but constrained by various factors including cost, infrastructure, public acceptance, and others [7]. To fully implement sustainable engineering solutions, engineers use the following principles in energy systems:

- Integrate environmental impact assessment tools.
- Conserve and improve natural ecosystems while protecting human health and well-being.
- Ensure that all material and energy inputs and outputs are as inherently safe and benign as possible.
- Minimize the depletion of natural resources and waste.
- Develop and apply engineering solutions in line with local geography, aspirations, and cultures.
- Actively engage with communities and stakeholders.
- Use material and energy inputs that are renewable.

Increasingly resource-intensive consumption across the world requires novel energy technologies with the economics, environmental, and societal context. Within the life cycle, inputs and output flows are either intermediate flows connecting processes or elementary flows between processes and their natural environments. The assessments of input and output flows are called life cycle inventories (LCIs) [28].

Sustainability is increasingly incorporated into energy policy and the strategies of companies. It is also within the context of ecology as a principle of Earth's natural balances. Environmental impact (I) is a function of three factors: population (P); affluence representing consumption (A); and technologies (T): $I = P \times A \times T$. As a concept, sustainable development implies limitations by technology, social perceptions on environmental resources, and by Earth's ability to absorb the effects of human activities [13].

The collective environmental regulations and technical advances, such as pollution control, waste minimization, and pollution prevention, have greatly diminished adverse environmental impacts of energy production processes. From a sustainability viewpoint, the most important factors that determine the suitability of processes in energy technology are [13, 18, 20]:

- Energy use per unit of economic value-added product
- Type of energy used (renewable or non-renewable)
- Materials use (or resource depletion)
- Fresh water use
- Waste and pollutants production
- Environmental impacts of product/process/service
- Assessment of overall risk to human health and the environment.

The risks to human health and the environment from probable exposures to a product, or emissions from a process constitute both the environmental and social aspects of the sustainability. Tools for hazard characterization of chemicals, exposure assessment models, health effect models, and risk assessment models need to be incorporated into sustainable technologies [18, 19]. Sustainability innovation considers commitment to the development of products and processes with superior environmental, social, and economic performances. Renewable energy sources have a completely different set of environmental costs stemming from capturing relatively low-intensity energy and hence the large area taken up by them.

Example 13.12 Cost of solar heat storage

A typical square two-story home with a roof surface area of 1260 ft^2 and a wall surface area of 2400 ft^2 is to be heated with solar energy storage using a salt hydrate as phase change material. It presently has as an insulation of 6 inches in the roof and 1 inch in the walls. Inside temperature will be held at 70 °F and expected outside low temperature is 10 °F. Average solar radiation is 650 Btu/ft^2 and the cost of solar air collector is $1.1/ft^2. The salt hydrate costs around $0.15/lb. Estimate the costs of salt hydrate and solar air collectors.

Solution:

Assume that the approximate thermal conductivity of the walls and roof is: $k = 0.025$ Btu/h °F ft^2.

Average solar radiation is 650 Btu/ft^2 and the cost of solar air collector is $1.1/ft^2. The salt hydrate costs around $0.15/lb and its heat of melting is 145 Btu/lb.

Heating requirement of building with present insulation:

Heat loss from roof ($\Delta x = 6/12 = 0.5$ ft); $\Delta T = (10{-}70) = -60$ °C

$$q_{loss, roof} = -kA \frac{\Delta T}{\Delta x} = -(0.025 \, \text{Btu/h} \, °F \, ft^2)(1260 \, ft^2)$$

$(-60 \, °F/0.5 \, ft) = 3780 \, \text{Btu/h}$

Heat loss from walls ($\Delta x = 1$ in $= 0.083$ ft)

$$q_{loss, wall} = -kA \frac{\Delta T}{\Delta x} = -(0.025 \, \text{Btu/h} \, °F \, ft^2)$$

$(2400 \, ft^2)(-60 \, °F/0.083 \, ft) = 43,373 \, \text{Btu/h}$

Total heat loss = Total heating requirement $\rightarrow q_{total \, loss}$

= 3780 Btu/h + 43,373 Btu/h = 47,153 Btu/h

Total heat storage material needed:

m_s = (47,153 Btu/h) (24 h)/145 Btu/lb = 7804.0 lb

Cost of salt hydrate = (7804.0 lb) ($0.15/lb) = **$1170.6**

Size and cost of collector to meet present heating requirements:

Area of collectors = (47,153 Btu/h) (24 h)/(650 Btu/ft^2) = 1740 ft^2

Cost of collectors: (1740 ft^2)($1.1/ft^2) = **$1914.0**

Total cost of solar energy capture and its storage:

Cost total = $1170.6 + $1914.0 = **$3084.6**

Annual cost of heating may be optimized by compromising between the two opposing effects of capital costs of thermal energy storage system and fuel as shown below.

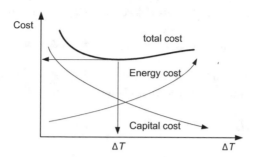

As the temperature difference between hot and cold space increases, the capital cost may decrease while the energy (fuel) cost increases. The total cost should be optimized for a required value for ΔT.

13.14 Process Intensification and Energy Systems

The power generation sector has traditionally involved the consumption of non-renewable resources, such as coal, natural gas, petroleum oil and naturally causing adverse environmental and ecological impacts. As a part of sustainability efforts, this tradition is being replaced worldwide by using renewable resources and clean technology with decarbonization in power generation. Sequestering the emitted CO_2 for long-term storage in deep underground, and conversion of CO_2 into value-added products may help mitigate climate change problems. In addition, suitable process intensification techniques based on equipment, material and process development strategies can play a key role at enabling the deployment of clean energy processes [1, 2, 32, 33].

13.14.1 Efficiency Optimization of Solar Concentrators

Solar concentrators can heat the working fluid to high temperatures that in turn increases the Carnot efficiency [23]. However, the absorber radiates a higher fraction of the incident energy at higher temperatures. Therefore, the temperature of the absorbing material should be optimized. If we consider a solar concentrator with concentration C, heating the working fluid to temperature T, and the absorber radiates as a black body at the same temperature T, Carnot efficiency becomes

$$\eta_C = \frac{T - T_o}{T} \tag{13.26}$$

where T_o is the ambient temperature. With the incident per unit area of I_o the power hitting the absorber is $P = C\,I_o\,A$, where A is the surface area of the absorber, while the power radiated by the absorber is

$P_{rad} = \sigma T^4 A$. Therefore, the available power $P_{available}$ becomes,

$$P_{available} = P_{incoming} - P_{radiated} = A(CI_o - sT^4)A \qquad (13.27)$$

where σ is the Stefan-Boltzmann constant, $\sigma = 5.67 \times 10^{-8}\,\text{W/m}^2\text{K}^4$. The output electric power per unit absorber area at Carnot efficiency is

$$P_{out}(T) = A(CI_o - sT^4)(1 - T_o/T) = A\left(CI_o - sT - T^3 - sT^4 - CI_oT_o/T\right) \qquad (13.28)$$

The overall conversion efficiency is

$$\eta(T) = \frac{P_{out}}{CI_oA} \qquad (13.29)$$

For a given rate of incoming and concentration, the theoretical optimum power becomes

$$\frac{1}{A}\frac{dP_{out}}{dT} = 3\sigma T_o T^2 - 4\sigma_o T^3 + CI_oT_o/T^2 = 0$$

or (13.30)

$$3\sigma T_o T^4 - 4\sigma_o T^5 + CI_oT_o = 0$$

If we consider a solar thermal electric plant using two-dimensional concentrators with $C = 200$ and molten salt as heat transfer medium, optimal efficiency for an ambient temperature of 300 K and solar insolation of 1000 W/m^2 can be estimated by solving numerically Eq. (13.30). The value of temperature 818 K obtained from Eq. (13.30) leads to an approximate Carnot efficiency of 63%. Maximum possible efficiency of 55% is obtained after about 13% of incident is reradiated [23].

Example 13.13 Cost saving from heat pump
A heat pump provides 60 MJ/h to a house. If the compressor requires an electrical energy input of 5 kW, calculate the COP. If electricity costs \$0.08 per kWh and the heat pump operates 100 h per month, how much money does the homeowner save by using the heat pump instead of an electrical resistance heater?

Solution:
The heat pump operates at steady state.
COP for a heat pump with a heat supply of 60 MJ/h $= 16.66\,\text{kW}$: and $W_{HP} = 5\,\text{kW}$
$$COP_{HP} = \frac{q_{out}}{W_{HP}} = 3.33$$

An electrical resistance heater converts all the electrical work supplied W_e into heat q_H. Therefore, in order to get 16.66 kW into your home, you must buy 16.66 kW of electrical power.

Cost of resistance heater:

Power = 16.66 kW

Cost ($/month) = Electricity ($0.08/kWh) Power (16.66kW)

Time $(100h/month)$ = 133.3 $/month

Cost of heat pump with a power of 5 kW:

Cost ($/month) = Electricity ($0.08/kWh) Power (5.0 kW)

Time $(100h/month)$ = 40.0 $/month

Therefore, monthly saving is (133.3 − 40.0) $/month = **93.3 $/month**.

Electrical resistance heaters are not very popular, especially in cold climates. The thermal efficiency of a heat pump drops significantly as the outside temperature falls. When the outside temperature drops far enough that the $COP_{HP} \sim 1$, it becomes more practical to use the resistance heater.

Example 13.14 Increasing the efficiency of a Rankine cycle by reducing the condenser pressure

A steam power plant is operating on the simple ideal Rankine cycle. The steam mass flow rate is 20 kg/s. The steam enters the turbine at 3500 kPa and 400 °C. Discharge pressure of the steam from the turbine is 78.5 kPa.

(a) Determine the thermal efficiency of the cycle.
(b) If the pressure of the discharge steam is reduced to 15 kPa determine the thermal efficiency.
(c) Determine the annual cost saving if the unit cost of electricity is $0.10/kWh and avoided emission from a coal-fired plant.

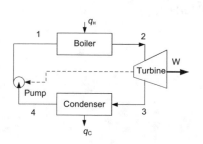
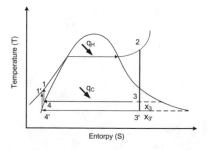

Solution:
Assume that the cycle is at steady-state flow and the changes in kinetic and potential energy are negligible. Efficiency of pump and turbine is 100%.

(a) $\dot{m}_s = 20.0$ kg/s.

Using the data from the steam tables
Superheated steam: $P_2 = P_1 = 3500$ kPa, $H_2 = 3224.2$ kJ/kg, $S_2 = 6.8443$ kJ/kg K,
$T_2 = 400\,°C$
Saturated steam: $P_3 = P_4 = 78.5$ kPa$(T_{sat} = 366.15$ K), $V_4 = 0.001038$ m^3/kg
$H_{3sat\,vap} = 2665.0$ kJ/kg, $H_4 = H_{3sat\,liq} = 389.6$ kJ/kg,
$S_{3sat\,vap} = 7.4416$ kJ/kg K, $S_{3sat\,liq} = 1.2271$ kJ/kg K
Basis: 1 kg/s steam. With a pump efficiency of $\eta_{pump} = 100\%$

$$W_{p,in} = V_1(P_1 - P_4) = (0.001038)(3500 - 78.5)\left(\frac{1\,kJ}{1\,kPa\,m^3}\right) = 3.55\,kJ/kg$$

$H_1 = H_4 + W_{p,in} = 393.1$ kJ/kg
Isentropic process $S_1 = S_4$ and $S_3 = S_2$.
The quality of the discharged wet steam $(S_2 < S_{3sat\,vap})$: $6.8443 < 7.4416$
$x_3 = (6.8463 - 1.2271)/(7.4416 - 1.2271) = 0.90$
$H_3 = 389.6(1 - 0.9) + 2665.0 \times 0.9 = 2437.5$ kJ/kg
Heat interactions: $q_{in} = H_2 - H_1 = 3224.2 - 393.1 = 2831.1$ kJ/kg
$q_{out} = -(H_4 - H_3) = -(389.6 - 2437.5) = 2048.0$ kJ/kg (heat received by the cooling medium that is the system in the condenser)

The thermodynamic efficiency of the cycle: $\eta_{th} = 1 - \dfrac{|q_{out}|}{q_{in}} = \mathbf{0.276}$ **or 27.6%**

Therefore, the plant uses only 27.6% of the heat it received in the boiler.
Turbine work output: $W_{out} = H_3 - H_2 = 3224.2 - 2437.5 = -786.7$ kJ/kg

(b)   Steam properties: $P_3 = P_4 = 15$ kPa, $T_{sat} = 327.15$ K, $V_4 = 0.001014$ m^3/kg.

$H_{3sat\,vap} = 2599.2$ kJ/kg, $H_4 = H_{3sat\,liq} = 226.0$ kJ/kg,

$S_{3sat\,vap} = 8.0093$ kJ/kg K, $S_{3sat\,liq} = 0.7550$ kJ/kg K

With a pump efficiency of $\eta_{pump} = 100\%$

$$W_{p,\text{in}} = V_1(P_1 - P_4) = (0.001014)(3500 - 15)\left(\frac{1\,\text{kJ}}{1\,\text{kPa m}^3}\right) = 3.53\,\text{kJ/kg}$$

$H_1 = H_4 + W_{p,\text{in}} = 226.0 + 3.53 = 229.5\,\text{kJ/kg}$

Isentropic process $S_1 = S_4$ and $S_3 = S_2$.

The quality of the discharged wet steam ($S_3 < S_{3\text{satvap}}$): $6.8443 < 8.0093$

$x_{3'} = (6.8443 - 0.7550)/(8.0093 - 0.7550) = 0.84$

$H_{3'} = 226.0(1 - 0.84) + 2599.2 \times 0.84 = 2219.5\,\text{kJ/kg}$

Heat interactions: $q_{\text{in}} = H_2 - H_1 = 3224.2 - 229.5 = 2994.7\,\text{kJ/kg}$.

$q_{\text{out}} = -(H_4 - H_{3'}) = -(226.0 - 2219.5.5) = 1993.5\,\text{kJ/kg}$ (heat received by the cooling medium that is the system in the condenser).

The thermal efficiency of the cycle: $\eta_{th} = 1 - \dfrac{|q_{\text{out}}|}{q_{\text{in}}} = \mathbf{0.334\ or\ 33.4\%}$

Therefore, the plant uses only 33.4% of the heat it received in the boiler:

Turbine work output: $W_{\text{out}} = H_{3'} - H_2 = 2219.5 - 3224.2 = -1004.7\,\text{kJ/kg}$ (heat produced by the turbine)

Cycle work out: $W_{\text{net}} = (q_{\text{out}} - q_{\text{in}}) = (1993.5 - 2994.7) = -1001.2\,\text{kJ/kg}$

Energy gain: $(20\,\text{kg/s})(1001.2 - 783.1)\,\text{kJ/kg} = 4362\,\text{kW}$

For a 360 days of operation = $(360)(24) = 8640$ h/year

The annual saving in the net power output:

$(20\,\text{kg/s})(1001.2 - 783.1)\,\text{kJ/kg}\,(8640)$ h/year $= 37,687,680\,\text{kWh/year}$

For a unit selling price of electricity of \$0.1/kWh, annual saving = **\$3,768,768/year**

Avoided CO_2 emissions from coal-fired plant with 0.37 kg CO_2/kWh from Table 11.2.

(0.37 kg CO_2/kWh) (37,687,680 kWh/year) = 13,944,441 kg CO_2/year

The thermal efficiency increased from 0.276 to 0.334 by reducing the condenser pressure from 78.5 to 15.0 kPa. However, the quality of the discharged steam decreased from 0.9 to 0.84, which is not desirable for the blades of the turbine. Savings are considerable.

Example 13.15 Economic and ecological gain of a Rankine cycle by increasing the boiler pressure

A steam power plant is operating on the simple ideal Rankine cycle. The steam mass flow rate is 20 kg/s. The steam enters the turbine at 3500 kPa and 400 °C. Discharge pressure of the steam from the turbine is 78.5 kPa.

(a) If the pressure of the boiler is increased to 9800 kPa while maintaining the turbine inlet temperature at 400 °C, determine the thermal efficiency.

(b) Determine the annual saving if the unit cost of electricity is \$0.10/kWh and avoided emission from a natural gas-powered plant.

Solution:
Assume that the cycle is at steady-state flow and the changes in kinetic and potential energy are negligible. Pump efficiency of $\eta_{pump} = 1$
Basis: 1 kg/s steam. Using the data from the steam tables
From Example 9.2 part a: $\eta_{th} = 0.276$ and $W_{net} = 783.1$ kJ/kg at 3500 kPa

(a) Superheated steam:$P_2 = P_1 = 9800$ kPa; $H_2 = 3104.2$ kJ/kg:
$S_2 = 6.2325$ kJ/kg K, $T_2 = 400°C$

Saturated steam: $P_3 = P_4 = 78.5$ kPa, $T_{sat} = 366.15$ K, $V_4 = 0.001038$ m^3/kg
$H_{3sat\,vap} = 2665.0$ kJ/kg; $H_4 = H_{3sat\,liq} = 389.6$ kJ/kg,
$S_{3sat\,vap} = 7.4416$ kJ/kg K, $S_{3sat\,liq} = 1.2271$ kJ/kg K

$$W_{p,in} = V_1(P_1 - P_4) = (0.001038)(9800 - 78.5)\left(\frac{1kJ}{1\ kPa\ m^3}\right) = 10.1\ kJ/kg$$

$H_1 = H_4 + W_{p,in} = 389.6 + 10.1 = 399.7$ kJ/kg
Isentropic process $S_1 = S_4$ and $S_3 = S_2$.
The quality of the discharged wet steam x_3 : $\left(S_3 < S_{3sat\,vap}\right)$: $6.2325 < 7.4416$.
$x_3 = (6.2325 - 1.2271)/(7.4416 - 1.2271) = 0.80$
$H_3 = 389.6(1 - 0.8) + 2665.0 \times 0.8 = 2210.0$ kJ/kg
Heat interactions: $q_{in} = H_2 - H_1 = 3104.2 - 399.7 = 2704.5$ kJ/kg
$q_{out} = -(H_4 - H_3) = -(389.6 - 2210.0) = 1820.4$ kJ/kg (heat received by the cooling medium in condenser).

The thermodynamic efficiency of the cycle: $\eta_{th} = 1 - \dfrac{q_{out}}{q_{in}} =$ **0.326 or 32.6%**

Therefore, the plant uses only 32.6% of the heat it received in the boiler.
The thermal efficiency increased from 0.276 to 0.326 by increasing the boiler pressure from 3500 to 9800 kPa. However the quality of the discharged steam decreased from 0.9 to 0.80, which is not desirable for the blades of the turbine.
Turbine work output: $W_{out} = H_3 - H_2 = (2210.0 - 3104.2)$ kJ/kg $= -894.2$ kJ/kg
Cycle work out: $W_{net} = (q_{out} - q_{in}) = (1820.4 - 2704.5)$ kJ/kg $= -884.1$ kJ/kg

(b) $\dot{m}_s = 20.0$ kg/s, and for a 360 days of operation $= (360)(24) = 8640$ h/year.

Annual gain power output:
$(20\,\text{kg/s})(884.1 - 783.1)\,\text{kJ/kg}\,(8640)\,\text{h/year} = 17,452,800\,\text{kWh/year}$
For a unit selling price of electricity of $0.1/kWh: annual saving = **$1,745,280**
**Avoided CO_2 emissions from natural gas-fired plant with 0.23 kg CO_2/kWh
from** Table 11.2
(0.23 kg CO_2/kWh) (17,452,800 kWh/year) = 4,014,144 kg CO_2/year

---

**Example 13.16 Economic and ecological gains from a Rankine cycle by
increasing the boiler temperature**
A steam power plant is operating on the simple ideal Rankine cycle. The steam
mass flow rate is 20 kg/s. The steam enters the turbine at 3500 kPa and 400 °C.
Discharge pressure of the steam from the turbine is 78.5 kPa.

(a) If the temperature of the boiler is increased to 525 °C while maintaining the
     pressure at 3500 kPa, determine the thermal efficiency.
(b) Determine the annual saving if the unit cost of electricity is $0.10/kWh and the
     avoided emission from a natural-gas- powered plant.

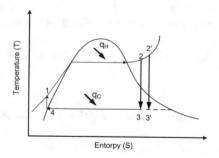

Entorpy (S)

Solution:
Assume that the cycle is at steady-state flow and the changes in kinetic and potential
energy are negligible.

(a) Basis: 1 kg/s steam. Using the data from the steam tables:

From Example 9.2 part a: $\eta_{\text{th}} = 0.276$ and $W_{\text{net}} = 783.1$ kJ/kg at 400 °C
Superheated steam: $P_2 = P_1 = 3500\,\text{kPa};$ $H_2 = 3506.9\,\text{kJ/kg};$ $S_2 = 7.2297\,\text{kJ/kg K},$
$T_2 = 525°C$
Saturated steam: $P_3 = P_4 = 78.5\,\text{kPa},$ $T_{\text{sat}} = 366.15\,\text{K},$ $V_4 = 0.001038\,\text{m}^3/\text{kg}$

$H_{3\text{sat vap}} = 2665.0 \text{ kJ/kg}, \quad H_{3\text{sat liq}} = 389.6 \text{ kJ/kg},$

$S_{3\text{sat vap}} = 7.4416 \text{ kJ/kg K}, \quad S_{3\text{sat liq}} = 1.2271 \text{ kJ/kg K}$

$W_{p,\text{in}} = V_1(P_1 - P_4) = (0.001038)(3500 - 78.5)\left(\frac{1 \text{ kJ}}{1 \text{ kPa m}^3}\right) = 3.55 \text{ kJ/kg}$

$H_1 = H_4 + W_{p,\text{in}} = 389.6 + 3.55 = 393.15 \text{ kJ/kg}$

Isentropic process $S_1 = S_4$ and $S_3 = S_2$.

The quality of the discharged wet steam ($S_2 < S_{3\text{sat vap}}$): $7.2297 < 7.4416$

$x_{3'} = (7.2297 - 1.2271)/(7.4416 - 1.2271) = 0.96$

$H_{3'} = 389.6(1 - 0.96) + 2665.0 \times 0.96 = 2574.0 \text{ kJ/kg}$

Heat interactions: $q_{\text{in}} = H_2 - H_1 = 3506.9 - 393.1 = 3113.8 \text{ kJ/kg}$

$q_{\text{out}} = -(H_4 - H_{3'}) = -(389.6 - 2574.0) = 2184.4 \text{ kJ/kg}$ (heat received by the cooling medium)

The thermodynamic efficiency of the cycle is $\eta_{\text{th}} = 1 - \frac{q_{\text{out}}}{q_{\text{in}}} = \mathbf{0.298 \text{ or } 29.8\%}$

Therefore, the plant uses only 29.8% of the heat it received in the boiler.

The thermal efficiency increased from 0.276 to 0.298 by increasing the boiler temperature from 400 to 525 °C. The quality of the discharged steam increased from 0.9 to 0.96, which is desirable for the protection of the turbine blades.

Turbine work out $W_{\text{out}} = H_{3'} - H_2 = (2574.0 - 3506.9) = -932.9 \text{ kJ/kg}$

Cycle work out: $W_{\text{net}} = (q_{\text{out}} - q_{\text{in}}) = (2184.4 - 3113.8) = -929.4 \text{ kJ/kg}$

(b) $\dot{m}_s = 20.0 \text{ kg/s}$ and for a 360 days of operation $= (360)(24) = 8640 \text{ h/year}$. Annual increase in the power output: $(20 \text{ kg/s})(929.4 - 783.1) \text{ kJ/kg} (8640)$ h/year $= 25.280,640 \text{ kWh/year}$

With a unit selling price of electricity of \$0.1/kWh: Annual saving = **\$2,528,064** **Avoided CO_2 emissions from natural gas-fired plant with 0.23 kg CO_2/kWh from** Table 11.2.

(0.23 kg CO_2/kWh) (25.280,640 kWh/year) = 5,814,547 kg CO_2/year

---

Example 13.17 Economic and ecological gains in a two-stage compression work by intercooling

Air with a flow rate of 2 kg/s is compressed in a steady state and reversible process from an inlet state of 100 kPa and 300 K to an exit pressure of 1000 kPa. Estimate the work for (a) polytropic compression with $\gamma = 1.3$, and (b) ideal two-stage polytropic compression with intercooling using the same polytropic exponent of $\gamma = 1.3$, (c) estimate conserved compression work by intercooling and electricity per year if the unit cost of electricity is \$0.15/kWh and the compressor is operated 360 days per year. Estimate the avoided emissions from a natural gas-fired plant.

Solution:

Assumptions: steady-state operation; air is ideal gas; kinetic and potential energies are negligible.

$P_2 = 1000 \text{ kPa}, \quad P_1 = 100 \text{ kPa},$

$T_1 = 300 \text{ K}, \quad \gamma = 1.3, \quad MW_{air} = 29 \text{ kg/kmol}$

Basis 1 kg/s air flow rate

(a) Work needed for polytropic compression with $\gamma = 1.3$

$$W_{comp} = \frac{\gamma R T_1}{MW(\gamma - 1)} \left[\left(\frac{P_2}{P_1} \right)^{(\gamma-1)/\gamma} \right] = \frac{1.3(8.314 \text{ kJ/kmol K})(300 \text{ K})}{29 \text{ kg/kmol } (1.3 - 1)} \left[\left(\frac{1000}{300} \right)^{(1.3-1)/1.3} - 1 \right] =$$

261.3 kJ/kg

(b) Ideal two-stage polytropic compression with intercooling ($\gamma = 1.3$).

$$P_x = (P_1 P_2)^{1/2} = (100 \times 1000)^{1/2} = 316.2 \text{ kPa}$$

$$W_{comp} = \frac{2\gamma R T_1}{MW(\gamma - 1)} \left[\left(\frac{P_x}{P_1} \right)^{(\gamma-1)/\gamma} \right]$$

$$= \frac{2(1.3)(8.314 \text{ kJ/kmol K})(300 \text{ K})}{29 \text{ kg/kmol } (1.3 - 1)} \left[\left(\frac{316.2}{100} \right)^{(1.3-1)/1.3} - 1 \right] = 226.8 \text{ kJ/kg}$$

Recovered energy = $(261.3 - 226.8)$ kJ/kg = 34.5 kJ/kg

Reduction in energy use: $\dfrac{261.3 - 226.8}{261.3} = 0.13 \text{ or } 13\%$

(c) Conservation of compression work = (2 kg/s) (261.3 − 226.8) kJ/kg = 69 kW.
Yearly conserved work: (69 kW) (8640 h/year) = 596,160 kWh/year
Saving in electricity (2,385,745 kWh/year) ($0.15/kWh) = **$89,424/year**
The compression work has been reduced by 13% when two stages of polytropic compression are used instead of single polytropic compression and conserved 596,160 kWh/year.
Avoided CO_2 emissions from natural gas-fired plant with 0.23 kg CO_2/kWh from Table 11.2.
(0.23 kg CO_2/kWh) (596,160 kWh/year) = 137,116 kg CO_2/year

Example 13.18 Economic and ecological gains from compression work
Ten adiabatic compressors are used to compress air from 100 kPa and 290 K to a pressure of 900 kPa at a steady-state operation. The isentropic efficiency of the compressor is 80%. The air flow rate is 0.4 kg/s. Determine the minimum and actual power needed by the compressor and the avoided emissions from a coal-fired plant.

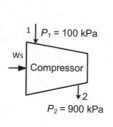

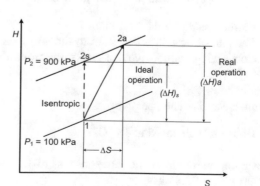

Solution:

Assume: Steady-state adiabatic operation. Air is ideal gas. The changes in kinetic and potential energies are negligible.

Enthalpies of ideal gas depends on temperature only.

The air mass flow rate = 0.4 kg/s, $\eta_C = 0.8$

Inlet conditions (Appendix D Table D1): $P_1 = 100\,kPa$, $T_1 = 290\,K$, $H_1 = 290.16\,kJ/kg$, $Pr_1 = 1.2311$

Exit conditions (Table D1): $P_2 = 900$ kPa

$$Pr_2 = Pr_1 \frac{P_2}{P_1} = 1.2311 \left(\frac{900\,kPa}{100\,kPa} \right) = 11.1$$

From Table D1: for $Pr_2 = 11.1$, we find $T_2 = 540\,K$ and $H_{2s} = 544.35\,kJ/kg$

$$\eta_C = \frac{H_{2s} - H_1}{H_{2a} - H_1} = 0.8 \rightarrow H_{2a}\ H_{2a} = 607.9\,kJ/kg \text{ and } T_2 = 600\,K \quad \text{(Table D1)}$$

(Approximate)

As seen from the TS diagram above H_{2a} is the actual enthalpy at the exit

Actual power required:

$$\dot{m}\Delta H_a = \dot{W}_{net,in} = \dot{m}(H_{2a} - H_1) = 0.4 \text{ kg/s } (607.9 - 290.16) \text{ kJ/kg} = \mathbf{127.1\ kW}$$

Minimum power required:

$$\dot{m}\Delta H_s = \dot{W}_{net,in} = \dot{m}(H_{2s} - H_1) = 0.4 \text{ kg/s } (544.3 - 290.1) \text{ kJ/kg} = \mathbf{101.7\ kW}$$

If the operation was ideal the rate of conserved energy would be 2540 kW for ten compressors.

Yearly gain: $2540\,kW\ (24\,h/day)(365\,days/year) = 22,250,400\,kWh/year$

Avoided CO_2 emissions from coal-fired plant with 0.37 kg CO_2/kWh from Table 11.2

(0.37 kg CO_2/kWh) (22,250,400 kWh/year) = 8,232,648 kg CO_2/year

---

Example 13.19 Economic and ecological gain from heating a house by heat pump

A heat pump is used to heat a house and maintain it at 18 °C. On a day where the outside temperature is −2 °C, the house is losing heat at a rate of 75,000 kJ/h. The heat pump operates with a coefficient of performance (COP) of 2.8. Determine:

(a) Power needed by the heat pump,
(b) The rate of heat absorbed from to the surrounding cold air,
(c) Cost saving and the avoided emissions from a natural-gas-powered plant.

Solution:

Assume: Steady-state operation

(a) House is maintained at 18 °C.

Heat pump must supply the same amount of lost heat from the cold source: $q_H = 75,000$ kJ/h = 20.83 kW

Power required by the heat pump:

$$\text{COP}_{\text{HP}} = \frac{\dot{q}_H}{\dot{W}_{\text{net,in}}} \rightarrow \dot{W}_{\text{net,in}} = \frac{\dot{q}_H}{\text{COP}_{\text{HP}}} = \frac{20.83 \text{ kW}}{2.8} = \textbf{7.44 kW}$$

(b) Energy balance: $\dot{q}_H - \dot{q}_C = \dot{W}_{\text{net,in}} = 7.44 \text{ kW}$.

$$\dot{q}_C = \dot{q}_H - \dot{W}_{\text{net,in}} = (20.83 - 7.44) \text{ kW} = \textbf{3.39 kW}$$

13.39 kW is extracted from the outside. The house is paying only for the energy of 7.44 kW that is supplied as electrical work to the heat pump.
If we have to use electricity in a resistance heater, we have to supply 20.83 kW
Energy conserved by using heat pump instead of electric heater is
$13.39/20.83 = 0.64$ (or 64%).
Energy conserved by using heat pump instead of electric heater is:
$(20.83 - 13.39) \text{ kW} = 7.44 \text{ kW}$
Yearly gain: 7.44 kW (24 h/day)(365 days/year) = 65,174.4 kWh/year
Avoided CO_2 emissions from natural gas-fired plant with 0.23 kg CO_2/kWh
from Table 11.2
(0.37 kg CO_2/kWh) (65,174.4 kWh/year) = 8,232,648 kg CO_2/year

---

Example 13.20 Economic and ecological gains from air conditioner
Estimate the cost of electricity saving for a 5000 Btu/h (1500 W) air-conditioning unit operating, with a SEER of 10 Btu/Wh. The unit is used for a total of 1500 h during an annual cooling season and the unit cost of electricity is $0.14/kWh. Also, estimate the total cost of electricity.

Solution:
Air conditioner sizes are often given as 'tons' of cooling where 1 ton of cooling is being equivalent to 12,000 Btu/h (3500 W).
The unit considered is a small unit and the annual total cooling output would be:
$(5000 \text{ Btu/h}) \times (1500 \text{ h/year}) = 7,500,000 \text{ Btu/year}$
With a seasonal energy efficiency ratio (SEER) of 10, the annual electrical energy usage:
$(7,500,000 \text{ Btu/year})/(10 \text{ Btu/Wh}) = 750,000 \text{ Wh/year} = 750 \text{ kWh/year}$
With a unit cost of electricity of $0.14/kWh, the annual cost:
(750 kWh/year)($0.14/kWh) = **$105/year**
The average power usage may also be calculated more simply by:
Average power = (Btu/h)/(SEER, Btu/Wh) = 5000/10 = 500 W = 0.5 kW
With the electricity cost of $0.14/kWh, the cost per hour:
(0.5 kW) ($0.14)/kWh = $0.07/h
For 1500 h/year, the total cost: ($0.07/h) (1500 h/year) = **$105**

Example 13.21 Economic and ecological gains of cooling with a unit at a higher SEER rating
Estimate the annual cost of electric power consumed by a 6-ton air-conditioning unit operating for 2000 h per year with a SEER rating of 10 and a power cost of $0.16/kWh.

Solution:
Air conditioner sizes are often given as 'tons' of cooling:
1 ton of cooling = 12,000 Btu/h (3500 W).
This is approximately the power required to melt one ton of ice in 24 h.
The annual cost of electric power consumption: (6) (12,000 Btu/h) = 72,000 Btu/h

$$\text{Cost} = \frac{(\text{Size, Btu/h})(\text{time, h/year})(\text{Cost of energy, \$/kWh})}{(\text{SEER, Btu/Wh})(1000\,\text{W/kW})}$$

(72,000 Btu/h)(2000 h/year)($0.16/kWh)/[(10 Btu/Wh)(1000 W/kW)] = **$2304/year**
For the temperatures of hot and cold sources:
$T_H = 273.15 + 20°C = 293.15\,K$, $T_C = 273.15 − 4°C$
$= 269.15\,K$, and $T_H − T_C = 24\,K$

$$EER_{Carnot} = 3.41\left(\frac{T_C}{T_H − T_C}\right) = 3.41\,(269.13/24) = \mathbf{38.24}$$

The maximum EER decreases as the difference between the inside and outside air temperature increases.

Example 13.22 Cost of coal in a steam power plant
An adiabatic turbine is used to produce electricity by expanding a superheated steam at 4100 kPa and 350 °C. The power output is 50 MW. The steam leaves the turbine at 40 kPa and 100 °C. If the combustion efficiency is 0.75 and the generator efficiency is 0.9, determine the yearly cost of coal and amount of emissions.

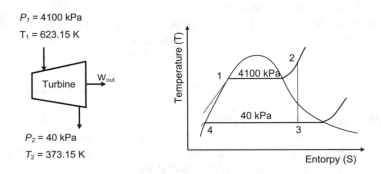

Solution:
Assume: Steady-state adiabatic operation. The changes in kinetic and potential energies are negligible.
Basis: stems flow rate = 1 kg/s
Combustion efficiency $\eta_{comb} = 0.75$, Generator efficiency $\eta_{gen} = 0.9$
Power output = 50 MW = 50,000 kW

From steam tables. Turbine inlet conditions:

$P_2 = 4100\,\text{kPa},\quad T_2 = 623.15\,\text{K},\quad H_2 = 3092.8\,\text{kJ/kg},\; S_2 = 6.5727\,\text{kJ/kg K}$

Turbine outlet conditions:

$P_4 = P_3 = 40\,\text{kPa}, T_3 = 373.15\,\text{K},\; V_4 = 1.027\,\text{cm}^3/\text{g} = 0.001027\,\text{m}^3/\text{kg}$

$S_{4\text{sat vap}} = 7.6709\,\text{kJ/kg K},\; S_{4\text{sat liq}} = 1.2026\,\text{kJ/kg K}, H_{4\text{sat vap}} = 2636.9\,\text{kJ/kg},$

$H_{4\text{sat liq}} = 317.6\,\text{kJ/kg}(40\,\text{kPa})$

$W_{p,\text{in}} = V_4(P_1 - P_4) = (0.001027\,\text{m}^3/\text{kg})(4100 - 40)\,\text{kPa}$

$(1\,\text{kJ}/1\,\text{kPa}\,\text{m}^3) = 4.2\,\text{kJ/kg}$

$H_1 = H_4 + W_{p,\text{in}} = 317.6 + 4.2 = 321.8\,\text{kJ/kg}$

For the isentropic operation $S_3 = S_2 = 6.5727$ kJ/kg K

Since $S_2 < S_{3\text{sat vap}}$: $6.5727 < 7.6709$ the steam at the exit is saturated liquid–vapor mixture.

Quality of that mixture: $x_3 = \dfrac{S_2 - S_{2\text{satliq}}}{S_{2\text{satvap}} - S_{2\text{satliq}}} = \dfrac{6.5727 - 1.2026}{7.6709 - 1.2026} = 0.83$

$H_3 = (1 - x_3)H_{3\text{satliq}} + x_3 H_{3\text{satvap}} = 2243.1\,\text{kJ/kg}$

Heat interactions: $q_{\text{in}} = \dot{m}(H_2 - H_1) = (3092.8 - 321.8)\,\text{kJ/kg} = 2771.0\,\text{kJ/s}$

$q_{\text{out}} = -\dot{m}(H_4 - H_3) = -(317.6 - 2243.1)\,\text{kJ/kg} = 1925.5\,\text{kJ/s}$ (heat received by the cooling medium that is the system in the condenser)

Thermal efficiency: $\eta_{\text{th}} = 1 - \dfrac{q_{\text{out}}}{q_{\text{in}}} = 0.305$ or 30.5%

Overall plant efficiency: $\eta_{\text{overall}} = \eta_{\text{th}}\eta_{\text{comb}}\eta_{\text{gen}} = (0.305)(0.75)(0.9) = \mathbf{0.205}$ **or 20.5%**

The rate of coal energy required: $\dot{E}_{\text{coal}} = \dfrac{|\dot{W}_{\text{net}}|}{\eta_{\text{overall}}} = \dfrac{50,000\,\text{kW}}{0.205} = 243,900\,\text{kJ/s}$

Energy of coal (bituminous) = 24,000 kJ/kg (Table 2.6).

$\dot{m}_{\text{coal}} = \dfrac{\dot{E}_{\text{coal}}}{\eta_{\text{overall}}} = \dfrac{243,900}{24,000}\left(\dfrac{1\,\text{ton}}{1000\,\text{kg}}\right) = 0.0101\,\text{ton/s} = 36.6\,\text{ton/h}$

Annual average price of bituminous coal = \$60/ton (2018)
Annual consumption of coal:
(36.6 ton/h) (24 h/day) (365 day/year) = 320,616 ton/year
Annual cost of coal: **(320,616 ton/year) (\$60 ton) = \$19,236,960/year**
Emission of coal: 2.3 kg CO_2/kg coal (Table 11.2); ton = 2000 lb = 907.1 kg
Annual emission of coal:
(320,616 ton/year) (907.1 kg/ton) (2.3 kg CO_2/kg coal) = 668,910,779 kg CO_2/year

---

Example 13.23 Economic and ecological gains with compact fluorescent bulbs
Assume that an average residential rate of electricity is \$0.14/kWh and a household consume about 10,000 kWh per year. If the lighting is provided by **compact fluorescent bulbs only**, estimate the yearly cost saving and avoided emissions from a natural gas-fired power plant.

Solution:

Assume that 11% of the energy budget of household is for lighting, and **compact fluorescent bulbs** consume about 75% less energy than standard lighting.

Annual energy use and cost with standard lighting:

(10,000 kWh/year) (0.11) = 1100 kW/year

(1100 kW/year)($0.14/kWh) = $154/year

Annual energy use and cost with compact fluorescent bulbs:

(1100 kW/year) (1– 0.75) = 275 kWh/year

(275 kWh/year) ($0.14/kWh) = $38.5/year

Annual energy and cost savings:

(1100 − 275) kWh = **825 kWh**

(154 − 38.5)$/year = **$115 5/year**

Avoided emission from coal-fired plant:

(825 kWh)(24 h/day)(365 day/year) (0.37 kg CO$_2$ 2/kg coal)

= 2,673,990 kg CO$_2$

Conservation of energy and cost with **compact fluorescent bulbs are considerable**.

13.14.2   Heat Integration in a Sustainable Refinery Operation

For a crude oil refinery (Fig. 13.17) separation by distillation is an energy-intensive process. This energy consumption is associated with significant greenhouse gas emissions. Reducing thermal energy consumption will help reduce CO$_2$ emissions [1]. Environmental pollution impact metrics are estimated from the 'Carbon Tracking' options with the data source of US-EPA-Rule-E9-5711 and using crude oil as a primary fuel source for the hot utilities. Energy can be recovered by heat integration (Fig. 13.17) in a conventional crude oil refinery by using the hot streams to heat the cold feed stream of crude oil in a retrofit of an existing conventional refinery [1]. Table 13.9 shows the estimated energy savings after the retrofits.

Summary

- In the business as usual, gas utilities are energy sources that strive to meet policy goals but there is pushback from environmentalists to decarbonize the energy industry. The advancement of clean energy technologies is becoming a global goal that is in line with how investments made in the energy sector to grow based on policy and reliability.
- Resilience in utilities goes beyond what is considered reliable and covers focuses on how quickly the system can become operational after failure.
- Coal-fired power plants are required to meet standards that limit the amounts of some of the pollutants that these plants release into the air. Placing a fee on CO$_2$ capture ($25 to $80/ton) may encourage the deployment of CCS technology.

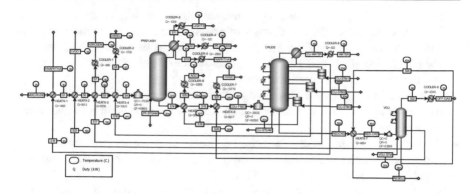

Fig. 13.17 Process flow diagram after using newly installed heat exchangers to match the available and required heats. All the heat duties required by the heat exchangers are in kW and inlet and outlet temperatures are in °C [1]

Table 13.9 Estimated efficiencies and exergy savings for the three columns [1]

	Base case			Modified case					
Unit	Ex_{min} (kW)	Ex_{loss}^a (kW)	%	ηEx_{min} (kW)	Ex_{loss} (kW)	η%	Saved Ex_{loss} (kW)	Change Ex_{loss} %	Electricity Saving[b] ($/year)
Preflash	14,836.6	3385.4	81.4	18,229.5	3385.5	84.3	−0.8	−0.005	−554.6
Crude	9749.9	9414.7	50.8	41,015.9	8768.0	82.4	646.7	6.8	427,009.4
VDU	−3813.3	8051.7	31.1	−6210.5	8085.9	43.4	−34.4	−0.3	−22,740.7
Total									**403,714.0**

[a]Ex_{loss}: Total column exergy loss from the converged simulation by Aspen Plus with the BK-10 method
[b]Electricity equivalent of energy saving is based on a unit cost of electricity of $0.0775/kW h

- Chemical-looping combustion with inherent CO_2 capture, if properly developed, can produce water and CO_2 stream, and helps reduce carbon capture and storage costs.
- In 2040, half of the total electricity will be from renewables. Adaptation of characteristics of the local resources in energy and efficiency, as well as dependence on nonrenewable sources should be analyzed.
- A sharp increase in power generations by solar PV and wind is expected, as well as by natural gas, while stagnation in power generation by coal and nuclear continues. Renewable power has become an affordable and green source of new power generation world wide. The global weighted-average levelized cost of electricity from renewable resources have all been within the range of fossil-fuel-fired power generation costs. Commercial solar PV cells may range up to 26% conversion and are connected to electricity grids. Efficiency is increased using **concentrating solar PV**. To mitigate the grid integration problem with solar PV, battery and other means are being developed.

- *Offshore wind operating and maintenance* costs are higher than those for onshore wind due to the higher costs of access to the site and perform maintenance on towers and cabling. The average O&M cost of hydropower is around 2% of total installed costs per year with below the 2% average for larger plants, and around 3% for smaller capacities.
- Geothermal O&M costs are high for geothermal projects, because of the need to work over production wells on a periodic basis to maintain fluid flow and hence production.
- Any type of organic materials including trees, crops, grass, woody material, animal waste, and agricultural waste from biological systems are called biomass, which can be used to produce biofuels.
- Starch is the main energy-containing molecules and cannot be directly fermented. First, in starch liquefaction, starch must be broken down to smaller units using the enzyme alpha-amylase through hydrolysis to break the bonds between adjacent glucose molecules.
- Biological systems use photosynthesis to capture and convert solar energy into chemical. The chlorophyll molecule absorbs light of the visible spectrum and excites electrons that powers photosynthesis to produce organic compounds mainly of carbohydrates such as sugars and starches.
- Biomass and waste biodegrade by bacteria in landfills and produce methane, which can be used for heat and power production. Biogas consisting of methane and CO_2 is produced from organic waste through anaerobic digestion that facilitates the breakdown of the waste without the presence of oxygen.
- Large centralized plants need reliable supply of a large amount of feedstock throughout the year. Distributed process technology may offer economic solutions to areas where the biomass feedstock is not readily transportable in large quantities as it can transport solid biomass materials from multiple sources more easily.
- Energy conservation focuses on reducing energy consumption and increasing efficiency in energy usage for the same useful energy output. Energy conservation and consequently higher energy efficiencies across most sectors of economy is becoming part of sustainable development.
- Energy is the main driver of technological development to deliver the prosperity of modern society by being able to produce all goods and services. The International Energy Outlook 2019 projects a 50% increase in global energy use and despite the increased use of renewable energy, fossil fuels will continue to supply global energy through 2050. Therefore, GHG emissions will continue to rise.
- Industrial and transportation sectors will keep being the largest world end-energy consumers in increasing trend. The increase in residential and commercial sectors will be gradual only. Liquid transportation fuels, natural gas, and electricity will increasingly dominate world end-use of energy types, while coal will remain stagnant and renewables will have the lowest share.

- Industrial, residential, and commercial sectors are the main users of electricity, while the transportation sector has the lowest share as electricity user currently. This will change as the electrification increases.
- *Environment Impact Assessment* identifies the short- and long-term impact of energy generation plant on the environment within the liability to local and federal government bodies. Possible impacts include natural physical resources, natural biological/resources, and quality-of-life values including aesthetic and cultural values. By 2050, the demand for energy will nearly double with water and food increase by over 50% globally.
- The relation between economic growth and energy consumption is the main factor for *energy policy* implication since the gross domestic product (one of the main indicators of economic growth) increases with the increasing usage of energy. Energy policy recognizes the importance of energy efficiency and the necessary transition to renewables to solve the risks and challenges of fossil fuel use.
- Expenditures of solar, wind, and storage are very close to each other, while the hydro expenditure is relatively low. Annual operating and maintenance costs are high and similar for wind and storage, while lower for solar and much lower for hydro energy. Energy economies should be handled with extreme care due to its impact in the ecosystem.
- *Levelized cost of electricity (LCOE)* is the price at which electricity must be produced from a specific source to break even. It is an economic assessment tool in estimating the costs of generation from different sources. Renewable energy generation has produced low-cost electricity and it continues to decrease with data over the past 10 years. Using this data, predictions can be made as to what the future costs may be based upon the trends.
- Wind total installed costs for both the onshore and offshore systems will keep falling. However, offshore installations will keep the costing more. This will be the case for the levelized cost of electricity from both the systems.
- Advances in solar power technology have increased the power generation of photovoltaic cells at a lower cost and industrial level installation projects are cheaper per unit than smaller scale installations. The levelized cost of electricity is expected to continue to decrease.
- The total installed costs for both the solar PV and concentrated solar powers will keep decreasing. However, the cost is relatively higher for concentrated solar power.
- The installed cost and LCOE from bioenergy will remain high and flat. The values of LCOE from bioenergy will be comparable with those of wind and solar-based LCOEs in 2022
- Geothermal power production is a mature technology and continues to grow at a lesser pace.
- Hydroelectric power generation is a reliable source of electricity. The installation costs have been driven up in recent years due to an increase in developing areas with a more challenging geographical and engineering challenges.

- *Bio break* model represents the regional feedstock supply system and biofuel biorefinery economics. Bio break model estimates willingness to pay (*WTP*) , willingness to accept (*WTA*) , and price gap (*PG*) between WTA and WTP:
- *Thermoeconomics* assigns costs to exergy-related variables by using the *exergy cost theory* and *exergy cost balances* and achieve cost accounting and optimization methods by minimizing the overall cost under financial, environmental, and technical constraints
- Wind power-based electrolysis for hydrogen production cost accounts energy and turbine costs, geographic factors, large-scale energy storage, compression, pipeline transport, and dispensing. The cost of electrolytic hydrogen depends on the cost of electricity as well as the capital cost of the electrolyzer systems and their operating efficiency.
- One of the main concerns is the storage of renewable energy for utility-scale power generation, integrating into the grid, as well as for transport applications.
- The *third-generation biofuels* derived from algae continue to be a promising new low-carbon economy for several reasons. Algae can be produced on non-arable areas such as lakes, oceans, or deserts, thus reducing competition with the food supply chain. Sustainable development of algal biofuels requires the favorable values for energy efficiency and emissions, water use, supply of nitrogen, phosphorus, CO_2, and appropriate land resources.
- *Risk analysis* consists of financial risk, environmental risk, technical risk, and social risk.
- Energy efficiency is capable of yielding energy and demand savings. Increased energy efficiency reduces the cost of energy, so consumers use it more. The direct rebound is the amount of energy savings from increased efficiency that are is offset by increased use of energy.
- The ratio of useful energy and energy required from the energy balances may be called the energy return on investment (EROI) that helps assess the energy cost on society and its impact on development.
- *Circular economy* describes strategies for waste prevention and resource efficiency toward a sustainable business model. Therefore, it is regenerative by intention and design and represents phased of closed flow of materials and energy in manufacturing. This can be achieved through design, maintenance, repair, and recycling
- *Bioeconomy* is based on all economic activity originated from invention, development, production, and use of biological products and processes in the production of food, energy, and material from renewable biological resources in a sustainable way. Energy efficiency of a conversion process for producing biofuels from a biomass affects the bioeconomy.
- The benefits of a circular economy overlap with the elements of sustainability that are economy, environment, and society; therefore, it should be aligned with the sustainable development policy.
- *Hydrogen economy* implies producing and using hydrogen as a clean fuel. Because of high energy density (kJ/kg), hydrogen is very attractive for a large

amount of energy storage or energy carrier to be used for various purposes, including for the transportation sector.

- An affordable and viable alternative can be *methanol economy*. Methanol has more hydrogen by mass in the same volume. Therefore, methanol is a safe carrier fuel for hydrogen and may solve many issues of climate change. Methanol can be used as a fuel for blending or producing other fuels, as well as a chemical and intermediate for large-volume chemicals. Dimethyl ether with methanol can produce various liquid fuels, such as gasoline, kerosene, gasoil.
- *Ecological cost* analysis may minimize the depletion of nonrenewable natural resources. Determining the extraction of raw materials and fossil fuels from natural resources is not sufficient in fully understanding the ecological impact of production processes. The production, conversion, and utilization of energy may lead to environmental problems, such as air and water pollution, impact on the use of land and rivers, thermal pollution due to mismanagement of waste heat, and global climate change.

Problems

13.1 An over used car may consume around 150 gallons gasoline per month. Estimate the cost of energy consumed by the car per year.

13.2 A city consumes natural gas at a rate of 500×10^6 ft^3/day. The volumetric flow is at standard conditions of 60 °F and 1 atm = 14.7 psia. If the natural is costing \$12/GJ of higher heating value what is the daily cost of the gas for the city.

13.3 A city consumes natural gas at a rate of 800×10^6 ft^3/day. The volumetric flow is at standard conditions of 60 °F and 1 atm = 14.7 psia. If the natural is costing \$10/GJ of higher heating value what is the daily cost of the gas for the city.

13.4 A car consumes about 6 gallons a day, and the capacity of a full tank is about 15 gallons. The density of gasoline ranges from 0.72 to 0.78 kg/l (Table 2.2). The lower heating value of gasoline is about 44,000 kJ/kg. Assume that the average density of gasoline is 0.75 kg/l. If the car was able to use 0.2 kg of nuclear fuel of uranium-235, estimate the time in years for refueling and avoided cost of gasoline.

13.5 A car consumes about 3 gallons a day, and the capacity of the full tank is about 11 gallons. The density of gasoline ranges from 0.72 to 0.78 kg/l (Table 2.2). The lower heating value of gasoline is about 44,000 kJ/kg. Assume that the average density of gasoline is 0.75 kg/l. If the car was able to use 0.1 kg of nuclear fuel of uranium-235, estimate the time in years for refueling and avoided cost of gasoline.

13.6 When a hydrocarbon fuel is burned, almost all the carbon in the fuel burns completely to form CO_2 (carbon dioxide), which is the principle gas causing the greenhouse effect and thus global climate change. On average, 0.59 kg of CO_2 is produced for each kWh of electricity generated from a power plant that burns natural gas. A typical new household uses about 7000 kWh of electricity per year. Determine the amount of CO_2 production and carbon fee in a city with 100,000 households.

13.7 When a hydrocarbon fuel is burned, almost all the carbon in the fuel burns completely to form CO_2 (carbon dioxide), which is the principle gas causing the greenhouse effect and thus global climate change. On average, 0.59 kg of CO_2 is produced for each kWh of electricity generated from a power plant that burns natural gas. A typical new household uses about 10,000 kWh of electricity per year. Determine the amount of CO_2 production and carbon fee in a city with 250,000 households.

13.8 A large public computer lab operates Monday through Saturday. There the computers are either being used constantly or remain on until the next user comes. Each computer needs around 240 W. If the computer lab contains 53 computers and each is on for 12 h a day, during the course of the year how much CO_2 will the local coal power plant have to release and carbon fee to be paid to keep these computers running?

13.9 The average university will have a large public computer lab open Monday through Saturday. There the computers are either being used constantly or remain on until the next user comes. Each computer needs around 13.10 W. If the computer lab contains 53 computers and each is on for 12 h a day, during the course of the year how much coal will the local coal power plant have to consume to keep these computers running?

13.10 A large public computer lab runs six days per week from Monday through Saturday. Each computer uses a power of around 120 W. If the computer lab contains 45 computers and each is on for 12 h a day, during the course of the year how much CO_2 will the local coal power plant have to release to the atmosphere in gram moles to keep these computers running?

13.11 If a car consumes 60 gallons gasoline per month. Estimate the energy consumed by the car per year.

13.12 A car having an average 22 miles/gal is used 32 miles every day. If the cost of a gallon fuel is $3.8 estimate the yearly cost of fuel.

13.13 A car having an average 22 miles/gal is used 32 miles every day. Estimate the yearly energy usage.

13.14 A 150-Watt electric light bulb is used on average 10 h per day. A new bulb costs $2.0 and lasts about 5000 h. If electricity cost is $0.15/kWh, estimate the yearly cost of the bulb.

13.15 A laptop consuming 90 W is used on average 5 h per day. If a laptop costs $500 and will be used for four years estimate the total electricity cost in four years for the laptop. Electricity cost is $0.15/kWh.

13.16 A laptop consuming 90 W is used on average 7 h per day. If a laptop costs $500 and will be used for four years estimate the total electricity cost in four years for the laptop. Electricity cost is $0.10/kWh.

13.17 A 20-hP electric motor is used to pump ground water into a storage tank four hours every day. Estimate the work done by the pump in kW every year and the cost of electricity every year. Assume that the electricity unit cost is \$0.1/kWh.

13.18 A city consumes natural gas at a rate of 250×10^6 ft³/day. The volumetric flow is at standard conditions of 60 °F and 1 atm = 14.7 psia. If the natural gas is costing \$6/GJ of higher heating value what is the daily cost of the gas for the city.

13.19 A home consumes natural gas at a rate of 4.3 ft³/day to heat the home. The volumetric flow is at standard conditions of 60 °F and 1 atm = 14.7 psia. If the natural gas is costing \$0.67/MJ of higher heating value what is the daily cost of the gas for the home?

13.20 A water heater consumes propane, which is providing 80% of the standard heat of combustion when the water produced after combustion is vapor. If the price of propane is \$2.2/gal measured at 25 °C. What is the heating cost in \$ per million Btu and in \$ per MJ?

13.21 An average video games system consumes 170 W of power during game-play. If a person were to play an hour a day for 80% of the year how many liters of gasoline would the person have burned and what is the cost? (Evaluated at HHV).

13.22 Assuming complete combustion of methanol and gasoline (assuming 100% octane) estimate the HHVs by using the oxidations and carbon dioxide intensity in $kgCO_2/MJ$.

13.23 Assuming complete combustion of ethanol and diesel (assuming 100% dodecane) estimate the HHVs by using the oxidations and carbon dioxide intensity in $kgCO_2/MJ$.

13.24 Assuming complete combustion of butanol and natural gas estimate the HHVs by using the oxidations and carbon dioxide intensity in $kgCO_2/MJ$.

13.25 Assuming complete combustion of ethanol and bituminous coal estimate the HHVs by using the oxidations and carbon dioxide intensity in $kgCO_2/MJ$.

13.26 Consider the exergy costing on a boiler and turbine system shown below

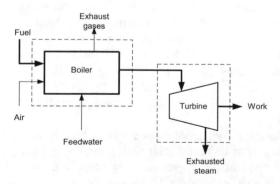

A turbine produces 100 MW of electricity per year. The average cost of the steam is \$0.05/(kWh) of exergy (fuel). The total cost of the unit (fixed

capital investment and operating costs) is 2.1×10^5. If the turbine exergetic efficiency increases from 80 to 85%, after an increase of 3% in the total cost of the unit, evaluate the change of the unit cost of electricity.

13.27　Consider the exergy costing on a boiler and turbine system shown in Pr. 3.26. A turbine produces 80 MW of electricity per year. The average cost of the steam is $0.05/(kWh) of exergy (fuel). The total cost of the unit (fixed capital investment and operating costs) is 1.51×10^5. If the turbine exergetic efficiency increases from 75 to 87%, after an increase of 2% in the total cost of the unit, evaluate the change of the unit cost of electricity.

13.28　Consider the exergy costing on a boiler and turbine system shown in Pr. 3.26. A turbine produces 200 MW of electricity per year. The average cost of the steam is $0.05/(kWh) of exergy (fuel). The total cost of the unit (fixed capital investment and operating costs) is 5.5×10^5. If the turbine exergetic efficiency increases from 80% to 88%, after an increase of 2% in the total cost of the unit, evaluate the change of the unit cost of electricity.

13.29　A typical square two-story home with a roof surface area of 1260 ft^2 and a wall surface area of 2400 ft^2 is to be heated with solar energy storage using a salt hydrate as phase change material. It presently has as an insulation of 6 inches in the roof and 1 inch in the walls. Inside temperature will be held at 70 °F and expected outside low temperature is 10 °F. Average solar radiation is 650 Btu/ft^2 and the cost of solar air collector is $3/ft^2. The salt hydrate costs around $0.15/lb. Estimate the costs of salt hydrate and solar air collectors.

13.30　A large building with a roof surface area of 3700 ft^2 and a wall surface area of 5000 ft^2 is to be heated with solar energy storage using a salt hydrate as phase change material. It presently has an insulation of 5 inches in the roof and 1 inch in the walls. Inside temperature will be held at 70 °F and expected outside low temperature is 20 °F. Average solar radiation is 700 Btu/ft^2 and the cost of solar air collector is $5.1/ft^2. The salt hydrate costs around $0.15/lb. Estimate the costs of salt hydrate and solar air collectors.

13.31　A typical square two-story home with a roof surface area of 1260 ft^2 and a wall surface area of 2400 ft^2 is to be heated with solar energy storage using paraffin as phase change material. It presently has as an insulation of 6 inches in the roof and 1 inch in the walls. Inside temperature will be held at 70 °F and expected outside low temperature is 10 °F. Average solar radiation is 300 Btu/ft^2 and the cost of solar air collector is $4.5/ft^2. The paraffin costs around $0.4/lb. Estimate the costs of salt hydrate and solar air collectors.

13.32　A large building with a roof surface area of 3200 ft^2 and a wall surface area of 4200 ft^2 is to be heated with solar energy storage using octadecane as phase change material. It presently has as an insulation of 6 inches in the roof and 1 inch in the walls. Inside temperature will be held at 70 °F and expected outside low temperature is 10 °F. Average solar radiation is 350 Btu/ft^2 and the cost of solar air collector is $3.4/ft^2. The octadecane costs around $0.45/lb. Estimate the costs of salt hydrate and solar air collectors.

13.33   A power plant is operating on an ideal Brayton cycle with a pressure ratio of $r_p = 9$. The fresh air temperature at the compressor inlet is 295 K. The air temperature at the inlet of the turbine is 1200 K. The cycle operates with a compressor efficiency of 75% and a turbine efficiency of 80%. The unit cost of fuel is \$0.10/kWh. The cycle operates 360 day per year.

   (a) Using the standard-air assumptions, determine the thermal efficiency of the cycle.
   (b) If the power plant operates with a regenerator with an effectiveness of 0.78, determine the thermal efficiency of the cycle and annual saving of fuel cost.

13.34   A steam power plant is operating on the simple ideal Rankine cycle. The steam mass flow rate is 25 kg/s. The steam enters the turbine at 3500 kPa and 400 °C. Discharge pressure of the steam from the turbine is 90 kPa.

   (a) Determine the thermal efficiency of the cycle.
   (b) If the pressure of the discharge steam is reduced to 20 kPa determine the thermal efficiency.
   (c) Determine the annual cost saving if the unit cost of electricity is \$0.10/kWh and avoided emission from a coal-fired plant.

13.35   A steam power plant is operating on the simple ideal Rankine cycle. The steam mass flow rate is 40 kg/s. The steam enters the turbine at 3500 kPa and 400 °C. Discharge pressure of the steam from the turbine is 85 kPa.

   (a) Determine the thermal efficiency of the cycle.
   (b) If the pressure of the discharge steam is reduced to 25 kPa determine the thermal efficiency.
   (c) Determine the annual cost saving if the unit cost of electricity is \$0.10/kWh and avoided emission from a coal-fired plant.

13.36   A steam power plant is operating on the simple ideal Rankine cycle. The steam mass flow rate is 25 kg/s. The steam enters the turbine at 3500 kPa and 400 °C. Discharge pressure of the steam from the turbine is 80 kPa.

   (a) If the pressure of the boiler is increased to 9000 kPa while maintaining the turbine inlet temperature at 400 °C, determine the thermal efficiency.
   (b) Determine the annual saving if the unit cost of electricity is \$0.10/kWh and avoided emission from a natural gas-powered plant.

13.37   A steam power plant is operating on the simple ideal Rankine cycle. The steam mass flow rate is 20 kg/s. The steam enters the turbine at 3500 kPa and 400 °C. Discharge pressure of the steam from the turbine is 78.5 kPa.

(a) If the temperature of the boiler is increased to 500 °C while maintaining the pressure at 3500 kPa, determine the thermal efficiency.

(b) Determine the annual saving if the unit cost of electricity is $0.12/kWh and the avoided emission from a natural-gas powered plant.

13.38 Ten adiabatic compressors are used to compress air from 100 kPa and 290 K to a pressure of 1000 kPa at a steady-state operation. The isentropic efficiency of the compressor is 80%. The air flow rate is 0.5 kg/s. Determine the minimum and actual power needed by the compressor and the avoided emissions from a coal-fired plant.

13.39 A heat pump system is used to heat a district with 1200 houses and maintain them at 18 °C. On a day where the outside temperature is −2 °C, the house is losing heat at a rate of 80,000 kJ/h. The heat pump operates with a coefficient of performance (COP) of 3.0. Determine:

(a) Power needed by the heat pump,

(b) The rate of heat absorbed from to the surrounding cold air

(c) Cost saving and the avoided emissions from a natural-gas powered plant.

13.40 A heat pump system is used to heat a district with 1200 houses and maintain them at 18 °C. On a day where the outside temperature is −2 °C, the house is losing heat at a rate of 80,000 kJ/h. The heat pump operates with a coefficient of performance (COP) of 3.0. Determine:

(a) Power needed by the heat pump,

(b) The rate of heat absorbed from to the surrounding cold air

(c) Cost saving and the avoided emissions from a natural-gas powered plant.

13.41 An air conditioner with a 5000 Btu/h (1500 W) operating with a SEER of 10 Btu/Wh. The unit is used for a total of 1700 h during an annual cooling season and the unit cost of electricity is $0.10/kWh. Estimate the cost of electricity saving and the avoided emissions from a coal-fired plant if a district uses 1500 units of the air conditioner.

13.42 An air conditioner with a 5000 Btu/h (1500 W) operating with a SEER of 22 Btu/Wh. The unit is used for a total of 2000 h during an annual cooling season and the unit cost of electricity is $0.10/kWh. Estimate the cost of electricity saving and the avoided emissions from a coal-fired plant if a district uses 1500 units of the air conditioner.

13.43 (a) Estimate the annual cost of electric power consumed by a 5 ton air conditioning unit operating for 2000 h per year with a SEER rating of 10 and a power cost of $0.10/kWh. (b) Estimate the emission avoided after replacing natural-gas powered plant with solar concentrated power plant.

13.44 (a) Estimate the annual cost of electric power consumed by a 6 ton air conditioning unit operating for 3000 h per year with a SEER rating of 22 and a power cost of $0.12/kWh. (b) Estimate the emission avoided after replacing natural-gas powered plant with solar concentrated power plant.

13.45 A 5 tons of current residential air conditioner operates with a seasonal energy efficiency ratio (SEER) rating of 10. This unit will be replaced with a newer unit operating with a SEER rating of 22. The unit operates 130 days with an average 10 h per day. Average inside and outside temperatures are 21 °C and 3 °C, respectively. The unit cost of energy is $0.16/kWh. Estimate the saving in the cost of electricity and the maximum energy efficiency ratio, as well as the avoided emission if you replace a natural gas fired plant with wind power.

13.46 An adiabatic turbine is used to produce electricity by expanding a superheated steam at 4100 kPa and 350 °C. The power output is 75 MW. The steam leaves the turbine at 90 kPa and 100 °C. If the combustion efficiency is 0.85 and the generator efficiency is 0.9, determine the yearly cost of coal and amount of emissions.

13.47 An adiabatic turbine is used to produce electricity by expanding a superheated steam at 4800 kPa and 350 °C. The power output is 100 MW. The steam leaves the turbine at 80 kPa and 100 °C. If the combustion efficiency is 0.70 and the generator efficiency is 0.9, determine the yearly cost of coal and amount of emissions.

13.48 An adiabatic turbine is used to produce electricity by expanding a superheated steam at 4500 kPa and 350 °C. The steam flow rate is 42 kg/s. The steam leaves the turbine at 40 kPa and 100 °C. If the combustion efficiency is 0.77 and the generator efficiency is 0.95, determine the yearly cost of coal and amount of emissions.

13.49 Assume that an average residential rate of electricity is $0.10/kWh and a household consume about 12,000 kWh per year. If the lighting is provided by **compact fluorescent bulbs only**, estimate the yearly cost saving and avoided emissions from a natural gas-fired power plant.

13.50 Assume that an average residential rate of electricity is $0.12/kWh and a district with 2000 households consumes about 20,000,0000 kWh per year. If the lighting is provided by **led bulbs (a light emitting diode) only**, estimate the yearly cost saving and avoided emissions from a natural gas-fired power plant.

References

1. Alhajji M, Demirel Y (2015) Energy and environmental sustainability assessment of crude oil refinery by thermodynamic analysis. Int J Energy Res 39:1925–1941
2. Adamu A, Russo-Abegão F, Boodhoo K (2020) Process intensification technologies for CO_2 capture and conversion—a review. BMC Chemical Eng 2:2

3. Allen JW, Unlu S, Demirel Y, Black P, Riekhof W (2018) Integration of biology ecology and engineering for sustainable algal based biofuel and bioproduct biorefinery. Bioresour Bioprocess 5(47):1–28

4. Bergstrom JC, Ty D (2017) Economics of carbon capture and storage. Intech Open. www.intechopen.com

5. Cuellar-Franca RM, Azapagic A (2015) Carbon capture, storage and utilisation technologies: a critical analysis and comparison of their life cycle environmental impacts. J CO_2 Utilization 9:82–102

6. Demirel Y, Matzen M, Winters C, Gao X (2015) Capturing and using CO_2 as feedstock with chemical-looping and hydrothermal technologies. Int J Energy Res 39:1011–1047

7. Demirel Y (2018) Biofuels. In: Dincer I (ed) Comprehensive energy systems. Elsevier, Amsterdam, vol 1, Part B, pp 875–908

8. Demirel Y (2018) Energy conservation. In: Dincer I (ed) Comprehensive energy systems. Elsevier, Amsterdam, vol 5, pp 45–90

9. Demirel Y (2018) Sugar versus lipid for sustainable biofuels. Int J Energy Res 42:881–884. https://doi.org/10.1002/er.3914

10. Demirel Y (2017) Lignin for sustainable bioproducts and biofuels. J Biochem Eng Bioprocess Technol 1:1–3

11. Ferroukhi R, Nagpal D, Lopez-Peña A, Hodges T, Mohtar RH, Daher B, Keulertz M (2015) Renewable energy in the water, energy & food nexus. IRENA, Abu Dhabi

12. Demirel Y, Gerbaud V (2019) Nonequilibrium thermodynamics transport and rate processes in physical, chemical and biological systems, 4th edn. Elsevier, Amsterdam

13. Galindo CP, Badr O (2007). Renewable hydrogen utilisation for the production of methanol. Energy Conversion and Management, 48:519–527

14. Geissdoerfer M, Savaget P, Bocken NMP, Hultink EJ (2017) The circular economy 'a new sustainability paradigm? J Cleaner Prod 143:757–768

15. Gumber S, Gurumoorthy AVP (2018) https://www.sciencedirect.com/science/article/pii/B9780444463903500025X?via%3Dihub. Chapter 25-Methanol economy versus hydrogen economy. Methanol Sci Eng 661–674

16. Hall CAS, Lambert JG, Balogh SB (2014) EROI of different fuels and the implications for society. Energy Policy 64:141–152

17. Hall CAS, Lambert JG, Balogh SB, Gupta A, Arnold M (2014) EROI and quality of life. Energy Policy 64153

18. IEA (2019) World energy outlook 2019. IEA, Paris. https://www.iea.org/reports/world-energy-outlook-2019

19. IEO (2019) International energy outlook 2019 with projections to 2050. September 2019, U. S. Energy Information Administration, U.S. Department of Energy, Washington, DC 20585. https://www.eia.gov/ieo

20. IRENA (2020) Renewable power generation costs in 2019. International Renewable Energy Agency, Abu Dhabi

21. IRENA (2020) Global renewables outlook: energy transformation 2050 (Edition: 2020). International Renewable Energy Agency, Abu Dhabi

22. Iaquaniello G, Centi G, Salladini A, Palo E (2018) Methanol economy: environment, demand, and marketing with a focus on the waste-to-methanol process. Methanol. Elsevier, Amsterdam

23. Jaffe RL, Taylor W (2018) The physics of energy. Cambridge Univ. Press, Cambridge

24. Küçük K, Tevatia R, Sorgüven E, Demirel Y, Özilgen M (2015) Bioenergetics of growth and lipid production in *Chlamydomonas reinhardtii*. Energy 83:503–510

25. Kumar A, Demirel Y, Jones DD, Hanna MA (2010) Optimization and economic evaluation of industrial gas production and combined heat and power generation from gasification of corn stover and distillers grains. Biores Technol 101:3696–3701

26. Lago C, Herrera I, Caldés N, Lechón Y (2019) Nexus bioenergy bioeconomy. In: Lago C, Caldés N, Lechón Y (eds) The role of bioenergy in the emerging bioeconomy. Amsterdam, Elsevier
27. McCardell SB (2014) Effective energy use: rewards and excitement. In: Anwar S (ed) Encyclopedia of energy engineering and technology, 2nd edn. CRC Press, Boca Raton
28. Matzen M, Demirel Y (2016) Methanol and dimethyl ether from renewable hydrogen and carbon dioxide: alternative fuels production and life-cycle assessment. J Cleaner Prod 139:1068–1077
29. Matzen M, Alhajji M, Demirel Y (2015) Chemical storage of wind energy by renewable methanol production: Feasibility analysis using a multi-criteria decision matrix. Energy 93:343–353
30. National Research Council (2011) Renewable fuel standard: potential economic and environmental effects of U.S. biofuel policy. The National Academies Press, Washington, DC
31. Nguyen N, Demirel Y (2013) Economic analysis of biodiesel and glycerol carbonate production plant by glycerolysis. J Sustain Bioenergy Syst 3:209–216
32. Nguyen N, Demirel Y (2011) A novel biodiesel and glycerol carbonate production plant. Int J Chemical Reactor Eng 9:1–25
33. Nguyen N, Demirel Y (2011) Using thermally coupled reactive distillation columns in biodiesel production. Energy 36:4838–4847
34. Olah GA, Goeppert A, Prakash GS (2011) Beyond oil and gas: The methanol economy. Wiley, Weinheim
35. Ptasinski KJ (2016) Efficiency of biomass energy. An Exergy Approach to Biofuels, Power, and Biorefineries. Wiley, New York, p 2016
36. Price L, Lu H (2011) Industrial energy auditing and assessments: a survey of programs around the world. In: European Council for an Energy Efficient Economy (ECEEE) 2011
37. Singh R, Tevatia R, White D, Demirel Y, Blum P (2019) Comparative kinetic modeling of growth and molecular hydrogen overproduction by engineered strains of *Thermotoga maritima*. Int J Hydrogen Energy 44:7125–7136
38. Theising TR (2016) Preparing for a successful energy assessment. AIChE, Chem Eng Prog May, pp 44–49
39. Theis J (2019) Quality guidelines for energy systems studies: cost estimation methodology for NETL assessments of power plant performance. In: Quality guidelines for energy systems studies: cost estimation methodology for NETL assessments of power plant performance (Technical Report)|OSTI.GOV
40. Wang X, Demirel Y (2018) Feasibility of power and methanol production by an entrained-flow coal gasification system. Energy Fuels 32:7595–7610
41. Withers J, Quesada H, Smith RL (2017) Bioeconomy survey results regarding barriers to the United States. Adv Biofuel Industry, Bio Resources 12:2846–2863

Appendix A
Physical and Critical Properties

See Tables A1 and A2.

Table A1 Physical properties of various organic and inorganic substances (1. Himmelblau DM, Riggs JB (2012) Basic Principles and calculations in chemical engineering 8th edn. Prentice Hall, Upper Saddle River; 2. Poling BE, Prausnitz J, O'Connel JP (2001) The properties of gases and liquids, 5th edn. McGraw-Hill, New York

Compound	Formula	MW	Sp Gr	T_m (K)	T_b (K)	ΔH_v (kJ/kg)	ΔH_m (kJ/kg)
Air		28.97					
Ammonia	NH_3	17.03	0.817	195.4	239.7	1374.0	322.4
Benzene	C_6H_6	78.11	0.879	278.7	353.3	394.3	126.0
Benzyl alcohol	C_7H_8O	108.13	1.045	257.8	478.4		
Butane	$n–C_4H_{10}$	58.12	0.579	134.8	272.7	383.6	80.3
Iso-Butane	$i–C_4H_{10}$	58.12	0.557	113.6	261.4	366.4	105.7
Carbon dioxide	CO_2	44.01	1.530	217.0			
Carbon disulfide	CS_2	76.14	1.261	161.1	319.4	351.9	
Carbon monoxide	CO	28.01	0.968	68.1	81.7	214.2	
Cyclohexane	C_6H_{12}	84.16	0.779	279.8	353.9	357.6	
Cyclopentane	C_3H_{10}	70.13	0.745	179.7	322.4	389.2	
Diethyl ether	$(C_2H_5)_2O$	74.12	0.708	156.9	307.8	352.1	
Ethane	C_2H_6	30.07	1.049	89.9	184.5	488.8	
Ethanol	C_2H_6O	46.07	0.789	158.6	351.7	837.8	109.0
Ethylene glycol	$C_2H_6O_2$	62.07	1.113	260.0	470.4	916.7	181.1
Glycerol	$C_3H_8O_3$	92.09	1.260	291.4	563.2		200.6
Heptane	C_7H_{16}	100.20	0.684	182.6	371.6	316.3	
Hexane	C_6H_{14}	86.17	0.659	177.8	341.9	335.3	
Hydrogen	H_2	2.02	0.069	14.0	20.4	445.5	59.5
Hydrogen chloride	HCl	36.47	1.268	158.9	188.1	444.2	
Hydrogen sulfide	H_2S	34.08	1.189	187.6	212.8	548.7	
Mercury	Hg	200.61	13.546				
Methane	CH_4	16.04	0.554	90.7	111.7	511.2	58.4
Methanol	CH_3OH	32.04	0.792	175.3	337.9	1101.7	99.2
Nitric acid	HNO_3	63.02	1.502	231.6	359.0	480.7	
Nitrogen	N_2	28.02	12.500	63.2	77.3	199.8	25.3
Nitrogen dioxide	NO_2	46.01	1.448	263.9	294.5	319.4	
Nitrogen oxide	NO	30.01	1.037	109.5	121.4	459.8	

(continued)

© Springer Nature Switzerland AG 2021
Y. Demirel, *Energy*, Green Energy and Technology,
https://doi.org/10.1007/978-3-030-56164-2

Table A1 (continued)

Compound	Formula	MW	Sp Gr	T_m (K)	T_b (K)	ΔH_v (kJ/kg)	ΔH_m (kJ/kg)
Nitrogen pentoxide	N_2O_5	108.02	1.630	303.0	320.0		
Nitrous oxide	N_2O	44.02	1.226	182.1	184.4		
Oxygen	O_2	32.00	1.105	54.4		212.5	13.7
n-Pentane	C_5H_{12}	72.15	0.630	143.5	309.2	357.5	
Iso-pentane	$i-C_5H_{12}$	72.15	0.621	113.1	300.9		
Propane	C_3H_8	44.09	1.562	85.5	18.8		80.0
Propylene	C_3H_6	1500	7.792	1.637	22.706	−6.915	
n-Propyl alcohol	C_3H_8O	60.09	0.804	146.0			
Iso-Propyl alcohol	C_3H_8O	60.09	0.785	183.5			
n-Propyl benzene	C_9H_{12}	120.19	0.862	173.7	38.2		
Sodium hydroxide	NaOH	40.00	2.130	592.0	1663.0		
Sulfur dioxide	SO_2	64.07	2.264	197.7	263.1	388.6	
Sulfur trioxide	SO_3	80.07	2.750	290.0	316.5	522.0	
Sulfuric acid	H_2SO_4	98.08	1.834	283.5			
Toluene	$C_6H_5CH_3$	92.13	0.866	178.2	383.8	363.6	
Water	H_2O	18.02	1.000	273.2	373.2	2253.0	333.7

Table A2 Critical properties

Species	MW	ω	T_c (K)	P_c (atm)	V_c (cm^3/gmol)
Air	28.97	0.035	132.5	37.2	88.3
H_2	2.02	−0.216	33.3	12.8	65.0
Air	28.97	0.035	132.0	36.4	86.6
N_2	28.01	0.038	126.2	33.5	90.1
O_2	32.00	0.022	154.4	49.7	74.4
CO	28.01	0.048	132.9	34.5	93.1
CO_2	44.01	0.224	304.2	72.8	94.1
NO	30.10	0.583	180.0	64.0	57.0
N_2O	44.01	0.141	309.7	71.7	96.3
SO_2	64.06	0.245	430.7	77.8	122.0
Cl_2	70.91	0.069	417.0	76.1	124.0
CH_4	16.04	0.012	191.1	45.8	98.7
C_2H_6	30.07	0.100	305.4	48.2	148.0
C_3H_6	42.08	0.140	365.0	45.5	181.0
C_3H_8	44.10	0.152	369.8	41.9	200.0
$n-C_4H_{10}$	58.12	0.200	425.2	37.5	255.0
$i-C_4H_{10}$	58.12	0.181	408.1	36.0	263.0
$n-C_5H_{12}$	72.15	0.252	469.5	33.2	311.0
$n-C_6H_{14}$	86.18	0.301	507.3	29.7	370.0
$n-C_7H_{16}$	100.20	0.350	540.1	27.0	432.0
$n-C_8H_{18}$	114.23	0.400	568.7	24.5	492.0
$n-C_9H_{20}$	128.26	0.444	594.6	22.6	548.0
Cyclohexane	84.16	0.210	553.0	40.0	308.0
Benzene	78.11	0.210	562.6	48.6	260.0

Appendix B
Heat Capacities

See Tables B1, B2, B3, and B4.

Table B1 Heat capacities in the ideal-gas state: $C_p^{ig}/R = A + BT + CT^2$; $T = 298$ K to T_{max} K

Chemical species	Formula	T_{max} (K)	C_p^{ig}/R	A	$10^3 B$	$10^6 C$
Methane	CH_4	1500	4.217	1.702	9.081	−2.164
Ethane	C_2H_6	1500	6.369	1.131	19.225	−5.561
Propane	C_3H_8	1500	9.011	1.213	28.785	−8.824
Acetylene	C_2H_2	26.04	0.906	191.7	191.7	672.0
Ethylene	C_2H_4	28.05	0.975	104.0	169.5	481.2
n-Butane	C_4H_{10}	1500	11.928	1.935	36.915	−11.402
n-Pentane	C_5H_{12}	1500	14.731	2.464	45.351	−14.111
n-Hexane	C_6H_{14}	1500	17.550	3.025	53.722	−16.791
n-Heptane	C_7H_{16}	1500	20.361	3.570	62.127	−19.486
n-Octane	C_8H_{18}	1500	23.174	4.108	70.567	−22.208
Benzene	C_6H_6	1500	10.259	−0.206	39.064	−13.301
Cyclohexane	C_6H_{12}	1500	13.121	−3.876	63.249	−20.928
Ethanol	C_2H_6O	1500	8.948	3.518	20.001	−6.002
Methanol	CH_4O	1500	5.547	2.211	12.216	−3.450
Toluene	C_7H_8	1500	12.922	0.290	47.052	−15.716
Air		2000	3.509	3.355	0.575	–
Ammonia	NH_3	1800	4.269	3.578	3.020	–
Carbon monoxide	CO	2500	3.507	3.376	0.557	–
Carbon dioxide	CO_2	2000	4.467	5.457	1.045	–
Chlorine	Cl_2	3000	4.082	4.442	0.089	–
Hydrogen	H_2	3000	3.468	3.249	0.422	–
Nitrogen	N_2	2000	3.502	3.280	0.593	–

(continued)

© Springer Nature Switzerland AG 2021
Y. Demirel, *Energy*, Green Energy and Technology,
https://doi.org/10.1007/978-3-030-56164-2

Table B1 (continued)

Chemical species	Formula	T_{max} (K)	C_p^{ig}/R	A	$10^3 B$	$10^6 C$
Nitrogen dioxide	NO_2	2000	4.447	4.982	1.195	–
Oxygen	O_2	2000	3.535	3.639	0.506	–
Sulfur dioxide	SO_2	2000	4.796	5.699	0.801	–
Sulfur trioxide	SO_3	2000	6.094	8.060	1.056	–
Water	H_2O	2000	4.038	3.470	1.450	–

Table B2 Heat capacities of liquids
constants for the equation $C_p/R = A + BT + CT^2$; $T = 298\,K$ to $T_{max}\,K$

Chemical species	C_p/R	A	$10^3 B$	$10^6 C$
Ammonia	9.718	22.626	−100.75	192.71
Aniline	23.070	15.819	29.03	−15.80
Benzene	16.157	−0.747	67.96	−37.78
Carbon tetrachloride	15.751	21.155	−48.28	101.14
Chlorobenzene	18.240	11.278	32.86	−31.90
Chloroform	13.806	19.215	−42.89	83.01
Cyclohexane	18.737	−9.048	141.38	−161.62
Ethanol	13.444	33.866	−172.60	349.17
Ethylene oxide	10.590	21.039	−86.41	172.28
Methanol	9.798	13.431	−51.28	131.13
n-Propanol	16.921	41.653	−210.32	427.20
Toluene	18.611	15.133	6.79	16.35
Water	9.069	8.712	1.25	−0.18

Table B3 Heat capacities of solids[+]
constants for the equation $C_p/R = A + BT + CT^{-2}$; $T = 298$ K to T_{max} K

Chemical species	T_{max} (K)	C_p/R	A	$10^3 B$	$10^{-5} C$
CaO	2000	5.058	6.104	0.443	−1.047
$CaCO_3$	1200	9.848	12.572	2.637	−3.120
$CaCl_2$	1055	8.762	8.646	1.530	−0.302
C (graphite)	2000	1.026	1.771	0.771	−0.867
Cu	1357	2.959	2.677	0.815	0.035
CuO	1400	5.087	5.780	0.973	−0.874
Fe	1043	3.005	−0.111	6.111	1.150
FeS	411	6.573	2.612	13.286	
NH_4Cl	458	10.741	5.939	16.105	
NaOH	566	7.177	0.121	16.316	1.948
SiO_2 (quartz)	847	5.345	4.871	5.365	−1.001

Table B4 Ideal-gas specific heats of various common gases (Cengel YA, Boles MA (2014) Thermodynamics: An engineering approach, 8th edn. McGraw-Hill, New York)

T (K)	C_p (kJ/kg K)	C_v (kJ/kg K)	γ	C_p (kJ/kg K)	C_v (kJ/kg K)	γ	C_p (kJ/kg K)	C_v (kJ/kg K)	γ
	Air			Carbon dioxide CO_2			Carbon monoxide CO		
250	1.003	0.716	1.401	0.791	0.602	1.314	1.039	0.743	1.400
300	1.005	0.718	1.400	0.846	0.657	1.288	1.040	0.744	1.399
350	1.008	0.721	1.398	0.895	0.706	1.268	1.043	0.746	1.398
400	1.013	0.726	1.395	0.939	0.750	1.252	1.047	0.751	1.395
450	1.020	0.733	1.391	0.978	0.790	1.239	1.054	0.757	1.392
500	1.029	0.742	1.387	1.014	0.825	1.229	1.063	0.767	1.387
550	1.040	0.753	1.381	1.046	0.857	1.220	1.075	0.778	1.382
600	1.051	0.764	1.376	1.075	0.886	1.213	1.087	0.790	1.376
650	1.063	0.776	1.370	1.102	0.913	1.207	1.100	0.803	1.370
700	1.075	0.788	1.364	1.126	0.937	1.202	1.113	0.816	1.364
750	1.087	0.800	1.359	1.148	0.959	1.197	1.126	0.829	1.358
800	1.099	0.812	1.354	1.169	0.980	1.193	1.139	0.842	1.353
900	1.121	0.834	1.344	1.204	1.015	1.186	1.163	0.866	1.343
1000	1.142	0.855	1.336	1.234	1.045	1.181	1.185	0.888	1.335
	Hydrogen, H_2			Nitrogen, N_2			Oxygen, O_2		
250	14.051	9.927	1.416	1.039	0.742	1.400	0.913	0.653	1.398
300	14.307	10.183	1.405	1.039	0.743	1.400	0.918	0.658	1.395
350	14.427	10.302	1.400	1.041	0.744	1.399	0.928	0.668	1.389
400	14.476	10.352	1.398	1.044	0.747	1.397	0.941	0.681	1.382
450	14.501	10.377	1.398	1.049	0.752	1.395	0.956	0.696	1.373
500	14.513	10.389	1.397	1.056	0.759	1.391	0.972	0.712	1.365
550	14.530	10.405	1.396	1.065	0.768	1.387	0.988	0.728	1.358
600	14.546	10.422	1.396	1.075	0.778	1.382	1.003	0.743	1.350
650	14.571	10.447	1.395	1.086	0.789	1.376	1.017	0.758	1.343
700	14.604	10.480	1.394	1.098	0.801	1.371	1.031	0.771	1.337
750	14.645	10.521	1.392	1.110	0.813	1.365	1.043	0.783	1.332
800	14.695	10.570	1.390	1.121	0.825	1.360	1.054	0.794	1.327
900	14.822	10.698	1.385	1.145	0.849	1.349	1.074	0.814	1.319
1000	14.983	10.859	1.380	1.167	0.870	1.341	1.090	0.830	1.313

Appendix C
Enthalpies and Gibbs Energies of Formation

See Table C1.

Table C1 Standard enthalpies and Gibbs energies of formation at 298.15 K

Chemical species	Formula	State	ΔH_{f298} (kJ/mol)	ΔG_{f298} (kJ/mol)
Methane	CH_4	(g)	−74.520	−50.460
Ethane	C_2H_6	(g)	−83.820	−31.855
Propane	C_3H_8	(g)	−104.680	−24.290
n-Butane	C_4H_{10}	(g)	−125.790	−16.570
n-Pentane	C_5H_{12}	(g)	−146.760	−8.650
Benzene	C_6H_6	(g)	82.930	129.665
Benzene	C_6H_6	(l)	49.080	124.520
Cyclohexane	C_6H_{12}	(g)	−123.140	31.920
Cyclohexane	C_6H_{12}	(l)	−156.230	26.850
Ethanol	C_2H_6O	(g)	−235.100	−168.490
Ethanol	C_2H_6O	(l)	−277.690	−174.780
Methanol	CH_4O	(g)	−200.660	−161.960
Methanol	CH_4O	(l)	−238.660	−166.270
Toluene	C_7H_8	(g)	50.170	122.050
Toluene	C_7H_8	(l)	12.180	113.630
Ammonia	NH_3	(g)	−46.110	−16.450
Calcium carbonate	$CaCO_3$	(s)	−1206.920	−1128.790
Calcium oxide	CaO	(s)	−635.090	−604.030
Carbon dioxide	CO_2	(g)	−393.509	−394.359
Carbon monoxide	CO	(g)	−110.525	−137.169
Hydrochloric acid	HCl	(g)	−92.307	−95.299
Hydrogen sulfide	H_2S	(g)	−20.630	−33.560
Iron oxide	FeO	(s)	−272.000	

(continued)

© Springer Nature Switzerland AG 2021
Y. Demirel, *Energy*, Green Energy and Technology,
https://doi.org/10.1007/978-3-030-56164-2

Table C1 (continued)

Chemical species	Formula	State	ΔH_{f298} (kJ/mol)	ΔG_{f298} (kJ/mol)
Nitric acid	HNO_3	(l)	−174.100	−80.710
Nitrogen oxides	NO	(g)	90.250	86.550
	NO_2	(g)	33.180	51.310
Sodium carbonate	Na_2CO_3	(s)	−1130.680	−1044.440
Sodium chloride	NaCl	(s)	−411.153	−384.138
Sodium hydroxide	NaOH	(s)	−425.609	−379.494
Sulfur dioxide	SO_2	(g)	−296.830	−300.194
Sulfur trioxide	SO_3	(g)	−395.720	−371.060
Sulfur trioxide	SO3	(l)	−441.040	
Water	H_2O	(g)	−241.818	−228.572
Water	H_2O	(l)	−285.830	−237.129

Appendix D
Ideal Gas Properties of Some Common Gases

See Tables D1, D2, and D3.

Table D1 Ideal gas properties of air

T (K)	H (kJ/kg)	P_r	U (kJ/kg)	V_r	S (kJ/kg K)
270	270.11	0.9590	192.60	808.0	1.596
280	280.13	1.0889	199.75	783.0	1.632
285	285.14	1.1584	203.33	706.1	1.650
290	290.16	1.2311	206.91	676.1	1.668
295	295.17	1.3068	210.49	647.9	1.685
300	300.19	1.3860	214.07	621.2	1.702
305	305.22	1.4686	217.67	596.0	1.718
310	310.24	1.5546	221.25	572.3	1.734
315	315.27	1.6442	224.85	549.8	1.751
320	320.29	1.7375	228.42	528.6	1.766
325	325.31	1.8345	232.02	508.4	1.782
330	330.34	1.9352	235.61	489.4	1.797
340	340.42	2.149	242.82	454.1	1.827
350	350.49	2.379	250.02	422.2	1.857
360	360.58	2.626	257.24	393.4	1.885
370	370.67	2.892	264.46	367.2	1.913
380	380.77	3.176	271.69	343.4	1.940
390	390.88	3.481	278.93	321.5	1.966
400	400.98	3.806	286.16	301.6	1.991
410	411.12	4.153	293.43	283.3	2.016
420	421.26	4.522	300.69	266.6	2.041
430	431.43	4.915	307.99	251.1	2.065
440	441.61	5.332	315.30	236.8	2.088
450	451.80	5.775	322.62	223.6	2.111

(continued)

Table D1 (continued)

T (K)	H (kJ/kg)	P_r	U (kJ/kg)	V_r	S (kJ/kg K)
460	462.02	6.245	329.97	211.4	2.134
470	472.24	6.742	337.32	200.1	2.156
480	482.49	7.268	344.70	189.5	2.177
490	492.74	7.824	352.08	179.7	2.198
500	503.02	8.411	359.49	170.6	2.219
510	513.32	9.031	366.92	162.1	2.239
520	523.63	9.684	374.36	154.1	2.259
530	533.98	10.37	381.84	146.7	2.279
540	544.35	11.10	389.34	139.7	2.299
550	555.74	11.86	396.86	133.1	2.318
560	565.17	12.66	404.42	127.0	2.336
570	575.59	13.50	411.97	121.2	2.355
580	586.04	14.38	419.55	115.7	2.373
590	596.52	15.31	427.15	110.6	2.391
600	607.02	16.28	434.78	105.8	2.409
610	617.53	17.30	442.42	101.2	2.426
620	628.07	18.36	450.09	96.92	2.443
630	683.63	19.84	457.78	92.84	2.460
640	649.22	20.64	465.50	88.99	2.477
650	659.84	21.86	473.25	85.34	2.493
660	670.47	23.13	481.01	81.89	2.509
670	681.14	24.46	488.81	78.61	2.525
680	691.82	25.85	496.62	75.50	2.541
690	702.52	27.29	504.45	72.56	2.557
700	713.27	28.80	512.33	69.76	2.572
710	724.04	30.38	520.23	67.07	2.588
720	734.82	32.02	528.14	64.53	2.603
730	745.62	33.72	536.07	62.13	2.618
740	756.44	35.50	544.02	59.82	2.632
750	767.29	37.35	551.99	57.63	2.647
760	778.18	39.27	560.01	55.54	2.661
780	800.03	43.35	576.12	51.64	2.690
800	821.95	47.75	592.30	48.08	2.717
820	843.98	52.59	608.59	44.84	2.745
840	866.08	57.60	624.95	41.85	2.771
860	888.27	63.09	641.40	39.12	2.797
880	910.56	68.98	657.95	36.61	2.823
900	932.93	75.29	674.58	34.31	2.848
920	955.38	82.05	691.28	32.18	2.873
940	977.92	89.28	708.08	30.22	2.897

(continued)

Table D1 (continued)

T (K)	H (kJ/kg)	P_r	U (kJ/kg)	V_r	S (kJ/kg K)
960	1000.55	97.00	725.02	28.40	2.921
980	1023.25	105.2	741.98	26.73	2.944
1000	1046.04	114.0	758.94	25.17	2.967
1020	1068.89	123.4	776.10	23.72	2.990
1040	1091.85	133.3	793.36	23.29	3.012
1060	1114.86	143.9	810.62	21.14	3.034
1080	1137.89	155.2	827.88	19.98	3.056
1100	1161.07	167.1	845.33	18.896	3.077
1120	1184.28	179.7	862.79	17.886	3.098
1140	1207.57	193.1	880.35	16.946	3.118
1160	1230.92	207.2	897.91	16.064	3.139
1180	1254.34	222.2	915.57	15.241	3.159
1200	1277.79	238.0	933.33	14.470	3.178
1220	1301.31	254.7	951.09	13.747	3.198
1240	1324.93	272.3	968.95	13.069	3.217
1260	1348.55	290.8	986.90	12.435	3.236
1280	1372.24	310.4	1004.76	11.835	3.255
1300	1395.97	330.9	1022.82	11.275	3.273
1320	1419.76	352.5	1040.88	10.747	3.291
1340	1443.60	375.3	1058.94	10.247	3.309
1360	1467.49	399.1	1077.10	9.780	3.327
1400	1515.42	450.5	1113.52	8.919	3.362
1460	1587.63	537.1	1168.49	7.801	3.412
1500	1635.97	601.9	1205.41	7.152	3.445
1560	1708.82	710.5	1260.99	6.301	3.492
1600	1757.57	791.2	1298.30	5.804	3.523
1640	1806.46	878.9	1335.72	5.355	3.553
1700	1880.1	1025	1392.7	4.761	3.597
1750	1941.6	1161	1439.8	4.328	3.633
1800	2003.3	1310	1487.2	3.994	3.668
1850	2065.3	1475	1534.9	3.601	3.702
1900	2127.4	1655	1582.6	3.295	3.735
1950	2189.7	1852	1630.6	3.022	3.767
2000	2252.1	2068	1678.8	2.776	3.799
2100	2377.7	2559	1775.3	2.356	3.860

Table D2   Ideal gas properties of carbon dioxide, CO_2

T (K)	H (kJ/kmol)	U (kJ/kmol)	S (kJ/kmol K)	T (K)	H (kJ/kmol)	U (kJ/kmol)	S (kJ/kmol K)
0	0.000	0.000	0.000	600	22.280	17.291	243.199
220	6.601	4.772	202.966	610	22.754	17.683	243.983
230	6.938	5.026	204.464	620	23.231	18.076	244.758
240	7.280	5.285	205.920	630	23.709	18.471	245.524
250	7.627	5.548	207.337	640	24.190	18.869	246.282
260	7.979	5.817	208.717	650	24.674	19.270	247.032
270	8.335	6.091	210.062	660	25.160	19.672	247.773
280	8.697	6.369	211.376	670	25.648	20.078	248.507
290	9.063	6.651	212.660	680	26.138	20.484	249.233
298	9.364	6.885	213.685	690	26.631	20.894	249.952
300	9.431	6.939	213.915	700	27.125	21.305	250.663
310	9.807	7.230	215.146	710	27.622	21.719	251.368
320	10.186	7.526	216.351	720	28.121	22.134	252.065
330	10.570	7.826	217.534	730	28.622	22.522	252.755
340	10.959	8.131	218.694	740	29.124	22.972	253.439
350	11.351	8.439	219.831	750	29.629	23.393	254.117
360	11.748	8.752	220.948	760	20.135	23.817	254.787
370	12.148	9.068	222.044	770	30.644	24.242	255.452
380	12.552	9.392	223.122	780	31.154	24.669	256.110
390	12.960	9.718	224.182	790	31.665	25.097	256.762
400	13.372	10.046	225.225	800	32.179	25.527	257.408
410	13.787	10.378	226.250	810	32.694	25.959	258.048
420	14.206	10.714	227.258	820	33.212	26.394	258.682
430	14.628	11.053	228.252	830	33.730	26.829	259.311
440	15.054	11.393	229.230	840	34.251	27.267	259.934
450	15.483	11.742	230.194	850	34.773	27.706	260.551
460	15.916	12.091	231.144	860	35.296	28.125	261.164
470	16.351	12.444	232.080	870	35.821	28.588	261.770
480	16.791	12.800	233.004	880	36.347	29.031	262.371
490	17.232	13.158	233.916	890	36.876	29.476	262.968
500	17.678	13.521	234.814	900	37.405	29.922	263.559
510	18.126	13.885	235.700	910	37.935	30.369	264.146
520	18.576	14.253	236.575	920	38.467	30.818	264.728
530	19.029	14.622	237.439	930	39.000	31.268	265.304
540	19.485	14.996	238.292	940	39.535	31.719	265.877
550	19.945	15.372	239.135	950	40.070	32.171	266.444
560	20.407	15.751	239.962	960	40.607	32.625	267.007

(continued)

Table D2 (continued)

T (K)	H (kJ/kmol)	U (kJ/kmol)	S (kJ/kmol K)	T (K)	H (kJ/kmol)	U (kJ/kmol)	S (kJ/kmol K)
570	20.870	16.131	240.789	970	41.145	33.081	267.566
580	21.337	16.515	241.602	980	41.685	33.537	268.119
590	21.807	16.902	242.405	990	42.226	33.995	268.670

Table D3 Ideal gas properties of hydrogen, H_2

T (K)	H (kJ/kmol)	U (kJ/kmol)	S (kJ/kmol K)	T (K)	H (kJ/kmol)	U (kJ/kmol)	S (kJ/kmol K)
0	0.000	0.000	0.000	1440	42.808	30.835	177.410
260	7.370	5.209	126.636	1480	44.091	31.786	178.291
270	7.657	5.412	127.719	1520	45.384	32.746	179.153
280	7.945	5.617	128.765	1560	46.683	33.713	179.995
290	8.233	5.822	129.775	1600	47.990	34.687	180.820
298	8.468	5.989	130.574	1640	49.303	35.668	181.632
300	8.522	6.027	130.754	1680	50.622	36.654	182.428
320	9.100	6.440	132.621	1720	51.947	37.646	183.208
340	9.680	6.853	134.378	1760	53.279	38.645	183.973
360	10.262	7.268	136.039	1800	54.618	39.652	184.724
380	10.843	7.684	137.612	1840	55.962	40.663	185.463
400	11.426	8.100	139.106	1880	57.311	41.680	186.190
420	12.010	8.518	140.529	1920	58.668	42.705	186.904
440	12.594	8.936	141.888	1960	60.031	43.735	187.607
460	13.179	9.355	143.187	2000	61.400	44.771	188.297
480	13.764	9.773	144.432	2050	63.119	46.074	189.148
500	14.350	10.193	145.628	2100	64.847	47.386	189.979
520	14.935	10.611	146.775	2150	66.584	48.708	190.796
560	16.107	11.451	148.945	2200	68.328	50.037	191.598
600	17.280	12.291	150.968	2250	70.080	51.373	192.385
640	18.453	13.133	152.863	2300	71.839	52.716	193.159
680	19.630	13.976	154.645	2350	73.608	54.069	193.921
720	20.807	14.821	156.328	2400	75.383	55.429	194.669
760	21.988	15.669	157.923	2450	77.168	56.798	195.403
800	23.171	16.520	159.440	2500	78.960	58.175	196.125
840	24.359	17.375	160.891	2550	80.755	59.554	196.837
880	25.551	18.235	162.277	2600	82.558	60.941	197.539
920	26.747	19.098	163.607	2650	84.368	62.335	198.229
960	27.948	19.966	164.884	2700	86.186	63.737	198.907

(continued)

Table D3 (continued)

T (K)	H (kJ/kmol)	U (kJ/kmol)	S (kJ/kmol K)	T (K)	H (kJ/kmol)	U (kJ/kmol)	S (kJ/kmol K)
1000	29.154	20.839	166.114	2750	88.008	65.144	199.575
1040	30.364	21.717	167.300	2800	89.838	66.558	200.234
1080	31.580	22.601	168.449	2850	91.671	67.976	200.885
1120	32.802	23.490	169.560	2900	93.512	69.401	201.527
1160	34.028	24.384	170.636	2950	95.358	70.831	202.157
1200	35.262	25.284	171.682	3000	97.211	72.268	202.778
1240	36.502	26.192	172.698	3050	99.065	73.707	203.391
1280	37.749	27.106	173.687	3100	100.926	75.152	203.995
1320	39.002	28.027	174.652	3150	102.793	76.604	204.592
1360	40.263	28.955	175.593	3200	104.667	78.061	205.181
1400	41.530	29.889	176.510	3250	106.545	79.523	205.765

Appendix E
Thermochemical Properties

See Tables E1, E2, E3, and E4.

Table E1 Saturated refrigerant-134a

		V (ft^3/lb)		U (Btu/lb)		H (Btu/lb)		S (Btu/lb R)	
T (°F)	P_{sat} (psia)	V_l	V_g	U_l	U_g	H_l	H_g	S_l	S_g
−40	7.490	0.01130	5.7173	−0.02	87.90	0.00	95.82	0.0000	0.2283
−30	9.920	0.01143	4.3911	2.81	89.26	2.83	97.32	0.0067	0.2266
−20	12.949	0.01156	3.4173	5.69	90.62	5.71	98.81	0.0133	0.2250
−15	14.718	0.01163	3.0286	7.14	91.30	7.17	99.55	0.0166	0.2243
−10	16.674	0.01170	2.6918	8.61	91.98	8.65	100.29	0.0199	0.2236
−5	18.831	0.01178	2.3992	10.09	92.66	10.13	101.02	0.0231	0.2230
0	21.203	0.01185	2.1440	11.58	93.33	11.63	101.75	0.0264	0.2224
5	23.805	0.01193	1.9208	13.09	94.01	13.14	102.47	0.0296	0.2219
10	26.651	0.01200	1.7251	14.60	94.68	14.66	103.19	0.0329	0.2214
15	29.756	0.01208	1.5529	16.13	95.35	16.20	103.90	0.0361	0.2209
20	33.137	0.01216	1.4009	17.67	96.02	17.74	104.61	0.0393	0.2205
25	36.809	0.01225	1.2666	19.22	96.69	19.30	105.32	0.0426	0.2200
30	40.788	0.01233	1.1474	20.78	97.35	20.87	106.01	0.0458	0.2196
40	49.738	0.01251	0.9470	23.94	98.67	24.05	107.39	0.0522	0.2189
50	60.125	0.01270	0.7871	27.14	99.98	27.28	108.74	0.0585	0.2183
60	72.092	0.01290	0.6584	30.39	101.27	30.56	110.05	0.0648	0.2178
70	85.788	0.01311	0.5538	33.68	102.54	33.89	111.33	0.0711	0.2173
80	101.370	0.01334	0.4682	37.02	103.78	37.27	112.56	0.0774	0.2169
85	109.920	0.01346	0.4312	38.72	104.39	38.99	113.16	0.0805	0.2167
90	118.990	0.01358	0.3975	40.42	105.00	40.72	113.75	0.0836	0.2165

(continued)

© Springer Nature Switzerland AG 2021
Y. Demirel, *Energy*, Green Energy and Technology,
https://doi.org/10.1007/978-3-030-56164-2

Table E1 (continued)

T (°F)	P_{sat} (psia)	V (ft³/lb)		U (Btu/lb)		H (Btu/lb)		S (Btu/lb R)	
		V_l	V_g	U_l	U_g	H_l	H_g	S_l	S_g
95	128.620	0.01371	0.3668	42.14	105.60	42.47	114.33	0.0867	0.2163
100	138.830	0.01385	0.3388	43.87	106.18	44.23	114.89	0.0898	0.2161
105	149.630	0.01399	0.3131	45.62	106.76	46.01	115.43	0.0930	0.2159
110	161.040	0.01414	0.2896	47.39	107.33	47.81	115.96	0.0961	0.2157
115	173.100	0.01429	0.2680	49.17	107.88	49.63	116.47	0.0992	0.2155
120	185.820	0.01445	0.2481	50.97	108.42	51.47	116.95	0.1023	0.2153
140	243.860	0.01520	0.1827	58.39	110.41	59.08	118.65	0.1150	0.2143
160	314.630	0.01617	0.1341	66.26	111.97	67.20	119.78	0.1280	0.2128
180	400.220	0.01758	0.0964	74.83	112.77	76.13	119.91	0.1417	0.2101
200	503.520	0.02014	0.0647	84.90	111.66	86.77	117.69	0.1575	0.2044
210	563.510	0.02329	0.0476	91.84	108.48	94.27	113.45	0.1684	0.1971

Table E2 Superheated refrigerant-R-134a

T (°F)	V (ft³/lb)	U (Btu/lb)	H (Btu/lb)	S (Btu/lb R)	V (ft³/lb)	U (Btu/lb)	H (Btu/lb)	S (Btu/lb R)	V (ft³/kg)	U (Btu/lb)	H (Btu/lb)	S (Btu/lb R)
	P = 10 psia (T_{sat} = −29.7 °F)				P = 15 psia (T_{sat} = −14.3 °F)				P = 20 psia (T_{sat} = −29.7 °F)			
Sat	4.358	89.3	97.3	0.226	2.974	91.4	99.7	0.224	2.266	93.0	101.3	0.222
0	4.702	94.2	102.9	0.239	3.089	93.8	102.4	0.230	2.281	93.4	101.8	0.223
20	4.929	97.6	106.8	0.247	3.246	97.3	106.3	0.238	2.404	96.9	105.8	0.232
40	5.154	101.2	110.7	0.255	3.401	100.9	110.3	0.246	2.524	100.6	109.9	0.240
60	5.375	104.8	114.7	0.263	3.553	104.5	114.4	0.254	2.641	104.2	114.0	0.248
80	5.596	108.5	118.8	0.271	3.703	108.2	118.5	0.262	2.757	108.0	118.2	0.256
100	5.814	112.3	123.0	0.278	3.852	112.1	122.8	0.270	2.870	111.9	122.5	0.264
120	6.032	116.2	127.3	0.286	3.999	116.0	127.1	0.278	2.983	115.8	126.8	0.272
140	6.248	120.1	131.7	0.293	4.145	120.0	131.5	0.285	3.094	119.8	131.3	0.279
160	6.464	124.2	136.2	0.301	4.291	124.1	136.0	0.292	3.204	123.9	135.8	0.286
180	6.678	128.4	140.7	0.308	4.436	128.2	140.5	0.300	3.314	128.1	140.4	0.292
200	6.893	132.6	145.4	0.315	4.580	132.5	145.2	0.307	3.423	132.4	145.0	0.301

T (°F)	V (ft³/lb)	U (Btu/lb)	H (Btu/lb)	S (Btu/lb R)	V (ft³/lb)	U (Btu/lb)	H (Btu/lb)	S (Btu/lb R)	V (ft³/lb)	U (Btu/lb)	H (Btu/lb)	S (Btu/lb R)
	P = 30 psia (T_{sat} = −29.7 °F)				P = 40 psia (T_{sat} = −29.7 °F)				P = 50 psia (T_{sat} = 40.3 °F)			
Sat	1.540	95.4	103.9	0.220	1.169	97.3	105.88	0.2197	0.942	98.71	107.4	0.218
40	1.646	99.9	109.1	0.231	1.206	99.3	108.26	0.2245				
60	1.729	103.7	113.3	0.239	1.272	103.2	112.62	0.2331	0.997	102.62	111.8	0.227
80	1.809	107.5	117.6	0.247	1.335	107.1	117.00	0.2414	1.050	106.62	116.3	0.236
100	1.888	111.4	121.9	0.255	1.397	111.0	121.42	0.2494	1.102	110.65	120.8	0.244
120	1.966	115.4	126.3	0.263	1.457	115.1	125.90	0.2573	1.152	114.74	125.3	0.252
140	2.042	119.5	130.8	0.271	1.516	119.2	130.43	0.2650	1.200	118.88	129.9	0.260
160	2.118	123.6	135.4	0.278	1.574	123.3	135.03	0.2725	1.284	123.08	134.6	0.267
180	2.192	127.8	140.0	0.285	1.631	127.6	139.70	0.2799	1.295	127.36	139.3	0.275
200	2.267	132.1	144.7	0.293	1.688	131.9	144.44	0.2872	1.341	131.71	144.1	0.282

(continued)

Table E2 (continued)

T (°F)	V (ft³/lb)	U (Btu/lb)	H (Btu/lb)	S (Btu/lb R)	V (ft³/lb)	U (Btu/lb)	H (Btu/lb)	S (Btu/lb R)	V (ft³/kg)	U (Btu/lb)	H (Btu/lb)	S (Btu/lb R)
	$P = 10$ psia ($T_{sat} = -29.7$ °F)				$P = 15$ psia ($T_{sat} = -14.3$ °F)				$P = 20$ psia ($T_{sat} = -29.7$ °F)			
220	2.340	136.5	149.5	0.300	1.744	136.3	149.25	0.2944	1.387	136.12	148.9	0.289
240					1.800	140.8	154.14	0.3015	1.432	140.61	153.8	0.296
260					1.856	145.3	159.10	0.3085	1.477	145.18	158.8	0.303
280					1.911	149.9	164.13	0.3154	1.522	149.82	163.9	0.310
	$P = 60$ psia ($T_{sat} = 49.9$ °F)				$P = 70$ psia ($T_{sat} = 58.4$ °F)				$P = 80$ psia ($T_{sat} = 65.9$ °F)			
Sat	0.788	100.0	108.7	0.218	0.677	101.0	109.83	0.2179	0.593	102.02	110.8	0.217
60	0.813	102.0	111.0	0.222	0.681	101.4	110.23	0.2186				
80	0.860	106.1	115.6	0.231	0.723	105.5	114.96	0.2276	0.621	105.03	114.2	0.223
100	0.905	110.2	120.2	0.239	0.764	109.7	119.66	0.2361	0.657	109.30	119.0	0.232
120	0.948	114.3	124.8	0.248	0.802	113.9	124.36	0.2444	0.692	113.56	123.8	0.241
140	0.990	118.5	129.5	0.255	0.839	118.2	129.07	0.2524	0.726	117.85	128.6	0.249
160	1.030	122.7	134.2	0.263	0.875	122.4	133.82	0.2601	0.758	122.18	133.4	0.257
180	1.070	127.1	138.9	0.271	0.910	126.8	138.62	0.2678	0.789	126.55	138.2	0.264
200	1.110	131.4	143.7	0.278	0.944	131.2	143.46	0.2752	0.820	130.98	143.1	0.272
220	1.148	135.9	148.6	0.285	0.978	135.6	148.36	0.2825	0.850	135.47	148.0	0.279
240	1.187	140.4	153.6	0.293	1.011	140.2	153.33	0.2897	0.880	140.02	153.0	0.286
260	1.225	145.0	158.6	0.300	1.044	144.8	158.35	0.2968	0.909	144.63	158.1	0.294
280	1.262	149.6	163.6	0.307	1.077	149.4	163.44	0.3038	0.938	149.32	163.2	0.301
300	1.300	154.3	168.8	0.313	1.109	154.2	168.60	0.3107	0.967	154.06	168.3	0.307
	$P = 90$ psia ($T_{sat} = 72.8$ °F)				$P = 100$ psia ($T_{sat} = 79.2$ °F)				$P = 120$ psia ($T_{sat} = 90.5$ °F)			
Sat	0.527	102.8	111.6	0.217	0.474	103.6	112.46	0.2169	0.394	105.06	113.8	0.216
80	0.540	104.4	113.4	0.220	0.476	103.8	112.68	0.2173				

(continued)

Table E2 (continued)

T (°F)	P = 10 psia (T_{sat} = −29.7 °F) V (ft³/lb)	U (Btu/lb)	H (Btu/lb)	S (Btu/lb R)	P = 15 psia (T_{sat} = −14.3 °F) V (ft³/lb)	U (Btu/lb)	H (Btu/lb)	S (Btu/lb R)	P = 20 psia (T_{sat} = −29.7 °F) V (ft³/kg)	U (Btu/lb)	H (Btu/lb)	S (Btu/lb R)
100	0.575	108.8	118.3	0.229	0.508	108.3	117.73	0.2265	0.408	107.26	116.3	0.221
120	0.607	113.1	123.2	0.238	0.538	112.7	122.70	0.2352	0.435	111.84	121.5	0.230
140	0.638	117.5	128.1	0.246	0.567	117.1	127.63	0.2436	0.461	116.37	126.6	0.238
160	0.667	121.8	132.9	0.254	0.594	121.5	132.55	0.2517	0.482	120.89	131.6	0.247
180	0.696	126.2	137.8	0.262	0.621	125.9	137.49	0.2595	0.508	125.42	136.7	0.255
200	0.723	130.7	142.7	0.269	0.646	130.4	142.45	0.2671	0.530	129.97	141.7	0.262
220	0.751	135.2	147.7	0.277	0.671	135.0	147.45	0.2746	0.552	134.56	146.8	0.270
240	0.777	139.8	152.7	0.284	0.696	139.6	152.49	0.2819	0.573	139.20	151.9	0.277
260	0.804	144.4	157.8	0.291	0.720	144.2	157.59	0.2891	0.593	143.89	157.0	0.285
280	0.830	149.1	162.9	0.298	0.743	148.9	162.74	0.2962	0.614	148.63	162.2	0.292
300	0.856	153.9	168.1	0.305	0.767	153.7	167.95	0.3031	0.633	153.43	167.5	0.299
320	0.881	158.7	173.4	0.312	0.790	158.5	173.21	0.3099	0.653	158.29	172.8	0.306

T (°F)	P = 140 psia (T_{sat} = 100.6 °F) V (ft³/lb)	U (Btu/lb)	H (Btu/lb)	S (Btu/lb R)	P = 160 psia (T_{sat} = 109.6 °F) V (ft³/lb)	U (Btu/lb)	H (Btu/lb)	S (Btu/lb R)	P = 180 psia (T_{sat} = 117.7 °F) V (ft³/kg)	U (Btu/lb)	H (Btu/lb)	S (Btu/lb R)
Sat	0.335	106.2	114.9	0.216	0.291	107.2	115.91	0.2157	0.256	108.18	116.7	0.215
120	0.361	110.9	120.2	0.225	0.304	109.8	118.89	0.2209	0.259	108.77	117.4	0.216
140	0.384	115.5	125.2	0.234	0.326	114.7	124.41	0.2303	0.281	113.83	123.2	0.226
160	0.406	120.2	130.7	0.242	0.347	119.4	129.78	0.2391	0.301	118.74	128.7	0.235
180	0.427	124.8	135.8	0.251	0.366	124.2	135.06	0.2475	0.319	123.56	134.1	0.244
200	0.447	129.4	141.0	0.259	0.384	128.9	140.29	0.2555	0.336	128.34	139.5	0.252
220	0.466	134.0	146.1	0.266	0.402	133.6	145.52	0.2633	0.352	133.11	144.8	0.260
240	0.485	138.7	151.3	0.274	0.419	138.3	150.75	0.2709	0.367	137.90	150.1	0.268
260	0.503	143.5	156.5	0.281	0.435	143.1	156.00	0.2783	0.382	142.71	155.4	0.275
280	0.521	148.2	161.7	0.288	0.451	147.9	161.29	0.2856	0.397	147.55	160.7	0.282

(continued)

Table E2 (continued)

T (°F)	V (ft³/lb)	U (Btu/lb)	H (Btu/lb)	S (Btu/lb R)	V (ft³/lb)	U (Btu/lb)	H (Btu/lb)	S (Btu/lb R)	V (ft³/kg)	U (Btu/lb)	H (Btu/lb)	S (Btu/lb R)
	$P = 10$ psia ($T_{sat} = -29.7$ °F)				$P = 15$ psia ($T_{sat} = -14.3$ °F)				$P = 20$ psia ($T_{sat} = -29.7$ °F)			
300	0.538	153.1	167.0	0.295	0.467	152.7	166.61	0.2927	0.411	152.44	166.1	0.289
320	0.555	157.9	172.3	0.302	0.482	157.6	171.98	0.2996	0.425	157.38	171.5	0.296
340	0.573	162.9	177.7	0.309	0.497	162.6	177.39	0.3065	0.439	162.36	177.0	0.303
360	0.589	167.9	183.2	0.316	0.512	167.6	182.85	0.3132	0.452	167.40	182.4	0.310
	$P = 200$ psia ($T_{sat} = 125.3$ °F)				$P = 300$ psia ($T_{sat} = 156.2$ °F)				$P = 400$ psia ($T_{sat} = 179.9$ °F)			
Sat	0.228	108.9	117.4	0.215	0.142	111.7	119.62	0.2132	0.096	112.77	119.9	0.210
160	0.263	117.9	127.7	0.232	0.146	112.9	121.07	0.2155				
180	0.280	122.8	133.2	0.241	0.163	118.9	128.00	0.2265	0.096	112.79	119.9	0.210
200	0.297	127.7	138.7	0.249	0.177	124.4	134.34	0.2363	0.114	120.14	128.6	0.223
220	0.312	132.6	144.1	0.257	0.190	129.7	140.36	0.2453	0.127	126.35	135.7	0.234
240	0.326	137.4	149.5	0.265	0.202	134.9	146.21	0.2537	0.138	132.12	142.3	0.243
260	0.340	142.3	154.9	0.272	0.213	140.1	151.95	0.2618	0.148	137.65	148.6	0.252
280	0.354	147.1	160.2	0.280	0.223	145.2	157.63	0.2696	0.157	143.06	154.7	0.261
300	0.367	152.1	165.6	0.287	0.233	150.3	163.28	0.2772	0.166	148.39	160.6	0.268
320	0.379	157.0	171.1	0.294	0.242	155.4	168.92	0.2845	0.174	153.69	166.5	0.276
340	0.392	162.0	176.6	0.301	0.252	160.5	174.56	0.2916	0.181	158.97	172.4	0.284
360	0.405	167.1	182.1	0.308	0.261	165.7	180.23	0.2986	0.189	164.26	178.2	0.291

Table E3 Saturated propane

T (°F)	P_sat (psia)	V (ft³/lb)		U (Btu/lb)		H (Btu/lb)		S (Btu/lb R)	
		V_l	V_g	U_l	U_g	H_l	H_g	S_l	S_g
−140	0.6	0.02505	128.0000	−51.33	139.22	−51.33	153.6	−0.139	0.501
−120	1.4	0.02551	58.8800	−41.44	143.95	−41.43	159.1	−0.109	0.481
−100	2.9	0.02601	29.9300	−31.34	148.80	−31.33	164.8	−0.080	0.465
−80	5.5	0.02653	16.5200	−21.16	153.73	−21.13	170.5	−0.053	0.452
−60	9.7	0.02708	9.7500	−10.73	158.74	−10.68	176.2	−0.026	0.441
−40	16.1	0.02767	6.0800	−0.08	163.80	0.00	181.9	0.000	0.433
−20	25.4	0.02831	3.9800	10.81	168.88	10.94	187.6	0.025	0.427
0	38.4	0.02901	2.7000	21.98	174.01	22.19	193.2	0.050	0.422
10	46.5	0.02939	2.2500	27.69	176.61	27.94	196.0	0.063	0.420
20	55.8	0.02978	1.8900	33.47	179.15	33.78	198.7	0.074	0.418
30	66.5	0.03020	1.5980	39.34	181.71	39.71	201.4	0.087	0.417
40	78.6	0.03063	1.3590	45.30	184.30	45.75	204.1	0.099	0.415
50	92.3	0.03110	1.1610	51.36	186.74	51.89	206.6	0.111	0.414
60	107.7	0.03160	0.9969	57.53	189.30	58.16	209.2	0.123	0.413
70	124.9	0.03213	0.8593	63.81	191.71	64.55	211.6	0.135	0.412
80	144.0	0.03270	0.7433	70.20	194.16	71.07	214.0	0.147	0.411
90	165.2	0.03332	0.6447	76.72	196.46	77.74	216.2	0.159	0.410
100	188.6	0.03399	0.5605	83.38	198.71	84.56	218.3	0.171	0.410
110	214.3	0.03473	0.4881	90.19	200.91	91.56	220.3	0.183	0.409
120	242.5	0.03555	0.4254	97.16	202.98	98.76	222.1	0.195	0.408
130	273.3	0.03646	0.3707	104.33	204.92	106.17	223.7	0.207	0.406
140	306.9	0.03749	0.3228	111.70	206.64	113.83	225.0	0.220	0.405
150	343.5	0.03867	0.2804	119.33	208.05	121.79	225.9	0.233	0.403

(continued)

Table E3 (continued)

T (°F)	P_{sat} (psia)	V (ft³/lb)		U (Btu/lb)		H (Btu/lb)		S (Btu/lb R)	
		V_l	V_g	U_l	U_g	H_l	H_g	S_l	S_g
160	383.3	0.04006	0.2426	127.27	209.16	130.11	226.4	0.246	0.401
170	426.5	0.04176	0.2085	135.60	209.81	138.90	226.3	0.259	0.398
180	473.4	0.04392	0.1771	144.50	209.76	148.35	225.3	0.273	0.394
190	524.3	0.04696	0.1470	154.38	208.51	158.94	222.8	0.289	0.387
200	579.7	0.05246	0.1148	166.65	204.16	172.28	216.5	0.309	0.376
206.1	616.1	0.07265	0.0726	186.99	186.99	195.27	195.3	0.343	0.343

Table E4 Superheated propane

T (°F)	V (ft³/kg)	U (Btu/lb)	H (Btu/lb)	S (Btu/lb R)	V (ft³/kg)	U (Btu/lb)	H (Btu/lb)	S (Btu/lb R)
	P = 0.75 psia (T_{sat} = −135.1 °F)				P = 1.5 psia (T_{sat} = −118.1 °F)			
Sat.	104.8	140.4	154.9	0.496	54.99	144.4	159.7	0.479
−110	113.1	146.6	162.3	0.518	56.33	146.5	162.1	0.486
−90	119.6	151.8	168.4	0.535	59.63	151.7	168.2	0.503
−70	126.1	157.2	174.7	0.551	62.92	157.1	174.5	0.520
−50	132.7	162.7	181.2	0.568	66.20	162.6	181.0	0.536
−30	139.2	168.6	187.9	0.584	69.47	168.4	187.7	0.552
−10	145.7	174.4	194.7	0.599	72.74	174.4	194.6	0.568
10	152.2	180.7	201.9	0.615	76.01	180.7	201.8	0.583
30	158.7	187.1	209.2	0.630	79.27	187.1	209.1	0.599
50	165.2	193.8	216.8	0.645	82.53	193.8	216.7	0.614
70	171.7	200.7	224.6	0.660	85.79	200.7	224.5	0.629
90	178.2	207.8	232.6	0.675	89.04	207.8	232.5	0.644
	P = 5.0 psia (T_{sat} = −83.0 °F)				P = 10.0 psia (T_{sat} = −58.8 °F)			
Sat.	18.00	153.0	169.6	0.454	9.47	159.0	176.6	0.441
−40	20.17	165.1	183.8	0.489	9.96	80.9	99.3	1.388
−20	21.17	171.1	190.7	0.505	10.47	86.9	106.3	1.405
0	22.17	172.2	197.7	0.521	10.98	93.1	113.4	1.421
20	23.16	183.5	205.0	0.536	11.49	99.5	120.8	1.436
40	24.15	190.1	212.5	0.552	11.99	106.1	128.3	1.452
60	25.14	196.9	220.2	0.567	12.49	113.0	136.1	1.467
80	26.13	204.0	228.2	0.582	12.99	120.0	144.1	1.482
100	27.11	211.3	236.4	0.597	13.49	127.3	152.3	1.497
120	28.09	218.8	244.8	0.611	13.99	134.9	160.7	1.512

(continued)

Table E4 (continued)

T (°F)	V (ft^3/kg)	U (Btu/lb)	H (Btu/lb)	S (Btu/lb R)	V (ft^3/kg)	U (Btu/lb)	H (Btu/lb)	S (Btu/lb R)
	$P = 0.75$ psia ($T_{sat} = -135.1$ °F)				$P = 1.5$ psia ($T_{sat} = -118.1$ °F)			
140	29.07	226.5	253.4	0.626	14.48	142.6	169.4	1.526
	$P = 20.0$ psia ($T_{sat} = -30.7$ °F)				$P = 40.0$ psia ($T_{sat} = 2.1$ °F)			
Sat.	4.971	166.2	184.6	0.430	2.594	174.6	193.8	0.422
20	5.648	182.4	203.3	0.471	2.723	180.6	200.8	0.436
40	5.909	189.1	211.0	0.487	2.864	187.6	208.8	0.453
60	6.167	195.9	218.8	0.502	3.002	194.6	216.9	0.469
80	6.424	203.1	226.9	0.518	3.137	201.8	225.1	0.484
100	6.678	210.5	235.2	0.533	3.271	209.4	233.6	0.500
120	6.932	218.0	243.7	0.548	3.403	217.0	242.2	0.515
140	7.184	225.8	252.4	0.562	3.534	224.9	251.1	0.530
160	7.435	233.9	261.4	0.577	3.664	232.9	260.1	0.545
180	7.685	242.1	270.6	0.592	3.793	241.3	269.4	0.559
200	7.935	250.6	280.0	0.606	3.921	249.8	278.9	0.574
	$P = 60.0$ psia ($T_{sat} = 24.1$ °F)				$P = 80.0$ psia ($T_{sat} = 41.1$ °F)			
Sat.	1.764	180.2	199.8	0.418	1.336	184.6	204.3	0.415
50	1.894	189.5	210.6	0.400	1.372	187.9	208.2	0.423
70	1.992	196.9	219.0	0.417	1.450	195.4	216.9	0.440
90	2.087	204.4	227.6	0.432	1.526	203.1	225.7	0.456
110	2.179	212.1	236.3	0.448	1.599	210.9	234.6	0.472
130	2.271	220.0	245.2	0.463	1.671	218.8	243.6	0.487
150	2.361	228.0	254.2	0.478	1.741	227.0	252.8	0.503
170	2.450	236.3	263.5	0.493	1.810	235.4	262.2	0.518
190	2.539	244.8	273.0	0.508	1.879	244.0	271.8	0.533

(continued)

Table E4 (continued)

T (°F)	V (ft³/kg)	U (Btu/lb)	H (Btu/lb)	S (Btu/lb R)	V (ft³/kg)	U (Btu/lb)	H (Btu/lb)	S (Btu/lb R)
	$P = 0.75$ psia ($T_{sat} = -135.1$ °F)				$P = 1.5$ psia ($T_{sat} = -118.1$ °F)			
210	2.626	253.5	282.7	0.523	1.946	252.7	281.5	0.548
230	2.713	262.3	292.5	0.537	2.013	261.7	291.5	0.562
250	2.800	271.6	302.7	0.552	2.079	270.9	301.7	0.577
	$P = 100$ psia ($T_{sat} = 55.1$ °F)				$P = 120$ psia ($T_{sat} = 67.2$ °F)			
Sat.	1.073	188.1	207.9	0.414	0.895	191.1	210.9	0.412
80	1.156	197.8	219.2	0.435	0.932	196.2	216.9	0.424
100	1.219	205.7	228.3	0.452	0.989	204.3	226.3	0.441
120	1.280	213.7	237.4	0.468	1.043	212.5	235.7	0.457
140	1.340	221.9	246.7	0.483	1.094	220.8	245.1	0.473
160	1.398	230.2	256.1	0.499	1.145	229.2	254.7	0.489

Index

© Springer Nature Switzerland AG 2021
Y. Demirel, *Energy*, Green Energy and Technology,
https://doi.org/10.1007/978-3-030-56164-2

Printed in the United States
by Baker & Taylor Publisher Services